Thr

BIOCOORDINATION CHEMISTRY:
Coordination Equilibria in
Biologically Active Systems

ELLIS HORWOOD SERIES IN INORGANIC CHEMISTRY

Series Editor: J. BURGESS, Department of Chemistry, University of Leicester

Inorganic chemistry is a flourishing discipline in its own right and also plays a key role in many areas of organometallic, physical, biological, and industrial chemistry. This series is developed to reflect these various aspects of the subject from all levels of undergraduate teaching into the upper bracket of research.

Alcock, N.	**Bonding and Structure: Structural Principles in Inorganic and Organic Chemistry**
Almond, M.J.	**Short-lived Molecules**
Beck, M.T. & Nagypal, I.	**Chemistry of Complex Equilibria**
Burger, K.	**Biocoordination Chemistry:**
	Coordination Equilibria in Biologically Active Systems
Burgess, J.	**Ions in Solution: Basic Principles of Chemical Interactions**
Burgess, J.	**Metal Ions in Solution**
Burgess, J.	**Inorganic Solution Chemistry**
Cardin, D.J., Lappert, M.F. & Raston, C.L.	
	Chemistry of Organo-Zirconium and -Hafnium Compounds
Constable, E.C.	**Metals and Ligand Reactivity**
Cross, R.J.	**Square Planar Complexes: Reaction Mechanisms and Homogeneous Catalysis**
Harrison, P.G.	**Tin Oxide Handbook**
Hartley, F.R., Burgess, C. & Alcock, R.M.	**Solution Equilibria**
Hay, R.W.	**Bioinorganic Chemistry**
Hay, R.W.	**Reaction Mechanisms of Metal Complexes**
Housecroft, C.E.	**Boranes and Metalloboranes: Structure, Bonding and Reactivity**
Kendrick, M.J., May, M.T., Plishka, M.J. & Robinson, K.D.	**Metals in Biological Systems**
Lappert, M.F., Sanger, A.R., Srivastava, R.C. & Power, P.P.	**Metal and Metalloid Amides**
Lappin, G.	**Redox Mechanisms in Inorganic Chemistry**
Maddock, A.	**Mössbauer Spectroscopy**
Massey, A.G.	**Main Group Chemistry**
McGowan, J. & Mellors, A.	**Molecular Volumes in Chemistry and Biology:**
	Applications Including Partitioning and Toxicity
Parish, R.V.	**NMR, NQR, EPR, and Mössbauer Spectroscopy in Inorganic Chemistry**
Rochel, L.P.	**The Chemical Elements: Chemistry, Physical Properties and Uses in Science and Industry**
Romanowski, W.	**Highly Dispersed Metals**
Snaith, R. & Edwards, P.	**Lithium and its Compounds: Structures and Applications**
Williams, P.A.	**Oxide Zone Geochemistry**

BIOCOORDINATION CHEMISTRY:
Coordination Equilibria in Biologically Active Systems

Editor

KÁLMÁN BURGER Ph.D., D.Sc.
c. Member of the Hungarian Academy of Sciences
Department of Inorganic and Analytical Chemistry
A. József University, Szeged, Hungary

Authors

KÁLMÁN BURGER, A. József University, Szeged, Hungary
JUNZO HIROSE, Nagoya City University, Nagoya, Japan
YOSHINORI KIDANI, Nagoya City University, Nagoya, Japan
TAMÁS KISS, L. Kossuth University, Debrecen, Hungary
HARRI LÖNNBERG, Turku University, Turku, Finland
LÁSZLÓ NAGY, A. József University, Szeged, Hungary
BÉLA NOSZÁL, L. Eötvös University, Budapest, Hungary
IMRE SÓVÁGÓ, L. Kossuth University, Debrecen, Hungary

ELLIS HORWOOD
NEW YORK   LONDON   TORONTO   SYDNEY   TOKYO   SINGAPORE

First published in 1990 by
ELLIS HORWOOD LIMITED
Market Cross House, Cooper Street,
Chichester, West Sussex, PO19 1EB, England

A division of
Simon & Schuster International Group
A Paramount Communications Company

Typeset by Ellis Horwood Limited
Printed and bound in Great Britain by Bookcraft (Bath) Limited, Midsomer Norton, Avon

British Library Cataloguing in Publication Data

Biocoordination chemistry: coordination equilibria in biologically active systems
1. Organic compounds. Chemical analysis
I. Burger, K. (Kalman), 1929–
547.3
ISBN 0–13–179912–6

Library of Congress Cataloging-in-Publication Data available

Table of contents

Preface 9

I. Introduction (Kálmán Burger) 11

II. Acid-base Properties of Bioligands (Béla Noszál) 18
  1. Introduction 18
    1.1 Nomenclature 19
  2. Acid–base properties at molecular level 19
    2.1 Macroconstants and their determination 19
    2.2 The application of protonation macroconstants 20
  3. Acid-base properties at submolecular level 22
    3.1 Microscopic protonation equilibria and microconstants 24
      3.1.1. Determination of microconstants 27
    3.2 Group constants and their determination 29
      3.2.1. Applicability of group constants 31
    3.3 Limiting cases for the relations of macro- and microequilibria . .32
    3.4 Submicroconstants 33
      3.4.1 The determination of rota-microconstants 35
  4. Microspeciation and rota-microspeciation 36
  5. Characteristic proton-binding sites of bioligands 39
    5.1 Nucleic bases 40
    5.2 Amino acids and peptide residues 41
    5.3 Miscellaneous 42
  6. The example of aspartic acid 49
  References 49

III. Complexes of amino acids (Tamás Kiss) 56
  1. Introduction 56
    1.1 General 56
    1.2 Acid–base properties of amino acids 59

      1.3 Aspects of metal ion coordination .62
2. Bidentate amino acids 68
      2.1 Aliphatic amino acids 68
      2.2 Proline and hydroxyproline .73
3. Tridentate amino acids with weakly coordinating side chains 75
      3.1 Serine and threonine 75
      3.2 Aspartic acid and glutamic acid. 77
      3.3 Asparagine and glutamine 81
4. Aromatic amino acids 82
      4.1 Phenylalanine and tyrptophan 82
      4.2 Tyrosine isomers 83
      4.3 3,4-Dihydroxyphenylalanine 84
5. Tridentate amino acids with side chain nitrogen donors. 88
      5.1 Diaminocarboxylic acids and arginine 88
      5.2 Histidine 91
6. Sulphur-containing amino acids 96
      6.1 Methionine and S-methylcysteine 96
      6.2 Cysteine and penicillamine .97
      6.3 Cystine and penicillamine disulphide 103
7. Complexes of amino acid derivatives. 104
      7.1 Aminophosphonic acids 104
      7.2 Aminohydroxamic acids 107
8. Mixed ligand complexes 110
9. Conclusions 113
References 116

IV. **Metal Complexes of peptides and their derivatives** (Imre Sóvágó) 135
1. Introduction 135
2. Acid–base properties and coordination ability of peptides 135
      2.1 General 135
      2.2 Acid–base properties 136
      2.3 Possible modes of coordination in peptides 137
3. Glycine oligopeptide complexes. .140
      3.1 Copper(II) complexes. 140
      3.2 Complexes of nickel(II) and palladium(II) 143
      3.3 Complexes of other metal ions 145
4. Complexes of peptides with non-coordinating side chains 147
      4.1 Stability and structure of complexes. 147
      4.2 Stereoselectivity in the metal complexes of peptides 150
      4.3 L-proline as as 'breakpoint' in the metal ion–peptide systems. . .151
5. Complexes of peptides containing oxygen or nitrogen donors
    in the side chain 154
      5.1 Peptides of serine and threonine 154
      5.2 Tyrosine-containing peptides 155
      5.3 Peptides with carboxylate or acid amide groups in the
            side chain. 157
      5.4 Lysine-containing peptides .158

    6.   Complexes of histidine-containing peptides 159
        6.1   The dipeptides His-X and X-His 159
        6.2   Tripeptides of histidine 162
        6.3   Other histidine-containing peptides 164
    7.   Complexes of sulphur-containing peptides 165
        7.1   Peptides containing the thioether group 166
        7.2   Thiol-containing peptides 168
        7.3   Coordination chemistry of glutathione 171
        7.4   Complexes of disulphides 172
        7.5   Complexes of thioamides 173
    8.   Conclusions 174
    References 176

V. **Thermodynamic and kinetic aspects of metalloenzymes and**
    metalloproteins (Junzo Hirose and Yoshinori Kidani). 185
    Introduction 185
    1.   Metal-binding and dissociation mechanism for metalloenzymes
        and proteins 186
        1.1   Introduction 186
        1.2   Carboxypeptidase A 187
        1.3   Carbonic anhydrase 196
        1.4   Cu, Zn-superoxide dismutase .199
        1.5   Human transferrin and ovotransferrin 205
    2.   Mechanism of binding and dissociation of small ligands (inhibitors
        and substances such as substrates) in metalloenzymes and
        metalloproteins 211
        2.1   Introduction 211
        2.2   Carboxypeptidase A 211
        2.3   Carbonic anhydrase 217
        2.4   Superoxide dismutase 227
        2.5   Transferrin 230
    References 230

VI. **Metal complexes of carbohydrates and sugar-type ligands**
    (Kálmán Burger and László Nagy) 236
    1.   General properties 236
    2.   Coordination equilibria 238
        2.1   Sugar (small molecular carbohydrate) complexes 238
        2.2   Glycofuranoside complexes 242
        2.3   Sugar acid complexes 243
        2.4   Amino sugar complexes 248
        2.5   Ion-binding properties of macromolecular carbohydrates 249
    3.   Structure and bonding 252
        3.1   Sugar (mono-, di- and oligosaccharide) complexes 252
        3.2   Sugar acid complexes 257
        3.3   Complexes of nitrogen-containing sugar derivatives 259
        3.4   Dimeric and oligomeric complexes 263

8   **Table of contents**

       3.5   Redox behaviour 267
  References 272

VII. **Proton and metal ion interaction with nucleic acid bases, nucleosides,**
     and nucleoside monophosphates (Harri Lönnberg) 284
    1.   Introduction 284
    2.   Complexing of purine and its 9-substituted derivatives 287
       2.1   Protolytic equilibria 287
       2.2   Metal ion complexes 289
    3.   Complexing of adenosine and adenine 290
       3.1   Protolytic equilibria 290
       3.2   Metal ion complexes of adenosine 291
       3.3   Metal ion complexes of adenine 295
    4.   Complexing of 6-oxopurines and their nucleosides 297
       4.1   Protolytic equilibria 297
       4.2   Metal ion complexes of guanosine and inosine 298
       4.3   Metal ion complexes of guanine and hypoxanthine 303
    5.   Complexing of cytidine and cytosine 303
       5.1   Protolytic equilibria 303
       5.2   Metal ion complexes 305
    6.   Complexing of 4-oxopyrimidines and their nucleosides 308
       6.1   Protolytic complexes 308
       6.2   Metal ion complexes 309
    7.   Binding at the sugar moiety of ribinucleosides 312
    8.   Complexing of nucleoside 5'-monophosphates 315
    9.   Thermodynamics of protolytic and complexing equilibria 322
   10.   Stability–basicity correlations .324
   11.   Interligand interactions 326
    Acknowledgement 328
    References 328

Index 347

Preface

Ellis Horwood suggested that I should write a monograph on the coordination chemistry of biologically active systems in a letter he sent to me in 1982. Unfortunately, for personal reasons (new job, new responsibilities), the work could not start before 1988. Even then, this undertaking was made possible only through the help of colleagues and friends who agreed to write the individual chapters.

Our aim was to survey in one volume the results and conclusions of coordination chemical studies of kinetically labile biologically active systems. Via a great number of selected typical examples, we have attempted to demonstrate how the most intriguing task of determining the compositions and concentrations of species formed in solution in labile equilibrium reactions can be, and has been, solved in different groups of bioactive molecules.

Primarily, equilibrium measurements have served this purpose, structural investigations contributing to the evaluation and interpretation of the equilibrium data. We have made an effort to illustrate how the different experimental methods complement each other in leading to the characterization of the systems. As a consequence of this type of treatment, the material covered in the various chapters reflects the principles and rules governing (or merely influencing) the compositions, structures, and stabilities of coordination compounds formed in biologically active systems. A knowledge of these interactions is crucial for an understanding of the role of metal ions in biochemistry.

A detailed reference list at the end of each chapter helps the reader to find his way around the corresponding literature.

We hope very much that this book will be of use to all those chemists, physicists, and biologists who are interested in the roles of metal ions in biologically active systems.

The authors express their gratitude to Professors D. L. Rabenstein and H. Kozlowski for their comments, to Dr D. Durham for the stylistic revision of some chapters, and to all those colleagues who have permitted reproduction of illustrations from work published elsewhere.

Kálmán Burger
Szeged, January 1990

I

Introduction:
Biocoordination chemistry: coordination
chemical interactions in biologically
active systems

Kálmán Burger
Department of Inorganic and Analytical Chemistry, A. József University, H-6701
Szeged, Hungary

Biologically active organic molecules (proteins, polypeptides, carbohydrates, nucleotides, nucleosides, alkaloids, etc.) and their derivatives or decomposition products all contain electron pair donor atoms (e.g. amino, histidine, guanidino, or peptide nitrogens, carboxylato, carbonyl, or hydroxo oxygens, sulphydryl or thioether sulphur atoms, etc.). They must therefore be considered to be potential ligands which may participate in metal ion coordination and protonation–deprotonation processes [73Si]†. Since biological fluids are aqueous solutions with well-defined pH, containing both bioactive organic compounds and metal ions, coordination chemical interactions in these systems cannot be neglected.

This recognition led to a new complex discipline, which makes use of the theoretical and experimental armoury or inorganic and coordination chemistry and molecular biology: biocoordination chemistry, which is concerned with coordination interactions in biologically active systems.

Protonation or metal ion coordination is known to decrease the electron density on the donor groups participating in this process. This change usually extend to other parts of the molecule, leading to significant changes in its electronic structure, which results in alteration in its chemical behaviour (e.g. decomposition and hence stability, metabolism *in vivo*, etc.).

As a result of H-bonding and the template effect of metal ion binding in systems

---

† Literature citations are in this from (73Si stands for Sigel 1973) throughout the book. Each chapter has its own list of references.

containing several donor groups (e.g. proteins, carbohydrates), coordination processes may change the configuration and/or conformation of the species in solution.

Changes in the protonation of macromolecular polypeptides or proteins may result, for example, in the formation or cleavage of intramolecular H-bonds, which stabilize the secondary and tertiary structure (helical, globular, etc.) of the molecule. Thus, shifts in the protonation–deprotonation equilibria may be connected with transformation of the peptide or protein structure. H-bonding can connect two donor groups only if they are situated at a suitable distance. Metal ion coordination numbers much higher than two allow the formation of complicated structural units, e.g. pockets housing the active sites of enzymes or centres in an entatic state [68Va] in a molecule serving as a biocatalyst, etc. Metal ions affect the stability of the helical tertiary structure of nucleic acids in solution, for instance [88Ko], most probably owing to coordination processes.

Protonation and metal complexation also change the charges on the species in solution, resulting in changes in their solvation; this influences their transport rate *in vivo*, among other effects.

The biological activity of molecules or ions is influenced by several different chemical and physical effects: the transport rate of the bioactive product to the receptor *in vivo*, the reaction rates of the formation of the active form and of its decomposition (metabolism) in the bioprocess, and the affinity of the active compound for the receptor (this latter depends on the electronic structure, configuration, and/or conformation of the molecule).

Metal ion coordination may affect all of the above processes and hence the biological activity of the system. The biological (pharmacological or medical) effects of bioactive compounds can be understood, therefore, only if the coordination chemical interactions are taken into consideration. On the other hand, the effects of metal ion coordination on the chemical reactions (e.g. decomposition) and therefore chemical stability of bioactive compounds, and on their metabolism *in vivo*, also influence their activity.

The effects of metal complex formation are utilized in pharmaceutical practice too. Metal complex formation may hinder the enzymatic decomposition of peptides. This effect is put to use in the employment of metal ions in protein-purification processes. The retarded action of some polypeptide-containing drugs (corticotropin, insulin) can be ensured through the administration of their zinc complexes, owing to the stabilizing effect of metal coordination on the protein.

The metal ion coordination of nucleic acids has been shown to interfere with the transformation of genetic information, leading to the preparation of anticarcinogenic drugs, for example [69Ro, 84Cr]. Besides the well-known anticarcinogenic platinum compounds, complexes of other metals exert analogous medical effects. Among them, mention must be made of ruthenium [87Sa], gold [87Mi2], and even copper [87Be] complexes.

The effect of metal coordination on the chemical stability of organic molecules can promote the production of stable pharmaceutical preparations. Anticarcinogenic bis-indole alkaloids (vincristine, vinblastine, etc.), for example, are very sensitive to hydrolytic decomposition. Drugs containing these alkaloids were therefore commercially available earlier only in powder ampoules. The compound had to be dissolved in a sterile isotonic aqueous solution just before its administration. Coordination chemical investigations have shown that the zinc and even the calcium complexes of these drugs are stable for several years in aqueous solution. This made possible the preparation of stable injections of vincristine (in aqueous solution) [86Bu].

The metal ion coordination of bioactive compounds may influence (in the optimum case decrease) the toxicity of the molecules. The iron(III) complexes of the anticarcinogenic anthracycline derivatives Duanorubicin and Andriamicin were found to be less toxic then the organic parent compounds, but to have the same pharmacological (medical) activity [85Ma, 87Fi]. Metal complexes of some antimalaria drugs exhibit similar behaviour [87Wa, 88Wa].

Complex formation has been shown to influence the pharmacological activity of a number of organic compounds used as drugs. As an example, the copper complexes of salycilic acid derivatives have stronger antipyretic effects than those of the metal-free parent compounds [87Ha]. The mixed ligand magnesium complex of theophylline and salicylic acid has much more pronounced antiflammatory and antipyretic effects than the individual components [81Ba].

The effects of metal ion coordination can also be utilized for synthetic purposes. This may be exemplified by the use of metal ions as activators and protecting groups in synthetic nucleic acid chemistry [86Mi].

Certain metal ions are vital for bioprocesses (iron, zinc, manganese, etc.). Enzyme-catalysed reactions which involve nucleic acid constituents, for example using nucleotides as coenzymes, require the presence of metal ions [87Mil, 87Ka]. DNA polymerase is activated by the magnesium ion [83Wu]. Metal ion coordination processes in general play important roles in the catalysis of different acid–base and redox processes in biological systems. This is well demonstrated by the role of metalloenzymes *in vivo*. In other cases, metal ions causing diseases have to be removed from living organisms. In both cases, metal ions are transported through biological membranes. Since metal ions are strongly hydrophilic because of their charges, and biomembranes are hydrophobic, metal ions can pass through these membranes only in the form of their hydrophobic complexes. The iron-containing drugs used in iron deficiency therapy are all coordination compounds of iron. The treatment of Wilson's disease with D-penicillinamine [70Wa] is based on the removal of copper by transport in the form of its penicillin–amine complex. Similarly, hydroxamic acid and aminohydroxamic acid derivatives are used in the chelation therapy of metal overload diseases, ensuring the transport of the metal in its complexes with the latter ligands [88Ra, 88Tu]. Radioactive metal isotopes have been introduced for diagnostic purposes. Their absorption is likewise possible only in the form of their complex compounds. Clarke & Podbielski [87Cl] have reviewed the technetium complexes used for the latter purpose.

The importance of metal compounds in bioprocesses is reflected by clinical investigations. A correlation has been found, for instance, between the metal ion concentration (trace amounts) and the risk factors involved in some cardiovascular diseases [88Aa]. The investigated human tissues contained the metal ions coordinated to donor groups in the system.

Metal ions can also be used as structural probes by means of which the accessibility and reactivity of potential donor groups on proteins can be tested. The coordination of the silver ion, for example, may be an indicator of the accessibility of sulphur-containing donor groups.

For the reasons described above, a detailed knowledge of the compositions, structures, and stabilities of metal ion complexes of the different types of biologically active molecules and/or ions in aqueous solution is necessary. There is a special need for a better understanding of the molecular biological phenomena governing biological activity, and for improvements in the use of metal complexation in the phamacological and pharmaceutical-technological applications of bioactive systems.

Coordination chemical studies of biologically active macromolecules, such as polypeptides, proteins, carbohydrates, etc., are therefore of equally great importance from biochemical, pharmacological, and pharmacotechnological aspects. For quantitative investigations of metal complex formation, a knowledge of protonation equilibria is similarly of fundamental importance.

Most bioactive molecule (proteins, carbohydrates, etc.) contain various numbers of donor groups, many of them having similar, but not identical, basicities and metal ion affinities. Numerous strongly overlapping protonation and metal complexation equilibria are, therefore, characteristic for such systems. These equilibria are also influenced by intramolecular interactions (e.g. H-bonds) due to the ordered conformation (helical, globular, etc.) of the macromolecule. All this makes an exact evaluation of the experimental equilibrium data on such systems difficult. The classical methods of equilibrium chemistry lead to the unambiguous determination of the number of binding sites in the protonation and complex formation processes. A computer evaluation of the experimental data even results in the determination of the compositions (metal–ligand and proton–ligand ratios) and concentration distribution of the species in solution. The macroscopic equilibrium constants derived from the measurements characterize the molecule, but (because of the overlapping processes) not the binding strengths of the individual donor groups in the different species.

Several research schools have attempted to overcome the above difficulties. The biologically important centres (acitve sites, metal binding moieties, etc.) have been imitated by means of small molecular model compounds. For modelling of the calcium ion-binding sites of the vitamin K-dependent blood coagulation factor protein, for example, γ-carboxyglutamic acid derivatives have been prepared and characterized via exact microscopic protonation and calcium ion coordination equilibrium constants [88Bul]. Besides the quantitative characterization of the functional groups, these investigations reflected the interactions of neighbouring groups in different protonation or complexation states. On the other hand, not even the most exact microscopic constants could give any indication of the influence of the structure of the macromolecule on the metal-binding processes. The calcium ion has been shown [81Bu, 88Bul] to have a much higher affinity toward the protein than toward the small molecular models of the binding site of the same protein. The information yielded by the study of model compounds relating to natural biological systems has proved to be limited. Nevertheless, small model compounds of the active sites of biomolecules are interesting ligands in their own right. Their coordination chemical study has contributed to the better understanding of ambidentate complexation processes.

Small molecular structural models may provide better information on the original system than do the small models devised for imitation of the equilibria in macromolecular systems. As an example, trinuclear carbohydrate-bridged iron(III)-L-amino acid complexes in which the amino-N does not participate in coordination are among the most attractive models of the structure of the iron core of ferritin [74Ho, 81Pu].

Since classical macroscopic equilibrium constants do not characterize the individual binding sites in polyfunctional molecules, and the great number of functional groups participating in overlapping coordination processes excludes the possibility of the determination of microscopic equilibrium constants, some simplifications were introduced to develop a novel evaluation method [79No, 86No] yielding 'group constants' characteristic of the individual binding sites in the macromolecule. The latter method reflects the effects of H-bonding between macromolecular functional groups situated near each other for configurational or conformational reasons; it also reveals hydrophobic or stacking interactions, but not the effects of electron shifts, from one donor to the other through the molecule (see Chapter II).

A further difficulty in the exact characterization of complexes *in vivo* is caused by the fact that in real biological fluids dozens of potential ligands compete for several ions. In *in vivo* systems, multi-ligand and multi-metal ion equilibria may lead not only to the parent complexes but also to mixed ligand complexes and polynuclear complexes.

The combination of analogous equilibrium studies on the macromolecule and its fragments and on other potential ligands possibly accompanying the active compounds in the system may contribute to the quantitative characterization of this type of system [88Bu2]. It must be emphasized, however, that even information given by the most extensive, well-planned equilibrium studies alone is strongly limited without some specific structural (NMR, vibration spectroscopy, etc.) investigation of the system. On the other hand, the exact evaluation of structural investigations of systems involving coordination chemical interactions in solution is impossible without equilibrium studies to determine the distribution of the various species in the solution. Thus, a combination of equilibrium and structural measurements, i.e. the use of different experimental methods, is essential in the characterization of bioactive coordination chemical systems.

The fundamental nature of coordination chemical interactions in biologically active systems makes their investigation of great importance, but the experimental and particularly the theoretical difficulties arising from the great complexity of these systems makes this work rather difficult. The systematic study of the coordination chemistry of polyfunctional bioactive molecules started only some 20 years ago, and in spite of the vast number of publications on this field, the results are still rather limited.

The complexes of the building units of bioactive macromolecules (e.g. amino acid or sugar complexes) and the small molecular weight model compounds of the binding sites of these macromolecule have been characterized with physicochemical exactness by means of equilibrium and structural investigations, as have the metal complexes of small bioactive compounds (e.g. pharmaceutical preparation). Much less is known about the macromolecular systems themselves. The most advanced experimental methods and accurate measurements are used for the quantitative characterization of such systems. In spite of the high accuracy of the experimental data, at the present state of the art their evaluation is possible only semiquantitatively on the basis of approximations that are more or less valid. Most of these are good enough to contribute to an understanding of the biochemistry of the macromolecules, and in ideal cases even to their practical application, e.g. in pharmacy. A complete and exact understanding of such systems is a task for the future. With the increasing complexity of the system studied, the exactness of its characterization decreases. Our current knowledge concerning the small molecular systems is already of help in the understanding of the more complex ones, and will be even more so in the future. In this respect, conclusions derived from the study of simple systems are more lasting and probably more useful for the future than those relating to complicated macrosystems, which are subject to rather rapid change. This is reflected in the material presented in this book.

Since the compositions and concentrations of species formed in the aqueous solutions containing kinetically labile systems (most biofluids are such solutions) form the basis of the most intriguing question of biocoordination chemistry, the following chapters demonstrate the methods used for the study of such systems and report on the results and conclusions of these investigations. Most of the primary information is based on equilibrium measurements, but the contribution made by structural studies to the final conclusions is likewise of great importance and is therefore also discussed.

The authors of the chapters have selected the most typical examples of their field for discussion and have included detailed reference lists to assist the further orientation of the reader. The limited length of the chapters has not permitted a bibliographically complete presentation of the material. We hope that nonetheless we have been able to present a realistic picture of the coordination chemistry of biological importance.

REFERENCES

68Va   Vallee, B. L. & Williams, R. J. P.: *Proc. Natl. Acad. Sci. USA* **59**, 498 (1968).

69Ro   Rosenberg, B., Van Camp, L., Trosko, J. E., & Mansour, V. H.: *Nature* **222**, 385 (1969).

70Wa   Walsha, J. M.: *Brit. J. Hospital Med.* **91**, 248, (1970).

73Si   Sigel, H. (ed.): *Metal Ions in Biological Systems*, Marcel Dekker, New York (starting in 1973).

74Ho   Holt, E. M., Holt, S. L., Tucker, W. F., Asphund, R. O., & Watson, K. J.: *J. Am. Chem. Soc.* **96**, 2621 (1974).

79No   Noszál, B., & Burger, K.: *Acta Chim. Acad. Sci. Hung.* **100**, 275 (1979).

81Ba   Barry, R. H., Rubin, H., Johnson, J. B., & Lazarus, J. H.: *J. Pharm. Sci.* **70**, 204 (1981).

81Bu   Burnier, J. P., Borowski, M., Furie, B. C., & Furie, B.: *Mol. Cell. Biochem.* **39**, 191 (1981), and references cited therein.

81Pu   Puri, R. N., Asphund, R. O., & Holt, S. L.: *J. Coord. Chem.* **11**, 125 (1981).

83Wu   Wu, F. Y. H. & Wu, C. W.: *Metal Ions in Biological Systems* **15**, 157 (1983).

84Cr   Crow, A. J., Smith, P. J., & Atassi, G.: *Inorg. Chim. Acta* **93**, 179 (1984).

85Ma   Martin, R. B.: *Metal Ions in Biological Systems* **19**, 19 (1985).

86Bu   Burger, K., Véber, M., Sipos, P., Galbács, Z., Horváth, I., Szepesi, G., Takácsi-Nagy, G., & Siemroth, J.: *Inorg. Chim. Acta* **124**, 175 (1986).

86Mi   Mizuno, Y.: *The organic chemistry of nucleic acids*, Kodanska, Tokyo (1986).

86No   Noszál, B.: *J. Phys. Chem.* **90**, 4104, 6345 (1986).

87Be   Berners-Price, S. J., Johnson, R. K., Mirabelli, C. K., Faucette, L. F., McCabe, F. L., & Sadler, P. J.: *Inorg. Chem.* **26**, 3383 (1987).

87Cl   Clarke, M. J. & Podbielski, L.: *Coord. Chem. Rev.* **78**, 253 (1987).

87Fi   Fiallo, M. M. L. & Garniers, A.: *Inorg. Chim. Acta* **137**, 119 (1987).

87Ha   Hac, E., & Gagalo, I.: *Pol. J. Pharm.* **39** 219 (1987).

87Ka   Kalbitzer, H. R.: *Metal Ions in Biological Systems* **22**, 81 (1987).

87Mi1   Mildvan, A. S.: *Magnesium* **6**, 28 (1987).

87Mi2   Mirabelli, C. K., Hill, D. T., Faucette, L. F., McCabe, F. L., Girard, G. R., Bryan, D. B., Sutton, B. M., Bartus, J. O., Crooke, S. T., & Johnson, R. K.: *J. Med Chem.* **30**, 2181 (1987).

87Sa   Sava, G., Zorzet, S., Perissin, L., Mestroni, G., Zassinovich, G., & Bontempi, A.: *Inorg. Chim. Acta* **137**, 69 (1987).

87Wa   Wasi, N., Singh, H. B., Gajamana, A., & Raichowdhary, A. N.: *Inorg. Chim. Acta* **135**, 133 (1987).

88Aa   Aalbers, T. G., Houtman, J. P. V., & Makkink, B.: *Trace Elements in Medicine* **5**, 114 (1988).

88Bu1   Burger, K., Sipos, P., Véber, M., Horváth, I., Noszál, B., & Löw, M.: *Inorg. Chim. Acta* **152**, 233 (1988).

88Bu2   Burger, K.: *Bioelectrochemistry and Bioenergetics* **20**, 33 (1988).

88Ko   Kornilova, S. V., Blagoi, Y. P., Moskalenko, I. P., Nikiforova, N. A., & Gladchenko, N. A.: *Stud. Biophys.* **123**, 77 (1988).

88Ra   Raymund, K. N. & Garrett, T. M.: *Pure Appl. Chem.* **60**, 1087 (1988).

88Tu   Turowski, D. N., Rodgers, S. J., Scarrow, R. C., & Raymond, K. N.: *Inorg. Chem.* **27**, 474 (1988).

88Wa   Wasi, N. & Singh, H. B.: *Inorg. Chim. Acta* **151**, 287 (1988).

II

Acid-base properties of bioligands

Béla Noszál
Department of Inorganic and Analytical Chemistry, L. Eötvös University, H-1518, Budapest, 112 Hungary

1. INTRODUCTION

Peptides, proteins, nucleic acids and their constituents are molecules of low symmetry and of two or more functional groups, which are able to associate with (dissociate from) protons. Accordingly, these multidentate bioligands exist in a great number of protonation forms and inherent fine structures. The physiological role (e.g. enzyme effect) of the polyfunctional biomolecules is strictly structure-dependent. Furthermore, the type and extent of metal–ligand interactions are also considerably influenced by the protonation stage of the ligand. Protons are ubiquitous in every biological medium. Consequently, the knowledge of proton-binding characteristics is of primary importance for a thorough understanding of effector–receptor and metal–ligand interactions.

Earlier, for the thermodynamic description of such equilibrium processes, macroconstants were the only useful tools. These parameters provide proton association (dissociation) data about the molecule as a whole, but do not define the basicity of individual functional groups. Nevertheless, macroconstants are very useful in calculating the pH-dependent concentration of all species containing different number of protons, the isoelectric point, and the net charge of the molecule at any pH.

Contrary to macroconstants, microconstants and two other recent parameters (group constants and submicroconstants) yield profound structure-related data on protonation thermodynamics. The exact quantification of highly specific biochemical and coordinative interactions can be based on such constants only.

The fundamental purpose of this chapter is to unify the thermodynamic and structural aspects of the acid–base properties of bioligands. First, the above introduced equilibrium parameters and their applications will be surveyed, followed by the characterization of the main proton-binding sites of bioligands. Last, but not least, the most elaborated example (aspartic acid rotamer populations and their protonation constants) will be discussed in detail.

1.1   Nomenclature
Macroconstants and macroscopic constants as well as microconstants and microscopic constants are equally used synonyms in the literature. Note that such terms as microconstant, microspecies, microform, microequilibrium (and their macrocounterparts) have nothing to do with size. In fact macromolecules (proteins, nucleic acids, polysaccharides, etc.) may take part in microequilibrium processes and exist in several microforms. The micro- or macro- prefix only specifies whether the site of the protonation as well as the protonation state at the rest of the molecule is defined or not.

2.   ACID-BASE PROPERTIES AT MOLECULAR LEVEL

2.2   Macroconstants and their determination†
The overwhelming majority of published equilibrium constants are macroconstants. This is why when the term 'equilibrium constant' is not specified, it always refers to macroconstant.

Macroconstants can be sorted into several classes. Their types and meanings are explained below.

Let B^{2-} designate a trivalent base. (Such trivalent bases of two negative charges can be exemplified by aspartate or glutamate anions, where the neutral amino and the two negatively charged carboxylate groups can bind protons.) Its proton-binding processes are as follows:

$$B^{2-} + H^+ \rightleftharpoons HB^- \tag{1}$$

$$HB^- + H^+ \rightleftharpoons H_2B \tag{2}$$

$$H_2B + H^+ \rightleftharpoons H_3B^+ \tag{3}$$

The extent of these processes can be quantified by the following terms:

$$K_1 = \frac{[HB^-]}{[B^{2-}][H^+]} \;(4) \qquad \beta_1 = K_1 \;(7) \qquad\qquad K_{d_1} = \frac{[H^+][H_2B]}{[H_3B^+]} \;(10)$$

$$K_2 = \frac{[H_2B]}{[HB^-][H^+]} \;(5) \qquad \beta_2 = K_1 K_2 = \frac{[H_2B]}{[B^{2-}][H^+]^2} \;(8) \qquad K_{d_2} = \frac{[H^+][HB^-]}{[H_2B]} \;(11)$$

$$K_3 = \frac{[H_3B^+]}{[H_2B][H^+]} \;(6) \qquad \beta_3 = K_1 K_2 K_3 = \frac{[H_3B^+]}{[B^{2-}][H^+]^3} \;(9) \qquad K_{d_3} = \frac{[H^+][B^{2-}]}{[HB^-]} \;(12)$$

K_1, K_2, and K_3 are stepwise (in other words successive) protonation constants, whereas β_1, β_2, and β_3 are cumulative protonation constants. Since all these parameters regard the processes from the viewpoint of association, the terms

---

† In several (earlier) papers acid-base equilibria at molecular level and microconstants are corresponding terms.

association constant or stability constants can also be used. K_{d_1}, K_{d_2}, K_{d_3} are stepwise dissociation constants. It can be seen that some stepwise protonation and dissociation constants are reciprocals, so, for example, $\log K_1 = -\log K_{d_3} = pK_{d_3}$.

In order to handle all the processes on the basis of the same principle, in this chapter the equilibria will be regarded from the viewpoint of association.

Determination of macroconstants

There are tens of thousands of chemical equilibrium systems depicted in terms of macroconstants. These data are compiled in tabular books [79Se, 64Si], including critical surveys [74Ma1, 75Sm].

Most of the macroconstants have been determined in the past 40 years, mainly between 1960 and 1980. Accordingly, several monographs have appeared on the determination of stability constants [85Le, 77In, 70Be, 64Bu, 61Ro]. Thus, concerning the methodology, only the absolutely fundamental principles will be discussed here.

In the solution of the above trivalent base two mass-balance equations are valid:

$$C_B = [B^{2-}] + [HB^-] + [H_2B] + [H_3B^+] \tag{13}$$

$$C_H = [H^+] + [HB^-] + 2[H_2B] + 3[H_3B^+] \tag{14}$$

where C_B and C_H are the total (analytical) base and proton concentrations, respectively. Concentrations of all the composite species (HB^-, H_2B, and H_3B^+) can be expressed by means of the component (H^+ and B^{2-}) concentrations and the cumulative protonation constants:

$$C_B = [B^{2-}](1 + \beta_1[H^+] + \beta_2[H^+]^2 + \beta_3[H^+]^3) \tag{15}$$

$$C_H = [H^+] + [B^{2-}](\beta_1[H^+] + 2\beta_2[H^+]^2 + 3\beta_3[H^+]^3) \tag{16}$$

C_B and C_H values are known analytical concentrations; $[H^+]$ can be measured by pH-metry. Thus, by using an appropriate mathematical apparatus, the equilibrium constants can be determined, where the minimum number of different pH values for the determination of n macroconstants is also n. However, in practice, to minimize the ambiguity of the constants one must form a redundant mathematical system, taking many more pH values than the number of macroconstants to be determined. For further practical details of the computations, see the pertinent monographs [85Le, 77In, 70Be].

2.2 The application of protonation macroconstants

Macroconstants are very useful for characterizing in solution the basicity of the molecule as a whole. The pH-dependent distribution of macrospecies can also be calculated by means of macroconstants. The relative concentration (α_{HB^-}) of HB^- species of a trivalent ligand can be expressed as:

$$\alpha_{HB^-} = \frac{[HB^-]}{[B^{2-}] + [HB^-] + [H_2B] + [H_3B^+]} = \frac{\beta_1[H^+]}{1 + \beta_1[H^+] + \beta_2[H^+]^2 + \beta_3[H^+]^3} \tag{17}$$

Equation (17) shows that α_{HB^-} (and all other analogous α values) depend on pH, but they are irrespective of total concentration. Such α-type relative concentrations (occurrence probabilities) can be converted into real concentrations if we multiply them by the ligand total concentration (C_B). A distribution diagram for all the 14 macrospecies of corticotropin (ACTH, a 32 amino acid-containing polypeptide) can be seen in Figs 2.1 and 2.2 [89No1]. Another application of macroconstants is the

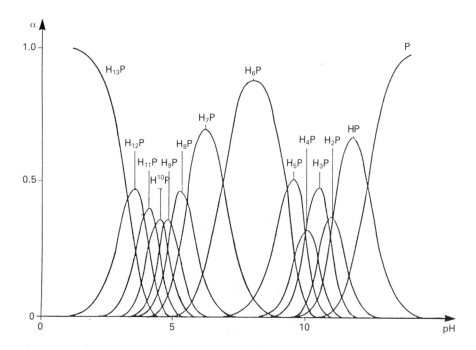

Fig. 2.1 — Distribution diagram for ACTH(1–32) macrospecies containing different number of hydrogen ions (H) on the peptide backbone (P). Charges are not indicated here. H_4P is the neutral species. This representation favours the major species.

calculation of the average number of protons bound to the ligand backbone. This number can be calculated by the classical complex formation function ($\bar{n}$) introduced by J. Bjerrum [41Bj]. The $\bar{n} - [H^+]$ function for our trivalent base in question can be given as:

$$\bar{n} = \frac{[HB^-] + 2[H_2B] + 3[H_3B^+]}{[B^{2-}] + [HB^-] + [H_2B] + [H_3B^+]} = \frac{\beta_1[H^+] + 2\beta_2[H^+]^2 + 3\beta_3[H^+]^3}{1 + \beta_1[H^+] + \beta_2[H^+]^2 + \beta_3[H^+]^3} \cdot$$

(18)

The right-hand side of equation (18) can be obtained from the middle and left-hand side, expressing all the proton-containing species concentrations by $[B^{2-}]$, $[H^+]$, and the β constants. The value of $\bar{n}$ in this case varies between 0 and 3, and it depends on pH, but it is independent of total concentrations. This function at the same time

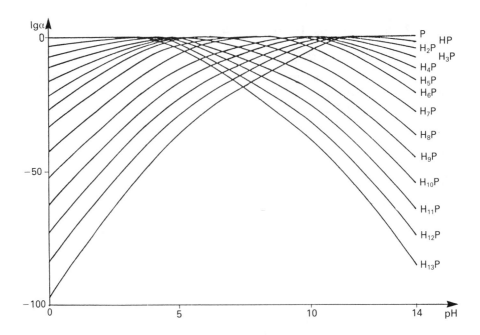

Fig. 2.2 — Distribution diagram for ACTH(1–32) macrospecies, where log α values are plotted against pH. The occurrence probability of minor species can be better seen in this type of plotting. Note that no parallel running curves occur here.

can be used to calculate the average charge of ligand at any arbitrarily chosen pH, which is equal to $\bar{n} - 2$ in this particular example.

The isoelectric point of the ligand can also be determined from this function if we rearrange equation (18) to calculate the pH which corresponds to zero net charge. Note, however, that (18) cannot be made explicit for $[H^+]$, so graphical or approximate maathematical treatment is necessary to evaluate the hydrogen ion concentration. Fig. 2.3 shows $\bar{n}$–pH and charge–pH functions of aspartic acid [89No2], where $\log \beta_1 = 9.53$, $\log \beta_2 = 13.28$ and $\log \beta_3 = 15.24$.

3.   ACID-BASE PROPERTIES AT SUBMOLECULAR LEVEL

Macroconstants are influenced not only by chemical but also by statistical factors (see section 3.3 for details). This is why macroconstants cannot be assigned to individual functional groups (except for monodentate ligands).

The submolecular acid–base properties can be characterized in terms of group constants, microconstants, and submicroconstants. The above order of these thermodynamic parameters is proportional to the thoroughness of their information about the molecule and the equlibrium process. Group constants reflect the basicity of the proton-binding site, but they disregard the protonation state of the rest of the molecule and can be used in special cases only. Microconstants provide information on the basicity of the proton-binding sites and their interactions with other proton-

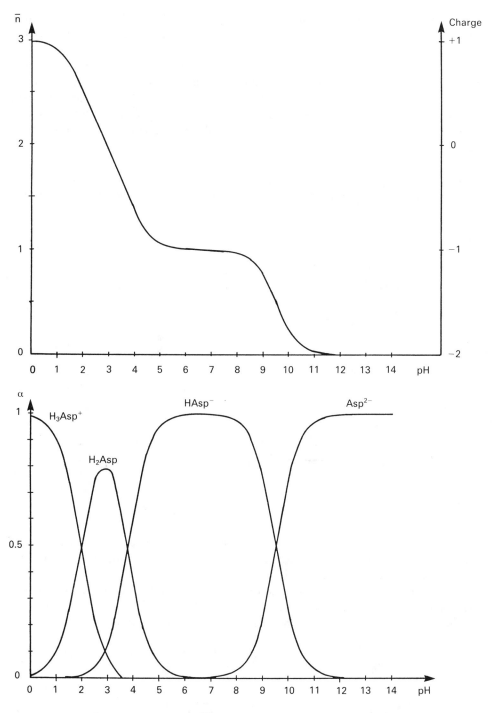

Fig. 2.3 — Number of bound protons($\bar{n}$) and the charge of aspartic acid as a function of pH, calculated by equation (18), using $\log \beta_1 = 9.53$, $\log \beta_2 = 13.28$, $\log \beta_3 = 15.24$. Bottom section shows the corresponding species distribution.

ating groups of the molecule. Submicroconstants, in addition, define the rotational, stacking, etc., state of other flexible moieties of the molecule, when protonation takes place.

In chronological order, microconstants were introduced first [23Bj]; whereas groups constants [79No, 86No1] and submicroconstants [74Fu, 89No2] are much more recently developed parameters. The literature of submolecular acid–base chemistry is incomparably less extensive than that of molecular chemistry. In fact, this review presents the first detailed analysis of all types of submolecular equilibrium constants and their applications.

3.1 Microscopic protonation equilibria and microconstants

Fig. 2.4 shows the protonation scheme of glycine. The unprotonated and doubly

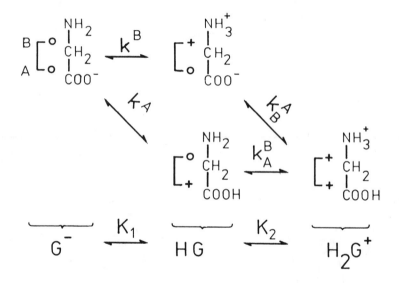

Fig. 2.4 — Protonation scheme of glycine. See text for abbreviations.

protonated species which predominate in basic and very acidic solutions can be seen on the left and right sides, respectively. The two one-proton-containing isomers are located in the middle, under one other.

Symbols of all the micro- and macrospecies, as well as the micro- and macroconstants, can also be seen in Fig. 2.4.

Microconstants can be expressed by microspecies concentrations as:

$$k^A = \frac{[{}_0^+]}{[{}_0^0][H^+]} \quad (19) \qquad k^B = \frac{[{}_0^+]}{[{}_0^0][H^+]} \quad (20)$$

$$k_B^A = \frac{[{}_+^+]}{[{}_0^+][H^+]} \quad (21) \qquad k_A^B = \frac{[{}_+^+]}{[{}_+^0][H^+]} \quad (22)$$

The superscript on microconstant k indicates the group protonating in the process in question, whereas the subscript (if any) refers to the group which holds the proton during the process.

Stepwise macroconstants can be expressed by macrospecies concentrations:

$$\beta_1 = K_1 = \frac{[HG]}{[G^-][H^+]} \tag{23}$$

$$K_2 = \frac{[H_2G^+]}{[HG][H^+]} \tag{24}$$

$$\beta_2 = K_1 K_2 = \frac{[H_2G^+]}{[G^-][H^+]^2} \ . \tag{25}$$

Relationships between the macro- and microconstants are as follows:

$$\beta_1 = k^A + k^B \tag{26}$$

$$\beta_2 = k^A k_A^B = k^B k_B^A \tag{27}$$

$$K_2^{-1} = (k_B^A)^{-1} + (k_A^B)^{-1} \ . \tag{28}$$

Note that only two of the last three equations (26)–(28) are independent. Microconstants k^A and k_B^A refer to the carboxylate basicity, whereas k^B and k_A^B refer to the amino basicity. The difference $\log k^A - \log k_B^A = \log k^B - \log k_A^B = \Delta \log k_{A-B}$ interactivity parameter quantifies the decrease in basicity at one site, when the other site takes up a proton (see also section 3.3).

The amino and carboxylate proton-binding sites of glycine are of significantly different basicity. Thus, microconstants belonging to the main pathway of protonation k^B and k_B^A are practically equal to the K_1 and K_2 macroconstants, respectively. It must be emphasized, however, that (1) in principle, values of macro- and microconstants can be identical for monodentate ligands only, and (2) the minor protonation isomer ($HOOC-CH_2-NH_2$) does exist (in low concentration) and may take part in biochemical and coordinative interactions.

Fig. 2.5 shows the schematic protonation diagram of a trifunctional ligand. Abbreviations and symbols are analogous with those for the bifunctional ligand. The relationships between the macro- and microconstants can be written as:

$$\beta_1 = k^A + k^B + k^C \tag{29}$$

$$\beta_2 = k^A k_A^B + k^A k_A^C + k^B k_B^C = k^B k_B^A + k^C k_C^A + k^C k_C^B \tag{30}$$

$$\beta_3 = k^A k_A^B k_{AB}^C = k^B k_B^A k_{AB}^C = \ldots \ . \tag{31}$$

The comparison of Figs 4 and 5 indicate that the number of microspecies and microconstants exponentially increases with the number of functional groups of the ligand. Table 2.1 gives relationships and corresponding numbers for groups, doubly-protonated protonation isomers, microspecies, and microconstants. $N_{group} = 13$ was chosen as the case of corticotropin ($ACTH_{1-32}$) discussed in detail in section 3.2.

The data in Table 2.1 prove that in solutions containing molecules of four or more functional groups an enormous number of differently protonated microspecies exist, and even more microconstants are necessary to characterize the system. In view of

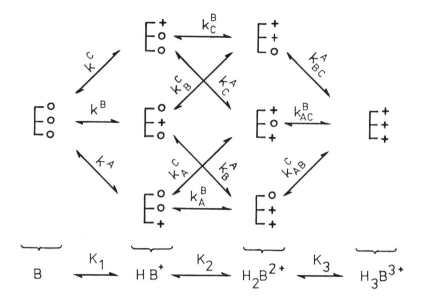

Fig. 2.5 — Protonation scheme of a basic ligand containing three functional groups.

Table 2.1 — Dependence of the number of doubly-protonated protonation isomers ($N_{2prot.isom.}$), microspecies ($N_{microspec.}$) and microconstants ($N_{microconst.}$) on the number of groups of the molecule (N_{group})

N_{group}	$N_{2prot.isom.}$	$N_{microspec.}$	$N_{microconst.}$
1	—	2	1
2	1	4	4
3	3	8	12
4	6	16	32
5	10	32	80
.	.	.	.
.	.	.	.
.	.	.	.
13	78	8192	53248
.	.	.	.
.	.	.	.
.	.	.	.
n	$\binom{n}{2}$	2^n	$n2^{n-1}$

this the difficulties of the determination of microspecies concentrations and micro-constants for molecules of three or more groups are obvious. Another source of difficulties is that the concentration ratio of protonation isomers is constant at any pH:

$$\frac{[\text{I}_0^+]}{[\text{I}_+^0]} = \frac{k^A[\text{I}_0^0][\text{H}^+]}{k^B[\text{I}_0^0][\text{H}^+]} = \frac{k^A}{k^B} \tag{32}$$

where concentrations of the protonation isomers were expressed by equations (19) and (20).

Microspecies are, in fast, continuous interconversion in solution, thus they occur exclusively in the presence of each other. The analytical determination of these coexisting species is difficult but important, since they give mutual spectroscopic (and other analytical) signals, but act individually in specific biochemical processes. The coexistence of microspecies allows only indirect determination of their concentrations. A *sine qua non* of such determinations is information about the relevant microconstants.

3.1.1 Determination of microconstants
For reasons surveyed above, the number of systems depicted in terms of reliable microscopic protonation constants is a few dozens only. Most of them refer to molecules of two groups. It is obvious from equations (26), (27), and (29)–(31) that besides pH-metric macroconstants, independent pieces of information are necessary to evaluate the microconstants.

There are two fundamental approaches for the determination of microconstants: (A) deductive methods and (B) combined spectroscopic–pH-metric methods.

(A) The classical deductive methods include determination of macroconstants for the studied ligand and for auxiliary ligand(s). The auxiliary ligand is usually a close derivate of the main moleclule, but it contains a reduced number of functional groups. Macroconstant(s) of the auxiliary molecule can then be handled as some microconstant(s) of the main compound. The first such determination was done for glycine by Ebert [26Eb], who assumed that the protonation constant of glycine-–methyl ester is identical with k_A^B (see Fig. 2.4) of glycine. This concept assumes that the effect of the $-\text{COCH}_3$ group on the amino basicity is equivalent with that of the $-\text{COOH}$. Such close similarlity was recently proved between $-\text{COOH}$ and $-\text{CONH}_2$ [89No2].

The other principle of deductive methods postulates that the interactions between two functional groups are the same in the small model molecule and in the analogous moiety of the large molecule in question. For example, the interactivity parameter between the two carboxylates can be calculated for the symmetrical malonic acid (see also section 3.3), where $k^A = k^B$ and $k_B^A = k_A^B$, owing to symmetry. The same interactivity parameter can be applied to the γ-carboxylate groups of γ-carboxy glutamic acid, which are also separated by one linking carbon atom [88Bu].

In the third type of case spectroscopic parameters of auxiliary ligands can also be utilized in the process of microconstant determinations for the main compound. Chemical modifications and deductions were applied to evaluate microconstants for glutamic acid [36Ne], cysteine [44Ry, 55Gr, 64Wr], citric acid [60Lo, 75Pe], tetracyclines [85Ma1, 65Ri], histidine [74Su], tyrosine [58Ma1, 76Is], DOPA [77Is], aspartic acid [33Ed, 89No2], γ-carboxy glutamic acid [88Bu], arginine [90No3], moieties of polypeptide fragments such as angiotensine [90Ny], corticotropin [89No1], thymopoiethine [90No4], cysteamine and selenocysteamine [70Ta], catecholamine derivatives [77Ga], and substituted piperazines [89Pa].

Deductive methods were the only available means for the determination of microconstants and microspecies concentrations before the widespread use of spectroscopic techniques. They are, however, indispensable even today in three cases:

(1) When the basicities of the functional groups are very different, so the contribution of the minor protonation isomer(s) to the spectroscopic (or other analytical) signal is negligible.

(2) When the protonation state of one (or some) of the proton-binding sites cannot be monitored selectively by spectroscopic methods, owing to their close vicinity or similar spectroscopic characters.

(3) When molecules are too complicated for spectral resolution.

(B) Combined spectroscopic–pH-metric methods involve pH-metric determination of macroconstants plus selective spectroscopic data on the protonation state of one (or some) of the functional groups at known pH values. The most widely used spectroscopic techniques are UV–VIS to study the phenolate and thiolate protonation and NMR to monitor the protonation state of virtually any kind of functional group at one of the adjacent NMR nuclei. Raman spectroscopy was used to investigate the microequilibria of cysteine [62El] and fluorescence spectrometry for doxorubicine [77St].

The protonation state of a group can be calculated from spectroscopic data as follows:

$$\alpha_{(\text{pH})} = \frac{A_{(\text{pH})} - A_b}{A_a - A_b} \tag{33}$$

where $\alpha_{(\text{pH})}$ is the protonation mole fraction of the group in question at the given pH. $A_{(\text{pH})}$, A_b, and A_a are light absorption values of the solution at the given pH: at (usually, but not necessarily) basic pH; where the group in question is almost completely unprotonated; and at (usually) acidic pH, or where the group in question is completely protonated, respectively. Another precondition of the validity of equation (33) is that the total concentration of the ligand must be the same in every solution. Equation (33) can be applied to NMR spectroscopy if A values are chemical shifts of an NMR nucleus reflecting selectively the protonation state of one of the groups at the indicated pH.

The protonation mole fraction of a site can be expressed in terms of microspecies concentrations. If group B protonates, $\alpha_{(\text{pH})}$ takes the following form:

$$\alpha_{B(\text{pH})} = \frac{[{}_0^+] + [{}_+^+]}{[{}_0^0] + [{}_0^+] + [{}_+^0] + [{}_+^+]} \tag{34}$$

Implicit in this procedure is the assumption that the proton-binding at group B causes the same spectral changes, whether or not group A holds a proton. Concentrations of one or two-proton-containing microspecies can be expressed by means of equations (19)–(27):

$$\alpha_{B(pH)} = \frac{k^{B}[H^{+}] + \beta_{2}[H^{+}]^{2}}{1 + \beta_{1}[H^{+}] + \beta_{2}[H^{+}]^{2}} . \tag{35}$$

The most important meaning of equation (35) is that k^{B} is the only unknown quantity, since $\alpha_{B(pH)}$ and the hydrogen-ion concentration are experimental values, while β_{1} and β_{2} macroconstants can be obtained from pH-metric measurements. The statistical treatment of experimental data for basically (35)-type equations provides reliable microconstants [71Ma, 76Ra]. Equations (26) and (27) show that in the knowledge of the macroconstants and one of the microconstants, all other microconstants can be calculated for molecules of two functional groups. Note that the independent determination of a microconstant related to the minor protonation pathway is more valuable, since conditions of the accurate calculation for the microconstants are then better.

Microconstants were determined by combined spectroscopic–pH-metric methods for tyrosine [58Ed1, 65Ya, 66Co, 72Ni, 82Ki2, 84Ki, 87Ki], tyrosine derivatives and metal complexes [65Ya, 82Ki2, 84Ki, 87Ki], DOPA [71Ma, 75Bo, 78Ja, 77Is, 89Ki], dopamine [78Is, 89Ki], catecholamines [62Ri, 76Gr, 89Ki], a variety of phenolic amines [65Ka], dihydroxyphenylpropionic acid and its derivatives [77Is], dihydroxyphenylbutiric and dihydroxybenzoic acids [79Is], sulphonamides [80Sa], aminobenzoic acids [67Ch], citric acid [61Ma], histidine [72Sh, 83Ta1], histidine derivatives [83Ta2, 83Ta3], selenoglyoxalic acid [70Ku], tetracycline [69Ke, 76As], lysine [76Ra, 80Su], hydroxylysine [80Su], glycylhistidyllysine [77Ra], glutathione [72Ja, 73Ra], daunorubicine [82Ki1] salbutamol [84Pa], spermidine [84On], morphine [72Ni], doxorubicine [77St], cephalosporines [76St], several diamines [73Cr], thiadiamines [83Hu], methyl-mercaptopiperidine [83Ba], cysteine [55Be, 60Go, 62El, 69Co, 71Cl, 74Wa, 76Re], homocysteine, cysteine-ethyl ester [76Re], penicillamine [71Wi], cysteine-containing hexapeptides [87Mi], pefloxacin and norfloxacin [90Ta], and tobramycine [90No3].

In practice, enormous difficulties arise in the determination of microconstants for molecules of over three functional groups, owing to the great number of microconstants and microspecies in solution and the proliferating experimental errors during the evaluation process. Fortunately, in several cases the problem of microconstants can be reduced to the question of group constants.

3.2 Group constants and their determination

Group constants can be deduced from microconstants by simplifications. These simplifications have definite chemical preconditions.

Table 2.1 showed that the number of microconstants is $n2^{n-1}$, when the number of functional groups is n. Thus each of the groups has 2^{n-1} different basicity values depending on the protonation of the other groups in the molecule. For example, group B in a trifunctional molecule (see Fig. 2.5 and equations (29)–(31)) has four different basicities characterized by k^{B}, k_{A}^{B}, k_{C}^{B}, and k_{AC}^{B}. The most usual reason (in rigid molecules the only one) why microconstants of the same group differ from one another is the electron-attracting effect, which occurs when the proximate group(s) (group A or C or both in this example) protonate. This causes a more or less decrease in basicity of the group in question. The effect is significant if the adjacent group

(A or C) is separated by a sufficiently low number of intervening carbon (or other) atoms from the given group (B); that is, the static inductive effect actually reaches the group under consideration. The more groups that hold protons in an intramolecular chemical environment, the stronger the electron-attracting effect, and the less the basicity of the group in question. Thus, among microconstants belonging to the same group the largest one, k^B, refers to proton association when no other sites are protonated. The smallest one, k^B_{AC}, on the other hand, quantifies the proton association ability when all other sites hold protons. If the difference between these two extreme microconstants is so small (owing to the relative remoteness of the other groups) that it does not exceed the error of the microconstant determination, all the microconstants of the same group are virtually equal. Accordingly, all of them can be quantified in one parameter: group constant. The unification of microconstants for the three groups of a tridentate ligand can be represented as

$$k^A = k^A_B = k^A_C = k^A_{BC} = k_A \tag{36}$$

$$k^B = k^B_A = k^B_C = k^B_{AC} = k_B \tag{37}$$

$$k^C = k^C_A = k^C_B = k^C_{AB} = k_C \tag{38}$$

where constants on the right with subscript only are the group constants [86No1].

Substituting (36)–(38) into (29)–(31) gives:

$$\beta_1 = k_A + k_B + k_C \tag{39}$$

$$\beta_2 = k_A k_B + k_A k_C + k_B k_C \tag{40}$$

$$\beta_3 = k_A k_B k_C \ . \tag{41}$$

In the system of equations (39)–(41) it is very important to note that the number of unknown group constants is equal to the number of macroconstants. Thus, group constants can be calculated from β macroconstants.

From the (39)–(41) system of equations group constants cannot be expressed explicitly, but the whole system of equations can be turned into a polynomial of the third degree:

$$\beta_3 - \beta_2 k + \beta_1 k^2 - k^3 = 0 \ . \tag{42}$$

No subscript is necessasry on the group constants, because starting with any of them, the same formula can be reached. Equation (42) can be generalized for an arbitrary number of functional groups (n):

$$\sum_{i=0}^{n} (-1)^i \beta_{n-i} \, k^i = 0 \ . \tag{43}$$

The three roots of polynomial (42) and the n roots of polynomial (43) are the group constant values, while the β-macroconstants can be determined in the traditional ways.

As stated above, microconstants may be treated as group constants if the effect of protonation on the adjacent site decreases to a negligible level along the intervening bonds. It has to be emphasized, however, that the group constant treatment does not exclude the interaction between any two groups. In fact, the group constant values are indicators of group–group couplings by hydrogen bonds. If H-bond formation is energetically favourable for the system, the functional groups do not attach to the proton individually; instead the attached proton belongs to both of the bridgehead groups. This means that for molecules of two functional groups, the number of singly-protonated species in the system reduces to one. (This is the case of *o*-tyrosine [82Ki2]). The scheme, where H-bond formation occurs between groups A and B, is shown in Fig. 2.6. The H-bond formation modifies the relations between the macro-

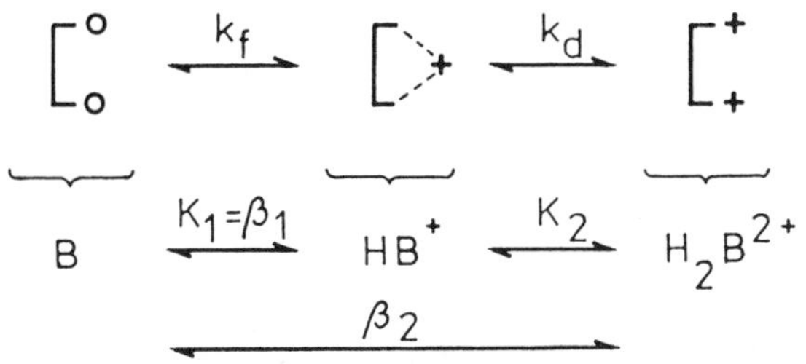

Fig. 2.6 — Protonation scheme of a molecule containing two functional groups coupled by an H-bond.

and group constants:

$$\beta_1 = k_f \tag{44}$$

$$\beta_2 = k_f \cdot k_d \tag{45}$$

where k_f is the group constant (more precisely group–pair constant) referring to the H-bond-forming protonation, and k_d is the group constant referring to the uptake of the second proton, when the H-bond decomposes.

The H-bond-forming protonation appears in the value of equilibrium constants. If the H-bond-forming protonation establishing linkage between groups A and B is more favourable for the system than any of the solitary protonations of A or B, the first proton association takes place at a pH higher then expected on the basis of individual basicities. On the other hand, the second protonation, which causes the rupture of the stable H-bond, requires a low pH relative to the individual protonations of either A or B.

Accordingly, if unusually high and low constants are determined, the molecule certainly contains H-bond-coupled groups.

3.2.1 *Applicability of group constants*
Before the introduction of group constants [79No, 86No1], relations between group-constant-like quantities and dissociation constants of polybasic acids containing independent groups had been analysed in a few papers [58Ed2, 30Mu, 26Si]. These works consider ideal systems which are free of interactions, so the real circumstances, when the application is appropriate, have not been studied. This is why it is crucial to know when group constants can be used.

The principle of group constants says that microconstants can be substituted by group constants if the interaction between the functional groups along the intervening atoms is negligible. What is the number and character of atoms and bonds between the protonation sites when this negligibility is virtually perfect?

^{1}H and ^{13}C NMR studies have shown that carboxyl and amine deprotonation downfield shifts are still sensitive at the third (γ) methylene moiety, but not any further one [77Sa]. This statement is supported by protonation macroconstants of symmetrical dicarboxylic acids (mucic acid; 3,3-dithiopropionic acid; 1,2-bis(carboxy-methyltio)benzene) [74Ma1], where the difference between the log of stepwise macroconstants is 0.6, which is the indicator of the independence of the groups (see 3.3) even in the case of mucic acid, where the number of intervening atoms is only four. Consequently, influencing of the proximate group's basicity along the chain takes place if not more than three atoms are situated between the protonation sites. This statement is not true for conjugated or cyclic systems in the molecule.

Group constants can be used for the characterization of protonation sites in non-cyclic macromolecules, mainly polypeptides. Barbucci *et al.* have stated that in the protonation processes of synthetic polymers built up of basic rigid monomomers the repeating moities behave like independent units, owing to the lack of interactions [78Ba]. In a thymopoietin pentapeptide fragment, where overlapping microconstants of the tyrosyl and the lysyl residues could be determined spectrophotometrically, both the phenolic and amino microconstants proved to be identical in every state of the other group [84So].

Typical polypeptide molecules are composites of individually protonating and coupled group-containing loci. The larger the molecule the greater the propensity for site-couplings and other intramolecular interactions. In a series of tri- and tetrapeptides [90No2] as well as of octapeptides [90Ny] the in-chain protonation sites could be characterized by individual group constants, and no specific secondary structure had to be hypothesized for the interpretation of the data. On the other hand, group constant values for longer polypeptide fragments of random coil type such as ACTH(15–32), ACTH(1–28), ACTH(1–32), provide evidence [89No1] that the functional groups may be influenced by several intramolecular interactions (H-bonding, macromolecular hydrophobic effects, etc.).

3.3   Limiting cases for the relations of macro- and microequilibria

A typical bioligand contains functional groups of different inherent basicity which are in interaction with each other. There are, however, some special cases that can be demonstrated in terms of limiting relationships between macro- and microconstants. In this aspect divalent bases (or dibasic acids) have the advantage of simplicity in treatment. It can be seen from equations (26) and (27) that there are at least three microconstants expressed by only two independent macroconstants. Thus, microconstants cannot be expressed from macroconstants only. The exceptions are as follows.

(1) The molecule is symmetrical like oxalic, malonic, succinic ... etc. acid, or oxalate, malonate, succinate ..., etc., bases. For these compounds $k^A = k^B$ and $k_B^A = k_A^B$. This is a new relationship, by which $k^A = k^B = \beta_1/2$ and $k_B^A = k_A^B = 2K_2$. Note that there is no definite relation between k^A and k_B^A or k^B and k_A^B and so between K_1 and K_2. It depends chemically on the extent of interaction (distance, etc.) of the groups. The interactivity parameter can be calculated easily in such cases:

$$\Delta \log k_{A-B} = \log K_1 - \log K_2 - 0.6 \ .$$

(2) The molecule is not symmetrical but the sites are separated by several intervening atoms. This is the case of group constants, which can be divided into two subcategories.

(a) There is no out-of-chain interaction between the groups.

$$k^A = k_B^A = k_A \ ; \qquad k^B = k_A^B = k_B \ .$$

Accordingly $\beta_1 = K_1 = k_A + k_B$; $\beta_2 = K_1 K_2 = k_A k_B$.

(b) The sites are separated by many intervening atoms, but they are coupled by the accepted proton. There is only one single type of monoprotonated species and equilibrium parameter.

$$\beta_1 = K_1 = k_f \ ; \qquad K_2 = k_d \ ; \qquad \beta_2 = K_1 K_2 = k_f k_d \ .$$

(3) The molecule is symmetrical and the sites are well separated. This is the 'statistical case': $k^A = k^B = k_B^A = k_A^B = K_1/2 = 2K_2$ [26Si, 30Mu, 86No1]. Therefore $K_1/K_2 = 4$; $\log K_1 - \log K_2 = 0.6$. This gives an explicit explanation why macro-constants (with the exception of monodentate ligands) in principle can never be assigned to functional groups. This also provides the minimum ratio of successive macroconstants, which is $K_1/K_2 = 4$ for bidentate ligands, $K_1/K_2 = 3$ and $K_2/K_3 = 3$ for tridentate ligands, etc. The statistical case (the perfect independence and symmetry) hardly occurs in practice; rigid, symmmetrical molecules with remote groups are necessary.

3.4 Submicroconstants

Microconstant values reflect basically the inherent electron density of the proton-binding site, which is modified by the protonation state of the surrounding groups. Another intramolecular modifying factor is the relative position of the flexible moieties in the molecule. When all these factors are defined, the proton-binding ability of the species is expressed in terms of submicroconstants. The most important intramolecular motion is rotation around the chemical bonds, thus submicro-constants and submicrospecies are practically identical with rota-microconstants and rota-microspecies, respectively. Microspecies are composites of rotameric forms.

Considering three staggered rotational positions around carbon–carbon single bonds, m such bonds in the molecule and 2^n microspecies, the number of rota-microspecies in solution is $2^2\,3^m$; which means an enormously great number of coexisting rota-microspecies. To date the concept for the determination of rota-microspecies concentrations and rota-microconstants has been elaborated in detail for amino acids only, where $m = 1$ (one rotational axis).

Fig. 2.7 shows amino acid rotameric forms and their proton-binding equilibria. R indicates the amino acid side chain. In rotamer t and g the bulkiest carboxyl(ate) and R groups are in trans and gauche position, respectively, whereas in rotamer h (which is often the most hindered one) all the three bulkiest groups are situated in gauche. Indices N and A refer to the amino nitrogen and alpha-carboxylate protonations, respectively. For simplicity in treatment and representation the minor protonation isomer (^N_Al$^0_+$), its rotamers, and their equilibria are neglected here.

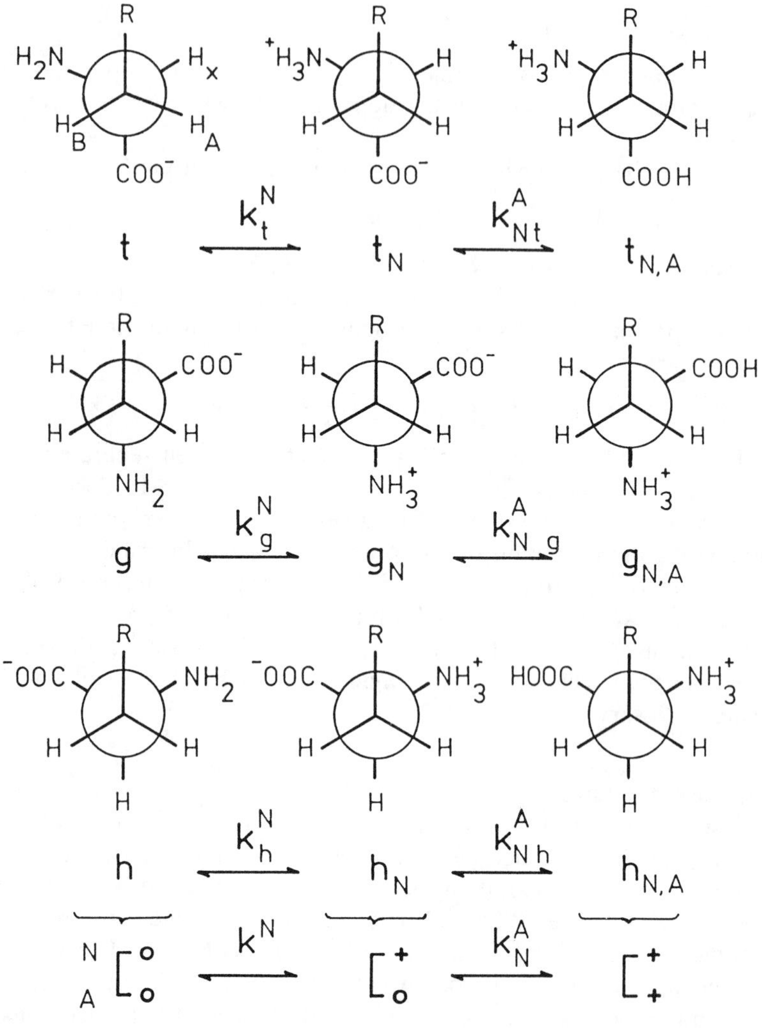

Fig. 2.7 — Protonation scheme of amino acid rotamers. The minor protonation isomers ($^{N[0}_{A[+}$), its rotamers and their equilibria are neglected here. See text for abbreviations.

Rota-microconstants can be deduced from microconstants and rotamer population values.

The microconstants belonging to the predominant protonation pathway can be expressed as:

$$k^N = \frac{[^+_0]}{[^0_0][H^+]} \quad (46) \qquad k^A_N = \frac{[^+_+]}{[^+_0][H^+]} \quad (47)$$

Populations for the unprotonated rotamers are as follows:

$$f_t = \frac{[t]}{[t]+[g]+[h]} = \frac{[t]}{[\ell_0^0]} \tag{48}$$

$$f_g = \frac{[g]}{[t]+[g]+[h]} = \frac{[g]}{[\ell_0^0]} \tag{49}$$

$$f_h = \frac{[h]}{[t]+[g]+[h]} = \frac{[h]}{[\ell_0^0]} \tag{50}$$

$$f_t + f_g + f_h = 1 \tag{51}$$

Examples for some other rotamer populations can be given analogously:

$$f_{g_N} = \frac{[g_N]}{[t_N]+[g_N]+[h_N]} = \frac{[g_N]}{[\ell_0^+]} \tag{52}$$

$$f_{g_{N,A}} = \frac{[g_{N,A}]}{[t_{N,A}]+[g_{N,A}]+[h_{N,A}]} = \frac{[g_{N,A}]}{[\ell_+^+]} \tag{53}$$

$$f_{t_N} + f_{g_N} + f_{h_N} = 1 \tag{54}$$

$$f_{t_{N,A}} + f_{g_{N,A}} + f_{h_{N,A}} = 1 \ . \tag{55}$$

The rota-microconstant for the *N*-protonation of rotamer g can be defined as:

$$k_g^N = \frac{[g_N]}{[g][H^+]} \ . \tag{56}$$

Expressing g_N and g concentrations from (52) and (49) gives:

$$k_g^N = \frac{f_{g_N}[\ell_0^+]}{f_g[\ell_0^0][H^+]} = \frac{f_{g_N}}{f_g} k^N \ . \tag{57}$$

As equation (57) shows, rota-microconstants fall near the relevant microconstant, modified by the appropriate rotamer populations.

3.4.1 The determination of rota-microconstants

Rotamer populations can be calculated from three-bond proton–proton (or proton–carbon) NMR coupling data [79Ma2, 75Ha, 62Ab, 64Pa]. For most amino acids the substituted ethane residue possesses a three-spin ABX pattern, owing to the non-equivalent two β- and one α-hydrogens. The observed $^3J_{(H_A - H_X)}$ and $^3J_{(H_B - H_X)}$ coupling 'constants' are independent of the magnetic field only, but they depend on pH, because of the change in rotamer populations. To elicit the rotamer mole fractions from the NMR coupling data, two-parameter formulations [72Ca, 64Pa, 62Ab, 76Es, 75Ha] and six-parameter formulations [76Fe, 76By] can be used. The more complicated six-parameter method may be marginally better only [79Ma2], thus we use former here:

$$J_{A,X} = f_t J_G + f_g J_T + f_h J_G \tag{58}$$

$$J_{B,X} = f_t J_T + f_g J_G + f_h J_G \tag{59}$$

where J_G and J_T are coupling parameters for the gauche and trans(anti) positions, respectively. Introducing (51) into (58) and (59), in the knowledge of J_T and J_G parameters as well as the experimental $J_{A,X}$ and $J_{B,X}$ values, the mole fractions can be calculated. Unambiguous f-values can be calculated only by this relatively simple method if one protonation form overwhelmingly predominates in the solution at the pH of the NMR measurement. For most amino acids this can be accomplished in very basic, neutral, and very acidic solutions. In cases of overlapping protonation equilibria, further considerations are necessary [89No2].

Concerning the NMR aspects of equations (58) and (59), the following problems occur. Since the H_A and H_B assignments are arbitrary, ^{1}H NMR coupling data alone do not provide sufficient evidence for the t and g assignments. This shortcoming can be overcome by using ^{13}C–^{1}H couplings as well [75Ha]. A further difficulty arises in acidic solutions, where the chemical shifts of the two β-methylene protons are very similar. Distinction between these two nuclei is possible by high-resolution NMR measurements only. The most widely accepted J_G and J_T constants for proton–proton couplings are 2.4 and 13.3 reported by Martin [79Ma2] as well as 2.56 and 13.6 reported by Hansen *et al.* [75Ha].

Several valuable contributions appeared on the rotamer analysis of aspartic acid [75Ha, 76Es, 73Is], valine [75Ha], phenylalanine [80Ve, 75Ha], tyrosine [80Ve], serine [67Og, 75Ha], threonine [75Ha], histidine [76Es, 73We], and numerous dipeptides [72Ba, 80Ve]. Nevertheless these papers did not contain rota-microconstants. Fujiwara *et al.* presented the first report on the determination of protonation constants for rotamers of serine and phenylalanine in basic, neutral, and acidic medium and for rotamers of aspartic acid and histidine in basic and neutral medium [74Fu]. The difference between the rota-microconstants and bulk constants varies in the range of 0.01–0.19 $\log k$ units. The last section of this chapter will show rota-microconstants for doubly isomeric species, where both protonation and rotational isomerism occur [89No2].

4. MICROSPECIATION AND ROTA-MICROSPECIATION

Speciation is a recent trend in clinical and environmental chemical analysis, using methods of equilibrium chemistry. Speciation works aim at determining not only the analytical concentration of a component in the sample, but also its distribution among different species. Microspeciation distinguishes between the different isomeric products of protonation [86No2], whereas rota-microspeciation, in addition, must specify the position of the flexible moieties of the molecule.

All these types of speciation need information about the appropriate equilibrium constants. Simple speciation uses thermodynamic macroconstants. Equation (17) in section 2.2 gives an example of how relative concentration of one of the species can be calculated at a given pH. A complete speciation calculates the distribution diagram for all the species throughout the entire pH range, as seen in Fig. 2.3 for a relatively simple and in Figs 2.1 and 2.2 for a more complicated binary (proton plus ligand) system. Such relative concentrations multiplied by the total ligand concentration give the real concentration of the species.

Microspeciation is based upon microconstants or group constants. The microconstant-based microspeciation can be done as follows. The relative concentration of one of the monoprotonated species ($\alpha_{[_0^+]}$) of a bidentate ligand can be expressed as:

$$\alpha_{[_0^+]} = \frac{[[_0^+]]}{[[_0^0]] + [[_0^+]] + [[_+^0]] + [[_+^+]]} . \tag{60}$$

Expressing the mono- and diprotonated microspecies concentrations by using equations (19)–(22), and then (26 and (27) gives:

$$\alpha_{[_0^+]} = \frac{k^B[H^+]}{1 + \beta_1[H^+] + \beta_2[H^+]^2} . \tag{61}$$

In a similar way, microspecies concentrations of more complicated molecules can be calculated from an analogous set of data. Microspecies distribution (together with rota-microspecies) for aspartic acid in the neutral–acidic pH region can be seen in Fig. 2.8.

For the group constant-based microspeciation first, we must classify groups. On this basis sites can be sorted into two classes:

(1) These groups are independent of the other protonating groups (but may take part in any kind of interaction with nonprotonating parts of the molecule). They protonate singly and exist in two possible stages of protonation: the unprotonated and protonated forms (A and AH, respectively, where the charges are not considered).

The relevant protonation equilibrium is characterized by the following group constant:

$$k_A = [AH]([A][H])^{-1} . \tag{62}$$

The probability of being protonated at such a group can be expressed by the ratio

$$P_{AH} = [AH]([A] + [AH])^{-1} . \tag{63}$$

Similarly the probability of being unprotonated is

$$P_A = [A]([A] + [AH])^{-1} . \tag{64}$$

Substituting (62) into (63) and (64) gives

$$P_{AH} = k_A[H](1 + k_A[H])^{-1} \tag{65}$$

$$P_A = (1 + k_A[H])^{-1} . \tag{66}$$

(2) These groups are divided into pairs with three stages of protonation: (a) the zero-proton-containing uncoupled stage: B,C; (b) the one-proton-containing coupled stage, holding the proton mutually: B-H-C; (c) the two-proton-containing uncoupled stage, holding one proton each: BH, CH.

Equilibrium between these positions can be depicted by the formation and the decomposition constants of the H-bond:

$$k_{fB,C} = [B\text{-}H\text{-}C]([B,C][H])^{-1} \tag{67}$$

$$k_{dB,C} = [BH,CH]([B\text{-}H\text{-}C][H])^{-1} . \tag{68}$$

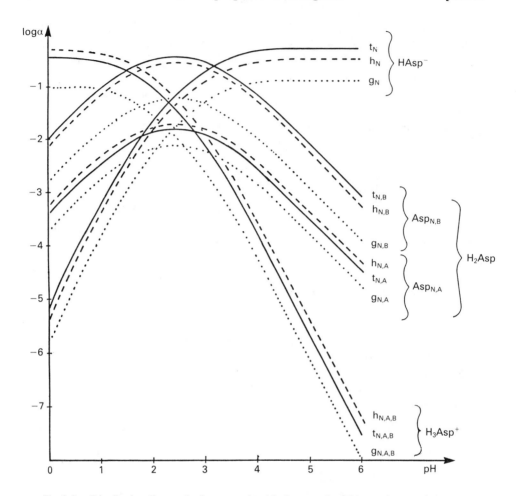

Fig. 2.8 — Distribution diagram for four aspartic acid micro-species ($HAsp^-$, $Asp_{N,B}$, $Asp_{N,A}$, H_3Asp^+) and their rotameters in the acidic and neutral pH region.

The probability of these successive stages is expressed by the group pair concentrations:

$$P_{B,C} = [B,C]([B,C] + [B\text{-}H\text{-}C] + [BH,CH])^{-1} \tag{69}$$

$$P_{B\text{-}H\text{-}C} = [B\text{-}H\text{-}C]([B,C] + [B\text{-}H\text{-}C] + [BH,CH])^{-1} \tag{70}$$

$$P_{BH,CH} = [BH,CH]([B,C] + [B\text{-}H\text{-}C] + [BH,CH])^{-1} . \tag{71}$$

By making use of equilibrium equations (67) and (68), (69)–(71) may be rewritten as (72)–(74):

$$P_{B,C} = (1 + k_{fB,C}[H] + k_{fB,C}k_{dB,C}[H]^2)^{-1} \tag{72}$$

$$P_{B\text{-}H\text{-}C} = k_{fB,C}[H](1 + k_{fB,C}[H] + k_{fB,C}k_{dB,C}[H]^2)^{-1} \tag{73}$$

$$P_{\text{BH,CH}} = k_{\text{fB,C}} k_{\text{dB,C}} [\text{H}]^2 (1 + k_{\text{fB,C}} [\text{H}] + k_{\text{fB,C}} k_{\text{dB,C}} [\text{H}]^2)^{-1} . \qquad (74)$$

Equations (65), (66), and (72)–(74) show that the probability of any stage of individual or coupled groups can be expressed by the appropriate groups constants and hydrogen ion concentrations. The above equations illustrate that an uncoupling group occurs at two levels of protonation and a coupling pair of groups exists in three stages. Accordingly, a molecule of h group pairs and $n - 2h$ uncoupling groups may occur in $2^{n-2h} 3^h$ different microscopic forms. The occurrence probability of the molecule in any protonating stage at a given pH is the product of probabilities belonging to (independent or coupled) groups in the appropriate protonation stage at that particular pH. For example, if a molecule contains two independent groups (A and D) and one group pair (B,C), it may occur in $2^4 = 2^2 3^1 = 12$ microscopic forms, which are combinations of the elementary forms of the groups.

The occurrence probability of the molecule when group A is bare, D is protonated, and the B,C pair binds one proton is

$$P_{\text{A,,B-H-C,DH}} = P_{\text{A}} P_{\text{B-H-C}} P_{\text{DH}} . \qquad (75)$$

This can be expressed in terms of group constants:

$$P_{\text{A,B-H-C,DH}} = (1 + k_{\text{A}} [\text{H}])^{-1} k_{\text{fB,C}} [\text{H}] \times$$
$$\times (1 + k_{\text{fB,C}} [\text{H}] + k_{\text{fB,C}} k_{\text{dB,C}} [\text{H}]^2)^{-1} k_{\text{D}} [\text{H}] (1 + k_{\text{D}} [\text{H}])^{-1} . \qquad (76)$$

The actual concentration of the molecule in any particular microscopic form can be calculated by multiplying the given probability by the total (analytical) concentration (C_{L}) of the substance (equation 77):

$$[\text{A,B-H-C,DH}] = P_{\text{A,B-H-C,DH}} C_{\text{L}} \qquad (77)$$

Fig. 2.9 shows the complete microspeciation diagram for glutathione (γ-glutamyl-cysteinyl-glycine) [86No2].

Rota-microspeciation may be based upon rota-microconstants or microconstants plus rotamer populations. For its simplicity, the latter will be shown here:

Equation (61) expresses the relative concentration $(\alpha_{\text{I}_0^+})$ for one of the microspecies. Microspecies are composites of rota-microspecies. The relative concentration of any rota-microspecies can be calculated by multiplying the appropriate microspecies relative concentration and rotamer population values. For example α_{g_N} (the relative concentration of a species protonated at the N-site in rotameric form g) can be given at any pH as:

$$\alpha_{\text{g}_N} = \alpha_{\text{I}_0^+} f_{\text{g}_N} . \qquad (78)$$

Note that both $\alpha_{\text{I}_0^+}$ and f_{g_N} are mole fraction-like quantities, of which $\alpha_{\text{I}_0^+}$ is pH-dependent, whereas f_{g_N} is pH-independent. A rota-microspeciation diagram for some aspartic acid species can be seen in Fig. 2.11 (see section 6).

5. CHARACTERISTIC PROTON-BINDING SITES OF BIOLIGANDS

Bioligands possess a wide variety of proton-binding sites in the sense of composition and basicity (which is one of the sources of the exuberant versatility of life on Earth), thus no kind of classification can be perfect.

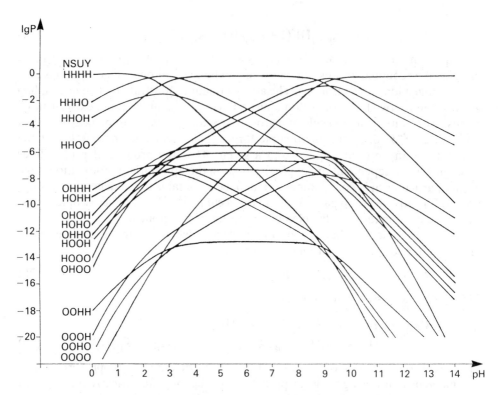

Fig. 2.9 — Complete distribution diagram of 16 glutathion microspecies. N, S, Y, and U are group legends indicating the amino, thiolate, glycyl carboxylate, and glutamyl carboxylate sites, respecively. H and O signals below them (regardless of charge) show whether the given group holds proton (H) or not (O). A set of such signals in a horizontal row defines a particular protonation stage of a glutathion molecule. The parallel running curves refer to protonation isomers.

Here the acid–base chemistry of major types of bioligands (nucleic bases as constituents of nucleic acids as well as amino acids and some peptide residues as constituents of proteins) will be discussed in detail. References for the basicity of other important bioligands are compiled in section (5.3).

5.1 Nucleic bases

Nucleic bases have a very correlative electron structure and take part in several tautomerisms, so intracyclic microconstants are very difficult to determine. The constants listed in Table 2.2 are considered either macroconstants or microconstants of the main protonation pathway, owing to the sufficient difference between the stepwise values. Since Fig. 2.10 shows the predominant tautomers in neutral solution, data in Table 2.2 refer to protonation, when $\log k < 7$ and proton-dissociation when $\log k > 7$. The pH-dependent charge of the molecule can be calculated accordingly.

Constants of the relevant nucleosides are 0.2–$0.5 \log k$ value smaller, owing to the

Table 2.2 — Protonation constants for nucleic bases

Compound	$\log k$	Temperature, ionic strength	References	Related literature
Adenine				
N9	9.7	25°C, 0.1	74Ma2	75Sm, 85Ma, 79Ma1
N1	4.1		72Zi	
Hypoxanthine				
N9	11.4	25°C, 0.1	76Li	74Ma2, 85Ma, 79Ma1
N1	8.7		79Ra	
N7	2.0			
Guanine				
N9	12.3	25°C, ?	48Ta	85Ma, 79Ma1
N1	9.2			
N7	3.3			
Cytosine				
N1	11.7	25°C, 0.1	69Re	85Ma, 79Ma1, 74Ma2
N3	4.4		67Ch	
Thymine				
N1	> 13	25°C, 0.1	70Ch	85Ma, 79Ma1, 74Ma2
N3	9.7		78Sr	
Uracyl				
N1	13.3	25°C, 0.001	78Sr	85Ma, 79Ma1, 74Ma2
N3	9.4	25°C, 0.1	61Gu	

electron-attractive effect of the hydroxyl groups on the sugar residue. Ribose and deoxyribose bind at N9 in purines and N1 in pyrimidines, so the N9 and N1 constants disappear in nucleosides, nucleotides, and nucleic acids. The *N*-basicities in the corresponding nucleoside phosphates are virtually the same as in nucleosides for sites with $\log k < 7$, and 0.1–$0.4 \log k$ units greater for sites with $\log k > 7$, owing to the basicity-increasing effect of the predominating twice negatively charged phosphate residue above pH 6.3. Effects of the state of protonation on the base-pairing ability of the nucleosides are surveyed in Martin's reviews (85Ma2, 79Ma1].

5.2 Amino acids and peptide residues
Table 2.3 contains protonation microconstants for basic sites in the given protonation state of the molecule in the 'compound or moiety' column. Protonation sites are always here in their basic form, while the surrounding moieties are formulated in their predominant form at the pH, when protonation of the site in question occurs. $\Delta \log k$ values quantify the effect of protonation or deprotonation of the indicated functional group on the basicity of the site in question. Negative $\Delta \log k$ values and the attached unprotonated form of the influencing group (for example, α-NH_2 or -COO^-) mean a $\log k$-decreasing effect.

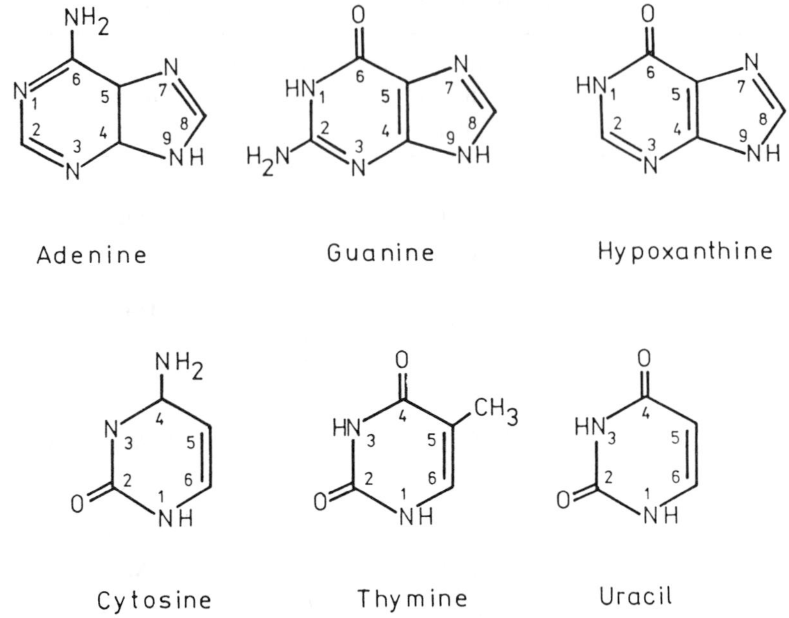

Adenine   Guanine   Hypoxanthine

Cytosine   Thymine   Uracil

Fig. 2.10 — Predominant protonation and tautomeric forms of five natural nucleic bases and hypoxanthine in neutral solution.

For example, the guanidino basicity of arginine is $\log k = 14.7$ when the α-amino and the α-carboxylate are both in their protonated (α-NH$_3^+$, α-COOH) form. This compilation contains data for those amino acids and peptide residues where microconstants were available. Abbreviations are as follows. TP3, TP4 are thymopoiethine fragments, HArg-Lys-AspOH, HArg-Lys-Asp-ValOH, respectively. Pe stand for $-$CO$-$NH$-$ peptide bond and the oncoming peptide sequences. The number of decimals is the same as in the original reference. The α-amino–α-carboxylate interaction is given for glycine only; nevertheless, it is approximately valid for all other amino acids.

Similarly, when a trifunctional amino acid is the *N*- or *C*-terminal residue of a peptide, the basicity of the two free (non-peptide-bonded) functional groups changes, but their interactivity parameter remains as in the monomeric amino acid. Imidazole $\log k$ values in peptides strongly depend on the surrounding aromatic side chains, owing to intramolecular stacking, hydrophobic interactions.

5.3 Miscellaneous

Microconstants or interactivity parameters are published between the phenolate and amino sites for dopamin and DOPA [71Ma]; the tertiary amino, en-amide, and phenolate sites for tetracyclines [85Ma1]; the selenolate and carboxylate for selenoglyoxalic acid [70Ku]; the carboxylates for citric acid [61Ma]; the aromatic amino and carboxylate for 4-aminobenzoic acid [81Va]; the phenolate and amino for salbuta-

Table 2.3 — Prontonation constants (log k) and interactivity parameters (Δlog k) for amino acid and peptide proton-binding sites

Site	Compound or moiety	Example	$\log k$	$\Delta \log k$	Ionic strength ($\mathrm{mol\,dm^{-3}}$) and temperature	Reference	Related literature
guanidino	NH₂—C=NH—NH—(CH₂)₃—H₂H—CH—COO⁻	arginine	15	α-NH₂ −0.2 / α-COO⁻ −0.1	2.0 / 25.0°C	90No4	
ε-amino	H₂N—(CH₂)₄—H₂N—CH—COO⁻	lysine	10.76	α-NH₂ −0.5 / α-COO⁻ −0.38	0.2–0.4 / 25.0°C	76Ra / 33Ed	76Sa / 80Su
	H₂N—(CH₂)₄—Pe—CH—COO⁻	gly-lys	10.57		0.2–0.4 / 25.0°C	76Sa	77Ra
	H₂N—(CH₂)₄—H₂N—CH—Pe	lys-gly	10.72		0.2–0.4 / 25.0°C	76Sa	

Table 2.3 — Prontonation constants (log k) and interactivity parameters ($\Delta\log k$) for amino acid and peptide proton-binding sites

Site	Compound or moiety	Example	log k	$\Delta\log k$	Ionic strength (mol dm^{-3}) and temperature	Reference	Related literature
	H$_2$N—(CH$_2$)$_4$ Pe—CH—Pe	TP3, TP4	10.6		0.5 25.0°C	90Sc	
δ-amino	H$_2$N—(CH$_2$)$_3$ H$_2$N—CH—COO$^-$	ornithine	10.55	α-NH$_2$ −0.76	0.2–0.4 25.0°C	76Sa	
α-amino	H$_2$N—CH$_2$—COO$^-$	glycine	9.78	α-COO$^-$ −2.18	0 25.0°C	58Ed2	
	H$_2$N—CH$_2$—Pe	triglicyne	7.89		0.1 25.0°C	73Fe 73Ya	
phenolate	O$^-$—⟨phenyl⟩—CH$_2$ H$_2$N—CH—COO$^-$	tyrosine	10.04	α-NH$_2$ −0.41	0.2 25.0°C	58Ma1 82Ki2	76Is 58Ed 65Ya 66Co 72Ni 84Ki

thiolate

Structure	Name	pK		I / Temp.	References
$O^- $—⟨benzene⟩—$CH_2$—$CH$(Pe)—Pe	ACTH, angiotensin	10.0		0.3 / 25.0°C	89No1 / 90Ny
O^-—⟨benzene⟩—$CH(NH_2)$—COO^-	4-hydroxy-phenyl-glycine	9.60	α-NH_2 -0.54	1.0 / 25.0°C	90No5
S^-—CH_2—$CH(NH_2)$—COO^-	cysteine	10.03	α-NH_2 -1.5 / α-COO^- -1.0	0–0.4 / 23.°C	55Be / 62El 64Wr 71Cl 60Go 74Wa 76Re
S^-—CH_2—CH(Pe)—Pe	gluta thione	8.93, 9.08		0.2–0.55 / 25.0°C	73Ra / 58Ma2 72Ja

Table 2.3 — Prontonation constants ($\log k$) and interactivity parameters ($\Delta \log k$) for amino acid and peptide proton-binding sites

Site	Compound or moiety	Example	$\log k$	$\Delta \log k$	Ionic strength $(\mathrm{mol\,dm^{-3}})$ and temperature	Reference	Related literature
imidazole		histidine	6.7	$\alpha\text{-NH}_3^+$ 0.7 $\alpha\text{-COO}^-$ -1.3	0.15 25.0°C	74Su	77Ra 83Ta1 83Ta2 83Ta3
		gly-his	6.82, 7.14		0.15–0.19 25.0°C	77Ra	
		angiotensin	5.89–6.46		0.3 25.0°C	90Ny	

δ-carboxylate	COO^-—$(CH_2)_2$—$CH(^+H_3N)$—COO^-	glutamic acid	4.32	α-NH_3^+ 0.77–0.80 α-COO^- −0.47	0 25.0°C	36Ne
β-carboxylate	COO^-—CH_2—$CH(^+H_3N)$—COO^-	aspartic acid	3.24	α-COO^- −0.39	2.0 25.0°C	89No2
	COO^-—CH_2—$CH(Pe)$—COO^-	TP3	4.41	−0.39	0.5 25.0°C	90Sc
	COO^-—CH_2—$CH(^+H_3N)$—Pe	angio-tensin	3.85	α-NH_3^+ 1.2	0.3 25.0°C	90Ny
α-carboxylate	^+H_3N—CH_2—COO^-	glycine	2.35	α–NH_3^+ 2.18 β-$COOH$ 0.39	0 25.0°C 2.0 25.0°C	58Ed2 89No2
	$COOH$—CH_2—$CH(^+H_3N)$—COO^-	aspartic acid	1.98			

Table 2.3 — Prontonation constants ($\log k$) and interactivity parameters ($\Delta \log k$) for amino acid and peptide proton-binding sites

Site	Compound or moiety	Example	$\log k$	$\Delta \log k$	Ionic strength $(\mathrm{mol\,dm^{-3}})$ and temperature	Reference	Related literature
	Pe—CH$_2$—COO$^-$	triglycine	3.20		0.1 25.0°C	73Fe 73Ya	
	COOH $\|$ CH$_2$ $\|$ Pe—CH—COO$^-$	TP3	3.33	β-COOH 0.39	0.5 25.0°C	90Sc	

mol [84Pa]; the different amino groups for aminoglycoside antibiotics [90No3]; the cyclic secondary (tertiary) amine and aromatic carboxylate for norfloxacine and pefloxacine [90Ta]; and the aminooxy and phenolate for benzyloxyamine [86Di]. Further references are given in 3.1.1.

6. THE EXAMPLE OF ASPARTIC ACID

The acid–base equilibria of aspartic acid are composite processes. As Fig. 2.11 shows, solutions of aspartic acid contain five protonation microspecies in considerable concentration, and thus fifteen rota-microspecies. The other 3 extremely minor microspecies and the relevant 9 rota-microspecies are neglected here. Despite difficulties in the determination of either the microconstants or the rotamer populations, aspartic acid is at present the most thoroughly elaborated example of bioligand protonations. Using pH-metry and NMR spectrometry for some auxiliary ligands as well (α-alanine, β-alanine, asparagine) all rotamer populations and rota-microconstants could be determined [89No2] see (Tables 2.4 and 2.5).

REFERENCES

23Bj	Bjerrum, J.: *Z. Physic. Chem.* **106**, 209 (1923).
26Eb	Ebert, L.: *Z. Physic. Chem.* **121**, 385 (1926).
26Si	Simms, H. S.: *J. Am. Chem. Soc.* **48**, 1239 (1926).
30Mu	von Muralt, A.: *J. Am. Chem. Soc.* **52**, 3518 (1930).
33Ed	Edsall, J. T. & Blanchard, M. H.: *J. Am. Chem. Soc.* **55**, 2337 (1933).
36Ne	Neuberger, A.: *Biochem. J.* **30**, 2085 (1936).
41Bj	Bjerrum, J.: *Metal ammine formation in aqueous solution*, Haase, Copenhagen, 1941.
44Ry	Ryklan, L. & Schmidt, C. L. A.: *Arch. Biochem.* **5**, 89 (1944).
48Ta	Taylor, H. F. W.: *J. Chem. Soc.* **765** (1948).
55Be	Benesch, R. E. & Benesch, R.: *J. Am. Chem. Soc.* **77**, 5877 (1955).
55Gr	Grafius, M. A. & Neilands, J. B.: *J. Am. Chem. Soc.* **77**, 3389 (1955).
58Ed1	Edsall, J. T., Martin, R. B., & Hollingworth, B. E.: *Proc. Natl. Acad. Sci. USA (Biochem.)* **44**, 505 (1958).
58Ed2	Edsall, J. T. & Wyman, J.: *Biophysical chemistry*; Academic, New York (1958) p. 487.
58Ma1	Martion, R. B., Edsall, J. T., Wetlaufer, D. B., & Hollingworth, B. R.: *J. Biol. Chem.* **233**, 1429 (1958).
58Ma2	Martin, R. B. & Edsall, J. T.: *Bull. Soc. Chim. Biol.* **40**, 1763 (1958).
60Go	Gorin, G. & Clary, C. W.: *Arch. Biochem. Biophys.* **90**, 40 (1960).
60Lo	Loevenstein, A. & Roberts, J. D.: *J. Am. Chem. Soc.* **82**, 2705 (1960).
61Gu	Gut, J., Prystas, M., Jonas, J., & Sorm, F.: *Collect. Czech. Chem. Commun.* **26**, 974 (1961).
61Ma	Martin, R. B.: *J. Phys. Chem.* **65**, 2053 (1961).
61Ro	Rossotti, F. J. C. & Rossotti, H. S.: *The determination of stability constants*; McGraw-Hill: New York (1961).
62Ab	Abraham, R. J. & McLaughlin, K. A.: *Mol. Phys.* **5**, 513 (1962).
62El	Elson, E. L. & Edsall, J. T.: *Biochemistry* **1**, 1 (1962).

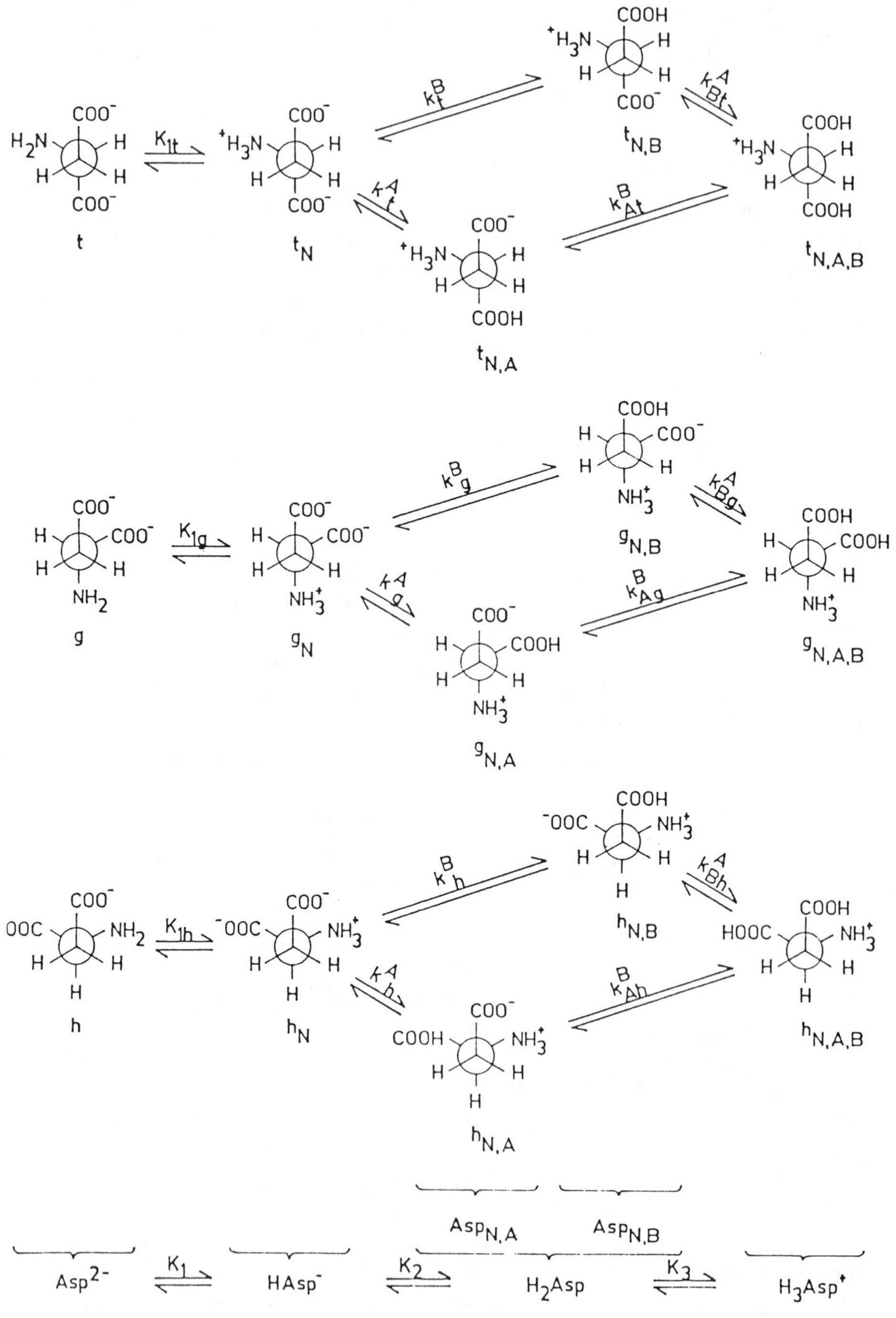

Acid-base properties of bioligands

Fig. 2.11 — Microspecies, rota-microspecies, and their equilibria of aspartic acid.

Table 2.4 — Populations of aspartic acid rotamers. See also Fig. 2.11 for abbreviations

Microform	Rotamer population		
	t	g	h
Asp^2	0.63	0.14	0.23
$HAsp^-$	0.49	0.12	0.39
$H_2Asp \begin{cases} Asp_{N,A} \\ Asp_{N,B} \end{cases}$	0.45	0.13	0.42
	0.42	0.14	0.44
H_3Asp^+	0.34	0.16	0.50

Table 2.5 — Bulk (macro- or micro-) constants and rota-microconstants of aspartic acid in $\log k$ units. The number of decimals is proportional to the accuracy

Macro- or microconstant		Rota-microconstants					
		t		g		h	
Symbol	Value	Symbol	Value	Symbol	Value	Symbol	Value
K_1	9.53	K_{1t}	9.42	K_{1g}	9.46	K_{1h}	9.76
k^B	3.63	k_t^B	3.6	k_g^B	3.7	k_h^B	3.7
k^A	2.37	k_t^A	2.3	k_g^A	2.4	k_h^A	2.4
k_A^B	3.24	k_{At}^B	3.1	k_{Ag}^B	3.3	k_{Ah}^B	3.3
k_B^A	1.98	k_{Bt}^A	1.9	k_{Bg}^A	2.1	k_{Bh}^A	2.0

62Ri Riegelman, S., Strait, L. A., & Fischer, E. Z.: *J. Pharm. Sci.* **51**, 129 (1962).

64Bu Butler, J. N.: *Ionic equilibrium: a mathematical approach*; Addison-Wesley: Reading, MA (1964).

64Pa Pachler, K. G. R.: *Spectrochim. Acta* **20**, 581 (1964).

64Si Sillén, L. G. & Martell, A. E.: *Stability constants of metal-ion complexes* Vol. I, (1964), Vol. II (1971).

64Wr Wrathall, D. C., Izatt, LR. M., & Christensen, J. J.: *J. Am. Chem. Soc.* **86**, 4779 (1964).

65Ka Kappe, T. & Armstrong, M.: *J. Med. Chem.* **8**, 368 (1965).

65Ri Rigler, N. E., Bug, S. P., Leyden, D. E., Sudmeier, J. L., & Reilley, C. N.: *Anal. Chem.* **37**, 872 (1965).

65Ya   Yamazaki, F., Fujidi, K., & Murata, Y.: *Bull. Chem. Soc. Japan* **38**, 8 (1965).

60Co   Coates, E., Gardam, P. G., & Rigg, B.: *J. Chem. Soc., Faraday Trans*. **62**, 2577 (1966).

67Ch   Christensen, J. J., Ryttig, J. H., & Izatt, R. M.: *J. Phys. Chem.* **71**, 3001 (1967).

67Og   Ogura, H., Arata, Y., & Fujiwara, S.: *J. Mol. Spectr.* **23**, 76 (1967).

69Co   Coates, E., Marsden, C. G., & Rigg, B.: *J. Chem. Soc. Faraday Trans*. **65**, 3032 (1969).

69Ke   Kesselring, U. W. & Benet, L. Z.: *Anal. Chem.* **41**, 1535 (1969).

69Re   Reinert, H. & Weiss, R.: *Z. Physiol. Chem.* **350**, 1310 (1969).

70Be   Beck, M. T.: *Chemistry of complex equilibria*; Akadémiai Kiadó, Budapest (1970).

70Ch   Christensen, J. J., Ryttig, J. H., & Izatt, R. M.: *J. Chem. Soc. (B)*, 1643 (1970).

70Ku   Kurz, J. L. & Harris, J. C.: *J. Org. Chem.* **35**, 3086 (1970).

70Ta   Tanaka, H., Sakurai, H., & Yokoyama, A.: *Chem. Pharm. Bull.* **18**, 1015 (1970).

71Cl   Clement, G. E. & Hartz, T. P.: *J. Chem. Educ.* **48**, 395 (1971).

71Ma   Martin, R. B.: *J. Phys. Chem.* **75**, 2657 (1971).

71Wa   Wilson, E. W. & Martin, R. B.: *Arch. Biochem. Biophys.* **142**, 445 (1971).

72Ba   Bartle, K. D., Jones, D. W., & L'Amie, R.: *J. Chem. Soc. Perkin* **II**. 650 (1972).

72Ca   Casey, J. P. & Martin, R. B.: *J. Am. Chem. Soc.* **94**, 6141 (1972).

72Ja   Jagt, D. L. W., Hansen, L. D., Lewis, E. A., & Han, L. B.: *Arch. Biochem. Biophys.* **153**, 55 (1972).

72Ni   Niebergall, P. J., Schnaare, R. L., & Sugita, E. T.: *J. Pharm. Sci.* **61**, 232 (1972).

72Sh   Shrager, R. J., Cohen, J. S., Heller, R. S., Sachs, D. H., & Shechter, A. N.: *Biochemistry* **11**, 541 (1972).

72Zi   Zimmer, S. & Biltonen, R.: *J. Sol. Chem.* **1**, 291 (1972).

73Cr   Creyf, H. S., van Poucke, L. C., & Eeckhaut, Z.: *Talanta* **20**, 973 (1973).

73Is   Ishizuka, H., Yamamoto, T., Arata, Y., & Fujiwara, S.: *Bull. Chem. Soc. Japan* **46**, 468 (1973).

73Fe   Feige, P., Mocker, D., Dreyer, R., & Munze, R.: *J. Inorg. Nucl. Chem.* **35**, 3269 (1973).

73Ra   Rabenstein, D. L.: *J. Am. Chem. Soc.* **95**, 2797 (1973).

73Ya   Yamanchi, O., Nakao, Y., & Nakahara, A.: *Bull. Chem. Soc. Japan* **46**, 2119 (1973).

73We   Weinkam, R. J. & Jorgensen, E. C.: *J. Am. Chem. Soc.* **95**, 6184 (1973).

74Fu   Fujiwara, S., Ishizuko, H., & Fudano, S.: *Chemistry Letters* 1281 (1974).

74Ma1   Maker, G. K. R. & Williams, D. R.: *J. Inorg. Nucl. Chem.* **36**, 1675 (1974).

74Ma2   Martell, A. E. & Smith, R. M.: *Critical stability constants* I., III., V. Plenum, New York, London (1974, 1977, 1982).

74Su   Sundberg, R. J. & Martin, R. B.: *Chemical Reviews* **74**, 471 (1974).

74Wa   Walters, D. B. & Leydens, D. E.: *Anal. Chim. Acta* **72**, 275 (1974).

75Bo   Bogess, R. K. & Martin, R. B.: *J. Am. Chem. Soc.* **97**, 3076 (1975).

75Ha   Hansen, P. E., Feeney, J., & Roberts, G. C. K.: *J. Magn. Reson.* **17**, 249 (1975).

75Pe   Pearce, N. K. & Creamer, L. K.: *Aust. J. Chem.* **28**, 2409 (1975).

75Sm   Smith, R. M. & Martell, A. E.: *Critical stability constants* II., IV. Plenum, New York, London (1975, 1976).

76As   Asleson, G. L. & Frank, C. W.: *J. Am. Chem. Soc.* **98**, 4745 (1976).

76By   Bystrov, V. F.: *Progr. NMR Spectr.* **10**, 41 (1976).

76Es   Espersen, W. G. & Martin, R. B.: *J. Phys. Chem.* **80**, 741 (1976).

76Fe   Feeney, J.: *J. Magn. Reson.* **21**, 473 (1976).

76Gr   Granot, J.: *FEBS Letters* **67**, 271 (1976).

76Is   Ishimitsu, R., Hirose, S., & Sakurai, H.: *Chem. Pharm. Bull.* **24** (12) 3195 (1976).

76Li   Linder, P. W., Stanford, M. J., & Williams, D. R.: *J. Inorg. Nucl. Chem.* **38**, 1847 (1976).

76Ra   Rabenstein, D. L. & Sayer, T. L.: *Anal. Chem.* **48**, 1141 (1976).

76Re   Reuben, D. M. E. & Bruice, T. C.: *J. Am. Chem. Soc.* **98**, 114 (1976).

76Sa   Sayer, T. L. & Rabenstein, D. L.: *Can. J. Chem.* **54**, 3392 (1976).

76St   Streng, W. H., Huber, H. E., De Young, J. L., & Zoglio, M. A.: *J. Pharm. Sci.* **65**, 1034 (1976).

77Ga   Ganellin, R.: *J. Med. Chem.* **20**, 579 (1977).

77In   Inczédy, J.: *Analytical application of complex equilibria*; Akadémiai Kiadó, Budapest (1977).

77Is   Ishimitsu, T., Hirose, S., & Sakurai, H.: *Talanta* **24**, 555 (1977).

77Ra   Rabenstein, D. L., Greenberg, M. S., & Evans, C. A. *Biochem.* **16** (5), 977 (1977).

77Sa   Sarneski, J. E., Reilley, C. N.: The determination of proton binding sites by NMR-titrations In: *Essays on analytical chemistry*; Wänninen, E. ed., Pergamon, Oxford (1977) p. 35.

77St   Stugeon, R. J. & Schulman, S. G.: *J. Pharm. Sci.* **66**(7), 958 (1977).

78Ba   Barbucci, R., Ferruti, P., Micheloni, M., Delfini, L., Segre, A. L., & Conti, F.: *Polymer* **19** 1329 (1978).

78Is   Ishimitsu, T., Hirose, S., & Sakurai, H.: *Chem. Pharm. Bull.* **26**, 74 (1978).

78Ja   Jameson, R. F., Hunter, G., & Kiss, T.: *J. Chem. Soc. CH*, (17), 768 (1978).

78Sr   Srivastava, R. C. & Srivastava, M. N.: *J. Inorg. Nucl. Chem.* **40**, 1493 (1978).

79Is   Ishimitsu, T., Hirose, S., & Sakurai, H.: *Chem. Pharm. Bull.* **27**, 247 (1979).

79Ma1   Martin, R. B. & Mariam, Y. H.: Interactions between metal ions and nucleic bases, nucleosides in solution. *In metal ions in biological systems* Vol. 8. p. 62 ed. Sigel, Marcel Dekker New York, Basel (1979).

79Ma2   Martin, R. B.: *J. Phys. Chem.* **83**, 2404 (1979).

79No   Noszál, B. & Burger, K.: *Acta Chim. Hung.* **100**, 275 (1979).

79Ra   Randhawa, S., Pannu, B. S., & Chopra, S. L.: *Thermochim. Acta* **32**, 111 (1979).

79Se   Serjeant, E. P. & Dempsey, B.: *Ionisation constants of organic acids in aqueous solution* Pergamon, Oxford (1979).

80Sa   Sakurai, H., & Ishimitsu, T.: *Talanta* **27**, 2931 (1980).

80Su   Surprenant, H. L., Sarneski, J. E., Key, R. R., Byrd, J. T., & Reilley, C. T.: *J. Magn. Reson.* **40**, 231 (1980).

80Ve   Vestues, P. I. & Martin, R. B.: *J. Am. Chem. Soc.* **102**, 7906 (1980).

81Va   Vandegraaf, B., Hoefnage, A. J., & Wepster, B. M.: *J. Org. Chem.* **46**(4), 653 (1981).

82Ki1   Király, R. & Martin, R. B.: *Inorg. Chim. Acta* **67**, 13 (1982).

82Ki2   Kiss, T. & Tóth, B.: *Talanta*, **29**, 539 (1982).

83Ba   Barrera, H., Bayon, J. C., Gonzales-Duarte, P., Sola, J., & Vives, J.: *Talanta* **30**, 537 (1983).

83Hu   Huys, C. T., Geominne, A. M., & Eeckhaut, Z.: *Thermochim. Acta* **65**, 19 (1983).

83Ta1   Tanokura, M.: *Biochim. Biophys. Acta* **742**, 576 (1983).

83Ta2   Tanokura, M.: *Biochim. Biophys. Acta* **742**, 586 (1983).

83Ta3   Tanokura, M.: *Biopolymers* **22**, 2563 (1983).

84Ki   Kiss, T. & Gergely, A.: *J. Chem. Soc., Dalton Trans.* 1951 (1984).

84On   Onasch, F., Aikens, D., Bunce, S., Schwartz, H., Nairn, D., & Hurwitz, C.: *Biophys. Chem.* **19**, 245 (1984).

84Pa   Paál, T. L.: *Acta Chim. Hung.* **115**, 185 (1984).

84So   Sóvágó, I., Kiss, T., & Gergely, A.: *Inorg. Chim. Acta* **93** (4), L53 (1984).

85Le   Leggett, D. J.: *Computational methods for the determination of formation constants* Plenum Press, New York, London (1985).

85Ma1   Martin, R. B.: *Tetracyclines and daunorubicin. In metal ions in biological systems* Vol 19, p. 19, ed. H. Sigel, Marcel Dekker, New York, Basel (1985).

85Ma2   Martin, R. B.: *Acc. Chem. Res.* **18**, 32 (1985).

86Di   Dittgen, M., Noszál, B., & Stampf, G.: *Drug Develop. Industr. Pharm.* **12** (11–13), 1663 (1986).

86No1   Noszál, B.: *J. Phys. Chem.* **90**, 4104 (1986).

86No2   Noszál, B.: *J. Phys. Chem.* **90**, 6345 (1986).

87Mi   Kiss, T., Balla, J., Nagy, G., Kozlowski, H., & Kowalik, J.: *Inorg. Chim. Acta* **138**, 25 (1987).

87Mi   Milburn, P. J., Konishi, Y., Meinwald, Y. C., Scheraga, & H. A.: *J. Am. Chem. Soc.* **109**, 4486 (1987).

88Bu   Burger, K., Sipos, P., Véber, M., Horváth, I., Noszál, B., & Lów, M.: *Inorg. Chim. Acta* **152**, 233 (1988).

89Ki   Kiss, T., Sóvágó, I., & Martin, R. B.: *J. Am. Chem. Soc.* **111**, 3611 (1989).

89No1   Noszál, B. & Osztás, E.: *Int. J. Pept. Prot. Res.* **33**, 162 (1989).

89No2   Noszál, B. & Sándor, P.: *Anal. Chem.* **61**, 2631 (1989).

89Pa   Pandit, N. K. & Sisco, J. M.: *Pharm. Res.* **6**, 177 (1989).

90No1   Noszál, B. & Tánczos, R.: *Int. J. Pept. Prot. Res.* (in preparation).

90No2   Noszál, B., Schón, I., Nyéki, O., Tánczos, R., & Nyiri, J.: *Int. J. Pept. Prot. Res.* (in preparation).

90No3   Noszál, B. & Ivancsics, R.: *J. Antibiot* (in preparation).

90No4   Noszál, B. & Tánczos, R.: *Int. J. Pept. Prot. Res.* (in preparation).

90No5   Noszál, B., Nyíri, J., & Sztaricskai, F.: *J. Antibiot.* (in preparation).

90Ny   Nyéki, O., Noszál, B., Osztás, E., & Burger, K.: *Int. J. Pept. Prot. Res.* **35**, 424 (1990).

90Sc   Schón, I., Nyéki, O., Nyíri, J., Tánczos, R., & Noszál, B.: *Int. J. Pept. Prot. Res.* (in preparation).

90Ta   Takács-Novák, K., Noszál, B., Hermecz, I., Keresztúri, G., & Szász, G.: *J. Pharm. Sci.* **79**, 1 (1990).

III

Complexes of amino acids

Tamás Kiss
Department of Inorganic and Analytical Chemistry, L. Kossuth University,
4010 Debrecen, Hungary

1. INTRODUCTION

1.1 General

Of all the possible model systems involving metal ions and biologically important ligands, those relating to metal–amino acid interactions have been the longest and the most extensively studied. The reason is that proteins contain the same types of donor groups, and they can therefore serve as simple models for metal–protein reactions and also for biological systems in which the properties of the proteins are modified by having metal ions attached to them. With the availability of various peptides of increasing complexity, and peptides which are active centres of larger metalloproteins, it has recently become possible to study much more complicated model systems too. The coordination chemistry of oligopeptides is reviewed in Chapter IV.

It should be borne in mind that amino acids are interesting ligands in their own right. Their *in vivo* interaction with transition metal ions is of immense biological importance; for example, the role of amino acid mixed ligand complexes in the transport of copper [81Sa1)], the treatment of Wilson's disease with D-penicillamine [70Wa], and the biosynthesis of methionine from methylcobalamin and homocysteine [75Ch2].

Our approach in this chapter is to discuss the most important amino acids commonly appearing as ligands in proteins and to examine the stoichiometry and structure of the complexes formed, mainly with transition metal ions. The complex-forming properties of some newly investigated amino acid derivatives, such as aminophosphonates and aminohydroxamates, will also be dealt with.

The chapter ends with a brief discussion of the most important factors influencing the formation of mixed ligand complexes of amino acids. As not all of the amino acids listed in Fig. 3.1 will be considered in the same detail, we refer to some recent

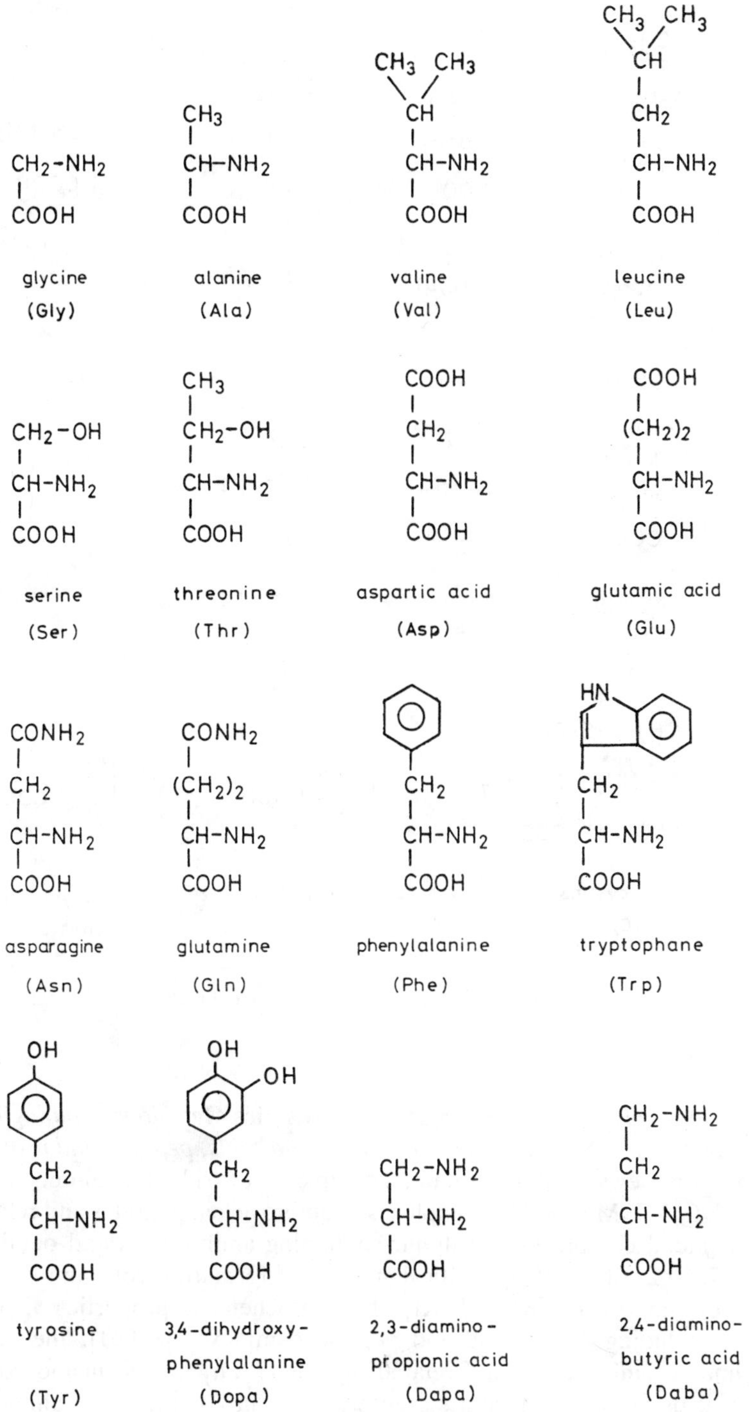

CH$_2$-NH$_2$
|
COOH

glycine

(Gly)

CH$_3$
|
CH-NH$_2$
|
COOH

alanine

(Ala)

CH$_3$ CH$_3$
\ /
CH
|
CH-NH$_2$
|
COOH

valine

(Val)

CH$_3$ CH$_3$
\ /
CH
|
CH$_2$
|
CH-NH$_2$
|
COOH

leucine

(Leu)

CH$_2$-OH
|
CH-NH$_2$
|
COOH

serine

(Ser)

CH$_3$
|
CH$_2$-OH
|
CH-NH$_2$
|
COOH

threonine

(Thr)

COOH
|
CH$_2$
|
CH-NH$_2$
|
COOH

aspartic acid

(Asp)

COOH
|
(CH$_2$)$_2$
|
CH-NH$_2$
|
COOH

glutamic acid

(Glu)

CONH$_2$
|
CH$_2$
|
CH-NH$_2$
|
COOH

asparagine

(Asn)

CONH$_2$
|
(CH$_2$)$_2$
|
CH-NH$_2$
|
COOH

glutamine

(Gln)

CH$_2$
|
CH-NH$_2$
|
COOH

phenylalanine

(Phe)

CH$_2$
|
CH-NH$_2$
|
COOH

tryptophane

(Trp)

OH

CH$_2$
|
CH-NH$_2$
|
COOH

tyrosine

(Tyr)

OH
OH

CH$_2$
|
CH-NH$_2$
|
COOH

3,4-dihydroxy-
phenylalanine

(Dopa)

CH$_2$-NH$_2$
|
CH-NH$_2$
|
COOH

2,3-diamino-
propionic acid

(Dapa)

CH$_2$-NH$_2$
|
CH$_2$
|
CH-NH$_2$
|
COOH

2,4-diamino-
butyric acid

(Daba)

Fig. 3.1 — α-amino acids reviewed in this chapter.

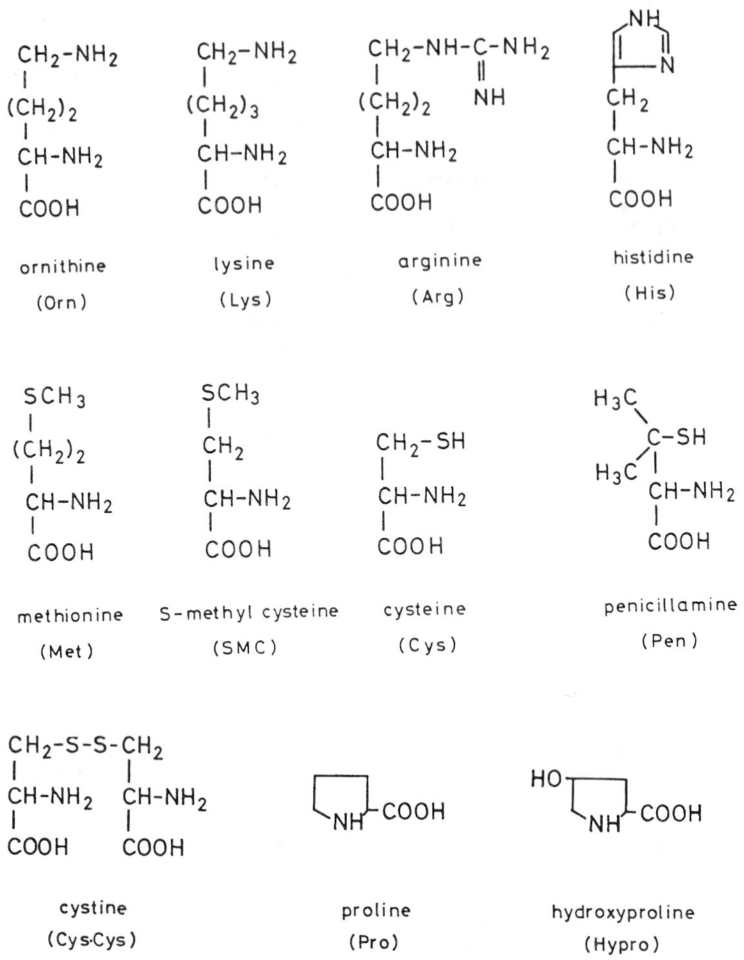

CH_2-NH_2
$|$
$(CH_2)_2$
$|$
$CH-NH_2$
$|$
$COOH$

ornithine

(Orn)

CH_2-NH_2
$|$
$(CH_2)_3$
$|$
$CH-NH_2$
$|$
$COOH$

lysine

(Lys)

$CH_2-NH-C-NH_2$
$|$ $\|$
$(CH_2)_2$ NH
$|$
$CH-NH_2$
$|$
$COOH$

arginine

(Arg)

histidine

(His)

SCH_3
$|$
$(CH_2)_2$
$|$
$CH-NH_2$
$|$
$COOH$

methionine

(Met)

SCH_3
$|$
CH_2
$|$
$CH-NH_2$
$|$
$COOH$

S-methyl cysteine

(SMC)

CH_2-SH
$|$
$CH-NH_2$
$|$
$COOH$

cysteine

(Cys)

penicillamine

(Pen)

$CH_2-S-S-CH_2$
$|$ $|$
$CH-NH_2$ $CH-NH_2$
$|$ $|$
$COOH$ $COOH$

cystine

(Cys·Cys)

proline

(Pro)

hydroxyproline

(Hypro)

Fig. 3.1 — α-amino acids reviewed in this chapter.

publications in this field: the excellent review series *Metal ions in biological systems*, edited by Sigel [74Si], the annual review *Amino acids, peptides and proteins* [69Sh], and other reviews dealing with metal complexes of tridentate amino acids [75Ch2, 79Ev, 79Gel, 79Ma], Pd(II) complexes of amino acids, peptides and related ligands [85Pe], metal complexes of sulphur-containing amino acids and peptides [72Mc, 74Mc, 79Ge2, 81So, 85Pe], critical surveys of formation constants of metal complexes of amino acids [84Pe, 90Ki], the photochemical properties of copper complexes, including amino acid and peptide complexes [81Fe1], the activation of coordinated amino acids and related ligands [79Ph], mixed ligand complexes of amino acids [73Si, 80Si], and stereoselectivity in complexes of amino acids and related compounds [67Gi, 74Be, 79Pe].

1.2   Acid base properties of amino acids

Figure 3.1 contains a list of the amino acids discussed in this chapter.

All amino acids undergo two reversible proton dissociation steps (1) in fairly well-separated pH ranges:

$$^+H_3NCH(R)COOH \xrightarrow{\text{pH 2–3}} {}^+H_3NCH(R)COO^- \xrightarrow{\text{pH 8–9}} H_2NCH(R)COO^-$$

$$(1)$$

Besides these two functional groups, most of the essential amino acids contain further acidic protons in the side-chain R. These side-chain groups play important roles in biological processes of peptides and proteins, as potential sites of interaction with protons, metal ions, or other charged biomolecules. These acidic functions and their pH ranges of deprotonation are presented in Table 3.1.

Table 3.1 — Acidic functions of α-amino acids and their deprotonation pH ranges

pH range		Acidic functions
2.0–3.0	α – COOH	C-terminal acid group
2.5–3.5	β – COOH	Aspartate
3.8–4.8	γ – COOH	Glutamate
5.5–6.5	(imidazolyl structure)	Imidazolyl of histidine
8.5–9.5	–SH	Sulfhydryl group of cysteine
9.0–10.0	(phenol structure)	Phenolic group of tyrosine
10.2–11.2	ω – NH$_3^+$	Lysine, Ornithine
8.0–9.0	α – NH$_3^+$	N-terminal amino group

It can be seen from Table 3.1 that the deprotonation of some of the side-chain groups occurs close to that of one of the terminal acidic groups. Hence, these proton dissociation processes overlap one another and take place in parallel, which can be characterized by microscopic acidity constants:

$$
\begin{bmatrix} 1 \\ 2H \end{bmatrix}
\quad
\begin{matrix} k_1 \nearrow & & k_{12} \searrow \end{matrix}
$$

$$
\begin{bmatrix} 1H \\ 2H \end{bmatrix}
\qquad\qquad\qquad\qquad
\begin{bmatrix} 1 \\ 2 \end{bmatrix}
\tag{2}
$$

$$
\begin{matrix} k_2 \searrow & & k_{21} \nearrow \end{matrix}
\quad
\begin{bmatrix} 1H \\ 2 \end{bmatrix}
$$

Note. Many different notations are accepted and used in the literature for the microscopic constants. Here, the subscripts 1 and 2 are assigned to the two acidic groups. The final subscript in a microscopic acidity constant refers to the group undergoing deprotonation.

The macroscopic acidity constant K_{HA} and K_{H_2A} are related to the microscopic acidity constants by equations (3) and (4):

$$
K_{HL} = k_1 + k_2 \quad (K_{H_2L})^{-1} = (k_{12})^{-1} + (k_{21})^{-1}
\tag{3}
$$

$$
\beta_{H_2L} = K_{HL} K_{H_2L} = k_1 k_{12} = k_{21}
\tag{4}
$$

Additional information is provided by the molar ratio of the two microspecies $R = [12/H]/[H12] = k_1/k_2$. The ratio R is characteristic of the extent of overlap of the two dissociation microprocesses, that is, the difference in the real acidities of the functional groups. The higher the deviation of ratio R from unity, the larger the difference in acidity of the two groups and the larger the separation of the two dissociation processes.

    The reciprocal effects of the two functional group deprotonations on each other (the interactivity parameter, as it is named in Chapter II) may be described by the ratio $S = k_1/k_{21} = k_2/k_{12}$ [58Ma]. The more separated the two acidic groups within the molecule, the less each affects the dissociation of the other group and the more closely the ratio S approaches unity. If the ratio $S = 1$, that is, there is no interaction between the acidic groups (this can be observed only in larger biomolecules, and never in the amino acids discussed in this chapter), the acidity of each donor group can be characterized by a single group constant [86No] instead of by microconstants.

The relation between the macroscopic acidity constants and the ratios R and S is $(R+1)^2/R = K_{HL}/K_{H_2L}S$ [71Ma].

The macroscopic acidity constants and the ratios R and S are suitable for the complete characterization of the acid–base chemistry of the tridentate amino acids with overlapping dissociations at both macroscopic and molecular levels. The data on the amino acids which will be discussed in this chapter are listed in Table 3.2.

Table 3.2 — Acid dissociation constants† of the essential α-amino acids

Amino acid		$pK_{\alpha\text{-COOH}}$	$pK_{NH_3^+}$	$pK_{sidechain}$	R	$\log S$
Glycine	Gly	2.36	9.56			
Alanine	Ala	2.31	9.70			
Valine	Val	2.26	9.49			
Leucine	Leu	2.27	9.57			
Serine	Ser	2.13	9.05			
Threonine	Thr	2.10	8.96			
Aspartic acid	Asp	1.94	9.62	3.70	50	0.39
Glutamic acid	Glu	2.18	9.59	4.20	18	0.47
Asparagine	Asn	2.15	8.72			
Glutamine	Gln	2.16	9.01			
Phenylalanine	Phe	2.17	9.11			
Tryptophane	Trp	2.34	9.32			
Tyrosine	Tyr	2.17	9.04	10.11	2.2	0.41
3,4-Dihydroxy-phenylalanine	Dopa	2.20	8.72	9.78 13.4	0.55	0.44
2,3-Diamino-propionic acid	Dapa	1.3	6.68	9.40	1.1	2.21
2,4-Diamino-butyric acid	Daba	1.7	8.19	10.21	5.0	1.18
Ornithine	Orn	1.9	8.78	10.54	7.6	0.68
Lysine	Lys	2.19	9.12	10.68	9.8	0.33
Arginine	Arg	2.3	9.02	15		
Histidine	His	1.7	9.09	6.02	0.02	1.3
Methionine	Met	2.10	9.05			
S-Methyl-cysteine	SMC	1.99	8.73			
Cysteine	Cys	1.91	8.16	10.29	0.48	1.79
Penicillamine	Pen	1.90	7.92	10.6	0.40	1.9
Cystine	Cys.Cys	<2, 2	8.03	8.80		
Proline	Pro	1.9	10.41			
Hydroxyproline	Hypro	1.8	9.47			

† Most values are taken from Ref. 82Ma1 at 25°C and 0.1 mol dm^{-3}, or from Refs given in the respective sections.

1.3 Aspects of metal ion coordination

Both in the solid state and in solution, simple amino acids can coordinate through either or both of the amino or carboxylate groups. The metal-binding capability of the terminal amino group is fairly strong, despite its high pK values. Coordination is favoured by the strong electron donor character and by the relatively strong ligand–field effect of NH_2 as compared with carboxylate. Monodentate coordination through the amino N donor atom is not so common, although it occurs not only in solid complexes, e.g. with Ag(I) or Pt(II), but also in solution, with kinetically inert metal ions, such as Cr(III), Co(III), Ir(II), Pt(II), and Rh(III). Monodentate interaction through the weaker ligand–field carboxylate O donor has been identified in a variety of crystal structures of complexes prepared at low pH. In these protonated complexes, carboxylate can coordinate via one or both O atoms. Carboxylate-coordinated complexes are always formed in solution too at pH 3–5, but it is difficult to detect them because they generally exist as minor species. However, the more common bonding mode is a bidentate chelate through the N and O donor atoms, which results in a thermodynamically stable five-membered ring for α-amino acids.

Besides the basic binding sites, that is, the α-amino and carboxylate groups, the donor atoms in the side chain offer further potential sites for interaction. The most obvious sign of side-chain involvement is an increase in the stability constants relative to those of simple bidentate amino acids (e.g. Ala). The pK values of the amino group and the stability constants for the first (K_{ML}) and the second (K_{ML_2}) steps of formation of the Co(II), Ni(II), Cu(II), Zn(II), and Cd(II) complexes of the amino acids discussed in this chapter are listed in Tables 3.3–3.7, respectively.

A simple comparison of stability constants in this way, however, is not a correct approach, because no allowance is made for the differences in ligand basicities. These differences are often ignored, although a greater proton affinity is expected to produce a greater metal ion-binding capability. Accordingly, a plot of $\log K_{ML}$ vs pK_{HL} should give a linear dependence for simple bidentate-coordinated amino acid complexes ML. On the basis of a careful analysis of the reported stability and structural results on the metal complexes of amino acid, Martin calculated the least square lines for Co(II), Ni(II), Cu(II), Zn(II) and Cd(II) [88Mal]. The ligands, in which the side chains were presumed not to interact with metal ions, or not to provide steric hindrance to chelation through the amino N and carboxylate O, were glutamine, phenylalanine, 2-amino-6-heptanoate, α-aminobutyrate, norleucine, and alanine. These ligands have a reasonable fit to a straight line with all five metal ions mentioned above. Any deviation from this straight line suggests some specific interaction with the side chain.

The equilibrium constant $\log K_{ML} - pK_{HL}$ for the reaction

$$M^{a+} + HL = ML^{(a-1)+} + H^+ \tag{5}$$

which expresses the competition between proton and metal ion for a ligand donor site, may be a realistic measure of the relative metal ion-binding strength. Martin

Table 3.3 — Stability data† on the Co(II) complexes of α-amino acids

Amino acid	$pK_{NH_3^+}$	$\log K_{ML}$	$\log K_{ML_2}$	% Closed form	$\log(K_{ML}/K_{ML_2})$
Gly	9.56	4.67	3.79	—	0.88
Ala	9.70	4.31	3.5	0	0.81
Ser	9.05	4.33	3.5	48	0.83
Thr	8.96	4.34	3.6	53	0.74
Asp	9.62	5.95	4.28	98	1.67
Glu	9.59	4.56	3.30	51	1.26
Asn	8.72	4.51	3.50	74	1.01
Gln	9.01	4.04	3.28	0	0.76
Phe	9.11	4.05	3.51	0	0.54
Tyr	9.20‡	4.24	4.0	28	0.24
DapaH	6.98‡	2.91	—	0	—
Dapa	6.98‡	6.28	5.08	100	1.20
DabaH	8.27‡	3.40	—	0	—
Daba	8.27‡	6.75	5.25	100	1.50
OrnH	8.83‡	3.65	3.11	0	0.54
Orn	8.83‡	5.01	3.48	90	1.53
Lys	9.16‡	3.84	3.23	0	0.61
Arg	9.02	3.87	3.18	0	0.69
His	9.09	6.87	5.51	99.8	1.36
Met	9.05	4.14	3.44	19	0.70
SMC	8.73	4.12	3.49	34	0.63
Cys	8.73	8.00	6.20	100	1.80
Pen	8.50§	8.98	7.90	100	1.08
Pro	10.41	5.05	4.22	68	0.83
Hypro	9.46	4.82	4.06	75	0.76

† Most values taken from Ref. 82Ma1 at 25°C and 0.1 mol dm^{-3}; values for Cys and Pen are taken from Ref. 83Ha1.
‡ Microscopic acid dissociation constant.
§ Estimated value from Ref. 88Ma1.

advocates the use of $\log K_{ML} - 0.7\, pK_{HL}$, in order not to over emphasize the role of the ligand basicity [79Ma].

Another approach to a quantitative characterization of the extent of side chain–metal ion interaction has been suggested by Martin [79Ma, 88Ma2] (the treatment is similar to that applied to the equilibrium description of dissociation microprocesses, see section 1.2). In this application, a metal ion may interact with a ligand such as an amino acid to give open (ML_o) and closed (ML_c) complexes, according to

Table 3.4 — Stability data† on the Ni(II) complexes of α-amino acids

Amino acid	$pK_{NH_3^+}$	$\log K_{ML}$	$\log K_{ML_2}$	% Closed form	$\log(K_{ML}/K_{ML_2})$
Gly	9.56	5.78	4.80	—	0.98
Ala	9.70	5.40	4.5	0	0.90
Val	9.49	5.42	4.30	22	1.12
Leu	9.57	5.45	4.26	24	1.19
Ser	9.05	5.40	4.50	44	0.90
Thr	8.96	5.46	4.55	54	0.91
Asp	9.62	7.15	5.24	98	1.91
Glu	9.59	5.60	4.15	45	1.44
Asn	8.72	5.68	4.55	77	1.13
Gln	9.01	5.16	4.26	0	0.90
Phe	9.11	5.15	4.44	0	0.71
Trp	9.32	5.48	4.92	40	0.56
Tyr	9.20‡	5.06	4.44	0	0.62
DapaH	6.98‡	4.04	3.54	0	0.50
Dapa	6.98‡	8.13	7.04	100	1.09
DabaH	8.27‡	4.66	4.03	0	0.63
Daba	8.27‡	8.97	7.37	100	1.60
OrnH	8.83‡	4.52	3.79	0	0.73
Orn	8.83‡	6.83	4.85	98	1.98
Lys	9.16‡	4.84	4.06	0	0.78
Arg	9.02	5.05	4.05	0	1.00
His	9.09	8.66	6.86	100	1.80
Met	9.05	5.33	4.56	34	0.77
SMC	8.73	5.26	4.56	37	0.70
Cys	8.73	8.7	10.9	100	− 2.2
Pen	8.50§	10.7	12.2	100	− 1.5
Pro	10.41	5.95	4.95	52	1.00
Hypro	9.46	5.94	5.01	77	0.93

† Most values taken from Ref. 82Ma1 at 25°C and 0.1 mol dm^{-3}; values for Dapa, Daba, Orn, Lys from
  Ref. 81Fa1, for Cys and Pen from Ref. 79So1.
‡ Microscopic acid dissociation constant.
§ Estimated value from Ref. 88Ma1.

(6)

Table 3.5 — Stability data† on the Cu(II) complexes of α-amino acids

Amino acid	$pK_{NH_3^+}$	$\log K_{ML}$	$\log K_{ML_2}$	% Closed form	$\log(K_{ML}/K_{ML_2})$
Gly	9.56	8.13	6.87	—	1.26
Ala	9.70	8.15	6.7	0	1.45
Val	9.49	8.11	6.79	13	1.32
Leu	9.57	8.2	6.8	22	1.4
Ser	9.05	7.89	6.6	17	1.29
Thr	8.96	8.00	6.69	42	1.31
Asp	9.62	8.89	7.04	83	1.85
Glu	9.59	8.33	6.51	40	1.82
Asn	8.72	7.83	6.53	37	1.30
Gln	9.01	7.76	6.46	0	1.30
Phe	9.11	7.86	6.91	0	0.95
Trp	9.63	8.29	7.19	32	1.10
Tyr	9.20‡	7.84	6.95	0	0.89
DapaH	6.98‡	6.21	5.10	0	1.11
Dapa	6.98‡	10.62	9.19	100	1.43
DabaH	8.27‡	7.04	5.74	0	1.30
Daba	8.27‡	10.62	7.99	100	2.63
OrnH	8.83‡	7.40	6.16	0	1.24
Lys	9.16‡	7.67	6.41	0	1.36
Arg	9.02	7.45	6.40	0	1.05
His	9.09	10.16	7.94	99.5	2.22
Met	9.05	7.86	6.7	11	1.16
SMC	8.73	7.88	6.84	45	1.04
Pro	10.41	8.84	7.52	47	1.32
Hypro	9.46	8.38	7.04	55	1.34

† Most values taken from Ref. 82Ma1 at 25°C and 0.1 mol dm^{-3}; values for Dapa, Daba, Orn, Lys are taken from Ref. 78Ge1, for Trp from Ref. 71We.
‡ Microscopic acid dissociation constant.

where ML_o represents a complex with bidentate chelation via the amino and carboxylate groups, while ML_c represents a complex with some degree of interaction between the amino acid side chain and the metal ion.

A dimensionless constant K_I is defined for the isomerization of the open to the closed complex

$$K_I = [ML_o]/[ML_c] \tag{7}$$

Table 3.6 — Stability data† on the Zn(II) complexes of α-amino acids

Amino acid	$pK_{NH_3^+}$	$\log K_{ML}$	$\log K_{ML_2}$	% Closed form	$\log(K_{ML}/K_{ML_2})$
Gly	9.56	4.96	4.23	—	0.73
Ala	9.70	4.56	3.99	0	0.57
Val	9.49	4.45	3.79	0	0.66
Leu	9.57	4.51	4.05	0	0.46
Ser	9.05	4.60	3.9	42	0.7
Thr	8.96	4.70	3.9	67	0.8
Asp	9.62	5.58	4.5	91	1.1
Glu	9.59	4.59	3.66	17	0.93
Phe	9.11	4.92	4.06	0	0.23
Trp	9.63	5.18	4.69	77	0.49
Tyr	9.20‡	4.28	3.99	0	0.29
DapaH	6.98‡	3.20	2.64	0	0.56
Dapa	6.98‡	6.31	5.35	100	0.96
DabaH	8.27‡	3.74	3.35	0	0.39
Daba	8.27‡	6.70	5.60	100	1.10
OrnH	8.83‡	3.37	3.08	0	0.29
Orn	8.83‡	6.17	—	99	—
Lys	9.16‡	4.06	3.47	0	0.59
Arg	9.02	4.15	3.95	0	0.20
His	9.09	6.51	5.53	99.4	0.98
Met	9.05	4.38	3.95	24	0.43
SMC	8.73	4.46	4.06	56	0.40
Cys	8.73‡	9.17	9.01	100	0.16
Pen	8.5§	9.5	9.9	100	−0.4
Pro	10.41	5.36	—	61	—
Hypro	9.46	5.03	4.35	74	0.68

† Most values taken from Ref. 82Ma1 at 25°C and 0.1 mol dm^{-3}; values for Dapa, Daba, Orn, Lys are taken from Ref. 81Fa1, for Trp from Ref. 71We, for Pro from 70Gi.
‡ Microscopic acid dissociation constant.
§ Estimated value from Ref. 88Ma1.

as is an equilibrium constant K_{ML_o} for the formation of the open complex

$$K_{ML_o} = [ML]/[M][L] \qquad (8)$$

The experimentally determined stability constant K_{ML} is given by

Table 3.7 — Stability data† on the Cd(II) complexes of α-amino acids

Amino acid	$pK_{NH_3^+}$	$\log K_{ML}$	$\log K_{ML_2}$	% Closed form	$\log(K_{ML}/K_{ML_2})$
Gly	9.56	4.24	3.47	—	0.77
Ala	9.70	3.80	3.30	0	0.50
Val	9.49	3.70	2.90	0	0.80
Leu	9.57	3.84	2.70	11	1.14
Ser	9.05	3.77	3.26	19	0.51
Asp	9.62	4.35	3.20	72	1.15
Glu	9.59	3.9	—	21	—
Phe	9.11	3.7	3.2	0	0.5
Trp	9.63	4.48	3.70	79	0.78
His	9.09	5.39	4.27	98	1.12
Met	9.05	3.69	3.41	2	0.28
SMC	8.73	3.79	3.25	35	0.54
Pen	8.50‡	10.6	—	100	—

† Most values taken from Ref. 82Ma1 at 25°C and 0.1 mol dm^{-3}; values for Trp are taken from Ref. 71We1, for Ser and SMC from Ref. 87So2.
‡ Estimated value from Ref. 88Ma1.

$$K_{ML} = \{[ML_o] + [ML_c]\}/[M][L] = K_{ML_o}(1 + K_I) \tag{9}$$

where $(1 + K_I)$ is the enhancement factor, which characterizes the stability enhancement due to the presence of the closed form. It can be calculated as a ratio of the experimentally observed stability constant and that of the corresponding open form complex. These latter constants K_{ML_o} can be estimated from the least square lines of the $\log K_{ML}$ vs pK_{HL} plots. The K_I values in turn allow the calculation of the percentage of the closed complex, i.e. the degree of interaction with the side-chain donor group. These values are also included in Tables 3.3–3.7.

Calculated percentages below 20–25% are not considered to be significant, because the uncertainty in the stability constants used here and determined by different authors in different laboratories (in part by different experimental and computational methods) does not allow this. (The reproducibility and reliability of the reported proton and metal complex formation constants have been critically surveyed for some simple aliphatic amino acids [90Ki] and several aromatic ones [84Pe].)

Tables 3.3–3.7 list 20–60% closed forms for Met, Glu, Ser, and Thr, which suggests weak interactions with the side-chain donor atoms. Much stronger metal ion side-chain coordination occurs in the metal complexes of Asp and Asn, with 74–98% closed forms. The practically 100% closed forms for His, Dapa, Daba, Cys, and Pen clearly show the tridentate bonding mode, with strong coordination of the side-chain imidazolyl, amino, and sulfhydryl groups.

The last column in Tables 3–7 lists the ratio of the stepwise stability constants, $\log(K_{ML}/K_{ML_2})$. Statistically, more coordination positions are available for the bonding of the first ligand to a given metal ion than for the second ligand. Depending on the geometry of the complex and the denticity of the ligand the statistically expected value for $\log(K_{ML}/K_{ML_2})$ will vary: for the coordination of bidentate ligands to a regular octahedral coordination sphere, it is 0.7 (as in many amino acid complexes of Co(II), Ni(II), Zn(II), and Cd(II)), while for the distorted coordination sphere of Cu(II) it is 1.2 [75Si]. The low $\log(K_{ML}/K_{ML_2})$ values found for Phe, Tyr, and Trp suggest a favourable interaction between the aromatic side chains in the bis complex [79Ma]. The high values for the Asp, Glu, and Asn complexes stem from the tridentate coordination of the first ligand molecule and from the electrostatic repulsion due to the coordination of the second ligand. The large negative values found for Cys (-2.20) and D-Pen (-1.50) are due to a change in geometry from octahedral to square planar during the coordination of a second ligand molecule.

Besides stability constant measurements, other evidence is required to confirm side-chain involvement in coordination. UV-Vis spectroscopy is very useful for transition metal complexes, and notably in the cases of Co(II), Ni(II), and Cu(II) complexes has proved to be a powerful tool for structure elucidation [68Sh]. For optically active ligands, circular dichroism (CD) provides significant information on molecular geometry and, in particular, on the nature of the interacting donor atoms [61Gi]. In many cases, CD spectra can differentiate between bidentate and tridentate coordination in solution for potentially tridentate amino acids [74Ma3]. NMR shifts in diamagnetic systems and selective line broadening in paramagnetic systems are of diagnostic use in detecting side-chain coordination [62Mi]. ESR and ENDOR spectroscopy also find significant application for metal complexes of biomolecules [87Si]. Besides X-ray crystallography [67Fr1], IR spectroscopy is one of the most frequently used techniques for determination of the number and nature of functional groups involved in the coordination of multidentate amino acids in the solid state [70Na1]. It must be remembered, however, that there is generally agreement between the structure in the solid state and that of the major species in solution only in the case of kinetically inert complexes. The same is not necessarily true for kinetically labile systems, and comparisons between the two phases must be performed carefully.

2.   BIDENTATE AMINO ACIDS

Besides simple aliphatic amino acids and the cyclic proline, which contain only two metal ion-binding sites and no functional group in the side chain, hydroxyproline is also dealt with here, although this contains an alcoholic hydroxy group as a further weak binding site. The reason is that the cyclic nature of Hypro results in an orientation of the hydroxy group whereby it is unable to coordinate directly to metal ions.

2.1   Aliphatic amino acids

The protonated form of the amino acids in question (Gly, Ala, Val, Leu, Pro, and Hypro) contains two ionizable protons, which dissociate stepwise in fully separated processes. Depending on the pH of the solution, the ligand can exist in three different forms: the cationic form H_2A^+, the zwitterionic form $HA^{\pm}$, and the anionic form A^-. Since the dissociation steps are well separated, the concentration of the neutral form HA is negligible [80Ha]. Besides these monomeric species various polymeric associates, for example HA_2^-, H_2A_2, $H_3A_2^+$, and $H_4A_2^{2-}$, can also be formed in concentrated (0.05–1.0 mol dm^{-3}) solution [78Vl]. The main reason for the association of the monomeric species is the formation of hydrogen bonds between them, whereas in dilute solution the monomeric species are stabilized through hydrogen bonds with water molecules.

The basic bonding mode of simple aliphatic amino acids in their metal complexes has long been established. These ligands readily form five-membered chelate complexes of types ML, ML_2, and ML_3 (the latter with metal ions which have a coordination number of at least 6) with most of the metal ions both in the solid state [67Fr1, 67Ja] and in neutral or slightly basic solutions [74Ma1, 82Ma1]. A few recent papers may be mentioned here, which prove this bonding mode with Pb(II) [82Bi1, 85Di], Tl(I) and Tl(III) [83Ja], V(III) [73Ka], VO(IV) [73To, 82Fa, 87So1, 88Pe], Cr(III) [84Ab, 84Co, 85Ya1], Mn(II) [87Go2], Co(II) [74Bh, 87Go1], Co(III) [84Pa1], Ni(II) [74Bh, 75Ge1, 87To], Cu(II) [87Go2, 878Hi, 88Sz], Zn(II) [79St, 81Al1, 85Di], and Cd(II) [82Ma2, 86So1]. Here, we focus mainly on complexes with less common stoichiometry and coordination modes.

Weak interactions can be detected with alkaline earth metal ions in solution [57Mu, 85Da, 87Ha]. The stability sequence of the complexes ML^+ is Mg(II) > Ca(II) > Sr(II) > Ba(II), and there is an almost linear relationship between the log K_{ML} values and the reciprocals of the ionic radii. This indicates that the complexes are mainly held together by electrostatic interactions between the carboxylate group and the metal ions. The crystal structure of CaGly·3H$_2$O [74Ei] supports this view.

Owing to the extensive hydrolysis of the lanthanide(III) ions even in neutral solution only monodentate carboxylate-coordinated complexes of these metal ions are formed [74Pr]; they can be prepared from acidic solution [84Le]. At the same time the formation constants of the UO_2(VI) and Th(IV) complexes point to the bidentate coordination of these simple aliphatic amino acids [83No, 85Ya1]. However, these data and thus the conclusion are rather uncertain as the role of hydrolysis of the actinide ions and their complexes was omitted.

A method has recently been developed for estimation of the stability constants of proton and metal complexes of ligands, including amino acids, that have not previously been measured [85m1, 85Sm2]. The method can be useful for a quantitative estimation of the weak interactions between amino acid and, for example, alkali metal or alkaline earth metal ions.

Ag(I) ion forms different complexes in the solid state without the formation of chelate rings. The characteristic coordination geometry is linear (or tetrahedral), which prevents the development of five- or six-membered chelate rings. A crystallographic study [71Ac] has shown that AgGly involves endless -Ag-NH$_2$CH$_2$COO-Ag-NH$_2$CH$_2$COO- chains (structure I), while in AgGly·1/2H$_2$O alternate Ag atoms are bonded to two N and two O donors, respectively (structure II). The complex AgGlyH·NO$_3$ prepared from acidic solution consists of carboxylate-coordinated dimeric units (structure III).

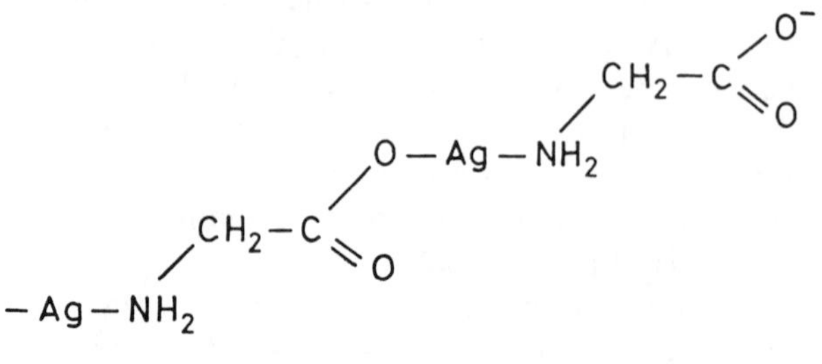

$$\underset{-\,Ag-NH_2}{\overset{CH_2-C\diagdown_O}{\diagup}} \qquad \overset{O-Ag-NH_2}{\diagup} \qquad \overset{CH_2-C\diagup^{O^-}}{\diagdown_O}$$

(I)

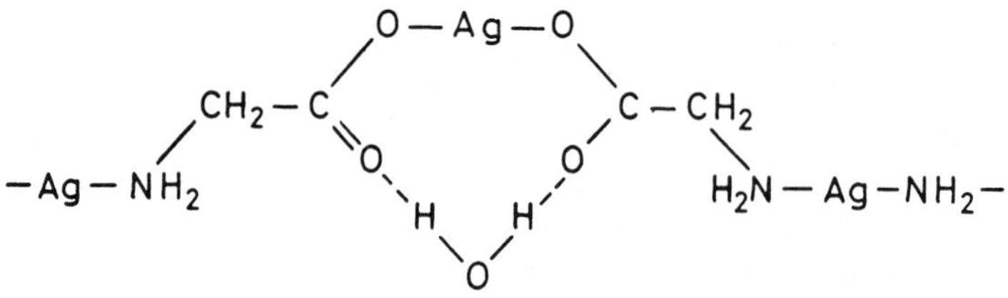

(II)

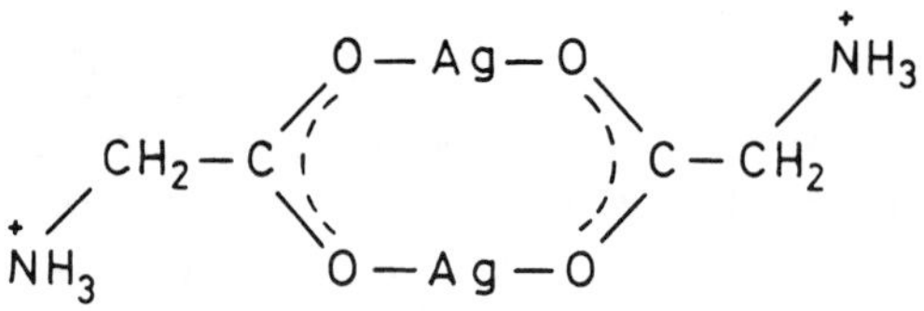

(III)

In aqueous solution and in solid complexes with a metal ion–ligand ratio of 1:2 Ag(I) ion binding occurs through the amino group; the electrostatic interaction between the positively charged Ag(I) ion and the negatively charged carboxylate is so weak that it does not contribute appreciably to stabilization of the amino-coordinated complexes [800h]. Another metal ion to which amino acids are frequently coordinated via their amino groups is Pt(II) [68Er, 88Pa1].

In addition to the usual 1:1 and 1:2 chelates, protonated complex formation with monodentate carboxylate-coordination has been proved for Pb(II) by potentiometry with a lead amalgam electrode [85Di] or a glass electrode [82Bi1] and by ^{207}Pb NMR spectrometry [83Na].

The stability data for the most investigated metal ions listed in Tables 3–7 show that the sequence of stability constants for divalent metal ions follows the Irving–Williams series, and that there is a certain periodicity in the $pK_{NH_3^+}$, $\log K_{ML}$ and $\log K_{ML_2}$ values of the aliphatic amino acids with increasing number of carbon atoms. As an illustration, the thermodynamic parameters for formation of the complexes NiL_2 with Gly, Ala, 2-aminobutanoate (Abu), Nva, and Nle [75So] are depicted in Fig. 3.2.

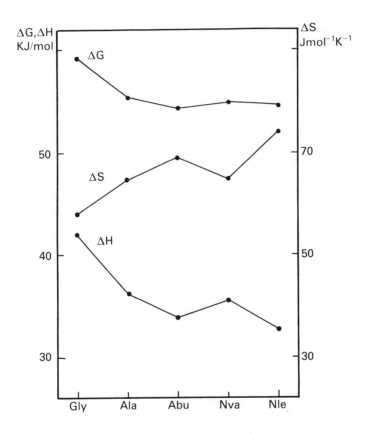

Fig. 3.2 — Thermodynamic parameters for the formation of NiL_2 complex with aliphatic amino acids as a function of the chain length.

The exceptionality of the stability of Gly complexes and the larger stability and more favourable enthalpy for amino acids containing odd numbers of carbon atoms are clearly seen in Fig. 3.2. (A similar periodicity is characteristic for the aliphatic carboxylic acids.) The high stability of the Gly complexes might be explained by steric reasons [63Ir], although a completely satisfactory explanation has not been found yet [88Ma1]. The main driving force for transition metal–amino acid complex formation is an enthalpy effect, as evidenced by the fact that the formation enthalpies of the divalent transition metal complexes also follow the Irving–Williams series [75Ch1].

The coordination geometry in the Cu(II) complexes is generally tetragonally distorted octahedral, and thus *cis-trans* isomerism can occur. The energy barrier to isomerization in the bidentate Gly–like structure is low even in the solid state [79De] and is therefore probably even lower in solution [80Fi, 85Go]. There is crystallographic [67Fr1] and ESR [74Je, 87Hi] evidence of the axial carboxylate-coordination of adjacent complexes and the formation of dimeric units.

Trinuclear carboxylate-bridged Fe(III)-L-amino acid complexes (structure IV) [74Ho, 81Pu] have been prepared with a variety of amino acids, in which the ligands exist as zwitterions and the amino N does not participate in the coordination. Such complexes provide one of the most attractive models for the structure of the ferritin iron core.

(IV)

Thermogravimetric studies of divalent metal chelates of simple aliphatic amino acids reveal that the thermal stabilities of the series of complexes examined do not correlate with the Irving–Williams series for the stabilities of complexes in solution; however, a reverse correlation is observed with the sequence of basicity of the ligands [70Bo, 87To].

Besides the usual bidentate complexes ML_n, there is a possibility for the formation of carboxylate-coordinated protonated complexes in acidic solution, e.g. with VO(IV) [73To, 82Fa, 88Pe], Mn(II) [87Go2], Fe(II) [85Fi], Fe(III) [86An], Co(II) [87Go2], Cu(II) [87Go2] and Zn(II) [81All, 85Di], and for the formation of hydroxo complexes in basic solution, e.g. with VO(IV) [82Fa, 88Pe], Cu(II) [80Gi], and Zn(II) [79Ge3, 85Di] even in the presence of excess ligand. The formation of CuL_3^-, with monodentate axial coordination of a third ligand molecule, has been detected with Gly and Ala at a very high ligand excess, by means of spectrophotometric [81Na, 82Na1] and NMR relaxation measurements [87Go2].

On the basis of a detailed pH-spectrophotometric study, the formation microconstants have been determined for the chelated and non-chelated forms of the complex $VOGly^+$

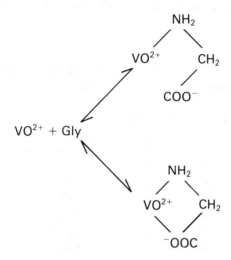

and it has been found that the ratio of the two microspecies is about 60:40 in favour of the non-chelated form.

It has been established that the protons of the coordinated amino groups in $Co(II)Gly_2$ and $Ni(II)Gly_2$ are sufficiently acidic to permit their successive removal in appreciably basic media. Thus, further deprotonation processes take place when these complexes are treated with potassium amide in liquid ammonia [67Wa, 68Wa, 69Wa].

The complexes cis- and trans-$Cl_2Pt(NH_2CHRCOOH)$ undergo reactions at the free carboxyl group without cleavage of the platinum–nitrogen bonds; these reactions have been widely studied. Owing to the stability of the Pt–NH_2 bond, platinum can be used as an effective amino-protecting group in peptide synthesis [88Be].

2.2 Proline and hydroxyproline

Pro and Hypro are in fact cyclic imino acids. Pro therefore contains a secondary amine group, and its pK_{HL} value (see Table 3.2) is higher than those for the amino acids containing a primary amino group, owing to the increased inductive effect on the nitrogen of the NH group, making its removal less easy. At the same time, the increased acidity of this group in Hypro can be explained by the electron-attracting effect of the alcoholic OH group. A similar effect of the hydroxy group can be observed in Ser and Thr (see section 3.1).

Since Pro does not contain a coordinating side-chain donor atom, it is capable only of bidentate coordination via the carboxylate and imino groups. Hypro has an alcoholic hydroxy group in position 2, but for steric reasons this is not capable of direct coordination to metal ions, and only some very weak indirect interaction can be assumed. Hence, it is very surprising that the data in Tables 3.3–3.7 show 47–72% closed form for Pro and 52–77% closed form for Hypro in the complex ML. However, if their basicity-adjusted stability constants ($\log K_{ML} - pK_{HL}$) are compared with that of Ala (see section 1.3), then the extra stabilization disappears in the case of Pro, but still exists for the Hypro complexes, which is now closer to expectations (see Table 3.8). If the corrected basicity-adjusted stability constant

Table 3.8 — Comparative stability data[†] on Co(II), Ni(II), Cu(II), and Zn(II) complexes of Ala, Pro, and Hypro

System	$pK_{NH_3^+}$	$\log K_{ML}$	% Closed form	$\log K_{ML} - pK_{NH_3^+}$	$\log K_{ML} - 0.7 pK_{NH_3^+}$
Co(II)-Ala	9.70	4.31	0	− 5.39	− 2.48
Co(II)-Pro	10.41	5.05	68	− 5.36	− 2.24
Co(II)-Hypro	9.46	4.82	75	− 4.64	− 1.80
Ni(II)-Ala	9.70	5.40	0	− 4.30	− 1.39
Ni(II)-Pro	10.41	5.95	52	− 4.46	− 1.34
Ni(II)-Hypro	9.46	5.94	77	− 3.52	− 0.68
Cu(II)-Ala	9.70	8.15	0	− 1.55	1.36
Cu(II)-Pro	10.41	8.84	47	− 1.57	1.55
Cu(II)-Hypro	9.46	8.38	55	− 1.08	1.76
Zn(II)-Ala	9.70	4.56	0	− 5.14	− 2.23
Zn(II)-Pro	10.41	5.36	61	− 5.05	− 1.93
Zn(II)-Hypro	9.46	5.03	74	− 4.43	− 1.59

† Values are taken from Ref. 82Ma1 at 25°C and 0.1 mol dm^{-3}.

($\log K_{ML} - 0.7 pK_{HL}$) suggested by Martin [79Ma] is considered, a slightly increased stability is obtained for Pro and a still considerable extra stability for Hypro. In fact, a basicity-adjusted stability constant corrected by a metal ion-dependent factor (which is less, than 0.7, however, for the metal ions listed in Tables 3.3–3.7) is used to calculate the percentages of the closed forms [88Ma2].

These results draw attention to the limitations of any kind of comparison; in this case, it seems that the correction factor used to give the best estimate of the effects of the differences in the acidities of the ligands on the metal complex formation constants is not only metal ion-dependent, but also a function of the type of ligand. Thus, the same factor cannot be used for the amino acids and for these cyclic imino acids.

Crystal structure analysis of $CuPro_2 \cdot 2H_2O$ [52Ma] proves that the environment of the Cu atom is tetragonally distorted octahedral, with two H_2O molecules occupying the more distant axial sites (there is no carboxylate-coordination of adjacent complexes in the apical positions, unlike the situation for simple aliphatic amino acids [67Fr1]). The structure in the solid state is *trans* and the visible absorption spectra in the solid state and in solution are very similar, and thus the structure in solution is probably *trans* too [66Gi]. A recent ESR study, however, suggested the equilibrium of *cis* and *trans* isomers [85Go]. Similarly to the simple aliphatic amino acids (see section 2.1), both protonated complexes involving only carboxylate-coordination and mixed hydroxo complexes have been detected with Mn(II), Co(II), Cu(II), and Zn(II) [69Ch, 87Go1].

The differences in the thermodynamic [73Ku1, 73Po] and kinetic [73Ku1] data on the formation of complexes between Co(II), Ni(II), and Cu(II) ions and L-Pro and L-Hypro indicate some interaction, probably hydrogen bond formation, between the water molecules coordinated to the metal ion and the hydroxy group of Hypro.

3.   TRIDENTATE AMINO ACIDS WITH WEAKLY COORDINATING SIDE CHAINS

3.1   Serine and threonine

Both ligands contain only two dissociable protons in the measurable pH range, as the alcoholic hydroxy group is so weakly acidic ($pK > 14$) that it does not undergo dissociation. If the pK values in Table 3.2 are compared with the data on the corresponding reference compound, Ala, it may be stated that the electron-attracting effect of the alcoholic hydroxy group lowers the basicity of the amino group by about the same extent as in the case of Hypro (see section 2.2).

The basic bonding properties of Ser and Thr in the solid state are simple; crystal structure determinations and IR studies show that only substituted Gly-type chelation occurs in their complexes with V(III) [86Ko], Ni(II) [69He1], Cu(II) [69He2, 69Fr1, 85Sz1], and Zn(II) [70He]. Only in Zn(L-Ser)$_2$ is there a weak non-chelating interaction between an alcoholic hydroxy group in one complex and the metal atom in a neighbouring complex [70He].

There is considerable disagreement in the literature, however, regarding the structures of Ser and Thr complexes in solution. The percentage values of the closed forms reported in Tables 3.3–3.7 indicate a weak chelation to Co(II), Ni(II), and Zn(II) ions, and even weaker interactions for Cu(II) and Cd(II). The various possible bonding modes suggested for the complexes ML$_2$ (M = Mn(II), Co(II), Ni(II), Cu(II), Zn(II), and Pb(II)) so far are depicted schematically below (structures V–VII). Structure V, in which there is no alcoholic hydroxy group coordination, is indicated by thermodynamic [67Sh1, 70Le1], Vis absorption and CD [70Ts, 73Gr, 74Ro] and ESR [83Sz] measurements. On the basis of stability and enthalpy comparison studies [67St, 72Ge1] and an NMR relaxation investigation [75Na] direct alcoholic hydroxy group coordination is suggested (Structure VI). Pettit & Swash [76Pe1] and Simeon *et al.* [82Kr] assume structure VII with the indirect coordination of hydroxy groups via water molecules. It should be admitted that structures VI and VII are similar in many respects, and it is therefore very difficult to differentiate between them experimentally. Nevertheless, structure VII seems to fit the experimental findings best, at least for the Cu(II) complexes: this type of outer-sphere coordination does not change the coordination sphere of the metal ions enough to be detected by spectroscopic (Vis, CD, and ESR) methods, but the strength of the hydrogen bond (ca. 15–20 kJ mol^{-1}) could account for the slight stereoselectivity found [76Pe1]. The partially occupied axial coordination sites make the proton-exchange processes followed by NMR slower [75Na].

(V)   (VI)   (VII)

Below pH ~ 4, protonated complexes CuLH^{2+} occur with both Ser and Thr, presumably owing to chelation via the alcoholic hydroxy and carboxylate groups [76Pe1].

In basic solution, the hydroxy-proton is ionized in the Cu(II) complexes of both Ser and Thr, with pK 10.2 and 11.2 [69Fr2, 87Ki1], and 10.3 and 11.3 [73Gr], respectively. Similar dissociation of the complex Co(II)Thr$_2$, with pK 9.7, has recently been reported [82Kr]. In these complexes CuL$_2$H$_{-1}$ and CuL$_2$H$_{-2}^{2-}$ Ser and Thr act as tridentate ligands; this is clearly indicated by the positive peaks in the visible CD spectra. The appreciable shift towards higher energies in the d–d transition confirms the deprotonation of at least one of the alcoholic hydroxy groups [87Ki1]. The other possibility is the deprotonation of a coordinated water molecule this would result in the formation of mixed hydroxo complexes. This is similar to what was recently inferred by Gillard *et al.* [80Gi] from pH-titration and CD measurements on the Cu(II)-Ala system at high pH. However, if the pH (= 12.17) of the complex CuAla$_2$ is compared with those obtained for CuSer$_2$ and CuThr$_2$, this possibility can be safely discarded for at least one of the deprotonation steps.

The enthalpy values obtained for these steps of deprotonation of CuSer$_2$ suggest that ionization of the alcoholic hydroxy group(s) and deprotonation of the coordinated water molecule(s) take place in parallel [87Ki1]. If structure VII is accepted for the complex CuL$_2$, the two processes seem to be indistinguishable.

The Cu(II) ion-promoted ionization of the alcoholic hydroxy group changes dramatically in the case of isoserine (3-amino-2-hydroxypropionic acid, Ise), where the hydroxy group is adjacent on both sides to further donor groups able to form strong coordinate bond with metal ions. The concentration distributions of the species formed in the Cu(II)-Ser and Cu(II)-Ise systems are depicted in Fig. 3.3. The species $Cu_2L_2H_{-2}$, which is formed exclusively in a wide pH range, has an alkoxo-bridged structure (VIII) [76Br1, 87Ki1].

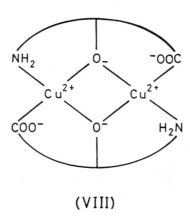

(VIII)

Hence, in the Cu(II)-Ise system ionization of the alcoholic hydroxy group occurs at as low as pH ∼ 5.

3.2 Aspartic acid and glutamic acid

Both Asp and Glu contain three functional groups that undergo acid–base reactions in the pH range 1 to 11. Literature values for the acid dissociation macroconstants are listed in Table 3.2. Owing to the partly overlapping processes, however, the acid–base chemistry of the two ligands at a molecular level can be described by 16 microscopic constants. These microconstants have been estimated from the pK values of partly esterified derivatives [36Ne, 67Jo, 73Br] and determined experimentally from ^{1}H NMR chemical shift titration data [89No1]. The ratios of the real acidity microconstants, which are characteristic of the extent of overlap between the dissociation of the individual acidic groups, $R_1 = k_{\alpha\text{-COOH}}/k_{\delta\text{-COOH}} = 50$ and $R_2 = k_{\delta\text{-COOH}}/k_{NH_3^+} = 1550$ for Glu, and $k_{\alpha\text{-COOH}}/k_{\gamma\text{-COOH}} = 18$ and $k_{\gamma\text{-COOH}}/k_{NH_3^+} = 1995$ for Asp, clearly show that the stepwise dissociation takes place through the major pathway sequence α-COOH, ω-COOH, NH_3^+.

Determination of the populations of the rotational isomers of Asp [73Is, 75Ha] has provided evidence that there is a close association between the two carboxylate groups and the ammonium group of mono-protonated Asp. In the case of Glu only the α-carboxylate is associated with the ammonium group [63Ll]. In an elegant way, even the dissociation microconstants of the acidic group for each rotational isomer of Asp have been calculated from ^{1}H NMR titration data [74Fu, 89No1]. The details of these submicroscopic dissociation constants and their significance are discussed in Chapter II.

Although many of the calcium-binding proteins contain very high proportions of aspartyl, glutamyl, and γ-carboxyl-glutamyl residues, relatively little attention has been devoted to the alkaline earth metal ion complexes of these amino acids.

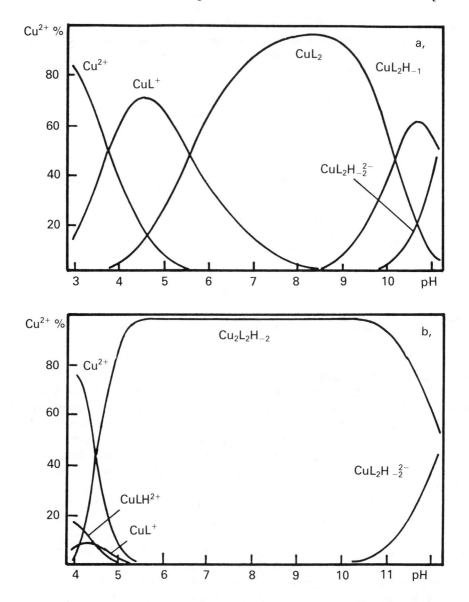

Fig. 3.3 — Concentration distribution of complexes formed in the (a) copper(II)-Ser and (b) copper(II)-Ise systems as a function of pH. $C_{\text{ligand}} = 0.004$, $C_{\text{Cu}} = 0.002$ mol dm^{-3}.

Stability constants have been reported for the Asp and Glu complexes of these metal ions [53Lu, 54Sc]. The closely linear relationship between log K_{ML} and the reciprocal ionic radius suggests predominantly ionic bonding, that is, coordination via the carboxylate group or both carboxylate groups of the ligands. As the stability constants are similar in magnitude to those for simple aliphatic amino acids such as

Gly [50Da, 76Ti] and Ala [50Da] the binding very probably occurs only through the Gly-like part of the Asp or Glu ligands, that is mainly via the -carboxylate group. The crystal structure of $CaGlu \cdot 3H_2O$ [74Ei] supports this view, although the lack of strong geometrical requirements in the ionic alkaline earth complexes allows ionic interactions with both carboxylate groups, or even the tridentate coordination of the ligands.

From ^1H NMR measurements, the interaction of the β-carboxylate and the α-amino groups of Asp with Mg(II) was concluded at high pH [81Ko].

γ-Carboxyglutamic acid (Glu-γ-COOH), in which the γ-carbon bears a carboxylate group additional to that already present in Glu, is a constituent of prothrombin and other blood-clotting proteins [81Bu1, 84Ne]. It has been found that Glu-γ-COOH itself forms a rather weak 1:1 complex with Ca(II), coordinating via the two γ-carboxylate groups as a substituted malonate [78Sp]. Small peptides containing two neighbouring Glu-γ-COOH moieties bind Ca(II) ions with similar affinity [79Ro] as prothrombin [75Ba]. The formation microequilibria of the proton, Ca(II), and Mg(II) complexes of Glu-γ-COOH and its derivatives have recently been studied [88Bu]. The calcium-binding constants obtained clearly indicate that additional stabilizing factors and/or binding of two or three dicarboxylate moieties are required for the formation of high-affinity binding sites in these blood-clotting proteins [84Ma].

The formation of 1:1 and 1:2 protonated and/or deprotonated complexes has been concluded from potentiometric and NMR titrations with other main group metal ions, such as Ga(III) [68Za, 76Bi1], In(III) [68Za, 76Bi1], Pb(II) [71Ko, 74Ko, 85Di], Tl(I) and Tl(III) [83Ja], and Zr(IV) [72Si1]. The stability data indicate monodentate coordination in the protonated complexes [76Bi1, 85Di] and the tridentate bonding mode in the complexes ML_n with Asp [68Za, 76Bi1, 71Ko, 83Ja, 85Di] and probably also with Glu [68Za, 85Di].

The interaction of the side-chain β- and γ-carboxylate groups of Asp and Glu with 3d transition metal ions, such as Co(II), Ni(II), and Cu(II) ions, is significant in solution; this is clearly shown by the percentage values of the closed forms: 83–98% for Asp and 40–51% for Glu (see Tables 3.3–3.5). As the side-chain carboxylate-coordination results in a six-membered chelate ring for Asp and a seven-membered ring for Glu, the greater extent of side-chain coordination in the case of Asp is in accordance with expectations. These data indicate that Glu coordinates to Zn(II) and Cd(II) only in a bidentate manner (see Tables 3.6 and 3.7).

Crystal structure studies have revealed that the complexes $MAsp \cdot 3H_2O$ of Co(II) [57Bo], Ni(II) [69He2, 82An], and Zn(II) [69He1] are isomorphous. The ligand is tridentate around the octahedral metal ions: coordination of the amino nitrogen atom yields a five-membered ring with respect to the α-carboxylate oxygen and a six-membered ring with respect to the β-carboxylate oxygen atoms. $ZnGlu \cdot 2H_2O$ and $CdGlu \cdot 2H_2O$ are also isomorphous, and their structures are very similar to that of $CuGlu \cdot 2H_2O$. The complexes are distorted octahedra; the Glu is bound to one metal ion in a bidentate manner via its glycinate ring, and to another metal ion via the side-chain carboxylate group. The magnetic and spectroscopic properties of the recently prepared $FeAsp \cdot 2H_2O$ and $FeGlu \cdot 2H_2O$ indicate that they have similar extended structures [85Fi].

All the possible geometrical isomers of the bis and tris complexes of the inert

Co(III) ion with Asp and Glu have been isolated and characterized [67Sh2, 69Gi, 70Ka2, 72Ka]. In the tris complexes, the side-chain carboxylates are free. In the bis complexes, Asp is tridentate [69Ho, 71Fr], while Glu is bidentate [75Oo]. With Cr(III), both bis and tris complexes of Asp have been isolated [73La, 74La, 76Gr]; the coordination mode is the same as in the Co(III) complexes.

Thermodynamic and spectral studies strongly suggest that the bonding modes existing in the solid state are maintained in solution too, at least in the gross details. A comparison of the formation enthalpy values of the Ni(II), Cu(II), and Zn(II) complexes of Asp with those of α-aminobutyric acid, and of those of Glu with those of Nva, confirms the tridentate coordination of Asp in both complexes ML and ML_2^{2-} (in the case of $CuAsp_2^{2-}$ only one β-carboxylate is coordinated in the apical position); for Glu, the stronger interaction appearing in the ΔG values is not reflected in the enthalpy values, suggesting only the very weak tridentate nature of Glu in these complexes [65Ri, 71Ba, 75Is, 74Ge, 74Na, 75Ri1]. Similar conclusions were reached from Vis absorption, CD and NMR measurements [69Ka, 70Ha, 70Ho, 70Ts, 70Wi1, 75Na]. No evidence of stereoselectivity has been found in the complexes ML_2^{2-} of Co(II) [71Ba], Ni(II) [65Ri, 73Bl], and Cu(II) [71Ba, 75Ri1, 77Br] with optically pure and racemic Glu or Asp. Thus, it seems that there is no marked preference for *cis* or *trans* coordination in the $x - y$ plane.

It deserves mention that a quite different coordination of Asp to Cu(II), through the α-amino and γ-carboxylate groups, has recently been concluded from the various metal ion–ligand nucleus distances calculated from the paramagnetically-induced relaxation [84Kh]. We consider, however, that the errors in the distances, due to the considerable uncertainty in the relaxation times, are somewhat too large to allow such an unambiguous conclusion.

Protonated complexes, mainly MLH^+, have been reported for both Asp and Glu with Co(III) [86Ba], Ni(II) [74Ge], Cu(II) [74Na, 75Ri1, 77Br, 80Cl], and Zn(II) [73Is]. Comparisons of the formation thermodynamic parameters [74Na, 75Ge2, 75Ri1] support the assumption that there is an equilibrium between two possible bonding modes for the complex MLH^+ (see structures IX and X, where $n = 1$: Asp, $n = 2$: Glu, $M =$ Ni(II), Cu(II)); however, this equilibrium is shifted significantly in the direction of the formation of structure IX in the case of Glu, where a seven-membered chelate ring would be formed through the coordination of the α- and γ-carboxylate groups [79Ev]. Other protonated species CuL_2H_2 and CuL_2H^-, have been reported by Brookes & Pettit [77Br].

(IX)   (X)

The results of the relatively few studies on other 3d transition metal ions, such as Cr(III) [70Mi, 82Ve], Mn(II) [84Kh], Fe(II) [59Pe], and Fe(III) [58Pe, 81Pu], are essentially in agreement with the binding properties of Asp and Glu discussed above.

Little work has been reported on the Asp and Glu complexes with 4d and 5d transition metal ions, for which the aqueous solution chemistry is often dominated by isopolyacid formation, ready hydrolysis, and/or kinetic inertness.

Mo(VI) and W(VI) exist in solution as isopoly compounds at low pH, and as the stable MoO_4^{2-} or WO_4^{2-} at high pH, and there is only a very narrow pH range within which complexation with amino acids can be presumed and studied. pH titrations [72Si2, 75Br, 77Ra, 82Ca], spectral studies [77Ra], and preparative work [76Bu] have indicated the formation of a complex MoO_3L^{2-} with both Asp and Glu, whereas no complexation has been detected in the W(VI)-ligand systems. The very similar formation constants obtained for the complex MoO_3L^{2-} for both Asp and Glu suggest that both ligands are tridentate.

Pd(II) forms 1:1 and 1:2 chelates with Asp and Glu. The stability constants reported earlier [71Su, 72Si3] are not consistent with those found more recently by Frye & Williams [79Fr], although the experimental conditions and methods of calculation differed. These data and those on the complexes of Gly, or Gln [73Fa] and Asn [74Te] suggest chelation via the α-amino and carboxylate groups. The side chains appear not to be coordinated, as expected for planar complexes. It should be mentioned, however, that the possibility of chloride ion complexation to the metal ion was neglected in the stability constant calculations, despite the fact that the palladium(II) was initially present as $PdCl_2$ or $PdCl_4^{2-}$.

Because of the more inert complex formation, mainly preparative work has been reported on the reactions of these ligands with Pt(II) [69Er], and Rh(III) [71Ka1]; monodentate and bidentate coordination of the ligands has been proposed.

The complexation of Asp and Glu with the lanthanides and actinides is similar to that with the alkaline earth metal ions. The interaction is mainly electrostatic, the binding sites being primarily the carboxylate groups. The amino group is not expected to interact strongly, if at all, with these metal ions [65Mo].

The pH titration curves of Ln(III) in the acidic range are essentially identical in the presence or absence of Asp, but the ligand prevents hydrolysis of the metal ion at higher pH, indicating the formation of reasonably stable and soluble complexes, probably mixed hydroxo complexes [74Pr]. Coordination of the α-amino and β-carboxylate groups has been assumed in the La(III)-Asp system in the solid state [78Le, 789Le].

Potentiometric and spectral studies of the reactions of Asp and Glu with Pr(III) [70Ka1], UO_2(VI) [63Fe, 74Se], and Th(IV) [72Si1] have revealed complexation through one or both carboxylate groups.

3.3 Asparagine and glutamine

The metal-binding ability of the amide group is generally weaker than that of a carboxylate group, and this is reflected in the weaker tendency of Asn and Gln to act as tridentate ligands as compared with Asp and Glu.

Crystal structure studies on the bis complexes of Asn with Cu(II) [75St], Zn(II) [77St1], and Cd(II) [73Fl] have demonstrated only bidentate Gly-like coordination and an interaction of the amide carbonyl group with a metal ion in a neighbouring complex. This linear-chain polymer structure has been confirmed by spectral and magnetic measurements on the Cu(II) complex [77Ri]. These results are in contradiction with those of Srivastava *et al.*, who assumed tridentate coordination in the complexes $CoAsn_2$ [74Te], $CuAsn_2$ and $CuGln_2$ [76Sr].

In solution, the γ-amide carbonyl group of Gln undergoes practically no interaction with transition metal ions; Gln was chosen for calculation of the least squares lines to be used in the estimation of the extent of side-chain interaction in the metal complexes of potentially tridentate amino acids [88Mal]. The percentage values of the closed forms listed in Tables 3.3–3.5 indicate a significant side-chain interaction in the complexes of Asn with Co(II) and Ni(II), but less so in the case of Cu(II). Calorimetric [74Ge, 75Ge2], Vis, CD [69Ka, 70Wi1, 73Ha], and NMR [80Je] studies indicate amide–carbonyl group coordination of Asn to Co(II), Ni(II), Cu(II), and Zn(II). This interaction, however, is weaker than in the case of Asp, as a third ligand molecule can displace the amide-carbonyls and 1:3 complexes may be formed with Co(II), Ni(II), Zn(II), and Cd(II) [74Ba, 74Wa1, 75Ge2].

Deprotonation and coordination of the amide group take place in the Cu(II)-Asn system, but not in the Cu(II)-Gln system in basic solution; the corresponding pK values are 10.5 and 12.0 [75Ge3]. The considerable shift in the d–d transition from 620 nm to 580 nm indicates the presence of four nitrogen donors in the equatorial plane [70Wi1], although a recent NMR study [88Sz] led to the conclusion of one equatorially and one axially bound $CONH^-$ group. Metal ion-induced deprotonation of the amide group takes place with both Asn and Gln in the complex $Pd(en)L^+$ [77Li], with pK 6.5 and 9.0, respectively. The difference in pK for these complexes reflects the difference between a six- and a seven-membered chelate ring, the latter being less stable.

4. AROMATIC AMINO ACIDS

4.1 Phenylalanine and tryptophan

Both Phe and Trp behave practically as bifunctional acids; the indole-NH group of Trp is extremely weak both as a base ($pK = -6.2$) [74We] and as an acid $pK = 16.8$ [67Ya]. The weak basicity of the indole group does not provide a basis for its coordination, hence these ligands are capable only of normal bidentate (N,O)-coordination. The 37–40% closed forms for the Ni(II) and Cu(II) complexes of Trp, and the significantly smaller $\log (K_{ML}/K_{ML_2})$ values for the metal complexes of both Phe and Trp (see Tables 3.3 and 3.4) indicate some additional interaction between the metal ion and the ligands and/or between the aromatic side chains in the bis complexes. Relatively large $\log K_{ML}$ value for Phe and Trp complexes have been noted for Cu(II) and other 3d metal ions [70Me, 71Ge, 74Ma2], but the difference in the enthalpy changes as compared with that for Ala is not very definite [61Iz, 70Me, 71Ge, 72Wi]. The unusual behaviour of the aromatic amino acid complexes of Cu(II) is also displayed in the visible CD spectra. The bis Cu(II) complexes of Phe (and also Tyr) exhibit nearly the same CD magnitudes. Instead of having a value of half this magnitude, the mixed complex MLB with optically inactive Gly is augmented by 30 to 80% above the expected value [71Wi1]. Complexes with aliphatic side chains do not produce similar augmentations. On the basis of these results, it appears that the bonding is Gly-like in the mono complexes, but at least one aromatic ring is associated with the Cu(II) in the bis complexes. This is probably a result of the smaller charge on the metal ion in the mono complex after one ligand has coordinated. This will make interaction between the d electrons of the Cu(II) ion and the aromatic system of a second ligand molecule more likely [84Pe].

A Cu(II)-aromatic ring interaction was not evident from a crystal structure analysis of CuPhe$_2$ [71He], but was apparent on CuTyr$_2$ [72He]; no crystal structure determination has been performed on CuTrp$_2$. Other evidence for an anomalous interaction comes from a comparison of the stability constants of the Cu(II) complexes with Trp and 1-methyl-Trp [74We]. Though the log K_{ML} values are almost identical, the log K_{ML_2} values differ by over 3 log units, suggesting considerable steric hindrance to the coordination of a second ligand. Such hindrance would be likely only if a Cu(II)-aromatic ring interaction were significant.

Visible, ESR [77Sp], and NMR [77Ko] spectral studies on Cu(II), Ni(II), and Pd(II) complexes of simple peptides containing Phe or Trp also suggest weak attractive interactions between these metal ions and the aromatic side chains.

Alternatively, most of these results can also be explained by an attractive interaction between the aromatic rings in the bis complexes [79Ma].

4.2 Tyrosine isomers

Besides the two acidic groups of simple amino acids, Tyr contains a third one, the phenolic hydroxy group on the aromatic side chain. This group dissociates in a process that overlaps strongly with the dissociation of the ammonium group (see R values in Table 3.2), the latter group being the more acidic. The same microscopic dissociation scheme with very similar acidity constants describes the stepwise proton losses from the Tyr isomer which contains the phenolic hydroxy group in the meta position [82Ki]. For the third isomer, o-Tyr, however, it may be assumed that the dissociation of $H_2L^\pm$ leads to the formation of the species with structure XI, which contains an intramolecular hydrogen bond, and thus the two microforms of HL are indistinguishable [82Ki]. Accordingly, as discussed in Chapter II, microscopic processes should not be reckoned with in this case. A detailed discussion of the acid–base chemistry of the Tyr isomers, including the microscopic enthalpy changes of the various microforms, is given in Ref. 82Ki.

$$CH_2 - CH - COO^-$$
$$\underset{\overset{|}{NH_2}}{}$$

(XI)

The phenolic hydroxy group in Tyr and m-Tyr is in an unfavourable position for direct coordination to the metal ions in their complexes. The reported thermodynamic data, spectral properties [70Le2, 71We, 72Ba, 73Li, 73Mc, 74We, 75Pl, 76Sa1, 82Pe, 84Ki1], and kinetic results [73Ba1] on the complexes of Tyr with various metal ions, such as Mn(II), Fe(II), Fe(III), Ni(II), Cu(II), Zn(II), Cd(II), Pb(II), Hg(II), some lanthanides and actinides, all point to the bidentate character of these ligands, there being considerable similarity with the results obtained for Phe. Besides the possible metal ion–aromatic ring [72He, 79Sa] and/or hydrophobic aromatic ring [80Ve] interactions, hydrogen bond formation has also been assumed between the phenolic hydroxy and the coordinated water molecule (see section 3.1) in the solid form [67La]. On the basis of the slight stereoselectivity found for the Cu(II)-L/D-Tyr system [84Al1] (in contradiction with earlier reports [71We]), similar weak interactions are assumed in solution too. This phenolic oxygen, however, does appear to coordinate directly to Cu(II) in some peptide complexes [78Su, 79Ga, 81He, 85Ya2, 86Ki], and it can act as a metal-binding site in a number of metalloproteins (see Chapter IV). At higher pH, following the formation of complexes $M(LH)_n$, the stepwise deprotonation of the non-coordinated phenolic hydroxy groups takes place. No spectral changes are observed during the deprotonation, indicating an unchanged bidentate bonding mode. In the case of Ni(II) and Cu(II), the thermodynamic quantities for these deprotonation processes agree well with those for the free ligand, and the ratio of the stepwise deprotonation constants corresponds to the statistical considerations, suggesting that there is no direct interaction between these groups [84Ki1]. With metal ions that are readily hydrolysable, such as Co(II) and Zn(II), the parallel dissociation of the phenolic hydroxy groups and the coordinated water molecule is expected, and is in fact indicated by the deprotonation thermodynamic quantities [82Pe, 84Ki1]. The microconstants characterizing these parallel processes of $ZnLH^+$ and $Zn(LH)_2$ have been determined via a combined pH-spectrophotometric method, one of the part processes, the deprotonation of the phenolic hydroxy groups, being followed via the UV band of the phenolate [84Ki1].

In contrast with Tyr and m-Tyr, in o-Tyr the phenolic hydroxy group is in a favourable position for coordination to the metal ion, with the simultaneous formation of a $5 + 6$-membered joined chelate ring system. At higher pH, the deprotonation of the complex $M(LH)_n$ (where M = Co(II), Ni(II) and Cu(II)) is accompanied by appreciable spectral changes, indicating the direct participation of the phenolate oxygen in the coordination, for example the medium-intensity band appearing at ~ 400 nm in the copper(II)-o-Tyr system, is clearly a charge transfer band characteristic of the Cu(II)-phenolate interaction [70Le1, 75Bo1]. The metal–phenolate interaction is barely manifested in the formation enthalpy data [84Ki1] on the Cu(II) complex. Thus, it is very probable that, even following deprotonation, the amino nitrogen and the carboxylate oxygen occupy the equatorial sites around the Cu(II) and the phenolate oxygen is coordinated only at the more distant and weaker apical sites. A detailed spectral (Vis, CD, and Raman) study of the Cu(II)-D-o-Tyr system [81Ga] has confirmed this assumption.

4.3   3,4-Dihydroxyphenylalanine

L-Dopa loses one proton from the carboxylic group at acidic pH (pK ~ 2.2), two protons from the ammonium group and a phenolic hydroxy group in overlapping processes in the intermediate pH range (pH 8–10), and one proton from the second phenolic hydroxy group in the highly basic pH range (pK ~ 13.4). The value of $R = [^+HNRO^-]/[NROH]$ indicates that the acidity sequence of the two groups is the reverse of that for Tyr; the phenolic hydroxy group is more acidic due to hydrogen bond formation between the phenolic groups in the ortho position (see Table 3.2).

As L-Dopa contains two separate chelate-forming donor group pairs, it can coordinate as a typical ambidentate ligand to a metal ion via both the amino acid side chain and the ortho phenolic hydroxy groups. Which donor group pair takes part predominantly in the coordination is determined primarily by the strength of the interaction between the metal ion and the binding sites, but also by the pH, as the extents of proton competition at the two separate binding sites are different, owing to the large differences in the overall basicity of the sites (log β_{H_2L} for the amino acid side chain is ~ 11.5, while for the catechol moiety it is ~ 22.5). Thus, depending upon the metal ion and the pH, either the (N,O) or the (O,O) or the mixed (N,O)(O,O) bonding mode occurs in the metal complexes. In general, (N,O)-chelation is favoured in the lower pH range, and (O,O)-coordination at higher pH, while at intermediate pH values the mixed bonding mode occurs. With the 'hard' metal ions, e.g. Mg(II) [72Go2, 78Ra], Ca(II) [78Ra], Al(III) [78Ra, 89Ki1], Cr(III) 75Da], Fe(III) [76Me1], VO_2(V) [74Ku], VO(IV) [78Ja], W(VI) [73Ku2], and Mo(VI) [76Gi] (O,O)-coordination is the preferred bonding mode, provided that the ligand is in excess. The amino acid-type (N,O)-coordination is rather weak with these metal ions.

Of the L-Dopa complexes of the 3d transition metal ions, those of Mn(II), Co(II), Ni(II), Cu(II), and Zn(II) are the best studied. Besides detailed pH-metric investigations [68Go, 72Go1, 72Go2, 74Gr, 75Bo1, 76Ge, 79Ge3, 83Am, 83Ki1] to determine the stoichiometry and stability of the complexes formed, kinetic [71Ka2, 73Ba2], Vis-UV [68Go, 74Kw, 75Bo1, 76Ge], and ESR [70Pi, 71Ca, 76Ge] spectral measurements and a preparative study [74Kw] have also been employed to clarify the bonding modes in these complexes. As an illustration, the concentration distribution of the complexes formed in the Cu(II)-L-Dopa system is depicted in Fig. 3.4.

At acidic pH, all these metal ions form amino acid type (N,O)-coordinated 1:1 complexes $M(LH_2)^+$; this bonding mode is so favoured with Ni(II) and Cu(II) that 1:2 complexes $M(LH_2)_2$ are also formed (in LH_2^- the two protons are on the non-coordinating phenolic hydroxy groups). Catecholate type (O,O)-coordinated 1:1 complexes MHL are formed only with Mn(II), Co(II), and Zn(II) (in HL^{2-} the proton is on the non-coordinating side chain amino group). The mixed bonding mode is preferred for all metal ions; this has been demonstrated by potentiometric and spectral studies of the metal-Ala-Cat mixed ligand systems, where the potential metal binding sites of L-Dopa occur in two different ligands [76Ge, 79Ge3, 83Ki1]. In these systems formation of the mixed ligand complexes $MAlaCat^-$ is preferred; their stability exceeds the statistical case by about 0.3–0.6 log unit. The occurrence of the (O,O)-coordinated complexes is usually indicated by spectral changes. As the pH is raised, the catecholate type coordination becomes dominant. Because of the overlapping processes of this bonding mode, rearrangement and the deprotonation of the donor groups not bonded to the metal ions, a given bonding mode cannot be ascribed to a given stoichimetric composition. Thus, the species $M(LH_2)(HL)^-$, which involves the mixed bonding mode, undergoes stepwise deprotonation by the different pathways given in Fig. 3.5.

The microconstants characterizing these processes have been determined [83Ki2] in a similar way as for the ligand dissociation microconstant [71Ma, 82Ki], by following the formation of the complexes containing (O,O) bonds via the UV band

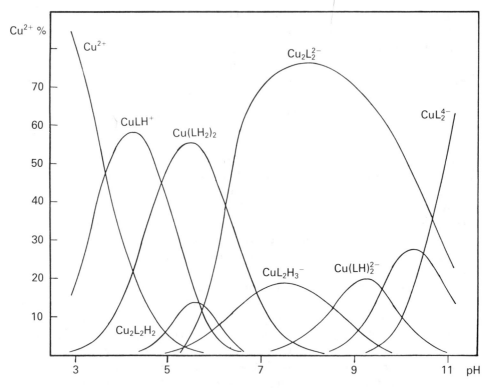

Fig. 3.4 — Concentration distribution of complexes formed in the copper(II)-L-Dopa system as a function of pH. $C_{\text{ligand}} = 0.004$, $C_{\text{Cu}} = 0.002$ mol dm^{-3}.

of the phenolate groups. The results show that the tendency to rearrange from the (N,O) or mixed bonding mode to entirely (O,O)-coordination follows the sequence Cu(II) ~ Zn(II) > Co(II) > Mn(II), Ni(II) [83Ki2].

Besides these monomeric species with various bonding modes, dimeric complexes too exist in the Cu(II)-L-Dopa system (see Fig. 3.5). In these binuclear species Cu(II) coordinates simultaneously at both binding sites of L-Dopa. Formation of chain-like dimers (structures XII and XIII) and a cyclic dimer (structure XIV) has also been assumed [72Go2] and confirmed by visible [75Bo1] and ESR [76Ge] spectral measurements.

(XII)   (XIII)

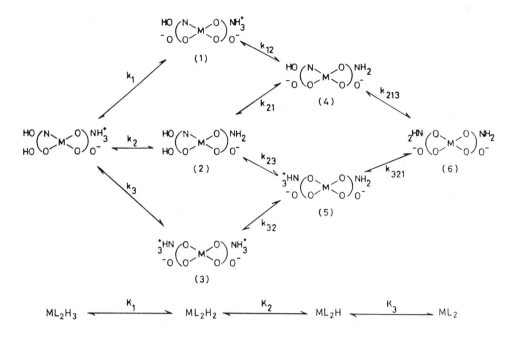

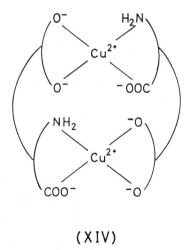

Fig. 3.5 — Parallel deprotonation pathways of the complex $CuL_2H_3^-$ formed in the copper(II)-L-Dopa system.

(XIV)

A detailed solution study [79Fa] of the Cu(II) complexes of methyldopa has led to substantially the same conclusions as discussed above for the complexes of L-Dopa.

In recent work [83Da], the heterobinuclear complex formation of L-Dopa with the Cu(II), Zn(II), or Cu(II), Cd(II) couples was investigated potentiometrically.

The results show that ternary mixed metal complexes of $CuML^+$ and $CuML(OH)$ are formed (M = Zn(II) or Cd(II)), indicating that both metal-binding sites of L-Dopa are involved in the coordination. Although the distribution of the metal ions at each site has not been studied, it can be assumed that it is not constant, but changes with the pH of the solution.

Besides the acid-base and complex-forming reactions of L-Dopa, the redox properties should also be mentioned, because of the biological importance [59He, 65He]. The ligand itself readily undergoes oxidation in the presence of oxidizing agents; the process is catalysed by certain metal ions such as Cu(II), Mn(III), VO(IV), Fe(III), etc. [89Ge1].

The complex-forming properties of semiquinone radicals generated by photolysis of aqueous solutions of L-Dopa have been studied with a number of metal ions (Mg(II), Zn(II), Ca(II), Cd(II), Sr(II)) over a wide pH range [81Fe2, 81Pi].

5.   TRIDENTATE AMINO ACIDS WITH SIDE CHAIN NITROGEN DONORS

5.1   Diaminocarboxylic acids and arginine

The α,ω-diaminocarboxylates, Dapa, Daba, Orn, and Lys, contain three dissociable protons, while Arg contains only two in the measurable pH range (see Table 3.2). The guanidino side chain of Arg, like guanidine itself, is only very weakly acidic; its pK_{HA} value is probably ~ 15 [89No2]. It should be mentioned, however, that there is a strong contradiction in the literature as regards the basicity constant of the guanidino group: the published values range between 11.5 and 15 [70Cl, 76Br2, 89No2]. The loss of the first proton can be ascribed to the carboxyl group and the pK_{COOH} changes in accordance with the electron-withdrawing effect of the ammonium group(s) and the electron-repelling effect of the growing alkyl chain. The next proton loss is normally assigned to the α-ammonium group, and the last one to the ω-ammonium group. This is only partly correct, however, as the shorter the alkyl chain between the two ammonium groups, the greater the overlap between the two dissociation processes [76Sa2]. The molar concentration ratio of the two micro-species, determined by the ratio of the real acidities of the two ammonium groups, $R = [\alpha NH_2,\omega NH_3^+]/[\alpha NH_3^+,\omega NH_2] = k_{\alpha-NH_3^+}/k_{\omega-NH_3^+}$, has been determined from a chemical shift analysis of the adjacent CH_2 or CH protons [76Sa2] and also in a very elegant deductive way [79Ma]. It can be observed from the values of R (see Table 3.2) that the percentage of the $\alpha NH_2,\omega NH_3^+$ Lys. It is also noteworthy that, though the difference between the two macroconstants ($pK_{HL} - pK_{H_2L}$) follows the sequence Dapa > Daba > Orn > Lys, the ratio R exhibits just the opposite sequence. This is because the R value is determined not only by K_{H_2L}/K_{HL}, but also by the interactions between the ammonium groups, as represented by S (see section 1.2).

For the α,ω-diaminocarboxylates, including Arg, there are alternatives between bidentate and tridentate chelation and between N and O donor atoms. All ligands can coordinate to metal ions in a bidentate manner, forming a five-membered Gly-like chelate ring. In these complexes the terminal amino group is protonated. With increasing number of the CH_2 groups between the two amino groups, the possibility of tridentate coordination via the formation of joined chelate systems invoving 5 + 5, 5 + 6, 5 + 7, or 5 + 8-membered rings decreases in the sequence Dapa > Daba > Orn

> Lys > Arg. At high pH, as the proton competition on the amino groups decreases, bidentate diamine-like bonding is also possible.

The data listed in Tables 3.3–3.7 clearly show that the bonding mode is Gly-like in the complexes MLH^{2+} formed in the acidic pH range. At higher pH, deprotonated complexes ML^+ are formed with Dapa, Daba, and Orn, in which the chelation is tridentate (90–100% closed forms in each case), although the data indicate that the $5+5$ and $5+6$ chelate systems are much more stable than the $5+7$ system that occurs with Orn. The stability of an 8-membered chelate ring is so low that there is practically no ML^+ formation with Lys and Arg.

In solution containing excess ligand, the predominant complex formed with Cu(II) at pH 5–6 is $Cu(LH)_2^{2+}$ [68Ha1, 76Br2, 78Ge1]. There is general agreement that the ligands are coordinated as substituted glycinates with protonated terminal ω-amino groups [66Br, 70Wi1, 76Bu, 78Ge1]. The crystal structure of $Cu(OrnH)_2^{2+}$ shows (N,O)-coordination in the equatorial plane, with remote, protonated δ-ammonium groups [77St2]. Vis [66Br], CD [70Wi2], ESR [71Ro], and NMR relaxation [78Ge1] studies reveal that, depending on the ligand, the stepwise deprotonation of $Cu(LH)_2^{2+}$ to CuL_2 is accompanied to different extents by a continuous rearrangement from the Gly-like (N,O)-coordination to the diamine-like (N,N) bonding mode. The possible structures of the complexes in equilibrium with each other are shown below. (As the extent of axial coordination varies for the different α,ω-diaminocarboxylates, this is not depicted in structures XV–XX.)

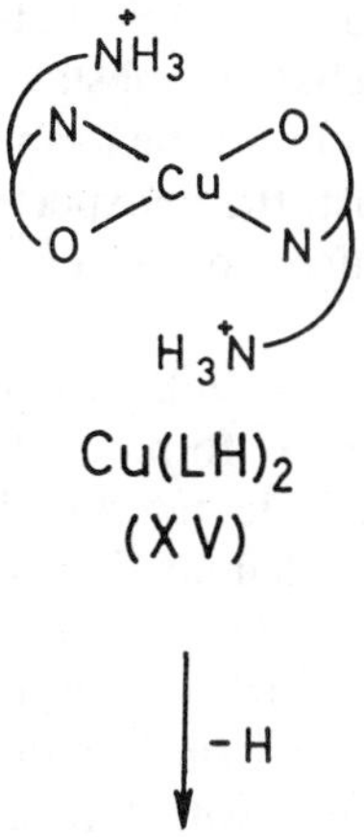

Cu(LH)₂
(XV)

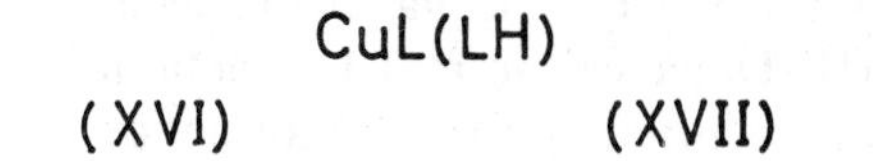

CuL(LH)

(XVI)   (XVII)

$$CuL_2$$

(XVIII)   (XIX)   (XX)

In the case of Dapa the equilibrium is shifted very strongly towards diamine-like coordination (structures XVII and XX), and at least one ligand is coordinated in a tridentate manner. Daba serves as a tridentate ligand, with significant amounts of both the species with (N,N) and the species with (N,O)-coordination in the equatorial plane and apical carboxylate and γ-amino coordination (structures XVII and XIX).

The exceptionally high ligand exchange rate constant [78Ge1] indicate very fast structural rearrangements between bonding modes XVI and XVII and XVIII and XIX in the Cu(II)-Orn system; a weak apical chelation is very probable in these complexes [70Wi]. Lys and Arg behave as substituted glycinates (structures XVI and XVIII) with insignificant interactions of the ω-amino groups in the apical positions [69Co, 70Wi2, 78Ge1]. In contrast, the unexpectedly large enthalpy values for the Cu(II)-Arg complexes were thought to be due to coordination of the guanidino group [60Pe].

With Ni(II) [66Br, 69Co, 71Ha1, 76Br2, 81Fa1], Co(II) [76Br2, 76Go], and Zn(II) [81Fa1], tris complexes too are formed, allowing both bidentate and tridentate coordination of the ligands. Bidentate Gly-like chelation occurs in the complexes $M(LH)_n$, while bidentate diamine-like coordination is suggested in the complexes ML_3^- of Dapa and Daba. In their deprotonated 1:1 and 1:2 complexes, Dapa and Daba and even Orn coordinate in a tridentate manner. For Cu(II), the tendency to tridentate bonding is very small with Orn and absent with Lys, but for Ni(II) it appears to be considerable with Orn and slight with Lys. Co(II) forms complexes similar to those of Ni(II) with all five ligands [76Bu, 76Go]. The tridentate and/or diamine-type bonding mode in the 1:1 and 1:2 Co(II) complexes of Dapa, Daba, and Orn is indicated by the reversible formation of a mono-bridged μ-dioxygen adduct $L_2CoO_2CoL_2$ [76Go], which necessitates the presence of at least 3N in the coordination sphere of each Co(II) ion [67Fa]. Zn(II) forms similar complexes with Dapa and Daba, but hydroxo complex formation occurs in preference to chelation of the δ-amino group of Orn and the ε-amino group of Lys [81Fa1].

A reversible equilibrium between the planar and octahedral forms of the complex NiL_2 occurs in the Ni(II)-Dapa system at a 1:2 metal ion:ligand ratio and at pH 9.0–9.5; this equilibrium is shifted towards the formation of the planar complex on cooling [81Fa1]. The Ni(II) in this complex is hexacoordinate at high temperature, but as the temperature is lowered, two carboxylate groups are released and four nitrogen donors take up positions in a plane around the Ni(II). The planar to octahedral conversion occurs because of the higher negative enthalpy change of the diaminoethane bonding mode over the Gly-type bonding mode of the ligand [81Fa2]. Visible spectral and magnetic susceptibility measurements served to determine the paramagnetic/diamagnetic Ni ratio.

Tetragonal Pd(II) chelates of Dapa and Daba, coordinated via two amino nitrogens with unbound carboxylate groups, were revealed by Vis absorption and CD results [70Wi2]. Lys is coordinated as a substituted glycinate. With Orn, equal amounts of five-membered (N,O) and seven-membered (N,N) chelates are formed in solution. The crystal structure of $PdOrn_2$ involves two seven-membered rings in the usual square planar arrangement, with both chelate rings in twisted-chair conformations [73Na]. Similar results were recently obtained for the Pt(II) complexes of Dapa, Daba, Orn, and Lys in the solid state [85Al].

5.2 Histidine

The His molecule contains four acidic protons, which can dissociate in the following sequence: carboxylic acid, imidazolium-N(3)-H, side-chain ammonium and imidazole-N(1)-H. The respective pK values are given in Table 3.2. The imidazole(1)-NH is very weakly acidic (pK 14.4) [74Su], and thus it does not dissociate in the measurable pH range (see later). From the complete deprotonation scheme, where all 12 microconstants were evaluated through the use of a classical deductive method (see section 3.1.1 in Chapter 2) it may be shown that the overlap between the dissociations of the acidic protons is negligible [74Su].

As concerns its metal ion-binding ability, His is a typical tridentate and ambidentate ligand. Besides the side chain carboxylate and amino groups, the imidazole ring can also be involved in chelation. There is a rapid tautomeric equilibrium between the two ring nitrogens for the single hydrogen in the neutral imidazole ring. In asymmetrically substituted imidazoles such as His, NMR studies [73Re, 77Bi] reveal a ratio of ≥ 4 in favour of the tautomer of N(1)-H (pyrrole nitrogen). Metal ion chelation occurs in the favoured tautomer with N(3) (the pyridine nitrogen), creating a six-membered (N3,N) or a seven-membered (N3,O) chelate ring. Thus, some analogues in structure and metal-binding capability can be anticipated between Dapa and His [79Ma].

Crystal structure determinations show that His coordinates as a tridentate ligand, giving rise to octahedral geometry in $Co(II)His_2$ [68Ha2], $Co(III)His_2^+$ [77Th], and $Ni(II)His_2$ [67Fr2], and to irregular octahedral geometry in $Cd(II)His_2$ [67Ca] and $Zn(II)His_2$ [63Ha1], with an approximately tetrahedral disposition of the four nitrogen donors and a further, weaker interaction of the two carboxylate oxygens. The crystals of composition $Cu(II)His_2(NO_3)_2(H_2O)_2$ obtained from acidic solution involve a typical strongly distorted octahedron, with *trans* Gly-like coordination in the equatorial plane and with distant water molecules at the axial sites [69Ev]; on the other hand, the solid $CuHis_2 \cdot 6H_2O$ has a *trans* histamine-like structure with no direct axial carboxylate bonding, but a weak interaction through hydrogen-bonding to the axial water molecules [78Ca). Mo(VI) forms diamagnetic oxygen-bridged dimers of composition $Mo_2O_4His_2$, with tridentate-coordinated His molecules [71De].

The aqueous Cu(II)-L-His system has been the subject of a large number of investigation [74Su, 79Ma, and Refs therein]. Potentiometric studies show a number of pH-dependent species (see structures XXI–XXVI).

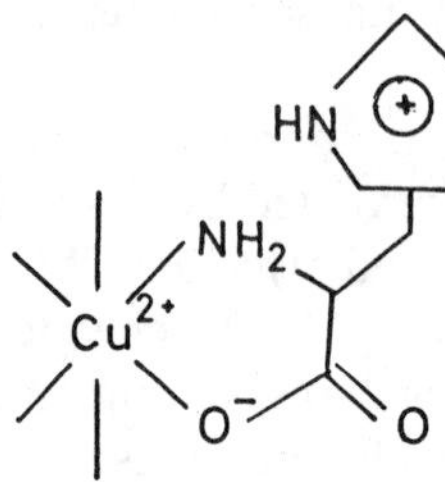

CuLH
(XXI)

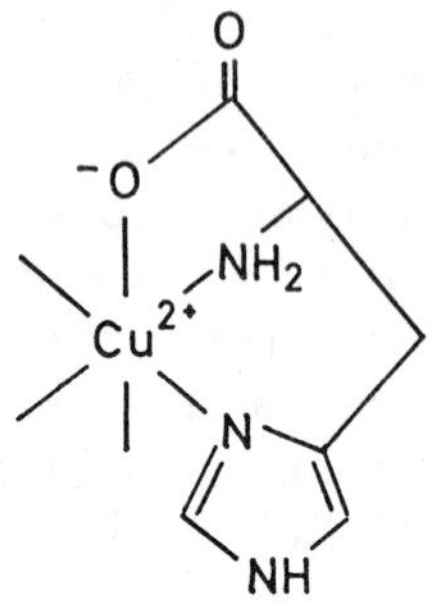

CuL
(XXII)

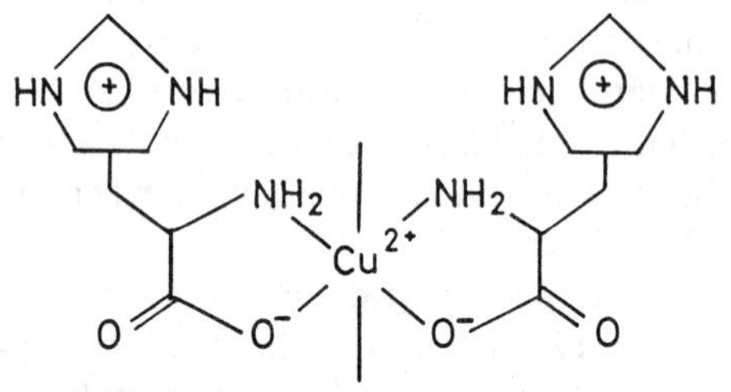

Cu(LH)₂
(XXIII)

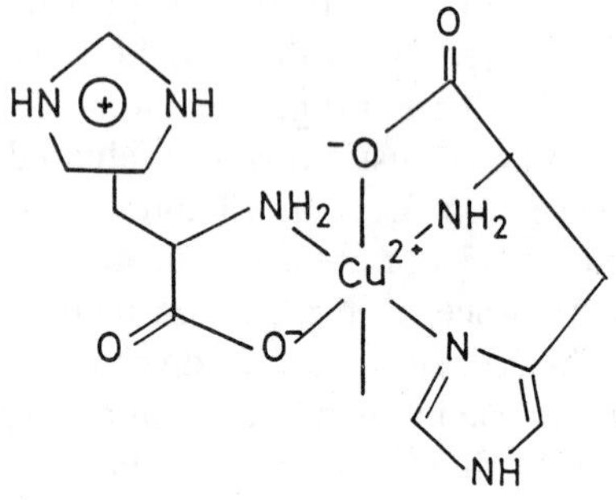

Cu(LH)L
(XXIV)

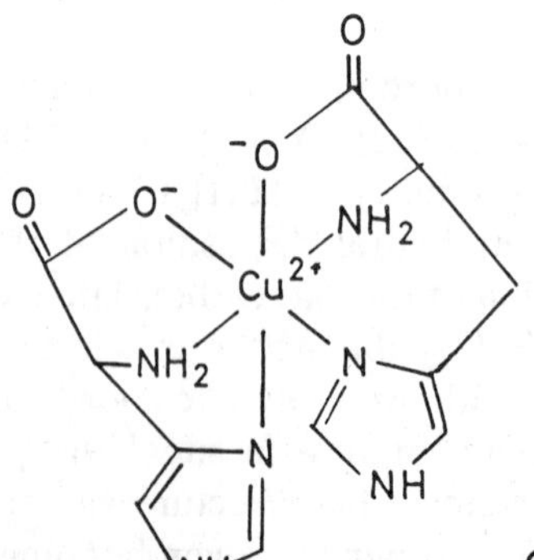

CuL₂

(XXV)

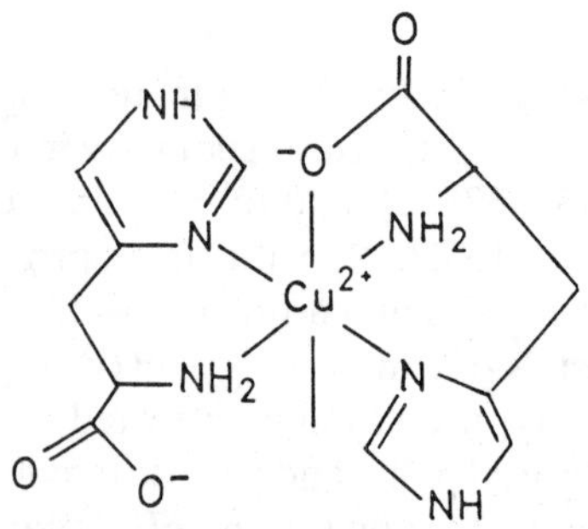

(XXVI)

Although the location of the proton in the complex $Cu(LH)^{2+}$ formed in the pH range 2–3 is still controversial, CD spectral results suggest that the most likely structure includes a Gly-like bound His with protonated imidazole ring (structure XXI) [83Ca]. This bonding mode is in agreement with stability constant comparisons [74Su, 84Pe] and a Raman spectral investigation [80It]. Earlier, the IR spectrum was interpreted by the formation of a seven-membered chelate ring with the participation of carboxylate and imidazole groups [66Ca].

It seems to be generally agreed that the species CuL^+ involves a tridentate ligand with two nitrogen donor atoms in the equatorial plane, with a weakly bound axial carboxylate group (structure XXII) [74Su]. The species $Cu(LH)_2^{2+}$ is formed in low concentration (it never exceeds 15%) in acidic solution containing excess His; similarly to the structure found in the solid state [69Ev], it presumably contains Gly-like bound ligand molecules (structure XXIII). Several features of the CD spectrum of $Cu(HL)L^+$ [83Ca], stability constant comparisons and optical studies [74Su] suggest that in this species anionic His^- is bound as a tridentate ligand, with an apical carboxylate, while neutral His is coordinated predominantly as a substituted Gly in the tetragonal plane, with a protonated imidazole ring (structure XXIV). There might be some electrostatic attracting interaction between the apical carboxylate and the non-coordinating imidazolium groups [81Go]. From pH 5.5 to about pH 11, a single complex, CuL_2, exists in solution, in which both His molecules are probably mainly tridentate [79Ma, 83Ca, 86He]. The general consensus of the many studies in which potentiometric, calorimetric, and spectroscopic (UV, Vis, CD, ESR, IR, NMR) techniques have been utilized is that the species CuL_2 is an equilibrium mixture of linkage isomers with three or four N atoms coordinated to the Cu(II) ion in the equatorial plane. If axial coordination occurs, then structures XXV and XXVI are the most favoured [87La]. These species, however, may also be in equilibrium with five- and six-coordinated structures. The hydrolysed species CuL(OH) and its dihydroxy-bridged dimer are also formed at a 1:1 metal ion:ligand ratio at pH > 6 [76Pe2].

Similarly as in the solid state, the His^- anion binds to 3d transition metal ions in a tridentate manner [74Su, 78Ki, 79Ma]. The protonated complexes formed with Co(II), Ni(II), Zn(II), Pb(II), V(IV), Hg(II), and Cd(II) in acidic solution contain either bidentate Gly-like or imidazolepropionic acid-like coordinated His molecules [84Pe and Refs therein].

Stereoselectivity has been demonstrated in metal ion complexes of His. The interaction of a metal ion with an optically active amino acid can result in three possible species: ML_2, MD_2, and MLD (where L and D denote the optical isomers of the amino acid). The first two complexes have identical stabilities, while the third contains ligands of opposite chiralities and is diastereomeric; it may therefore differ in thermodynamic properties from the optically pure complexes. Stereoselectivity may originate from differences in either the entropy or the enthalpy changes of complex formation or both. The absence of detectable stereoselectivity has been demonstrated in the formation of bis complexes of most bidentate amino acids and a number of potentially tridentate amino acids, such as Asn, Gln, Asp and Glu [79Pe and Refs therein]. His, on the other hand, appears to show a marked stereo-selectivity in its bis complexes of Co(II), Ni(II) and Zn(II) [63Mc, 69Ri, 70Mo1, 76Pe2, 76Ri]. The preference for formation of the diastereomeric MLD is shown by the following percentage values: Ni(II) 72%, Co(II), and Zn(II) 62% (instead of the 50% MLD corresponding to the statistical distribution) [88Ma1]. Stereoselectivity appears to be insignificant in the His complexes of VO(IV), Cd(II), and Pb(II) [76Pe2]. When present, it can arise from stereoselectivity in either the enthalpy changes alone (Co(II) and Ni(II)) or a combination of both the enthalpy and entropy changes (Cu(II) and Zn(II)) [76Pe2]. The origin of the stereoselectivity is discussed in detail in different reviews [74Be, 74Su, 79Pe]; here it will be mentioned only that stereoselectivity is generally favoured when a tridentate ligand with a bulky side chain is chelated to a small metal ion. Accordingly, the percentage of the mixed species NiDL, for instance, is as high as 85% for N-methyl-His and 94% for N,N-dimethyl-His [76Ri].

The above mentioned results are in interesting contrast to those of X-ray structure determinations on crystals obtained from solutions containing racemic His. The crystals of both the Ni(II) [67Fr2] and Zn(II) [63Ha1] complexes contained equal amounts of the two pure complexes ML_2 and MD_2. The crystals are therefore composed solely of species that account for less than one-fifth of the complexes present in solution [74Su]. The stereoselectivity found for $ZnHis_2$, together with stability constant comparisons, suggests that the complex in solution must be more octahedral then the nearly tetrahedral geometry found in the solid state.

Bis octahedral complexes of tridentate His can exist in three isomeric forms for optically pure ligands. All three geometric isomers of inert Co(III)-L-His_2 have been obtained by ion-exchange chromatography [69Zo], and the structures have been assigned on the basis of their electronic spectra [69Sc].

An NMR spectroscopic study has dealt with Pt(II)-L-His complexation [68Er]. The chemical shifts indicate that, as expected, the coordination is planar and that the donor atoms are amino and imidazole nitrogens over the entire pH range $1 < pH < 12$. For Pd(II)-L-His_2, the electronic and CD spectra indicate similar planar geometry, with the same four N donors in the coordination sphere [70Wi2].

The complexation of His with other metal ions, such as Be(II) [70Ch], Cu(I) [70Zu], Fe(II) [70Wi3], Fe(III) [58Pe], Hg(II) [74Li], Mn(II) [70Wi3, 75Le], UO_2(VI) [57Li], and some lanthanides [70Jo, 71Jo, 75Pl], has also been studied in solution.

Potentiometric and calorimetric studies have been carried out on the Cu(II)-M(II)-His systems (where M(II) = Ni(II), Zn(II), and Cd(II)) to detect heterobinuclear complex formation [79Am, 81Am]. On the basis of the thermodynamic parameters, it may be assumed that in the cyclic heterobinuclear species the M(II) ion is bound to the imidazole-N(3) nitrogen of one His molecule and to the carboxylate group of the other His molecule. Simultaneously, Cu(II) is bound Gly-like to one ligand and histamine-like to the other.

The acidity of the very weak imidazole-N(1)H (pyrrole-NH) of His increases upon metal ion coordination and its dissociation takes place in the measurable pH range. This deprotonation may occur by promotion across the ring or by substitution. In both cases metal ion coordination is required at the imidazole-N(3) (pyridine nitrogen). Pd(II) lowers the pK most from that of the unbound ligand by about 4 log units [72Pi]. 3d metal ions promote the ionization in the following sequence: Cu(II)$>$ Co(II) $>$ Ni(II) [70Mo2, 78So]. Besides these metal ions, the presence of another ligand in the coordination sphere can also affect the metal ion-induced ionization significantly. In a variety of mixed Cu(II) complexes of His, the deprotonations occur with pK 10.6 to 12.9. Owing to the back-coordination between Cu(II) and 2,2'-bipyridyl (bipy), the Cu(II) → ligand electron shift promotes ionization, while the electron shift in the opposite direction between Cu(II) and tiron (3,5-disulphocate-chol) appreciably suppresses the dissociation process as compared with that in the Cu(II)-His parent complexes [78So]. With Co(II), the ionization is associated with a change from octahedral to tetrahedral geometry, which is indicated by a characteristic change and intensification of the colour from pale pink–yellow to deep violet [70Mo2].

Cases are also known where N(1)-H deprotonation occurs through substitution of the proton by a metal ion. Such deprotonation cannot be achieved by His. However, in the tridentate-coordinated Ni(II), Cu(II) and Pd(II) complexes of GlyHis, where the ligand coordinates via the terminal amino group, a deprotonated peptide–amide group and the imidazolyl-N(3) group, all three complexes undergo concentration-dependent deprotonation at pH ~ 9.6 [71Mo]. The changes in the absorption spectra are indicative of the replacement of water by a nitrogen donor atom in the fourth tetragonal coordination position. The resulting complex contains an anionic imidazole ring that bridges two metal ions, one at each nitrogen. Interest in bridging imidazolate rings has heightened since the discovery that His-61 in the enzyme superoxide dismutase bridges a Cu(II) and Zn(II) through its imidazole ring [75Ri2]. For divalent metal ions, the deprotonation of N(1)-H due to substitution by a metal ion generally occurs at pH < 10, while ionization induced by a metal ion across the imidazole ring takes place at pH > 10 [74Su].

The reversible oxygenation of high-spin octahedral Co(II)His$_2$ in neutral and basic solutions to yield a brown, diamagnetic, binuclear oxygenated complex was discovered more than 40 years ago [46Bu]. Similar reversible oxygen uptake was observed for Co(II) complexes with Lys, Arg, Trp, and Asn and with many other synthetic molecules containing N and O donors. Recently, the mixed ligand complexes of bipy and 1,10-phenanthroline, for instance with various amino acids (Gly, Ala, Nval, Nle, Val, Leu, and Ile) have been studied in this respect [85Pa, 88Pa2].

Kinetic data [67Si] show that the mechanism of the oxygen uptake is a two-step process, with the rapid formation of a mononuclear oxygen adduct CoHis$_2$O$_2$, followed by formation of the oxygen-bridged dimer CoHis$_2$O$_2$His$_2$Co (structure XXVII). Most of an electron is transferred from each Co(II) to the bridging oxygen, so that the oxygenated complex is more precisely described as a Co(III)-peroxo complex [74Su]. For His, the equilibrium constant for formation of the binuclear complex is log $K = 6.4 \pm 0.1$ [76Ha, 83Fa], and the process is characterized by large negative enthalpy and entropy changes [72Po]. It is interesting that the violet tetrahedral complex Co(HisH$_{-1}$)$_2^{2-}$ formed in highly alkaline solution is able to absorb molecular oxygen [69Bu, 74Wa2].

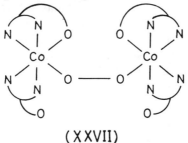

(XXVII)

6. SULPHUR-CONTAINING AMINO ACIDS

6.1 Methionine and S-methylcysteine

These thioether-containing ligands, like simple amino acids, contain two dissociable protons. The pK values of both the ammonium group and the carboxylic group are lower than those of the aliphatic analogues (see Table 3.2) owing to the electron-withdrawing effect of the RS groups. The thioether group is an extremely weak base, which undergoes protonation only in strongly acidic solutions, with a pK of -6.8 [72Bo]. The weak basicity of the thioether group does not provide a basis for its coordination, but its good π-electron acceptor character may result in fairly strong interactions with soft metal ions and weaker interactions with borderline metal ions.

The percentage values of the closed forms, 2–34% for Met and 34–56% for SMC (Tables 3.3–3.7), indicate that both Met and SMC chelate primarily as substituted glycinates with these borderline metal ions, with some involvement of the thioether group. This interaction is negligible with Met but significantly stronger with SMC, which is capable of forming a $5+5$-membered joined chelate system, while in the case of Met the second chelate ring is the less stable six-membered one. Nevertheless, the equilibrium data for the octahedral metal ions also demonstrate that the presence of the thioether group cannot prevent the formation of 1:3 complexes, that is, both SMC and Met act only as bidentate ligands in the tris complexes [76Sw, 86So1, 87So2]. The results of potentiometric [69Mc, 87So2] and NMR line broadening [76Es] studies reveal Gly-like chelation in the equatorial plane of Cu(II), with weak apical coordination of the thioether group of SMC. The weak, apical interaction is supported by the absence of an absorption band in the 400 nm region, characteristic of a Cu(II)-thioether interaction in the tetragonal plane [76Mi].

The crystal structures indicate no thioether interaction to borderline metal ions in the Met complexes of Cu(II) [69Ve, 78Ou], Zn(II) [77Wi] and Cd(II) [73Fr] and the SMC complexes of Cu(II) [86Du], Zn(II) [77Me], and Cd(II) [77Me].

Soft metal ions, such as Pd(II), Pt(II), Ag(I), Hg(II), and $CH_3Hg(I)$, are capable of strong interactions with the thioether group, forming bidentate (S,N)-coordinated or monodentate S-coordinated complexes, depending on the coordination geometry of the metal ion. Absorption [70Wi2, 82De, 83Ko], NMR [79Je], and IR [68Li] spectra and crystal structures [79Ni, 80Al, 82Ha, 86Du] indicate that the thioether sulphur is the major binding site and that the amino nitrogen is the second site for Pd(II) and Pt(II) with both Met and SMC. The (N,S)-coordination leads to the formation of a five- (SMC) or six-membered (Met) chelate ring.

An interesting feature of the coordination by the thioether donor is the creation of a chiral centre at the tetrahedral sulphur atom, leading to the formation of two diastereomeric forms. For many metal ions, inversion at the sulphur atom is rapid, but for Pt(II) snd Pd(II) the inversion process is slow enough for separate peaks to be observed in the NMR spectra [68Er, 80Ko, 82Ha]. Although Hg(II) is well known to have a high affinity for sulphur, its complexation with thioether-containing amino acids seems to be contraversial [85De]. The results of NMR shift measurements are consistent with monodentate thioether coordination in acidic solution [70Na2, 71Na]. Increase of the pH to 7 results in a change in the binding sites: amino and carboxylate donors will fully or partly displace the thioether group [71Na, 77Je]. In accordance with this tendency, in $Hg(II)(MetH)_2(ClO_4)_2 \cdot H_2O$ [76Ca] and in the analogous complex of ethionine (Eth) [81Bo] isolated from acidic solution, Hg(II) binds to the thioether group with additional intermolecular interactions to the carboxylate group of neighbouring molecules. However, crystals obtained via the reaction of HgO and SMC show that the ligand coordinates in a Gly-like manner [75Sz, 81Bo].

On the basis of detailed potentiometric and NMR measurements the linear coordination of Ag(I) to the sulphur donors of SMC, Met, and Eth is assumed in acid solution, while at higher pH each metal ion is bound via one NH_2 and one thioether group with resulting S-Ag-NH_2 linear coordination in the mononuclear and binuclear complexes formed [81Pe, 84To]. In contrast, the IR spectrum indicates that only the Met sulphur atoms are involved in the coordination in the solid complex $AgMet_2^-$, leaving the amino and carboxylate groups free to form chelates with other metal ions. Heterobinuclear complexes can be produced in this way with Cr(III), Fe(III), Co(II), Ni(II) and Cu(II) [66Mc].

It is worthwhile to mention that Co(III), which is classified as a typical hard metal ion, binds the thioether group in Met and SMC to form tridentate-coordinated complexes [76Me2].

As demonstrated above, the interaction of the thioether group with the biologically most important Cu(II) ion is relatively weak in these small biomolecules. In biological systems, for example, in the naturally occurring metalloproteins, where the formation of small chelates is less favourable, thioether sulphur serves as an important binding site. For instance, in the blue copper protein, plastocyanin, Cu(II) is situated in a distorted tetrahedral environment with side chain Met, Cys, and two His donors [78Co].

6.2   Cysteine and penicillamine

Cys and Pen each contain three dissociable protons (see Table 3.2); the last two of these deprotonations, those of the ammonium and the thiol groups, take place in overlapping processes. The R values (2.1 for Cys and 2.5 for Pen) expressing the ratios of the two monoprotonated forms, indicate that the thiol group is more acidic than the ammonium group. Thus, in weakly basic solution both Cys and Pen occur predominantly as ammoniumthiolates rather than as aminothiols. However, it must be mentioned that the R values display rather a significant temperature-dependence [69Co] and the acidity sequence of the two groups is reversed at high temperature. Both amino acids undergo oxidation to the disulphide or, in the presence of strong oxidants, to the sulphonic acid:

$$RSH \rightarrow RSSR \rightarrow RSO_2H \rightarrow RSO_3H \tag{11}$$

The reaction is catalysed by metal ions, Cu(II) [71Ha3, 81Zw, 82Fe, 83Ha2] and Fe(III) [85Ya3] being the most effective.

As a result of the presence of the three complex-forming functional groups (COO^-, NH_2, and S^-) Cys and Pen are typical ambidentate ligands. The possible metal binding sites are different in nature; COO^- is rather hard, S^- is fairly soft, and NH_2 is borderline. The SH and NH_3^+ groups have relatively high pK values, and the sulphur atoms may behave as bridging ligands. Thus, a great variety of metal complexes may be formed, including protonated and polynuclear complexes with monodentate, and different bidentate and tridentate bonding modes. The observed bonding modes of Cys in its metal complexes are given in Table 3.9. It can be seen

Table 3.9 — Bonding modes of thiol-containing amino acids in their metal complexes

Bonding modes	Metal ions
Monodentate (S)	Ag(I), Cu(I), Hg(II), MeHg(I), Pt(II)
Bidentate (N,O)	Co(III), Cr(III)
Bidentate (S,O)	Co(III), Fe(III), Zn(II), Cd(III), In(III)
Bidentate (S,N)	Co(II), Co(III), Fe(III), Ni(II), Zn(II), Pd(II), Pt(II)
Tridentate (S,N,O)	Co(II), Co(III), Cr(III), Zn(II), Mo(V), Mo(IV), Cd(II), In(III), Pb(II), Sn(II)

from Table 3.9 that the transition metal ions are the best examples on which to demonstrate the ambidentate nature of these thiol-containing amino acids.

The Cys complexes of Fe(II) and Fe(III) are especially important because of their possible relevance to the study of some non-heme iron proteins and ferredoxins containing Fe–S bonds. Fe(III), however, catalyses the oxidation of Cys, which makes the study of its complex formation very difficult. Extremely labile complexes with different stoichiometries, such as 1:1 (blue), 1:2 (red), and 1:3 (violet and green forms, thermally interconvertible), have been isolated in frozen solutions. On the basis of the absorption, ORD and CD spectra, (O,S)-coordination has been suggested in the blue, red, and violet forms, and (N,S)-coordination in the green form [68To1]; the complexes may be monomeric and/or polymeric. In the related Fe(II)-Cys complexes, the coordination is through N and S donors. Their magnetic and spectral properties indicate that the complexes have a polymeric structure, the sulphur behaving as a bridging ligand [75Mu1]. The thiol group has a greater affinity for Ag(I) than for Fe(II), and the Ag(I) ion converts it into a heteronuclear complex, in which Fe(II) is bound to Cys via the amino and carboxylate groups [68To2]. Very similar results have been obtained for the interaction of Fe(II) and Fe(III) with Pen [72St].

Both Co(II) and Co(III) form stable octahedral mono and bis complexes with Cys and Pen, where the ligands are notably tridentate [72Mc, 75Bo2, 83Ha1]. Various Co(III)-Cys tris complexes with different colours have been isolated from solutions at high pH and ligand excess; in these complexes, Cys generally acts as a bidentate (S,N) or (S,O) ligand. Besides the monomeric complexes, the formation of poly-nuclear species of $Co_2L_3^{2-}$ and $Co_3L_4^{2-}$ has also been proved [72Ga, 83Ha1] in the Co(II)-Cys and Co(II)-Pen systems. The spectral data clearly indicate the presence of both octahedral and tetrahedral Co(II) ions in $Co_3L_4^{2-}$, where two octahedral CoL_2^{2-} units are linked by a Co(II) in tetrahedral geometry via bridging thiolate groups [83Ha1].

Ni(II) prefers the planar (S,N)-coordination with both Cys and Pen [79Ge2]. As the data in Table 3.4 show the $\log (K_{NiA}/K_{NiA_2})$ values are negative. This unusual stability sequence results from the additional ligand field stabilization when the coordination of a second ligand molecule to the octahedral, paramagnetic 1:1 species yields the planar, diamagnetic 1:2 complex [79So1]. As the concentration distribution of the complexes formed in the Ni(II)-L-Cys and Ni(II)-D-Pen systems shows, little 1:1 complex exists in these solutions (see Fig. 3.6). It can also be seen

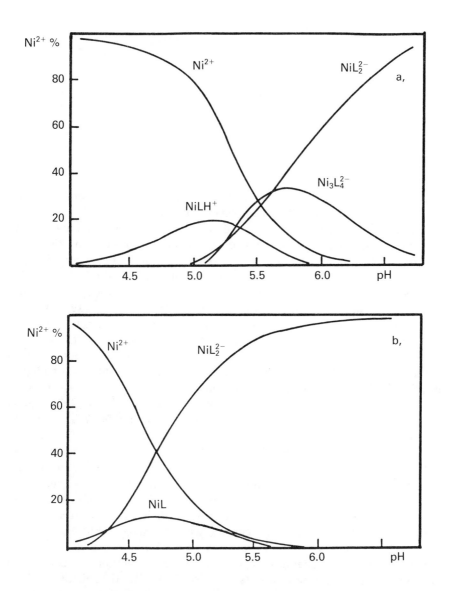

Fig. 3.6 — Concentration distribution of complexes formed in the (a) nickel(II)-L-Cys and (b) nickel(II)-D-Pen systems as a function of pH. $C_{ligand} = 0.010$, $C_{Cu} = 0.005$ mol dm^{-3}.

from Fig. 3.6 that the polynuclear complex of composition $Ni_3L_4^{2-}$ is characteristic only of the Ni(II)-Cys system. The absorption band at 380 nm indicates complex formation through sulphur bridging in $Ni_3L_4^{2-}$ (structure XXVIII) [79Ge2]. Because of the steric hindrance to bridging, due to the presence of the β-methyl groups in Pen, a polynuclear complex cannot be formed [79So1].

(XXVIII)

The absorption and NMR spectra indicate the preference for bidentate (S,N)-coordination in the Pd(II) and Pt(II) complexes of Cys [79Ge2, 85Th].

In the absence of a crystal field effect, Zn(II), as a borderline metal ion, readily forms complexes of octahedral or tetrahedral geometry with bidentate (S,O)- or (S,N)-coordination, but the tridentate (S,N,O) bonding mode is also very likely in the octahedral complexes [79Ge2]. Besides the complexes ZnL and ZnL_2^{2-}, which contain (S,N)- or (S,N,O)-coordinated Cys or Pen (cf. with the ~ 100% of the closed form of ZnL in Table 3.6), the formation of protonated complexes $ZnLH^+$, $Zn(LH)_2$, ZnL_2H^-, and polynuclear species $Zn_2L_3^{2-}$, $Zn_3L_4H^-$, $Zn_3L_4^{2-}$ has also been assumed in solution [68Pe, 76Co, 79So1]. In agreement with the solution studies, crystal structure and IR investigations [73Fr] indicate that Cys coordinates to Zn(II) through N and S donors in the complexes isolated from solutions at high pH, and through O and S donors in the complexes obtained at low pH. Cd(II) forms very similar complexes to those of Zn(II) with both Cys and Pen [72Mc, 79Ge2]. Hg(II) prefers linear or tetrahedral geometry in its complexes; thus, besides the bidentate (S,N) bonding mode, Cys and Pen are likely to coordinate in a monodentate manner via the sulphur atom [73Li, 75Sz]. Cys and Pen have been reported to act as tridentate ligands with Hg(II) in solution too [64Le, 70Su]. This is not entirely obvious if the preferred coordination number and geometry of Hg(II) are taken into account. RHg(I) readily forms 1:1 organomercury complexes with Cys and Pen in which exclusively Hg–S bonding occurs. In the case of a ligand excess, $(RHg)_2L$ may also be formed, in which the second mercury atom is coordinated to the amino group [75Ra, 76Ho, 81Ja1, 81Re].

The therapeutic and biochemical importance of the interaction between Cu(II) and D-Pen has already been mentioned (see section 1.1) as has the role of Cu(II) in the oxidation of these thiol-containing amino acids. (E^0 for the cystine/cysteine redox system is $- 0.22$ V [67Jo], while that for Cu^{2+}/Cu^+ in aqueous solution is 0.15 V, clearly indicating the redox reaction between the two couples.) Study of the Cu(II)-thiol interaction is appreciably hampered by the fact that redox and complex-formation processes occur in parallel, while both the metal ion and the ligand may form complexes that are stable in either the oxidized or the reduced form. Solution equilibrium, preparative, and crystal structure studies on the systems of Cu(II) with D-Pen and L-Cys and some other thiol derivatives have shown that, depending on the experimental conditions (metal:ligand ratio, pH, and ionic strength) different valence states of copper (Cu(I), Cu(II), or mixed) can be stabilized. Fig. 3.7

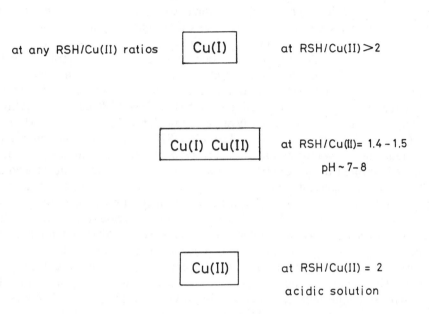

in the absence of halides   in the presence of halides

at any RSH/Cu(II) ratios   Cu(I)   at RSH/Cu(II) > 2

Cu(I) Cu(II)   at RSH/Cu(II)= 1.4 – 1.5
pH ~ 7– 8

Cu(II)   at RSH/Cu(II) = 2
acidic solution

Fig. 3.7 — Possible valence states of copper in the presence of thiol-containing ligands.

summarizes the possible valence states of copper in the presence of thiol-containing ligands

At any metal ion/ligand ratio in the absence of halide ions and at ratios of RSH/Cu(II) > 2 or RSH/Cu(II) < 1 in the presence of halide ions, the redox reaction between Cu(II) and the thiol compound is complete, and at higher concentrations a pale yellow, water-insoluble compound is obtained. The elemental analysis [78Ge2] and numerous observations on several Cu(I) mercaptides [72Mc] suggest a polymeric structure $Cu(I)_n(LH)_{n+1}^-$. Equilibrium study of the reaction between Cu(I) and D-Pen in dilute aqueous sholution shows that a mononuclear species $Cu(LH)_2^-$ and a polynuclear species $Cu_5A_4^{3-}$ are formed (the latter might be the most important member of a series of polynuclear species of the type $Cu(CuL)_n^{(n-1)-}$) [79Ös]. Although the formula of the solid compound and that of the polymeric species found in solution are different, both are Cu-S-bonded chain polymers. Depending on the metal ion/ligand ratio, Cu(II)-disulphide complexes (see section 6.3) are formed besides this Cu(I)-thiol polymer at RSH/Cu(II) < 1, while the thiol compound remains in excess at RSH/Cu(II) > 2.

When Cu(II) reacts with D-Pen at a given metal ion/ligand ratio in the presence of halide ion, in neutral solution a deep violet colour develops, the extent of formation of the colour being dependent on a number of factors: the solution pH, the D-Pen/Cu(II) ratio, the nature and concentration of the halide ion, and the O_2 content [75Mu2, 75Wr, 78Ge2]. A crystalline Tl(I) salt isolated from neutral solution has been shown to be a mixed valence complex of composition $Tl_5[Cu(I)_8Cu(II)_6Pen_{12}Cl]$ [76Bi2, 77Bi]. The structure consists of discrete, centrosymmetric clusters in which an eight-coordinated chloride ion in the central position is surrounded by a skeleton consisting of six Cu(II), each coordinated by two D-Pen molecules in (S,N) bonding mode, and linked to 8 Cu(I) via three-coordinate mercaptide ions. Detailed Vis and ESR spectral, magnetic, and electric conductivity studies have shown that the same complex is the major species in solution too at neutral pH, at D-Pen/Cu(II) = 1.40–1.45 and at various halide ion concentrations, depending on the nature of the halide ion [75Mu2, 75Wr, 78Ge2, 79La]. Although this mixed valence complex is fairly stable, slow thermal and oxidative decomposition processes occur even in neutral solution [87De]. In acidic solution it decomposes much faster, via formation of a short-lived intense blue complex [75Wr]. This product is formed in the Cu(II)-D-Pen system too at 1:2 ratio and at acidic pH [78Ge2], and its spectral properties are reminescent of those of the blue copper proteins [75Fe], containing Cu(II) in nearly tetrahedral geometry. The structural components of the mixed valence cluster may undergo various replacements. It was recently shown that Ag(I) will replace Cu(I) and Ni(II), or Pd(II) will replace Cu(II) [81Bi]. A more recent paper has examined the possibility of ligand modification, for example the formation of mixed ligand clusters in order to improve their stability [83Co]. The analogous mixed valence complex can also be obtained with β,β'-dimethylcysteamine [80Sc], α-mercaptoisobutyric acid [79Bi, 81Ke], L-Cys or L-cysteine methyl ester [80So], but the lifetime of the complex is substantially shorter with thiols not containing β substituents. The role of the β substituents is probably that their bulky, hydrophobic character hinders the hydration of Cu(I) in the cluster structure [76Sc]. The synthesis and X-ray structure of a new type of tetranuclear heterometallic complex, $Na_2[Au(I)_2Ni(II)_2(D-Pen)_4]$, have also been reported [82Bi2]. in this complex, both Ni(II) ions are *cis*-bidentate (N,S)-coordinated by two D-Pen ligands, while each Au(I) is linearly bound by two D-Pen S atoms. Each S atom forms a bridge between a Au(I) and a Ni(II) ion.

As concerns the L-Cys and D-Pen complexes of the early transition metal ions especially those of chromium and molybdenum deserve mention. It is known that glucose tolerance factor contains Cr(III), and it is interesting that a synthetic substance obtained by refluxing chromium acetate with nicotinic acid, glycine, cysteine, and glutamic acid has similar properties [77To]. At the same time chromium compounds, and particularly those of Cr(VI), are very toxic [78Da]. Cr(III) forms (N,O)-chelated complexes with L-Cys [85Ab] in acidic solution, while a complex Cr(III)-D-Pen$_2^-$ involving the tridentate coordination of the ligand is formed in a slow reaction at higher pH [72Su, 77Ho]. However, if CrO_4^{2-} and D-Pen are reacted, the formation of the same product proceeds in a relatively fast process via parallel redox and complexation reactions. In nitrogenase, molybdenum is bound to a Cys residue of the protein and thus there is a considerable interest in molybdenum-Cys complexes. The dinuclear dioxo-bridged Mo(V)-Cys complex

$Mo_2O_4Cys_2^{2-}$, produced by the interaction of MoO_4^{2-} with Cys as a result of redox and complexation reactions, has been widely studied [68Ka, 69Me, 70Ka3, 83Kr] and its X-ray structure has been determined (structure XXIX) [69Kn]. This complex is reported to have nitrate reductase, xanthine oxidase, and sulphite oxidase acitivities, and, as in the enzymes, these activities are stimulated by ferredoxin, but inhibited by KCN, Na_2AsO_3, and NaN_3 [85Qi]. The existence of a Mo(IV)-Cys complex $Mo(OH)_2Cys_2^{2-}$ has recently been demonstrated [77La] and spectroscopic evidence has been presented for new binuclear Mo(III)-Cys complexes $Mo_2O_2Cys_2^{2-}$ and $Mo_2O_2CysH_4^{2-}$ [81La1]. The interaction between molybdenum and D-Pen basically leads to the formation of similar complexes [77Su].

(XXIX)

With regard to the Cys and Pen complexes of the remaining metal ions (chiefly the main group elements), only to reviews are referred to here [72Mc, 79Ge2], and some recent papers on complexes with vanadium [81Sa2, 85Ko], Ag(I) [86Kr], Au(I) [84Al2], Tc(V) [86Gr], Bi(III) [85Hu], Sn(II), and Sn(IV) [82Cu], organotin(IV) [81Mo, 85Hu], and Se(IV) [85Sa] should also be mentioned.

Protonation constants and $CH_3Hg(I)$ formation constants have been determined for a number of selenohydryl compounds including selenocysteine, selenopenicill-amine, and selenomethionine [85Is, 86Ar].

6.3 Cystine and penicillamine disulphide

Both cystine (Cys Cys) and penicillamine disulphide (PDS) contain two amino acid moieties. They therefore have four dissociable protons, in the two ammonium and two carboxylic acid groups. The pK values clearly show that the two symmetrical functions are not fully separated from each other, as the differences in pK of the two ammonium groups and the two carboxylic groups are larger than the 0.6 log unit expected from statistical considerations.

The disulphide bridge is an even weaker base than an ether sulphur. Thus, mainly the (N,O) bonding mode is involved in the coordination with metal ions, but the two partly separated donor group pairs make dimer formation very likely. Potentio-metric measurements indicate that the dominant species formed between Co(II), Ni(II), Cu(II), and Zn(II) and D-penicillamine disulphide [81La2, 88Va] and between Cu(II) and cystine [71Ha2, 81Bl] in solution is a dimer of composition M_2L_2, in which two different cystinates chelate to each Cu(II) ion in a cyclic system. Atomic models demonstrate that simultaneous attachment of both ends of the molecule to the same Cu(II) ion is sterically impossible. An X-ray study of

$[Cu(II)PDS_2]_2 \cdot 9H_2O$ has confirmed this cyclic (N,O)-coordinated structure, and a weak axial Cu-S interaction has also been suggested [74Th]. No evidence for such a metal–disulphide interaction has been found in solution. Besides these predominant dimeric species, the formation of protonated mono and various bis complexes has also been assumed in solution [63Ha2, 71Ha2, 81Bl, 81La2, 88Ma3, 88Va]. The existence of the species proposed in basic solution, however, is questionable, as the interaction of metal ions with PDS is not reversible at pH > 9, where various transition metal ions, e.g. Co(II), Ni(II), Cu(II) and Zn(II) [88Va], promote the disproportionation of disulphides:

$$2\,RSSR + 2\,OH^- = 2\,RS^- + 2\,RSOH$$

$$2\,RSOH = RSH + RSO_2H$$

$$\overline{2\,RSSR + 2\,OH^- = 2\,RS^- + RSH + RSO_2H} \tag{12}$$

in which thiol and sulphinic acid are formed [80Os2]. The same has been found for Ag(I) with cystine [73Fr]. With soft metal centres, e.g. Pd(II), there is evidence from CD and ^{1}NMR studies that a strong metal–disulphide bond can form [77Bu, 78Ko].

The coordination chemistry of the disulphide bond in small peptide disulphides is reviewed in Chapter IV.

7 COMPLEXES OF AMINO ACID DERIVATIVES

In this section the complex-forming properties of two biologically important derivatives of α-amino acids will be discussed: the aminophosphonates (structure XXX) in which the COO^- group of amino acids is replaced by a phosphonate group, PO_3^{2-}, and aminohydroxamates (structure XXXI) in which it is replaced by a hydroxamate group, CO(NHOH).

(XXX)   (XXXI)

7.1 Aminophosphonic acids

Aminophosphonates are of considerable interest because of their occurrence in many living organisms and their biological activity. They were considered as possible constituents of living systems as early as the 1940s, but the first such compound, 2-aminoethanephosphonic acid (a phosphono derivative of β-Ala; 2-AEP), was isolated from sheep rumen protozoa only in 1959 [59Ho]. The discovery of 2-AEP in living systems was the first indication that the carbon-to-phosphorus bond exists in nature. Other aminophosphonic acids have since been identified in biological systems. Although they are found as free compounds in tissues, their most frequent forms of occurrence are their derivatives, such as phosphonopeptides, phosphonolipids, phosphonoglycolipids. The few hundred known natural and mainly synthetic aminophosphono acids and their derivatives include neuroactive compounds, growth factors, antibiotics, and herbicides. The occurrence, properties, reactions, and biological activities of aminophosphono compounds have recently been reviewed [84Ka]. In many cases, their biological activity is displayed through their inhibition of (metallo)enzymes with amino acid or peptide substrates [87Gi]. It seems obvious that this enzyme regulator ability of aminophosphonates is connected with the differences in size, shape, basicity, and charge of the carboxylate and phosphonate groups.

Simple aminophosphonic acids contain three dissociable protons; the first is very acidic, $pK < 1.0$, and is characteristic of the dissociation of PO_3H_2. The second and the third protons are involved in the dissociation of the PO_3H^- and NH_3^+ groups respectively. The pK values, together with those of possible side-chain acidic groups of several aminophosphonic acids, are listed in Table 3.10. As with Tyr and Dopa,

Table 3.10 — Dissociation constants of some aminophosphonic acids

Ligand	$pK_{PO_3H^-}$	$pK_{NH_3^+}$	$pK_{sidechain}$	Reference
Gly-P	5.39	10.05		78Wo
α-Ala-P	5.55	10.11		87Ki2
β-Ala-P	6.21	10.92		87Ki2
Phe-P	5.43	9.62		87Ki2
Tyr-P	5.50	9.45	10.42	87Ki2
Dopa-P	5.37	9.08	10.50	90Ba
			13.4	
Asp-α-P	5.52	10.07	3.44	89Ki2
Asp-β-P	6.07	10.79	2.40	89Ki2
Glu-γ-P	6.90	10.21	2.51	89Ki2

the dissociation processes of the phenolic hydroxy group and the ammonium group of Tyr-P and Dopa-P overlap one another, the R values characteristic of the concentration ratio of the two microforms ($= [^+HNRO^-]/[NROH]$) are 1.7 and 4.9, while the S values characteristic of the reciprocal effects of the two functional group deprotonations on each other are 0.34 and 0.57, respectively. In both ligands, the phenolic hydroxy group is more acidic. This change in the acidity sequence as compared to that for Tyr is attributed to the stronger electron-releasing effect of the PO_3^{2-} group than that of COO^-. Because of the higher basicity of PO_3^{2-} than that of COO^-, the overlap between the first two dissociation processes of Asp-P and Glu-P is much less than in the case of Asp and Glu; that is, it is practically negligible (cf. section 3.2).

Potentiometric and Vis and ESR spectral measurements [78Mo, 78Wo, 79Wo, 87Ki2] have revealed that the basic bonding mode of the simple aliphatic and aromatic aminophosphonates (α-Ala-P, β-Ala-P, and Phe-P) to Ca(II), Mg(II),

Mn(II), Co(II), Ni(II), Cu(II), and Zn(II) ions is bidentate (N,O)-coordination, but, owing to the more basic character of the PO_3^{2-} group, there is a possibility of the monodentate coordination of the ligands via their phosphonate group, which leads to the formation of protonated species. A stability comparison study [87Ki2] has clearly indicated that the differences in the complex-forming properties of aminophosphonic acids and their aminocarboxylic analogues are due to the significant differences in basicity, charge, electron-releasing effect, and size of the phosphonate and carboxylate groups. The stability increase of the metal–aminophosphonate complexes arising from the higher basicity of the PO_3^{2-} group is overcompensated by steric and electrostatic effects, and thus the basicity-adjusted (or relative) stability constants (see section 1.3) are less for the aminophosphonate complexes. The data show that the relative stability sequence for aminophosphonates and aminocarboxylates forming five- or six-membered chelate rings is as follows:

$$5(NH_2,COO^-) > 6(NH_2,COO^-) > 5(NH_2,PO_3^{2-}) > 6(NH_2,PO_3^{2-}).$$

The ^{31}P, ^1H and ^{13}C NMR spectra indicate the formation of (N,O)-chelated complexes Pd(II)L_2 and Pd(II)LX_2 (X = Cl$^-$, H_2O or OH$^-$) with Gly-P and Ala-P. In highly basic solution, monodentate amino-coordinated mixed hydroxo species are also formed. The monodentate coordination of the ligands via the phosphonate group, however, cannot be detected at any pH [89Ma].

There is spectral evidence that, in the case of the phosphonic acid analogues of Glu and Asp, the PO_3^{2-}/COO$^-$ substitution does not change the basic bonding modes in their transition metal complexes; that is, Glu-γ-P coordinates in a bidentate manner, while Asp-α-P and Asp-β-P coordinate as tridentate ligands. Electrostatic and steric effects, however, hinder the coordination of the second aminophosphonate ligand significantly [89Ki2]. For the two phosphono analogues of Asp, there is a possibility of ambidentate (aminocarboxylate or aminophosphonate) coordination of the ligands, which results in the co-existence of species with the same stoichiometric composition, but with different bonding modes. This behaviour is more marked when the PO_3^{2-}/COO$^-$ substitution is in the α-position. For example, equilibrium of the two microforms of CuL$_2^{4-}$ (structures XXXII and XXXIII) is assumed in the Cu(II)-Asp-α-P system [89Ki2].

(XXXII)   (XXXIII)

The metal ion and pyridoxal-5′-phosphate (PLP) catalysed transamination and dephosphonylation reactions of Asp-β-P, and thus metal complex formation between Asp-β-P, PLP, and PLP-Asp-β-P Schiff base and Al(III), Ga(III), Zn(II), and Cu(II), have been widely studied [77Ma, 84Sz, 85Ma, 85Sz2, 86Sz]. It is interesting that no enzyme has been reported so far that catalyses this dephosphony-lation reaction in an analogous way to the decarboxylation of Asp by aspartate-β-decarboxylase. The analogy between the mechanism of the two reactions, however, suggests that enzymes which catalyse such β-dephosphonylation reactions, leading to C-P bond fission in biological systems probably do exist [85Ma].

The complex-forming properties of the phenolic hydroxy-containing aminophos-phonates, Tyr-P and Dopa-P, do not differ considerably from those of their aminocarboxylate analogues (cf. sections 4.2 and 4.3) [87Ki2]. The only marked difference is the less favoured formation of the mixed bonding mode containing both catecholate- and aminophosphonate-chelated ligands with Dopa-P, and thus the significantly lower concentration of the open-chain and cyclic dimers (see structures XII–XIV) containing such types of bonding [90Ba].

7.2 Aminohydroxamic acids

There has been a great deal of interest in recent years in the chemistry of hydroxamic acid derivatives, because of their biological significance, and analytical and industrial applications. The hydroxamic acid moiety is known to be a constitutent of antibio-tics, tumour inhibitors, antifungal agents, food additives, and growth factors [82Ke]. Certain naturally occurring hydroxamic acid derivatives, the siderophores, are involved in the microbial transport of iron [73Ne]. Several of these activities depend on the strong metal-complexing ability of hydroxamates. There is current interest in utilizing these compounds in the design of metal chelates as suitable sources of trace elements essential in animal nutrition [86Br], or in using them in the chelation therapy of metal overload diseases [88Ra, 88Tu]. One group of hydroxamic acid derivatives, the amino- or peptide-hydroxamic acids, demonstrate wide-ranging biological activity as enzyme inhibitors [72Sm, 86Po], therapeutic drugs [80Mu], and antibacterial agents [66Ga].

Simple aminohydroxamic acids, which do not contain side-chain acidic groups, liberate two protons: one from the ammonium group (NH_3^+) and one from the hydroxamic (NHOH) group. There are contradictions in the literature regarding the acid–base chemistry of these ligands. According to Kurzak et al. [86Ku, 87Ku] and Paniago & Carvalho [84Pa2] the NH_3^+ group is more acidic than the hydroxamic group whereas Brown et al. [80Br, 83Br] and Leporati [86Le1] consider that the acidity sequence of the two groups is just the opposite. As the pK values of the ammonium group of an amino acid and the NHOH group of a hydroxamic acid are very similar, it is obvious to assume that the two protons dissociate in overlapping processes. Recently, a ^{13}C NMR titration study revealed that the overlap between the two dissociations is fairly little ($R = 12.5$) and the ammonium group is the more acidic [90Fa1].

As can be seen in structure XXXI, aminohydroxamates contain two partly separated metal binding sites. Metal ions can coordinate at the hydroxamate end of the molecule to form a five-membered (O,O) chelate ring, or via the amino and the hydroxamic groups to form six-membered (N,O) or five-membered (N,N) chelate rings. The simultaneous coordination of the amino, carbonyl, and hydroxamic groups to the same metal ion is sterically hindered.

The complex-formation equilibria of several simple aminohydroxamic acids (Gly-NHOH, Ala-NHOH, Nva-NHOH, Leu-NHOH, and Nle-NHOH) have been studied with Mn(II) [87Pa, 89Fa], Co(II) [86Le1, 87Pa, 88Le, 89Le, 89Fa], Ni(II) [86El1, 86Le1, 88Le, 89Le, 89Fa], Cu(II) [84Pa2, 86Ku, 86Le1, 87Ku, 87Mi, 88Le, 89Fa, 89Le], Zn(II) [86Le2, 87Pa, 89Le, 89Fa], Cd(II) [87Pa, 89Fa], Al(III) [89Fa], and Fe(III) [80Br, 85El, 89Fa] ions. The stability and spectral parameters of the complexes formed reveal that the hard metal ions such as Fe(III), Al(III), and Zn(II) prefer hydroxamic-type (O,O)-coordination, while Ni(II) and Co(II) prefer mixed amine-hydroxamate-type (N,N)- or (N,O)-coordination. X-ray crystallography clearly shows that in the complex NiL_2 of Gly-NHOH, besides the amino group the nitrogen, and not the oxygen atom of the NHOH group is involved in the coordination [82Br, 87El]. In the case of the Mn(II) and Cd(II) complexes, it is difficult to decide on the coordination mode unambiguously, because of the rather weak complexation. Cu(II) is able to coordinate at both metal binding sites, and thus it forms (O,O) complexes at low pH and (N,N) complexes at higher pH. In the intermediate pH range (between 3 and 6) a dimeric complex $Cu_2L_2H^+_{-1}$ is formed, which is presumed to be a monohydroxo-bridged dimer (structure XXXIV). A recent CD study, however, indicated that two different hydroxamic nitrogens, NOH and NO^-, are bound to Cu(II) ions, thus the hydroxamate-bridged dimer (structure XXXV) seems to be more likely [89Fa].

(XXXIV)   (XXXV)

From among the hydroxamic acid derivatives of aromatic amino acids the reactions of Tyr-NHOH and Trp-NHOH with Co(II), Ni(II), and Cu(II) ions have been studied [87Le1, 88Le] by pH-metric and spectral methods, but no interaction with the aromatic moiety has been detected. Similar results have been obtained for the interactions of Met-NHOH with Fe(III) [86El2] and Ser-NHOH with Ni(II) [83Br]. Thus, these aminohydroxamic acids behave as simple aliphatic derivatives (see above) as regards their complexation ability with metal ions.

The β-hydroxamic derivative of Asp (structure XXXVI) contains four metal binding sites; it can bind to metal ions in Gly-like, hydroxamic acid-like, or diamine-like manners, but there is also the possibility of tridentate (O,N,O)- or (O,N,N)-coordination. In a recent paper, coordination via the N atom of the amino group and the N or O atom of the NHO^- group of the ligand has been suggested in the Co(II), Ni(II), and Cu(II) complexes [88Le]. The formation of open-chain polymeric species $M_2L_3^{2+}$ and $M_3L_4^{2+}$ has also been assumed, in which an (N,N)-coordinated ML_2^{2-}

unit is linked via its α-COO⁻ group(s) to one or two (N,N)-coordinated ML units, respectively. More recently, however, monomeric species involving tridentate coordination of the ligand seemed to provide a satisfactory description of the equilibrium conditions in the Co(II), Ni(II), Zn(II), and Fe(III)-Asp-β-NHOH systems [90Fa2]. A polynuclear species with the composition $Cu_4L_4H_{-2}^{2-}$ (structure XXXVII) was assumed to be formed only for Cu(II). The spin-exchange interaction between the Cu(II) centres via the bridging NO⁻ groups may explain the ESR inactivity of the species [90Fa2].

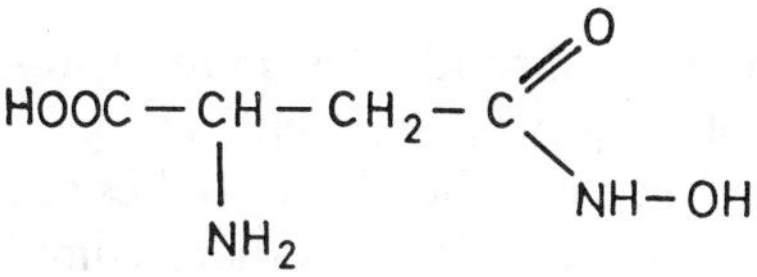

$$HOOC-CH-CH_2-C \overset{O}{\underset{NH-OH}{}}$$

$$\underset{NH_2}{|}$$

(XXXVI)

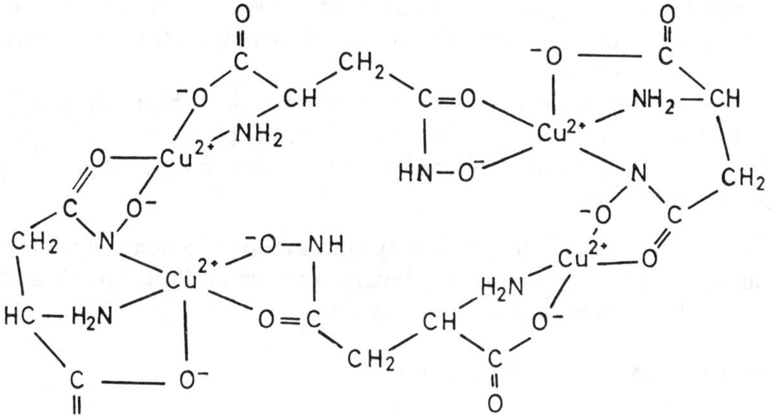

(XXXVII)

$$\overset{O}{\underset{NH_2}{}} CH_2-CH-C \overset{O}{\underset{NH-OH}{}}$$

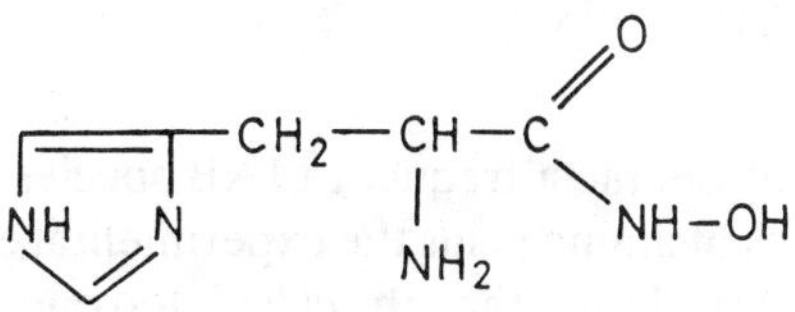

(XXXVIII)

His-hydroxamic acid (structure XXXVIII) is again a typical ambidentate ligand. The similar arrangement of the possible metal binding sites as in Asp-β-NHOH suggests complexes with similar bonding modes. The results published so far on complexation with Co(II) [87Le2], Ni(II) [87Le2], Cu(II) [87Le2, 88Ku], and Fe(III) [84Br, 87Sh] are partly contradictory [87Le2, 88Ku], and not enough to permit unambiguous conclusions on the complex-forming properties of this ligand. What is certainly true is that, with the exception of Fe(III), which as a hard metal ion prefers hydroxamic acid-like coordination, participation of the side-chain imidazolyl-N in bidentate or tridentate coordination of the ligand is very likely.

8. MIXED LIGAND COMPLEXES

Under biological and environmental conditions, multiligand–multimetal ion equilibria dominate, and not simple binary systems with a single metal ion and a ligand. Hence, in recent years attention has turned from studies of the equilibria of binary metal ion–amino acid complexes to those of ternary complex formation between amino acids, peptides, or other smaller or larger biomolecules. Besides the hundreds of papers published each year, excellent reviews have appeared in this field too [73Si, 75Si, 80Os1, 80Si, 85Sa]. Stability data on binary and ternary systems, together with a knowledge of the general features and factors characterizing and influencing mixed ligand complex formation, give the possibility of the speciation of real biological fluids, where hundreds of potential ligands are likely to compete for metal ions found *in vivo*.

This section briefly discusses the general factors which basically govern the formation of ternary species: (1) statistical reasons, (2) charge neutralization, (3) steric factors, (4) π-bond formation, and (5) intramolecular ligand–ligand interactions in ternary complexes.

(1) *Statistical reasons* — The probability of formation of a mixed ligand species MA_rB_s is always larger than that of any binary complex [73Ma]. The equilibrium constant (K_{stat}) for the general formation reaction

$$r MA_n^{a+} + s MB_n^{b-} = n MA_r B_s^{(ra-sb)+} \tag{13}$$

can be calculated on the basis of statistical considerations as follows [80Os1]:

$$K_{stat} = \frac{[MA_r B_s]^n}{[MA_n]^r [MB_n]^s} = \left(\frac{n!}{r! s!}\right)^n. \tag{14}$$

($K_{stat} = 4$ for the formation of the most frequent MAB species.) It is interesting to note that for mixed complexes of amino acids the experimental equilibrium constant (K_M) for reaction (13) is generally larger than the calculated one. Thus, the enhanced formation of the mixed complex cannot be explained on purely statistical grounds, and other effects usually also play an important role. The statistics of ternary complex formation has recently been discussed in several papers [83La, 83La2].

(2) *Charge neutralization* — The reaction of two oppositely charged binary complexes in eq (13) leads to a preferential formation of the mixed ligand complex, due to coulombic effects (enthalpy term) and the release of oriented solvent

molecules (entropy term). However, these effects are usually small, as is confirmed by the measured constants for a greater number of mixed ligand systems, including those of amino acids [75Si, 80Os1].

(3) *Steric factors* — Bulky groups, the size of the chelate rings, or a surplus of binding sites as compared with the available coordination positions may also influence mixed complex formation (see Table 3.11). For example, the bulky methyl

Table 3.11 — Mixed ligand complex formation constants† of some transition metal(II)-Gly-ligand B systems

Metal		Ligand B	$\log \beta_{MAB}$	$\log K_M$
Cu(II)	Gly	en	17.69	0.86
	Gly	Me$_2$en	17.29	2.76
	Gly	pn	16.91	2.16
Ni(II)	Gly	en	12.47	1.02
	Gly	Me$_2$en	11.94	2.72
	Gly	pn	11.60	2.18
Zn(II)	Gly	en	10.53	1.31
	Gly	Me$_2$en	9.80	1.84

† Values are taken from Refs. 76So and 79So1.

groups of N,N'-dimethylethylenediamine (Me$_2$en) favour the coordination of Gly over a second ligand of the same kind. In the case of the small ethylenediamine (en) molecule, the stabilization is much less. The data given in Table 3.11 may suggest that, as a tentative rule, the mixed ligand chelates containing a five- and a six-membered ring (Gly + propanediamine) are favoured compared to those containing two rings of equal size (Gly + en). However, results obtained for other types of ligands indicate that this rule is not generally valid and the effect may be metal ion-dependent [79So2]. These charge and steric effects are reflected in the $\log(K_{ML}/K_{ML_2})$ values of the binary complexes too. Thus, the larger the difference in the $\log(K_{ML}/K_{ML_2})$ values of the parent complexes, the larger the value of $\log K_M$ for the mixed complexes. In this way, the effect of the difference in the stability conditions of the binary complexes on the stability of the mixed ligand complex can be estimated [79So2]. The stabilization exceeding this estimated stability constant originates from some specific metal ion–ligand pair or ligand–ligand interaction in the ternary system.

(4) *π-bond formation* — The π-electron-accepting properties of heteroatomic N bases such as bipy, histamine (Hm), or His enhance the stability of ternary complexes tremendously, especially if the second ligand has O donor sites, e.g. catecholates (Cat) (see Table 3.12).

Table 3.12 — Mixed ligand complex formation constants (K_M) representing the effect of π-bond formation

Ligand A	Ligand B	Ni(II)	Cu(II)	Zn(II)
2,2′-Bipyridyl†	Gly	1.75	3.05	0.72
	Cat	3.71	6.15	2.98
Histamine‡	Gly	1.21	3.10	0.63
	Cat	—	4.86	—
	Tiron	—	3.96	—
Histidine§	Gly	1.16	2.20	0.92
	Tiron	—	2.30	—
1,2-diaminoethane¶	Gly	1.02	0.86	1.31
	Cat	—	2.65	—

† Ref. 71Gr; ‡ Refs. 69Hu, 78So; § Ref. 78So; ¶ Refs. 69Hu, 79Sol.

(5) *Intramolecular ligand–ligand interactions* — Ionic bonds, hydrogen bonds, and hydrophobic or aromatic ring-stacking interactions between suitable groups of the coordinated ligands also lead to enhanced mixed complex formation. Besides the usual stabilization factors, like K_M, a similar approach to open–closed equilibria can be used to characterize this enhancement of mixed complex formation, as has been shown for the estimation of the extent of tridentate coordination in amino acid binary complexes (see section 1.3). In Table 3.13, the first entries are examples of the

Table 3.13 — Mixed ligand complex formation constants (K_M) representing the effect of ligand–ligand interactions

System	$\log K_M$	% closed form	Reference
Cu(II)-L-His-L-OrnH	2.18	44	77Br
Cu(II)-L-His-L-ArgH	2.02	37	77Br
Cu(II)-L-Trp-ATP	2.10	41	76Si
Zn(II)-L-Trp-ATP	3.55	76	76Si
Co(II)-Phe-Nva	0.67	21	72Ge2

electrostatic interaction between the non-coordinating carboxylate of L-His and the ammonium group of L-OrnH of the guanidinium group of L-ArgH. In the case of the next two entries, the aromatic ring-stacking interaction between the aromatic π-electron system of the indole ring in Trp and the purine moiety of ATP results in favoured formation of the mixed complex. The last entry shows the stabilizing effect of the hydrophobic interaction between the aromatic moieties and the aliphatic side-chains.

To illustrate the wide variety of investigations on mixed complex formation involving amino acids and to demonstrate the importance of the factors influencing mixed complex formation, some recent studies are listed in Table 3.14.

9. CONCLUSIONS

The basic characteristic of the metal ion-binding ability of α-amino acids is chelation via the amino and α-carboxylate groups. The nature of the metal ion and the experimental conditions, however, can modify this general feature even for simple amino acids containing only these two donor groups, resulting in monodentate coordination via the carboxylate or the amino groups (see, for example, the various bonding modes occurring in Ag(I)-amino acid complexes or the formation of a carboxylate-bridged polynuclear complex with Fe(III)).

About half of the essential amino acids commonly occurring in proteins contain side-chain donor atoms that are at least potentially capable of forming other chelate rings with a metal ion bound at the amino and/or the α-carboxylate groups. As regards the possibility of participation of the side-chain donor groups in coordination, the decisive factors are the affinity of the third donor atom for metals and its location within the amino acid molecule. Accordingly, the alcoholic hydroxy group in Ser or Thr, the amide in Asn, and the thioether in S-methylcysteine are capable only of weak interactions with most metal ions, owing to their low basicity. Further, the unfavoured location of the same donor groups may additionally hinder the tridentate coordination of amino acids such as Hypro, Glu, Gln, or Arg. The phenolic hydroxy group is a fairly strong binding site, but it is separated from the (N,O) function in Tyr and thus is not capable of direct coordination to metal ions. When the phenolic hydroxy group is in the ortho position, as in o-Tyr, or in a chelate-forming pair, as in L-Dopa, its metal-binding ability increases dramatically. The same is true for the side-chain amino group in Orn, Lys, Dapa, and Daba. In His, Cys, and Pen, the metal-binding strength of the third donor group is strong enough and its location is also favourable for these tridentate amino acids to be very effective metal-binding biomolecules.

Besides the strength of the metal ion-donor group interaction and the steric effects, the pH can likewise greatly affect the participation of the donor groups in coordination in solution and hence in biological fluids. The pH-dependent hydrogen ion competition at the different binding sites results in rearrangements of the bonding modes in many metal complexes of L-Dopa, His, Cys, etc.

Numerous examples discussed in this chapter show that all side-chain donor atoms of potentially tridentate amino acids can behave as metal-binding sites, although they sometimes have to be forced to do so. In oligopeptides, in which the essential donors of amino acids are less available for metal binding, the coordination ability of the side-chain donor groups increases significantly. As will be seen in the next chapter, all side-chain donor atoms readily coordinate to metal ions under 'normal' conditions even in dipeptides. In proteins, these side-chain functional groups become the exclusive metal-binding sites, which are sterically brought close together by the tertiary structure of the proteins. Each functional group acts

Table 3.14 — Examples of recently investigated mixed ligand systems

Amino acid	Metal ion	Other ligand	Decisive factors/Remarks	Ref.
Gly, Ala, Val, Tyr, Ser, Asp, Glu	Cu(II)	1,10-phenanthroline, 2,2′-bipyridyl	Cu(II)-aromatic π-bond, ring-stacking interaction	80Kw
Cys, His, Glu, Thr	Zn(II)	histamine	Computer simulation of species distribution in blood plasma	80Ka
Val, Thr, Ile, Asp	Cu(II)	malonic acid	Charge neutralization	80Sh
L-His	Zn(II)	L/D-Thr, L-Ser, L/D-Ala	Steric factors/ stereoselectivity	81Ge
Lys, Phe, Arg	Zn(II)	Cys, Thrs, His, Arg	Computer simulation of species distribution in nutritive mixtures	81Al2
Ala, Ser, Thr, Pro	Zn(II)	1.10-phenanthroline	Steric effects	81Bu2
Val	Ni(II)	Gly, Ala	Steric effects	81En
His	Cu(II)	Dapa	Computer simulation of species distribution involved in rheumatoid arthritis	81Ja2
His	Pd(II)	Asn. Glu, Thr, homoserine	Ligand-ligand H-bond interaction	81Od
Gly, Ala, Ser, β-Ala	Cu(II)	1,2-diaminoethane 1,3-diaminopropane	Chelate ring size	81Ra
L/D-Ala, β-Ala, L/D-Asp, L/D-Orn	Cu(II)	GlySer, GlyAsn GlyAsp, GlyGlu	Ligand-ligand electrostatic interaction	82Ge
His, Glu, Thr	Cu(II)	histamine	Computer simulation of species distribution in blood plasma	82Ka
Gly, Ala, Abu	Cd(II)	Gly, Ala, Ahb	Inter-ligand interaction via central Cd(II)	82Ma2
Abu, Dapa, Daba, Orn	Cu(II)	imidazole	Chelate ring size, bond formation, surplus of binding sites	82Na2
Abu, Dapa, Daba, Orn	Cu(II)	histamine	Chelate ring size, bond formation, surplus of binding sites	82Na3
L/D-Glu	Pd(II)	L-Ala, L-Arg	Electrostatic ligand-ligand interaction	82Od
Ala, Gly, β-Ala, Val, Ser, Thr, Asn, Asp, Glu, Tyr	Cu(II)	Gly-Phe	Chelate ring size, steric effects, Cu(II)-aromatic interaction	82Sh
Ala, Trp	Cu(II), Zn(II)	ATP	Ring-stacking interaction	83Ar
Gly, His	Cu(II)	GlyHis, HisGly, carnosine	π-bond formation	83Fa
Tyr	Mn(II), Co(II) Ni(II), Cu(II) Zn(II)	AMP, ADP, ATP	Ring-stacking interaction	83Ma
Ser, Thr, Val, Glu	Co(III)	1,2-diaminoethane 1,10-phenenthroline, 2,2′-bipyridyl	Steric effect, polar	83Pa
Gly, Ala, β-Ala, Val, Ser, Thr, Asn, Asp, Glu, Tyr	Cu(II)	GlySer	Chelate ring size, steric effects, Cu(II)-aromatic π-interaction	83Sh
Ala, Abu, Nva, Nle, Leu, Ile	Mn(II), Cu(II)	ATP	Hydrofobic interaction	83Si
Gly, Ala, Ser, Val, Nva, Leu, Nle, Pro, Phe, Tyr, Trp	Cu(II)	4-carboxy-uracil	Ring-stacking interaction Hydrofobic interaction	83Vl
L-His, histamine	Cu(II), Zn(II)	ATP	π-bond formation, ring-stacking interaction	84Ar1
Phe, Tyr	Cu(II), Zn(II)	ATP	Ring-stacking interaction	84Ar2
His	Cu(II)	Leu, Glu, Met, Trp, Ala	Computer simulation of species distribution in nutritive mixtures	84Be1
His	Cu(II)	Thr, Lys, Gly, Phe, Val, Cys.Cys	Computer simulation of species distribution in nutritive mixtures	84Be2

Table 3.14 — Contd.

Amino acid	Metal ion	Other ligand	Decisive factors/Remarks	Ref.
Gly, Ala, Arg, Orn, Lys, Asp, Glu	Cu(II)	L-noradrenaline	Ligand-ligand interactions	84Da
Gly, Ala, Phe, Tyr, Trp	Cu(II), Ni(II)	5-methyl-uracil (thymin)	Ring-stacking interaction	84Fr
L-Dopa	Co(II), Ni(II) Cu(II), Zn(II)	Ala, His, GlyGly, ATP	π-bond formation, ring-stacking interaction	84Ki2
Ala, Gly, Leu, Phe, Pro	Co(III)	2,2′-bipyridyl, 1,10-phenanthroline	π-bond formation, ligand-ligand interaction	84Mo
Ala, Val, Phe, Tyr, Trp	Cu(II)	1,2-diaminoethane, and N-substituted derivatives	Cu(II)-aromatic ring interaction, decreased hydration	84Od
Ala, Asp, Glu, His	Cu(II)	Arg, Lys, Orn, Ser, Val, Glu	Ligand-ligand electrostatic interaction, hydrogen bond formation	84Ya
Gly, Ala, Val, Ser, Phe, Trp	Cu(II)	Edta	Effect of Edta on blood plasma model	85Ar
Gly, Ala, Ser, Val, Thr	Cu(II)	iminodiacetic acid	Surplus of binding sites	85Ch
Ala, Val, Ser, Lys, Trp, His	Cu(II)	gluconic acid, Vitamin B_{15}		85Fr
Phe, Tyr, Dopa	Cu(II), Ni(II)	2,2′-bipyridyl, 1,10-phenanthroline	π-bond formation, ring-stacking interaction	85Ki
His, Lys, Asn, Phe, Tyr, Trp	Pt(II)	2,2′-bipyridyl	Potential anticancer activity	85Ku
L/D-Val, N-alkyl-derivatives	Cu(II)	L-Pro, L-Hypro and N-alkyl derivatives	Enantioselective effect	85Ul
Val, Phe, Tyr, Trp, 5-hydroxy-Trp	Cu(II)	1,2-diaminoethane, histamine, 2,2′-bipyridyl, 1,10-phenanthroline	Ring-stacking interaction	85Ya4
L/D-His	Cu(II)	Lys, Tyr, Trp, Phe, Ala, Val, Arg, Glu, Asn, Glu, Ser, Thr	Electrostatic and hydrofobic ligand-ligand interaction	85Ya5
His	Cu(II)	Glu, Asn, Ser	Computer simulation of species distribution in blood plasma	86Be
Gly, Ala, Arg, Orn, Lys, Asp, Glu	Cu(II)	L-Dopa, dopamine	Ligand-ligand interaction	86Da
Gly, Ala, Val, Leu, Asp, Glu, Phe	Cu(II)	salicylic acid		86Di
Gly, Ala, Ser, His, Glu	Cu(II)	ascorbic acid		86Fr
Gly, Ala, Nor, Phe, Trp	Cu(II)	ethylenediamine-N-acetate	Cu(II)-aromatic ring interaction	86Le2
Asp	Cr(II)	Met, Eth	Surplus of binding sites	86Ma
His, Dapa, Daba, Orn	Zn(II)	histamine	Chelate ring size, π-bond formation	86Na
Cys, Asp, Glu, Met, His, Arg, Phe, Tyr	Pd(II)	2,2′-bipyridyl	Potential anticancer activity	86Pu
Gly, Dapa, His	Cu(II)	glycinamide, Gly-glycineamide, N-acetyl-His	π-bond formation	86So2
Asp, His, Glu	Zn(II), Co(II) Ca(II)	ATP	Surplus of binding site	87Da
Val, Ile, Asp, Glu, Gln, Pro, SMC	Pt(II)	2,2′-bipyridyl	Potential anticancer activity	87Ja
Gly, Ile, Val, Pro, Ala, Phe	Pd(II), Pt(II)	inosine, guanosine	Hydrofobic interaction	87Ka

Table 3.14 — Contd.

Amino acid	Metal ion	Other ligand	Decisive factors/Remarks	Ref.
Asp, Asn, Glu, His, Arg, Lys, Dopa	Cu(II)	2,2′-bipyridyl	π-bond formation, ring-stacking interaction	87Pr1
Ala, Gly, Val, Leu, Phe, Trp, Met, Eth	Cu(II)	2,2′-bipryridyl	π-bond formation	87Pr2
Pro, Thr, Ser, Val, Gly, Ile, Abu	Cu(II)	N-acetylglycine		88Li
Gly, Ala, Val, Leu, Phe, Trp, Met, Asp, His, Lys, Dopa	Cu(II)	bis(imidazol-2-yl)- methane, bis(imidazol- 2-yl)nitromethane	π-bond formation, ring-stacking interaction	88Pr
Gly, Ala, Val, Leu, Ser, Thr, Met, Tyr Phe, Trp	Cu(II)	Ala, Ser, Thr, Tyr Phe, Val, Met	Hydrophobic interaction	88Ta
Val, Leu, Phe, Thr	Cu(II)	L-Dopa, dopamine	Ligand-ligand interactions	88Ze

relatively independently of the others, and the specificity of these metal-binding active centres is determined to a large extent by the immediate vicinity of the protein chain and the possible secondary electrostatic and hydrophobic interactions between the constitutents of the active sites.

Thus, a knowledge of the complex-forming properties of the amino acids as the building units of peptides and proteins is indispensable for an understanding of the metal-binding properties of peptides, but it would be risky to extrapolate unreservedly from metal–amino acid or metal–peptide interactions to metal–protein interactions.

REFERENCES

36Ne   Neuberger, A.: *Biochem. J.* **30**, 2085 (1936).
46Bu   Burk, D., Hearon, J. Z., Caroline, L., & Schade, A. L.: *J. Biol. Chem.* **165**, 723 (1946).
50Da   Davies, C. W. & Waind, G. M.: *J. Chem. Soc.* 301 (1950).
52Ma   Mathieson, A. McL. & Welsh, H. K.: *Acta Cryst.* **5**, 599 (1952).
53Lu   Lumb, R. F. & Martell, A. E.: *J. Phys. Chem.* **57**, 690 (1953).
54Sc   Schubert, J.: *J. Am. Chem. Soc.* **76**, 3442 (1954).
57Bo   Boyne, T., Pepinsky, R., & Watanabe, T.: *Acta Cryst.* **10**, 438 (1957).
57Li   Li, N. C. & Doddy, E.: *J. Am. Chem. Soc.* **79**, 5859 (1957).
57Mu   Murphy, C. B. & Martell, A. E.: *J. Biol. Chem.* **226**, 37 (1957).
58Ma   Martin, R. B., Edsall, J. T., Wetlaufer, D. B., & Hollingworth, B. R.: *J. Biol. Chem.* **223**, 1429 (1958).
58Pe   Perrin, D. D.: *J. Chem. Soc.* 3125 (1958).
59He   Heacock, R. A.: *Chem. Rev.* **59**, 181 (1959).
59Ho   Horiguchi, M. & Kandatsu, M.: *Nature* **184**, 901 (1959).
59Pe   Perrin, D. D.: *J. Chem. Soc.* 290 (1959).
60Pe   Pelletier, S.: *J. Chem. Phys.* **57**, 318 (1960).
61Iz   Izatt, R. M., Wrathall, J. W., & Anderson, K. P.: *J. Phys. Chem.* **65**, 1914 (1961).

62Mi   Milner, R. S. & Pratt, L.: *Discuss. Faraday Soc.* **34**, 88 (1962).
63Fe   Feldman, I. & Koval, L.: *Inorg. Chem.* **2**, 145 (1963).
63Ha1   Harding, M. M. & Cole, S. J.: *Acta Cryst.* **16**, 643 (1963).
63Ha2   Hawkins, C. J. & Perrin, D. D.: *Inorg. Chem.* **2**, 843 (1963).
63Ir   Irving, H. & Pettit, L. D.: *J. Chem. Soc.* 1546 (1963).
63Ll   Llopis, J. & Ordonez, D.: *J. Electroanal. Chem.* **5**, 129 (1963).
63Mc   McDonald, C. C. & Phillips, W. D.: *J. Am. Chem. Soc.* **85**, 3736 (1963).
64Le   Lenz, G. R. & Martell, A. E.: *Biochemistry* **3**, 745 (1964).
65He   Heacock, R. A.: *Adv. Heterocyclic Chem.* **5**, 205 (1965).
65Mo   Moeller, T., Martin, D. F., Thompson, L. C., Ferrus, R., Feistel, G. R., &
        Randall, W. J.: *Chem. Rev.* **65**, 1 (1965).
65Ri   Ritsma, J. H., Wiegers, G. A., & Jellinek, F.: *Rec. Trav. Chim.* **84**, 1577
        (1965).
66Br   Brubaker, G. R. & Busch, D. H.: *Inorg. Chem.* **5**, 2110 (1966).
66Ca   Carlson, R. H. & Brown, T. L.: *Inorg. Chem.* **5**, 268 (1966).
66Ga   Gale, G. L. & Hynes, J. B.: *Can. J. Microb.* **12**, 73 (1966).
66Gi   Gillard, R. D., Irving, H. M., Parkins, R. M., Payne, N. C., & Pettit, L.
        D.: *J. Chem. Soc. (A)* 1159 (1966).
66Mc   McAuliffe, C. A., Qualiano, J. V., & Vallarino, L. M.: *Inorg. Chem.* **5**,
        1996 (1966).
67Ca   Candlin, R. & Harding, M. M.: *J. Chem. Soc. (A)* 421 (1967).
67Fa   Fallab, S.: *Angew. Chem. (Internat. Edit.)* **6**, 496 (1967).
67Fr1   Freeman, H. C.: *Adv. Protein Chem.* **22**, 257 (1967).
67Fr2   Fraser, K. A. & Harding, M. M.: *J. Chem. Soc. (A)* 421 (1967).
67Gi   Gillard, R. D.: *Inorg. Chim. Acta* **1**, 69 (1967).
67Ja   Jackowitz, J. F., Durkin, J. A., & Walter, J. L.: *Spectrochim. Acta* **23A**, 67
        (1967), and the references therein
67Jo   Jocelyn, P. C.: *Eur., J. Biochem.* **2**, 327 (1967).
67La   Laurie, S. H.: *Aust. J. Chem.* **20**, 2609 (1967).
67Si   Simplico, J. & Wilkins, R. G.: *J. Am. Chem. Soc.* **89**, 6092 (1967).
67Sh1   Sharma, V. S.: *Biochim. Biophys. Acta* **148**, 37 (1967).
67Sh2   Shibata, M., Nishikawa, H., & Hosaka, K.: *Bull. Chem. Soc. Japan* **40**, 236
        (1967).
67St   Stack, W. F. & Skinner, H. A.: *Trans. Faraday Soc.* **65**, 2044 (1967).
67Wa   Watt, G. W. & Knifton, J. F.: *Inorg. Chem.* **6**, 1010 (1967).
67Ya   Yagil, G.: *Tetrahedron* **23**, 2855 (1967).
68Er   Erickson, L. E., McDonald, J. W., Howie, J. K., & Clow, R. P.: *J. Am.
        Chem. Soc.* **90** 6371 (1968).
68Gi   Gillard, R. D.: In *Physical methods in advanced inorganic chemistry*, Hill,
        H. A. O., Day, P. eds., Interscience, London (1968).
68Go   Gorton, J. E. & Jameson, R. F.: *J. Chem. Soc. (A)* 2615 (1968).
68Ha1   Hay, R, W., Morris, P. J., & Perrin, D. D.: *Aust. J. Chem.* **21**, 1073 (1968).
68Ha2   Harding, M. M. & Long, H. A. *J. Chem. Soc. (A)* 2554 (1968).
68Ka   Kay, A. & Mitchell, P. C. H.: *Nature* **219**, 267 (1968).
68Li   Livingston, S. E. & Nolan, J. D.: *Inorg. Chem.* **7**, 1447 (1968).
68Pe   Perrin, D. D. & Sayce, I. G.: *J. Chem. Soc. (A)* 53 (1968).
68Sh   Shimura, Y.: *Spectroscopy and structure of metal chelate compounds,*

           Nakamoto, K., McCarty, P. J. eds., Wiley, New York (1968).
68To1   Tomita, A., Hirai, H., & Makishima, S.: *Inorg. Chem.* **7**, 760 (1968).
68To2   Tomita, A., Hirai, H., & Makishima, S.: *Inorg. Nucl. Chem. Lett.* **4**, 715
           (1968).
68Wa   Watt, G. W. & Knifton, J. F.: *Inorg. Chem.* **7**, 1159 (1968).
68Za   Zaharova, E. A. & Kymok, V. N.: *Zh. Obshch. Khim.* **38**, 1922 (1968).
69Bu   Bugger, S.: *Acta Chem. Scand.* **23**, 975 (1969).
69Ch   Childs, C. W. & Perrin, D. D.: *J. Chem. Soc. (A)* 1039 (1969).
69Co   Coates, E., Marsden, C. G., & Rigg, B.: *Trans. Faraday Soc.* **65**, 3032
           (1969).
69Er   Erikson, L. E., Dappe, A. J., & Uhleuhopp, J. C.: *J. Am. Chem. Soc.* **91**,
           2510 (1969).
69Ev   Evertsson, B.: *Acta Cryst.* **B25**, 30 (1969).
69Fr1   Freeman, H. C., Guss, J. M., Healy, M. J., Martin, R-P., Nockolds, C. E.,
           & Sarkar, B.: *Chem. Commun.* 229 (1969).
69Fr2   Freeman, H. C. & Martin, R-P.: *J. Biol. Chem.* **244**, 4823 (1969).
69Gi   Gillard, R. D. & Payne, N. C. *J. Chem. Soc. (A)* 1197 (1969).
69He1   van der Helm, D. & Hossian, M. B.: *Acta Cryst.* **B25**, 457 (1969).
69He2   van der Helm, D. & Franks, A.: *Acta Cryst.* **B25**, 451 (1969).
69Ho   Hosaka, K., Nishikawa, H., & Shibata, M.: *Bull. Chem. Soc. Japan* **42**, 277
           (1969).
69Hu   Huber, P. R., Griesser, R., Prijs, B., & Sigel, H.: *Eur. J. Biochem.* **10**, 238
           (1969).
69Ka   Katzin, L. I. & Gulyas, E.: *J. Am. Chem. Soc.* **91**, 6940 (1969).
69Kn   Knox, J. R. & Prout, C. K.: *Acta Cryst.* **B25**, 1857 (1969).
69Mc   McCormick, D. B., Sigel, H., & Wright, L. S.: *Biochem. Biophys. Acta*
           184, 318 (1969).
69Me   Melby, L. R.: *Inorg. Chem.* **8**, 349 (1969).
69Ri   Ritsma, J. H., van der Grampel, J. C., & Jellinek, F.: *Rec. Trav. Chim.* **58**,
           411 (1969).
69Sc   Schmidtke, H. H.: *Chem. Phys. Lett.* **4**, 4451 (1969).
69Sh   Sheppard, R. C. ed. *Amino acids, peptides and proteins,* Royal Society of
           Chemistry, Specialist Periodical Reports, London, 1969 Vol 1, 1988 Vol 18
69Ve   Veidis, M. V. & Palinek, G. J.: *Chem. Commun.* 1277 (1969).
69Wa   Watt, G. W. & Gillow, E. W.: *Inorg. Nucl. Chem. Lett.* **5**, 669 (1969).
69Zo   Zompa, L. J.: *Chem. Commun.* 783 (1969).
70Bo   Bottei, R. S. & Schneggenburger, R. G.: *J. Therm. Anal.* **2**, 11 (1970).
70Ch   Chawla, I. D. & Andrews, A. D.: *J. Inorg. Nucl. Chem.* **32**, 91 (1970).
70Cl   Clarke, E. R. & Martell, A. E.: *J. Inorg. Nucl. Chem.* **32**, 911 (1970).
70Gi   Girdhar, K. K., Vaidya, K. K., & Relam, P. S.: *J. Ind. Chem. Soc.* **47**, 715
           (1970).
70Ha   Harrison, M. R. & Rossotti, F. J. C.: *Chem. Commun.* 175 (1970).
70He   van der Helm, D., Nicholas, A. F., & Fisher, C. G.: *Acta Cryst.* **B26**, 1172
           (1970).
70Ho   Ho, F. F. L., Erickson, L. E., Watkins, S. R., & Reilley, C. N.: *Inorg.
           Chem.* **9**, 1139 (1970).
70Hu   Huang, T. J. & Haight, G. P.: *J. Am. Chem. Soc.* **92**, 2336 (1970).

70Jo   Jones, A. D. & Williams, D. R.: *J. Chem. Soc. (A)* 3138 (1970).
70Ka1   Katzin, L. I.: *Coord. Chem. Rev.* **5**, 279 (1970).
70Ka2   Kawasaki, K., Yoshii, J. & Shibata, M.: *Bull. Chem. Soc. Japan* **43**, 3819 (1970).
70Ka3   Kay, A. & Mitchell, P. C. H.: *J. Chem. Soc. (A)* 2421 (1970).
70Le1   Letter, J. E. & Bauman, J. E.: *J. Am. Chem. Soc.* **92**, 437 (1970).
70Le2   Letter, J. E. & Bauman, J. E.: *J. Am. Chem. Soc.* **92**, 443 (1970).
70Me   Meyer, J. L. & Bauman, J. E.: *J. Chem. Eng. Data* **15**, 404 (1970).
70Mi   Mizuochi, H., Shirakata, S., Kyuno, E., & Tsuchiya, R.: *Bull. Chem. Soc. Japan* **43**, 397 (1970).
70Mo1   Morris, P. J. & Martin, R. B.: *Inorg. Chem.* **9**, 2891, (1970).
70Mo2   Morris, P. J. & Martin, R. B.: *J. Am. Chem. Soc.* **92**, 1543 (1970).
70Na1   Nakamoto, K.: *Infrared spectra of inorganic and coordination compounds*, Wiley, Interscience, New York (1970).
70Na2   Natusch, F. S. & Porter, L. J.: *Chem. Commun.* 596 (1970).
70Pi   Pilbrow, J. R., Carr, S. G., & Smith, T. D.: *J. Chem. Soc. (A)* 723 (1970).
70Su   Sugiura, Y., Yokayama, A., & Tanaka, H.: *Chem. Pharm. Bull.* **18**, 693 (1970).
70Ts   Tsangaris, J. M. & Martin, R. B.: *J. Am. Chem. Soc.* **92**, 4255 (1970).
70Wa   Walsha, J. M.: *Brit. J. Hospital Med.* **91**, 248 (1970).
70Wi1   Wilson, E. W., Kasperian, M. H., & Martin, R. B.: *J. Am. Chem. Soc.* **92**, 5365 (1970).
70Wi2   Wilson, E. W. & Martin, R. B.: *Inorg. Chem.* **9**, 528 (1970).
70Wi3   Williams, D. R.: *J. Chem. Soc. (A)* 1550 (1970).
70Zu   Zuberbüchler, A.: *Helv. Chim. Acta* **53**, 669 (1970).
71Ac   Acland, C. B. & Freeman, H. C.: *Chem. Commun.* 1016 (1971).
71Ba   Barnes, D. S. & Pettit, L. D.: *J. Inorg. Nucl. Chem.* **33**, 2177 (1971).
71Ca   Carr, S. G., Smith, T. D., & Pilbrow, J. R.: *J. Chem. Soc. (A)* 2569 (1971).
71De   Delbare, L. T. J. & Prout, C. K.: *Chem. Commun.* 162 (1971).
71Fr   Froebe, L. R., Yamada, S., Hidaka, J., & Douglas, B. E.: *J. Coord. Chem.* **1**, 183 (1971).
71Ge   Gergely, A., Nagypál, I., & Király, B.: *Acta Chim. Acad. Sci. Hung.* **68**, 285 (1971).
71Gr   Griesser, R. & Sigel, H.: *Inorg. Chem.* **10**, 2229 (1971).
71Ha1   Hay, R. W. & Morris, P. J.: *J. Chem. Soc. (A)* 3562 (1971).
71Ha2   Hallman, P. S., Perrin, D. D., & Watt, A. E.: *Biochem. J.* **121**, 549 (1971).
71Ha3   Hanaki, A. & Kamide, H.: *Chem. Pharamacol. Bull.* **19**, 1006 (1971).
71He   van der Helm, D., Lawson, M. B., & Enwall, E. L.: *Acta Cryst.* **B27**, 2411 (1971).
71Jo   Jones, A. D. & Williams, D. R. *J. Chem. Soc. (A)* 3159 (1971).
71Ka1   Kalberer, H. & Frye, H.: *Inorg. Nucl. Chem. Lett.* **7**, 215 (1971).
71Ka2   Karpel, R. L., Kustin, K., Kowalik, A., & Pasternack, R. F.: *J. Am. Chem. Soc.* **93**, 1085 (1971).
71Ko   Kodama, M. & Takahashi, S.: *Bull. Chem. Soc. Japan* **44**, 697 (1971).
71Ma   Martin, R. B.: *J. Phys. Chem.* **75**, 2657 (1971).
71Mo   Morris, P. J. & Martin, R. B.: *J. Chem. Soc. (A)* 3159 (1971).
71Na   Natusch, F. S. & Porter, L. J.: *J. Chem. Soc. (A)* 2527 (1971).
71Ro   Rotilo, G. & Calabrese, L.: *Arch. Biochem, Biophys.* **143**, 218 (1971).

71Su   Sunar, O. P. & Trevedi, C. P.: *J. Inorg. Nucl. Chem.* **33**, 2177 (1971).
71We   Weber, O. A. & Simeon, Vl.: *Biochim. Biophys. Acta* **244**, 94 (1971).
71Wi1   Wilson, E. W. & Martin, R. B. *Inorg. Chem.* **10**, 1197 (1971).
71Wi2   Wilson, E. W. & Martin, R. B. *Arch. Biochim. Biophys.* **142**, 1006 (1971).
72Ba   Batayev, I. M. & Fogileva, R. S.: *Zh. Neorg. Khim.* **17**, 391 (1972).
72Bo   Bonvicini, P., Levi, A., Lucchini, V., & Scorrano, G.: *J. Chem. Soc. Perkin 2 Trans.* 2267 (1972).
72Ga   Garbet, K., Partridge, G. W., & Williams, R. J. P.: *Bioinorg, Chem.* **1**, 309 (1972).
72Ge1   Gergely, A., Mojzes, J., & Kassai-Bazsa, Zs.: *J. Inorg. Nucl. Chem.* **34**, 1277 (1972).
72Ge2   Gergely, A., Sóvágó, I., Nagypál, I., & Király, R.: *Inorg. Chim. Acta* **6**, 435 (1972).
72Go1   Gorton, J. E. & Jameson, R. F.: *J. Chem. Soc., Dalton Trans.* 310 (1972).
72Go2   Gorton, J. E. & Jameson, R. F.: *J. Chem. Soc., Dalton Trans.* 304 (1972).
72He   van der Helm, D. & Tatsch, C. E.: *Acta Cryst.* **B28**, 2307 (1972).
72Ka   Kawasaki, K. & Shibata, M.: *Bull. Chem. Soc. Japan* **45**, 3100 (1972).
72Mc   McAuliffe, C. A. & Murray, S. G.: *Inorg. Chem. Acta Rev.* **6**, 103 (1972).
72Pi   Pitner, T. E., Wilson, E. W., & Martin, R. B.: *Inorg. Chem.* **11**, 738 (1972).
72Po   Powell, H. K. J. & Nancollas, G. H.: *J. Am. Chem. Soc.* **94**, 2664 (1972).
72Si1   Singh, M. K. & Srivastava, M. N.: *Talanta* **19**, 699 (1972).
72Si2   Singh, M. K. & Srivastava, M. N.: *J. Inorg. Nucl. Chem.* **34**, 2081 (1972).
72Si3   Singh, M. K. & Srivastava, M. N.: *J. Inorg. Nucl. Chem.* **34**, 2067 (1972).
72Sm   Smissman, E. E. & Warner, W. D.: *J. Med. Chem.* **15**, 681 (1972).
72St   Stadtherr, L. G. & Martin, R. B.: *Inorg. Chem.* **11**, 92 (1972).
72Su   Sugiura, Y., Hojo, Y., & Tanaka, H.: *Chem. Pharm. Bull.* **20**, 1362 (1972).
72Wi   Williams, D. R. & Yeo, P. A.: *J. Chem. Soc., Dalton Trans.* 1988 (1972).
73Ba1   Barr, M. L., Baumgartner, E., & Kustin, K.: *J. Coord. Chem.* **2**, 263 (1973).
73Ba2   Barr, M. L., Kustin, K., & Liu, S-T.: *Inorg. Chem.* **12**, 2362 (1973).
73Bl   Blackburn, J. R. & Jones, M. M.: *J. Inorg. Nucl. Chem.* **35**, 1605 (1973).
73Br   Bridgart, G. J., Fuller, M. W., & Wilson, I. R.: *J. Chem. Soc., Dalton Trans.* 1274 (1973).
73Fa   Farooq, O., Ahmad, N., & Malik, A. V.: *J. Electroanal. Chem.* **48**, 475 (1973).
73Fl   Flook, R. J., Freeman, H. C., Moore, C. J., & Scudder, M. L.: *Chem. Commun.* 753 (1973).
73Fr   Freeman, H. C.: In *Inorganic biochemistry*, Eichorn, G. L. ed., Vol 1, Chap 4, Elsevier, Amsterdam (1973).
73Gr   Grenouillet, P., Martin, R-P., Rossi, A., & Ptak, M.: *Biochim. Biophys. Acta* **322**, 185 (1973).
73Ha   Haines, R. A. & Reimer, M.: *Inorg. Chem.* **12**, 1482 (1973).
73Is   Ishizuka, H., Yamamoto, T., Arata, Y., & Fujiwara, S.: *Bull. Chem. Soc. Japan* **46**, 468 (1973).
73Ka   Karwecka, Z. & Pajdowski, L.: *Adv. Mol. Rel. Proc.* **5**, 45 (1973).

73Kr   Kroneck, P. & Spence, J. T.: *Inorg. Nucl. Chem. Lett.* **9**, 177 (1973); *J. Inorg. Nucl. Chem.* **35**, 3391 (1973).
73Ku1   Kustin, K. & Liu, S. T.: *J. Chem. Soc., Dalton Trans.* 278 (1973).
73Ku2   Kustin, K. & Liu, S. T.: *Inorg. Chem.* **12**, 1486 (1973).
73La   Lassociuska, A.: *Rocz. Chem.* **47**, 889 (1973).
73Li   van der Linden, W. E. & Beers, C.: *Anal. Chim. Acta* **68**, 143 (1973).
73Ma   Martin, P-P., Petit-Ramel, M. M., & Scharff, J. P.: *Met. Ions Biol. Syst.* **2**, 1 (1973).
73Mc   McAuliffe, C. A. & Murray, S. G.: *Inorg. Chim. Acta* **7**, 171 (1973).
73Na   Nakayama, Y., Matsumoto, K., Ooipud, S., & Kurova, H.: *Chem. Commun.* 170 (1973).
73Ne   Neilands, J. B.: In *Inorganic biochemistry*, Eichorn, G. L. ed., Vol 1, Chap 5, Elsevier, Amsterdam (1973).
73Po   Po, I. T., Burke, J., Meyer, J. L., & Nancollas, G. H.: *Thermochim. Acta* **5**, 463 (1973).
73Re   Reynolds, W. F., Peat, I. R., Freedman, M. H., & Lyerla, J. R.: *J. Am. Chem. Soc.* **95**, 328 (1973).
73Si   Sigel, H. ed.: *Metal ions in biological systems*, Vol 2, Marcel Dekker, New York (1973).
73To   Tomiyasu, H. & Gordon, G.: *J. Coord. Chem.* **3**, 47 (1973).
74Ba   Baxter, A. C. & Williams, D. R.: *J. Chem. Soc., Dalton Trans.* 1117 (1974).
74Be   Bernauer, K.: *Met. Ions Biol. Syst.* **2**, 118 (1974).
74Bh   Bhagwat, V., Khadikar, P. V., Kerke, M. G., & Sharma, V.: *J. Inorg. Nucl. Chem.* **36**, 942 (1974).
74Ei   Einspahr, H. & Bugg, C. E.: *Acta Cryst.* **B30**, 1037 (1974).
74Fu   Fujiwara, S., Ishizuka, H., & Fudano, S.: *Chem. Lett.* 1281 (1974).
74Ge   Gergely, A., Nagypál, I., & Farkas, E.: *Magyar Kém. Foly.* **80**, 56 (1974).
74Gr   Grgas-Kuznar, B., Simeon, Vl., & Weber, O. A.: *J. Inorg. Nucl. Chem.* **36**, 2151 (1974).
74Ho   Holt, E. M., Holt, S. L., Tucker, W. F., Asplund, R. O., & Watson, K. J.: *J. Am. Chem. Soc.* **96**, 2621 (1974).
74Je   Jezowska-Trzebiatowska, B., Antonow, A., & Kozlowski, H.: *Bull. Acad. Pol. Sci.* **22**, 499 (1974).
74Ko   Kodama, M.: *Bull. Chem. Soc. Japan* **47**, 1547 (1974).
74Ku   Kustin, K., Liu, S-T., Nicolini, C., & Toppen, D. L.: *J. Am. Chem. Soc.* **96**, 7410 (1974).
74Kw   Kwik, W-L., Purdy, E., & Stiefel, E. J.: *J. Am. Chem. Soc.* **96**, 1638 (1974).
74La   Lassociuska, A.: *Rocz. Chem.* **48**, 967 (1974).
74Li   van der Linden, W. G. & Beers, C.: *Anal. Chim. Acta* **68**, 143 (1974).
74Ma1   Martell, A. E. & Smith, R. M.: *Critical stability constants*, Vol 1, Plenum, New York (1974).
74Ma2   Martin, R. B. & Prados, R.: *J. Inorg. Nucl. Chem.* **36**, 1665 (1974).
74Ma3   Martin, R. B.: *Met. Ions Biol. Syst.* **1**, 129 (1974).
74Mc   McCormick, D. B., Griesser, R., & Sigel, H.: *Met. Ions Biol. Syst.* **2**, 214

(1974).

74Na   Nagypál, I., Gergely, A., & Farkas, E.: *J. Inorg. Nucl. Chem.* **36**, 699 (1974).

74Pr   Prados, R., Stadtherr, L. G., Donato, H., & Martin, R. B.: *J. Inorg. Nucl. Chem.* **36**, 689 (1974).

74Ro   Rossi, A., Ptak, M., Grenouillet, P., & Martin, R-M.: *J. Chim. Phys.* **71**, 1371 (1974).

74Se   Sergeev, G. M. & Korsunov, I. A.: *Radiokhim.* **16**, 787 (1974).

74Si   Sigel, H. ed.: *Metal ions in biological systems*, Dekker, New York (1974) Vol 1, 1988 Vol 24

74Su   Sundberg, R. J. & Martin, R. B.: *Chem. Rev.* **74**, 471 (1974).

74Te   Tewari, R. C. & Srivastava, M. N.: *Acta Chim. Acad. Sci. Hung.* **83**, 259 (1974).

74Th   Thich, J. A., Mastropaolo, D., Potenza, J., & Schugar, H. J.: *J. Am. Chem. Soc.* **96**, 726 (1974).

74Wa1   Walker, M. D. & Williams, D. R.: *J. Chem. Soc., Dalton Trans.* 1186 (1974).

74Wa2   Watters, K. L. & Wilkins, R. G.: *Inorg. Chem.* **13**, 752 (1974).

74We   Weber, O. A.: *J. Inorg. Nucl. Chem.* **36**, 1341 (1974).

75Ba   Bajaj, S. P., Butkowski, R. J., & Mann, K. G.: *J. Biol. Chem.* **250**, 2150 (1975).

75Bo1   Bogges, R. K. & Martin, R. B.: *J. Am. Chem. Soc.* **97**, 3076 (1975).

75Bo2   Bogges, R. K. & Martin, R. B.: *J. Inorg. Nucl. Chem.* **37**, 359 (1975).

75Br   Brown, D. H. & Neumann, D.: *J. Inorg. Nucl. Chem.* **37**, 330 (1975).

75Ch1   Christensen, J. J., Eatough, D. J., & Izatt, R. M.: *Handbook of metal ligand heats,* Marcel Dekker, New York (1975).

75Ch2   Chow, S. T. & McAuliffe, C. A.: In *Progress in inorganic chemistry,* Lippard, S. J. ed., Vol 19, Wiley, New York (1975).

75Da   Davidov, J. P., Voronik, N. I., & Skripcova, A. V.: *Radiokhim.* 597 (1975).

75Fe   Fee, J. A.: *Structure and Bonding* **23**, 1 (1975).

75Ge1   Gelbaum, L. & Eugel, R.: *J. Inorg. Nucl. Chem.* **37**, 793 (1975).

75Ge2   Gergely, A. & Farkas, E.: *Magyar Kém. Foly.* **81**, 471 (1975).

75Ge3   Gergely, A., Nagypál, I., & Farkas, E.: *J. Inorg. Nucl. Chem.* **37**, 551 (1975).

75Ha   Hansen, P. E., Felney, J., & Roberts, G. C. K.: *J. Magn. Res.* **17**, 2449 (1975).

75Le   Led, J. J. & Grant, D. M.: *J. Am. Chem. Soc.* **97**, 6962 (1975).

75Mu1   Murray, K. S. & Neuman, P. J.: *Aust. J. Chem.* **28**, 773 (1975).

75Mu2   Musker, W. K. & Neagley, C. H.: *Inorg. Chem.* **14**, 1728 (1975).

75Na   Nagypál, I., Farkas, E., & Gergely, A.: *J. Inorg. Nucl. Chem.* **37**, 2145 (1975).

75Oo   Oonishi, I., Sato, S., & Saito, Y.: *Acta Cryst.* **B31**, 1318 (1975).

75Pl   Plejnschev, V. E., Nadezhdina, G. V., Loseva, G. S., Melnikova, V. V., & Parfenova, T. S.: *Zh. Neorg. Khim.* **20**, 60 (1975).

75Ra   Rabenstein, D. L. & Fairhurst, M. T.: *J. Am. Chem. Soc.* **97**, 2086 (1975).

75Ri1   Ritsma, J. H.: *Rec. Trav. Chim.* **94**, 210 (1975).

75Ri2	Richardson, J. S., Thomas, K. A., Rubin, B. H., & Richardson, D. C.: *Proc. Natl. Acad. Sci. U.S.A.* **72**, 1349 (1975).

75Si	Sigel, H.: *Angew. Chem. (Internat. Edit.)* **14**, 394 (1975).

75So	Sóvágó, I., Gergely, A., & Posta, J.: *Acta Chim. Acad. Sci. Hung.* **85**, 153 (1975).

75St	Stephens, F. S., Vagg, R. S., & Williams, P. A.: *Acta Cryst.* **B31**, 841 (1975).

75Sz	Sze, Y. K., Davies, A. R., & Neville, G. A.: *Inorg. Chem.* **4**, 163 (1975).

75Wr	Wright, E. W. & Frieden, E.: *Bioinorg. Chem.* **4**, 163 (1975).

76Bi1	Bianco, P., Haladjian, J., & Pilard, R.: *J. Chim. Phys.* **73**, 280 (1976).

76Bi2	Birker, P. J. M. W. L. & Freeman, H. C.: *Chem. Commun.* 312 (1976).

76Br1	Braibanti, A., Mori, G., & Dallavalle, F.: *J. Chem. Soc., Dalton Trans.* 826 (1976).

76Br2	Brookes, G. & Pettit, L. D.: *J. Chem. Soc., Dalton Trans.* 42 (1976).

76Bu	Butcher, R. J., Powell, H. K. J., Wilkins, C. J., & Young, S. H.: *J. Chem. Soc., Dalton Trans.* 356 (1976).

76Ca	Carty, A. J. & Taylor, N. J.: *Chem. Commun.* 214 (1976).

76Co	Corrie, A. M., Walker, M. D., & Williams, D. R.: *J. Chem. Soc., Dalton Trans.* 1012 (1976).

76Es	Espersen, W. G. & Martin, R. B.: *J. Am. Chem. Soc.* **98**, 40 (1976).

76Ge	Gergely, A. & Kiss, T.: *Inorg. Chim. Acta* **16**, 51 (1976).

76Gi	Gilbert, K. & Kustin, K.: *J. Am. Chem. Soc.* **98**, 5502 (1976).

76Go	Gold, M. & Powell, H. K. J.: *J. Chem. Soc., Dalton Trans.* 1418 (1976).

76Gr	Grouchi-Witte, G. & Weiss, E.: *Z. Naturforsch.* **31B**, 1190 (1976).

76Ha	Harris, W. R., McLendon, G., & Martell, A. E.: *J. Am. Chem. Soc.* **98**, 8378 (1976).

76Ho	Hojo, Y., Sugiura, Y., & Tanaka, H.: *J. Inorg. Nucl. Chem.* **38**, 641 (1976).

76Me1	Mentasti, E., Relizzetti, E., & Saini, G.: *J. Inorg. Nucl. Chem.* **38**, 785 (1976).

76Me2	de Meester, P. & Hodgson, D. J.: *J. Chem. Soc., Dalton Trans.* 1012 (1976).

76Mi	Miskowski, W. M., Tich, J. A., Solomon, R., & Schugar, H. J.: *J. Am. Chem. Soc.* **98**, 8344 (1976).

76Pe1	Pettit, L. D. & Swash, J. L. M.: *J. Chem. Soc., Dalton Trans.* 2416 (1976).

76Pe2	Pettit, L. D. & Swash, J. L. M.: *J. Chem. Soc., Dalton Trans.* 588 (1976).

76Ri	Ritsma, J. H.: *J. Inorg. Nucl. Chem.* **38**, 907 (1976).

76Sa1	Sandhu, R. S.: *Thermochim. Acta* **17**, 270 (1976).

76Sa2	Sayer, T. L. & Rabenstein, D. L.: *Can. J. Chem.* **54**, 3392 (1976).

76Sc	Schugar, H. J., Ou, C-C., Thich, J. A., Potenza, J. A., Lalancette, R. A., & Furey, W.: *J. Am. Chem. Soc.* **98**, 3047 (1976).

76Si	Sigel, H. & Naumann, C. F.: *J. Am. Chem. Soc.* **98**, 730 (1976).

76So	Sóvágó, I. & Gergely, A.: *Inorg. Chim. Acta* **20**, 27 (1976).

76Sr	Srivastava, M. N., Tewari, R. C., Srivastava, V. C., Bhargava, G. B., & Vishnoi, A. N.: *J. Inorg. Nucl. Chem.* **38**, 1897 (1976).

76Sw	Swash, J. L. M. & Pettit, L. D.: *Inorg. Chim. Acta* **19**, 19 (1976).

76Ti	Tihomirov, V. I. & Gornovskaya, N. K.: *Zh. Neorg. Khim.* **21**, 1970

(1976).

77Bi	Birker, P. J. M. W. L. & Freeman, H. C.: *J. Am. Chem. Soc.* **99**, 6890 (1977).

77Bl	Blomberg, F., Maurer, W., & Ruterjans, H.: *J. Am. Chem. Soc.* **99**, 8149 (1977).

77Br	Brookes, G. & Pettit, L. D.: *J. Chem. Soc., Dalton Trans.* 1918 (1977).

77Bu	Bunel, S., Ibarra, C., & Urbana, A.: *Inorg. Nucl. Chem. Lett.* **13**, 259 (1977).

77Ho	Hojo, Y., Sugiura, Y., & Tanaka, H.: *J. Inorg. Nucl. Chem.* **39**, 1859 (1977).

77Je	Jezowska-Trzebiatowska, B., Kovalik, T., & Kozlowski, H.: *Bull. Acad. Pol. Sci.* **25**, 57 (1977).

77Ko	Kozlowski, H., Formicka-Kozlowska, G., & Jezowska-Trzebiatowska, B.: *Org. Magn. Res.* **10**, 146 (1977).

77La	Lamache-Duhameaux, M.: *J. Inorg. Nucl. Chem.* **39**, 2081 (1977).

77Li	Lim, M-C.: *J. Chem. Soc., Dalton Trans.* 1398 (1977).

77Ma	Martell, A. E. & Langohr, M. F.: *Chem. Commun.* 342 (1977).

77Me	de Meester, P. & Hodgson, D. J.: *J. Am. Chem. Soc.* **99**, 6884 (1977).

77Ra	Rabenstein, D. L., Greenberg, M. S., & Saetre, R.: *Inorg. Chem.* **16**, 1241 (1977).

77Ri	Richardson, H. W., Wasson, J. R., Estes, W. E., & Hatfield, W. E.: *Inorg. Chim. Acta* **23**, 205 (1977).

77Sp	Sportelli, L., Neubacher, H., & Lohmann, W.: *Z. Naturforsch.* **32C**, 643 (1977).

77St1	Stephens, F. S., Vagg, R. S., & Williams, P. A.: *Acta Cryst.* **B33**, 433 (1977).

77St2	Stephens, F. S., Vagg, R. S., & Williams, P. A.: *Acta Cryst.* **B33**, 438 (1977).

77Su	Sugiura, Y., Kikuchi, T., & Tanaka, H.: *J. Inorg. Nucl. Chem.* **25**, 345 (1977).

77Th	Thorup, N.: *Acta Chem. Scand.* **A31**, 203 (1977).

77To	Toepfer, E. W., Mertz, W., Polansky, M. M., Roginski, E. E., & Wolf, W. R.: *J. Agr. Food Chem.* **25**, 162 (1977).

77Wi	Wilson, R. B., de Meester, P., & Hodgson, D. J.: *Inorg. Chem.* **16**, 1498 (1977).

78Ca	Camerman, N., Fawcett, J. K., Kruck, T. P. A., Sarkar, B., & Camerman, A.: *J. Am. Chem. Soc.* **100**, 2690 (1978).

78Co	Coleman, P. M., Freeman, H. C., Guss, J. M., Murata, M., Morris, V. A., Ramshaw, J. A. M., & Venkatappa, M. P.: *Nature* **272**, 319 (1978).

78Da	Davies, J. M.: *Lancet* 384 (1978).

78Ge1	Gergely, A., Farkas, E., Nagypál, I., & Kas, E.: *J. Inorg. Nucl. Chem.* **40**, 1709 (1978).

78Ge2	Gergely, A. & Sóvágó, I.: *Bioinorg. Chem.* **9**, 47 (1978).

78Ja	Jameson, R. F. & Kiss, T.: (unpublished results).

78Ki	Kitagawa, S., Yoshikawa, K., & Morishima, I.: *J. Phys. Chem.* **82**, 89 (1978).

78Ko Kozlowski, H., Formicka-Kozlowska, G., & Jezowska-Trzebiatowska, B.: *Bull. Acad. Pol. Sci.* **26**, 153 (1978).

78Le Legendziewicz, J., Kozlowski, H., & Burzala, E.: *Inorg. Nucl. Chem. Lett.* **14**, 409 (1978).

78Mo Mohan, M. S. & Abbott, E. H.: *J. Coord. Chem.* **8**, 175 (1978).

78Ou Ou, C., Powers, D. A., Thich, J. A., Felthouse, T. R., Hendrickson, D. N., Potenza, J. A., & Schugar, H. J.: *Inorg. Chem.* **17**, 34 (1978).

78Ra Rajan, K. S., Mainer, S., & Davis, J. M.: *Bioinorg. Chem.* **9**, 187 (1978).

78So Sóvágó, I., Kiss, T., & Gergely, A,.: *J. Chem. Soc., Dalton Trans.* 964 (1978).

78Sp Sperling, R., Furie, B. C., Blumenstein, B., Keyt, B., & Furie, B.: *J. Biol. Chem.* **253**, 3898 (1978).

78Su Sugiura, Y. & Hirayama, Y.: *Bioinorg. Chem.* **9**, 521 (1978).

78Vl Vliegen, J. & van Poucke, L. C.: *Bull. Soc. Chim. Belg.* **87**, 837 (1978).

78Wo Wozniak, M. & Nowogrocki, G.: *Talanta* **25**, 633 (1978).

79Am Amico, G., Daniele, P. G., Arena, G., Ostacoli, G., Rizzarelli, E., & Sammartano, S.: *Inorg. Chim. Acta* **35**, L383 (1979).

79Bi Birker, P. J. M. W. L.: *Inorg. Chem.* **18**, 3502 (1979).

79De Delf, B. W., Gillard, R. D., & O'Brien, P.: *J. Chem. Soc., Dalton Trans.* 1301 (1979).

79Ev Evans, C. A., Guevremont, R., & Rabenstein, D. L.: *Met. Ions Biol. Syst.* **9**, 41 (1979).

79Fa Fazakerley, G. V., Linder, P. W., Torrington, R. G., & Wright, M. R. W.: *J. Chem. Soc., Dalton Trans.* 1872 (1979).

79Fr Frye, H. & Williams, G. H.: *J. Inorg. Nucl. Chem.* **41**, 591 (1979).

79Ga Garnier, A. & Tosi, L.: *J. Inorg. Biochem.* **10**, 147 (1979).

79Ge1 Gergely, A. & Kiss, T.: *Met. Ions Biol. Syst.* **9**, 143 (1979).

79Ge2 Gergely, A. & Sóvágó, I.: *Met. Ions Biol. Syst.* **9**, 77 (1979).

79Ge3 Gergely, A., Kiss, T., & Deák, Gy.: *Inorg. Chim. Acta* **36**, 113 (1979).

79Je Jezowska-Trzebiatowska, B., Allain, A., & Kozlowski, H.: *Inorg. Nucl. Chem. Lett.* **15**, 279 (1979).

79La Laurie, S. H. & Prime, D. M.: *J. Inorg Biochem.* **11**, 229 (1979).

79Le Legendziewicz, J., Kozlowski, H., Jezowska-Trzebiatowska, B., & Huskovska, E.: *Inorg. Nucl. Chem. Lett.* **15**, 349 (1979).

79Ma Martin, R. B.: *Met. Ions Biol. Syst.* **9**, 1 (1979).

79Ni Nicholls, L. J. & Freeman, W. A.: *Acta Cryst.* **B35**, 2392 (1979).

79Ös Österberg, R., Ligaarden, R., & Persson, D.: *J. Inorg. Biochem.* **10**, 341 (1979).

79Pe Pettit, L. D. & Hefford, R. J. W.: *Met. Ions Biol. Syst.* **9**, 173 (1979).

79Ph Phipps, D. A.: *J. Mol. Cat.* **5**, 81 (1979).

79Ro Robertson, P., Koehler, K. A., & Miskey, R. G.: *Biochem. Biophys, Res. Commun.* **86**, 265 (1979).

79Sa Sabat, M., Jezowska, M., & Kozlowski, H.: *Inorg. Chim. Acta* **37**, L511 (1979).

79So1 Sóvágó, I., Gergely, A., Harman, B., & Kiss, T.: *J. Inorg. Nucl. Chem.* **41**, 1629 (1979).

79So2   Sóvágó, I. & Gergely, A.: *Inorg. Chim. Acta* **37**, 233 (1979).
79St   Stönzi, H. & Perrin, D. D.: *J. Inorg. Biochem.* **10**, 309 (1979).
79Wo   Wozniak, M. & Nowogrocki, G.: *Talanta* **26**, 381 (1979); **26** 1135 (1979).
80Al   Allain, A., Kubiak, M., & Jezowska-Trzebiatowska, B.: *Inorg. Chim. Acta* **46**, L25 (1980).
80Br   Brown, D. A., Chidambaram, M. W., & Glennon, J. D.: *Inorg. Chem.* **19**, 3260 (1980).
80Cl   Claridge, R. F. C., Kilpatrick, J. J., & Powell, H. K. J.: *Aust. J. Chem.* **33**, 2757 (1980).
80Fi   Fischer, B. E. & Sigel, H.: *J. Am. Chem. Soc.* **102**, 2998 (1980).
80Gi   Gillard, R. D., Lancashire, R. J., & O'Brien, P.: *Transition Met. Chem.* **5**, 340 (1980).
80Ha   Haberfield, P.: *J. Chem. Educ.* **57**, 346 (1980).
80It   Itabashi, M. & Itoh, K.: *Bull. Chem. Soc. Japan* **53**, 3131 (1980).
80Je   Jezowska-Trzebiatowska, B. & Latos-Grazynski, L.: *J. Inorg. Nucl. Chem.* **42**, 1079 (1980).
80Ka   Kayali, A. & Berthon, G.: *J. Chem. Soc., Dalton Trans.* 2374 (1980).
80Ko   Kozlowski, H., Siatecki, Z., & Jezowska-Trzebiatowska, B.: *Inorg. Chim. Acta* **46**, L25 (1980).
80Kw   Kwik, W. L., Ang, K. P., & Chen. G.: *J. Inorg. Nucl. Chem.* **42**, 303 (1980).
80Mu   Munakata, K., Kobashi, K., Takebe, S., & Hase, J.: *J. Pharm. Dyn.* **3**, 451 (1980).
80Oh   Ohtaki, H., Zama, M., Koyama, H., & Ishiguro, S.: *Bull. Chem. Soc. Japan* **53**, 2865 (1980).
80Os1   Ostacoli, G.: In *Bioenergetics and thermodynamics: model systems*, D. Reidl Publ. (1980).
80Os2   Ostern, M., Pelczar, J., Kozlowski, H., & Jezowska-Trzebiatowska, B.: *Inorg. Nucl. Chem. Lett.* **16**, 251 (1980).
80Sc   Schugar, H. J., Ou, C-C., Thich, J. A., Potenza, J. A., Felthouse, T. R., Haddad, M. S., Hendrickson, D. N., Furey, W., & Lalancette, R. A.: *Inorg. Chem.* **19**, 543 (1980).
80Sh   Shah, S. K. & Gupta, C. M.: *Talanta* **27**, 823 (1980).
80Si   Sigel, H.: In *Coordination chemistry-20*, Banarjea, D. ed., Pergamon Press, New York (1980) p27
80So   Sóvágó, I., Harman, B., & Gergely, A.: *Inorg. Chim. Acta* **46**, L107 (1980).
80Ve   Vestus, P. I. & Martin, R. B.: *J. Am. Chem. Soc.* **102**, 7906 (1980).
81Al1   Alemdaroglu, T. & Berthon, G.: *Bioelectrochem. Bioenergetics* **8**, 49 (1981).
81Al2   Alemdaroglu, T. & Berthon, G.: *Inorg. Chim. Acta* **56**, 51 (1981); **56**, 115 (1981).
81Am   Amico, P., Arena, G., Daniele, P. G., Ostacoli, G., Rizzarelli, E., & Sammartano, S.: *Inorg. Chem.* **20**, 772 (1981).
81Bi   Birker, P. J. M. W. L., Reedijk, J., & Verschoor, G. C.: *Inorg. Chem.* **20**, 2877 (1981).
81Bl   Blais, M-J., Kayali, A., & Berthon, G.: *Inorg. Chim. Acta* **56**, 5 (1981).

81Bo   Book, L., Carty, A. J., & Chung Chieh: *Can. J. Chem.* **59**, 144 (1981).
81Bu1   Burnier, J. P., Borowski, M., Furie, B. C., & Furie, B.: *Mol. Cell. Biochem.* **39**, 191 (1981).
81Bu2   Bunel, S., Larrazábel, G., & Decinti, A.: *J. Inorg. Nucl. Chem.* **43**, 2781 (1981).
81En   Enea, O., Berthon, G., & Scharff, J-P.: *Thermochim. Acta* **50**, 147 (1981).
81Fa1   Farkas, E., Gergely, A., & Kas, E.: *J. Inorg. Nucl. Chem.* **43**, 1591 (1981).
81Fa2   Farkas, E., Homoki, E., & Gergely, A.: *J. Inorg. Nucl. Chem.* **43**, 624 (1981).
81Fe1   Ferrandi, G. & Mualidharan, S.: *Coord. Chem. Rev.* **36**, 45 (1981).
81Fe2   Felix, C. C. & Sealy, R. C.: *J. Am. Chem. Soc.* **103**, 2831 (1981).
81Ga   Garnier-Suillerot, A., Albertini, J-P., Collet, A., Faury, L., Pashr, J. M., & Tosi, L.: *J. Chem. Soc., Dalton Trans.* 2544 (1981).
81Ge   Gergely, A.: *Inorg. Chim. Acta* **56**, L75 (1981).
81Go   Goodman, B. A., McPhail, D. B., & Powell, H. K. J.: *J. Chem. Soc., Dalton Trans.* 822 (1981).
81He   Hefford, R. J. W. & Pettit, L. D.: *J. Chem. Soc., Dalton Trans.* 1331 (1981).
81Ja1   Jawaid, M. & Ingman, F.: *Talanta* **28**, 137 (1981).
81Ja2   Jackson, G. E., May, P. M., & Williams, D. R.: *J. Inorg. Nucl. Chem.* **43**, 825 (1981).
81Ke   van Kempen, H., Perenboom, J. A. A. J., & Birker, P. J. M. W. L: *Inorg. Chem.* **20**, 917 (1981).
81Ko   Kozlowski, H., Swiatek, J., & Siatecki, Z.: *Acta Biochim. Pol.* **28**, 1 (1981).
81La1   Lamache-Duhameaux, M.: *J. Inorg. Nucl. Chem.* **43**, 208 (1981).
81La2   Laurie, S. H., Mohammed, E. S., & Prime, D. M.: *Inorg. Chim. Acta* **56**, 135 (1981).
81Mo   Molloy, K. C., Zuckerman, J. J., Domazetis, G., & James, B. D.: *Inorg. Chim. Acta* **54**, L217 (1981).
81Na   Nagypál, I., Debreczeni, F., & Connick, R. E.: *Inorg. Chim. Acta* **48**, 225 (1981).
81Od   Odani, A. & Yamauchi, O.: *Bull. Chem. Soc. Japan* **54**, 3773 (1981).
81Pe   Pettit, L. D., Siddiqui, K. F., Kozlowski, H., & Kowalik, T.: *Inorg. Chim. Acta* **55**, 87 (1981).
81Pi   Pierpont, C. G. & Buchanan, R. M.: *Coord. Chem. Rev.* **38**, 45 (1981).
81Pu   Puri, R. N., Asplund, R. O., & Holt, S. L.: *J. Coord. Chem.* **11**, 125 (1981).
81Ra   Ramanujam, V. V. & Krishnan, U.: *J. Inorg. Nucl. Chem.* **43**, 3407 (1981).
81Re   Reid, R. S. & Rabenstein, D. L.: *Can. J. Chem.* **59**, 1505 (1981).
81Sa1   Sarkar, B.: *Met. Ions Biol. Syst.* **12**, 233 (1981).
81Sa2   Sakurai, H., Shimomura, S., & Ishizu, K.: *Inorg. Chim. Acta* **55**, L67 (1981).
81So   Sóvágó, I. & Gergely, A.: *Agent and Actions Supplements* **8**, 289 (1981).
81Zw   Zwart, J. van Wolput, J. H. M. C., van der Cammen, J. C. J. M., & Koningsberger, D. C.: *J. Mol. Cat.* **11**, 82 (1981); Zwart, J., van Wolput, J. H. M. C., & Koningsberger, D. C.: *ibid,* **12**, 85 (1981).

82An   Antolini, L., Menabue, L., Pellaacani, G. C., & Marcotrigano, G.: *J. Chem. Soc., Dalton Trans.* 2541 (1982).

82Bi1   Bizri, Y., Cromer-Morin, M., & Scharff, J.: *J. Chem. Research (S)* 192 (1982).

82Bi2   Birker, P. J. M. W. L. & Verschoor, G. C.: *Inorg. Chem.* **21**, 990 (1982).

82Br   Brown, D. A., Roche, A. L., Pakkanen, T. A., & Smolander, K.: *Chem. Commun.* 676 (1982).

82Ca   Cavaleiro, A. V. S. V., de Jesus, J. D. P., & Gil, V. M. S.: *Transition Met. Chem.* **7**, 75 (1982).

82Cu   Cusack, P. A., Smith, P. J., & Donaldson, J. D.: *J. Chem. Soc., Dalton Trans.* 439 (1982).

82De   Decock-Le Révérend, B., Loucheux, C., Kowalik, T., & Kozlowski, H.: *Inorg. Chim. Acta* **66**, 205 (1982).

82Fa   Fábián, I. & Nagypál, I.: *Inorg. Chim. Acta* **62**, 193 (1982).

82Fe   Feldman, S. L., Hunter, J. S. V., Zgirski, A., Chidambaram, M. V., & Frieden, E.: *J. Inorg. Biochem.* **17**, 51 (1982).

82Ge   Gergely, A. & Farkas, E.: *J. Chem. Soc., Dalton Trans.* 381 (1982).

82Ha   Hadjiliadis, N., Theodoru, V., Photaki, I., Gellert, R. W., & Bau, R.: *Inorg. Chim. Acta* **60**, 1 (1982).

82Ka   Kayali, A. & Berthon, G.: *Polyhedron* **1**, 371 (1982).

82Ke   Kehl, H. ed.: *Chemistry and biology of hydroxamic acids,* S. Karger, Basel (1982).

82Ki   Kiss, T. & Tóth, B.: *Talanta* **29**, 539 (1982).

82Kr   Kralj, Z., Paulic, N., Raos, N., & Simeon, Vl.: *Croat. Chem. Acta* **55**, 125 (1982).

82Ma1   Martell, A. E. & Smith, R. M.: *Critical stability constants,* Vol 5, Plenum, New York (1982).

82Ma2   Matsui, H. & Ohtaki, H.: *Bull. Chem. Soc. Japan* **55**, 461 (1982).

82Na1   Nagypál, I., Debreczeni, F., & Erdődi, F.: *Inorg. Chim. Acta* **57**, 125 (1982).

82Na2   Nair, M. S., Santappa, M., & Murugan, P. K.: *Inorg. Chem.* **21**, 142 (1982).

82Na3   Nair, M. S., Venkatachalapathi, K., Santappa, M., & Murugan, P. K.: *Inorg. Chem.* **21**, 2418 (1982).

82Od   Odani, A. & Yamauchi, O. *Inorg. Chim. Acta* **66**, 163 (1982).

82Pe   Pettit, L. D. & Swash, J. L. M.: *J. Chem. Soc., Dalton Trans.* 485 (1982).

82Sh   Shelke, D. N.: *J. Coord. Chem.* **12**, 35 (1982).

82Ve   Venkatachalapathi, K., Nair, M. S., Ramaswamy, D., & Santappa, M.: *J. Chem. Soc., Dalton Trans.* 291 (1982).

83Am   Amico, P., Daniele, P. G., Ostacoli, G., & Zelano, V.: *Ann. Chim. (Rome)* **73**, 253 (1983).

83Ar   Arena, G., Cali, R., Cucinotta, V., Musumeci, S., Rizzarelli, E., & Sammartano, S.: *J. Chem. Soc., Dalton Trans.* 1271, (1983).

83Br   Brown, D. A. & Roche, A. L.: *Inorg. Chem.* **22**, 2199 (1983).

83Ca   Casella, L. & Gullotti, M.: *J. Inorg. Biochem.* **18**, 19 (1983).

83Co   Cooke, M. E., McDaniel, M. E., James, S. R., Jones, S. L., Trobak, N.,

Craytor, B. C., Bushman, D. R., & Wright, J. R.: *J. Inorg. Biochem.* **18** 313 (1983).

83Da   Daniele, P. G., Amico, P., Ostacoli, G., & Zelano, V.: *Ann. Chim. (Rome)* **73**, 199 (1983).

83Fa   Farkas, E., Sóvágó, I., & Gergely, A.: *J. Chem. Soc., Dalton Trans.* 1545 (1983).

83Ha1   Harman, B. & Sóvágó, I.: *Inorg. Chim. Acta* **80**, 75 (1983).

83Ha2   Hanaki, A. & Kamide, H.: *Bull. Chem. Soc. Japan* **56**, 2065 (1983).

83Ja   Jakovlev, J. B. & Ushakova, V. G.: *Zh. Neorg. Khim.* **28**, 2142 (1983).

83Ki1   Kiss, T. & Gergely, A.: *Acta Chim. Hung.* **114**, 249 (1983).

83Ki2   Kiss, T. & Gergely, A.: *Inorg. Chim. Acta* **78**, 247 (1983).

83Ko   Kozlowski, H., Decock-Le Révérend, B., Delarulle, J. L., Loucheux, C., & Ancian, B.: *Inorg. Chim. Acta* **78**, 31 (1983).

83La1   Laurie, S. H. & James, C.: *Inorg. Chim. Acta* **78**, 225 (1983).

83La2   Laurie, S. H.: *Inorg. Chim. Acta* **80**, L27 (1983).

83Ma   Manorik, P. A. & Davidenko, N. K.: *Zh. Neorg. Khim.* **28**, 2292 (1983).

83Na   Nakashima, T. T. & Robertstein, D. L.: *J., Magn. Res.* **51**, 223 (1983).

83No   Nourmand, M. & Meissami, N.: *J. Chem. Soc., Dalton Trans.* 1529 (1983).

83Pa   Pasini, A.: *Gazz. Chim. Ital.* **113**, 793 (1983).

83Sh   Shelke, D. N.: *Inorg. Chim. Acta* **80**, 255 (1983).

83Si   Sigel, H., Fischer, B. E., & Farkas, E.: *Inorg. Chem.* **22**, 925 (1983).

83Sz   Szabó-Plánka, T. & Horváth, L. I.: *Acta Chim. Hung.* **114**, 15 (1983).

83Vl   Vlasova, N. N. & Davidenko, N. K.: *Koord. Khim.* **9**, 1470 (1983).

84Ab   Abdullah, M. A., Barrett, J., & O'Brien, P.: *J. Chem. Soc., Dalton Trans.* 1647 (1984).

84Al1   Al-Ami, N. & Olin, A.: *Chem. Scripta* **23**, 161 (1984).

84Al2   Al-Saady, A. K. H., Mass, K., McAuliffe, M. C., & Parish, R. V.: *J. Chem. Soc., Dalton Trans.* 1609 (1984).

84Ar1   Arena, G., Cali, R., Cucinotta, V., Musumeci, S., Rizzarelli, E., & Sammartano, S.: *J. Chem. Soc., Dalton Trans.* 1651 (1984).

84Ar2   Arena, G., Cali, R., Cucinotta, V., Musumeci, S., Rizzarelli, E., & Sammartano, S.: *Thermochim. Acta* **74**, 77 (1984).

84Be1   Berthon, G., Piktas, M., & Blais, M-J.: *Inorg. Chim. Acta* **93**, 117 (1984).

84Be2   Berthon, G., Blais, M-J., Piktas, M., & Houngbossa, K.: *J. Inorg. Biochem.* **20**, 113 (1984).

84Br   Brown, D. A. & Sekhon, B. S.: *Inorg. Chim. Acta* **91**, 103 (1984).

84Co   Cooper, J. A., Blackwell, L. F., & Buckley, P. D.: *Inorg. Chim. Acta* **92**, 23 (1984).

84Da   Daniele, P. D., Amico, P., & Ostacoli, G.: *Ann. Chim. (Rome)* **74**, 105 (1984).

84Fr   Fridman, J. D.: *Koord. Khim.* **10**, 1034 (1984).

84Ka   Kafarski, P. & Mastalerz, P.: *Beitr. Wirkst. Forsch.* **21**, 1 (1984).

84Kh   Khazaeli, S. & Viola, R. E.: *J. Inorg. Biochem.* **22**, 33 (1984).

84Ki1   Kiss, T. & Gergely, A.: *J. Chem. Soc., Dalton Trans.* 1951 (1984).

84Ki2   Kiss, T., Deák, Gy., & Gergely, A.: *Inorg. Chim. Acta* **91**, 269 (1984).

84Le   Legendziewicz, J., Huskowska, E., Argay, Gy., & Waskowska, A.: *Inorg.*

Chim. Acta **95**, 57 (1984).

84Ma   Martin, R. B.: *Met. Ions Biol. Syst.* **17**, 1 (1984).

84Mo   Monroe, M. B., Boone, D. R., & Kust, R. G.: *Polyhedron* **3**, 49 (1984).

84Ne   Nelsestuen, G. L.: *Met. Ions Biol. Syst.* **17**, 353 (1984).

84Od   Odani, A. & Yamauchi, O.: *Inorg. Chim. Acta* **93**, 13 (1984).

84Pa1   Pavelcik, F.: *J. Coord. Chem.* **13**, 299 (1984).

84Pa2   Paniago, E. B. & Carvalho, S.: *Inorg. Chim. Acta* **92**, 253 (1984).

84Pe   Pettit, L. D.: *Pure and Appl. Chem.* **56**, 247 (1984).

84Sz   Szpoganicz, B. & Martell, A. E.: *J. Am. Chem. Soc.* **106**, 5513 (1984); *Inorg. Chem.* **22**, 4442 (1984).

84To   Tombeux, J. J., Schaubroeck, J., Huys, C. T., de Brabander, H. F., & Goeminne, A. M.: *Z. Anorg. Allg. Chem.* **517**, 235 (1984).

84Ya   Yamauchi, O.: *J. Mol. Cat.* **23**, 255 (1984).

85Ab   Abdullah, M., Barrett, J., & O'Brien, P.: *Inorg. Chim. Acta* **96**, L35 (1985).

85Al   Altman, J. & Wilchek, M.: *Inorg. Chim. Acta* **101**, 171 (1985); Altman, J., Wilchek, M., & Warshawsky, A.: *ibid*, **107**, 165 (1985).

85Ar   Arena, G., Musumeci, S., Rizzarelli, E., & Sammartano, S.: *Transition Met. Chem.* **10**, 399 (1985).

85Ch   Chitale, P. K., Mandloi, S. N., Kothari, N., & Verma, M. S.: *Pol. J. Chem.* **59**, 1001 (1985).

85Da   Daniele, P. G., de Robertis, A., de Stefano, C., Sammartano, S., & Rigano, C.: *J. Chem. Soc., Dalton Trans.* 2353 (1985).

85De   Decock-Le Révérend, B. & Kozlowski, H.: *J. Chim. Phys.* **82**, 883 (1985).

85Di   Diez-Caballero, R. J. B., Valentin, J. P. A., Garcia, A. A., & Batanero, P. S.: *Bull. Soc. Chim. France* 688 (1985).

85El   El-Ezaby, M. S. & Hassan, M. M.: *Polyhedron* **4**, 429 (1985).

85Fi   Fitzsimmons, B. W., Hume, A., Larkworthy, L. F., Turnbull, M. H, & Yavari, A.: *Inorg. Chim. Acta* **106**, 109 (1985).

85Fr   Fridman, J. D., Dzhusueva, M. S., & Dolgashova, N. V.: *Zh. Neorg. Khim.* **30** 2286 (1985).

85Go   Goodman, B. A. & McPhail, D. B.: *J. Chem. Soc., Dalton Trans.* 1717 (1985).

85Hu   Huber, F., Roge, G., Carl, L., Atassi, G., Spreafico, E., Filippeschi, S., Barbieri, R., Silverstri, A., Rivarola, E., Ruisi, G., di Bianca, F., & Alonzo, G.: *J. Chem. Soc., Dalton Trans.* 523 (1985).

85Is   Isab, A. A. & Arnold, A. P.: *J. Coord. Chem.* **14**, 73 (1985).

85Ki   Kiss, T. & Gergely, A.: *J. Inorg. Biochem.* **25**, 247 (1985).

85Ko   Konstantatos, J., Kalatzis, G., Vrachnou-Astra, E., & Katakis, D.: *J. Chem. Soc., Dalton Trans.* 2461 (1985).

85Ki   Kimar, L., Kandasamy, N. R., Srivastava, T. S., Amonkar, A. J., Adwankar, M. K., & Chitnis, M. P.: *J. Inorg. Biochem.* **23**, 1 (1985).

85Ma   Martell, A. E., Langohr, M. F., & Tatsumoto, K.: *Inorg. Chim. Acta* **108**, 105 (1985).

85Pa   Palade, D. M., Linkova, V. S., & Shapovalov, V. V.: *Zh. Neorg. Khim.* **30**, 384 (1985).

85Pe   Pettit, L. D. & Bezer, M.: *Coord. Chem.* **61**, 97 (1985).

85Qi   Qian, L., Chen, D., & Fang, D.: *Chem. Abstr.* **102**, 200203e (1985).
85Sa   Sakurai, H., Tanaka, H., Nakayama, M., & Chickuma, M.: *Chem. Abstr.* **103**, 81092b (1985).
85Sm1   Smith, R. M., Martell, A. E., & Motekaitis, R. J.: *Inorg. Chim. Acta* **99**, 207 (1985).
85Sm2   Smith, R. M., Motekaitis, R. J., & Martell, A. E.: *Inorg. Chim. Acta* **103**, 73 (1985).
85Sz1   Szabó-Plánka, T.: *Acta Chim. Hung.* **120**, 143 (1985).
85Sz2   Szponganicz, B. & Martell, A. E.: *Inorg. Chem.* **24**, 2414 (1985).
85Th   Theodorou, V. & Hadjiliadis, N.: *Polyhedron* **4**, 1283 (1985).
85Ul   Ulanovski, I. L., Kurganov, A. A., & Davankov, V. A.: *Inorg. Chim. Acta* **104**, 63 (1985).
85Ya1   Yadava, H. L., Singh, S., Prasad, P., Singh, R. K. P., Yadava, P. C., & Yadava, K. L.: *Bull. Soc. Chim. France* I-314 (1985).
85Ya2   Yamauchi, O., Tsujide, K., & Odani, A.: *J. Am. Chem. Soc.* **107**, 659 (1985).
85Ya3   Yabukov, K. M., Vinnichenko, G. M., Offengenden, E. Ya., & Astanina, A. N.: *Zh. Neorg. Khim.* **30**, 2018 (1985).
85Ya4   Yamauchi, O. & Odani, A.: *J. Am. Chem. Soc.* **107**, 5938 (1985).
85Ya5   Yamauchi, O. & Odani, A.: *Inorg. Chim. Acta* **100**, 165 (1985).
86An   Anderegg, G.: *Inorg. Chim. Acta* **121**, 229 (1986).
86Ar   Arnold, A. P., Sin Tan, K., & Rabenstein, D. L.: *Inorg. Chem.* **25**, 2433 (1986).
86Ba   Balzamo, S., Bucci, R., Carunchio, V., & Vinci, G.: *Transition Met. Chem.* **11**, 316 (1986).
86Be   Berthon, G., Hacht, B., Blais, M-J., & May, P. M.: *Inorg. Chim. Acta* **125**, 219 (1986).
86Br   Brown, D. A., Geraty, R., Glennon, J. D., & Choileain, N. N.: *Inorg. Chem.* **25**, 3792 (1986) and references therein
86Da   Daniele, P. G., Amico, P., Zerbinati, O., & Ostacoli, G.: *Ann. Chim. (Rome)* **76**, 393 (1986).
86Di   Diez-Caballero, R. J. B., Valentin, J. F. A., Garcia, A. A., & Batanero, P. S.: *Bull. Soc. Chim. France* 376 (1986).
86Du   Dubler, E., Cathomas, N., & Jameson, G. B.: *Inorg. Chim. Acta* **123**, 99 (1986) and references therein
86El1   El-Ezaby, M. S., Marafie, H. M., & Abu-Soud, H. M.: *Polyhedron* **5**, 973 (1986).
86El2   El-Ezaby, M. S., Marafie, H. M., Hassan, M. M., & Abu-Soud, H. M.: *Inorg. Chim. Acta* **123**, 53 (1986).
86Fr   Fridman, J. D., Alikejeva, S. V., Dolgashova, N. V., & Nemalcheva, T. G.: *Zh. Neorg. Khim.* **31**, 1232 (1986).
86Gr   Grases, F., Palou, J., & Amat, E.: *Transition Met. Chem.* **11**, 253 (1986).
86He   Henry, B., Boubel, J-C., & Delpuech, I-J.: *Inorg. Chem.* **25**, 623 (1986).
86Ki   Kiss, T. & Szücs, Z.: *J. Chem. Soc., Dalton Trans.* 2443 (1986).
86Ko   Kovala-Demertzi, D., Demertzis, M., & Tsangaris, J. M.: *Bull. Soc. Chim. France* 558 (1986).
86Kr   Krzewska, S. & Podsicidly, H.: *Polyhedron* **5**, 937 (1986).

86Ku   Kurzak, B., Kurzak, K., & Jezierska, J.: *Inorg. Chim. Acta* **125**, 77 (1986).
86Le1   Leporati, E.: *J. Chem. Soc., Dalton Trans.* 2587 (1986).
86Le2   Leporati, E.: *J. Chem. Soc., Dalton Trans.* 199 (1986).
86Ma   Maslowska, J. & Chruscinski, L.: *Polyhedron* **5**, 1131 (1986).
86Na   Nair, M. S., Pillai, M. S., & Ramalingam, S. K.: *J. Chem. Soc., Dalton Trans.* 1 (1986).
86No   Noszál, B.: *J., Phys. Chem.* **90**, 6345 (1986).
86Po   Powers, J. C. & Harper, J. W.: In *Proteinase inhibitors*, Barrett, A. J. & Salvesen, G. eds., Elsevier, Amsterdam (1986).
86Pu   Puthráya, K. H., Srivastava, T. S., Amonkar, A. J., Adwankar, M. K., & Chitnis, M. P.: *J. Inorg. Biochem.* **26**, 45 (1986).
86So1   Sóvágó, I., Várnagy, K., & Bényei, A.: *Magyar Kém. Foly.* **92**, 114 (1986).
86So2   Sóvágó, I., Harman, B., Gergely, A., & Radomska, B.: *J. Chem. Soc., Dalton Trans.* 235 (1986).
86Sz   Szpoganicz, B. & Martell, A. E.: *Inorg. Chem.* **25**, 327 (1986).
87Da   Davidenko, N. K. & Raspopina, V. A.: *Zh. Neorg. Khim.* **32**, 1140 (1987).
87De   Deuschle, V. & Weser, U.: *Inorg. Chim. Acta* **135**, 5 (1987).
87El   El-Issa, B. D., Makhyoun, M. A., & Salsa, B. A.: *J. Chim. Phys.* **84**, 843 (1987).
87Gi   Giannousis, P. P. & Bartlett, P. A.: *J. Med. Chem.* **30**, 1603 (1987).
87Go1   Gotsis, E. D. & Fiat, D.: *Polyhedron* **6**, 2037 (1987).
87Go2   Gotsis, E. D. & Fiat, D.: *Polyhedron* **6**, 2053 (1987).
87Ha   Harju, L.: *Talanta* **34**, 817 (1987).
87Hi   Hitchman, M. A., Kwan, L., Engelhardt, L. M., & White, A. H.: *J. Chem. Soc., Dalton Trans.* 457 (1987).
87Ja   Jain, N., Mital, R., Rau, K. S., Srivastava, T. S., & Battacharya, R. K.: *J. Inorg. Biochem.* **31**, 57 (1987).
87Ka   Kasselouri, S., Garoufis, A., & Hadjiliadis, N.: *Inorg. Chim. Acta 135*, L23 (1987).
87Ki1   Kiss, T., Simon, Cs., & Vachter, Zs.: *J. Coord. Chem.* **16**, 225 (1987).
87Ki2   Kiss, T., Balla, J., Nagy, G., Kozlowski, H., & Kowalik, J.: *Inorg. Chim. Acta* **138**, 25 (1987).
87Ku   Kurzak, B., Kurzak, K., & Jezierska, J.: *Inorg. Chim. Acta* **130**, 189 (1987).
87La   Laurie, S. H.: In *Comprehensive coordination chemistry*, Wilkinson, G. ed. Vol 2, Pergamon Press, New York, (1987).
87Le1   Leporati, E.: *J. Chem. Soc., Dalton Trans.* 1409 (1987).
87Le2   Leporati, E.: *J. Chem. Soc., Dalton Trans.* 435 (1987).
87Mi   de Mirande-Pinto, C. O. B., Paniago, E. B., Carvalho, S., Tabak, M., & Mascarenhas, Y. P.: *Inorg. Chim. Acta* **137**, 145 (1987).
87Pa   Paniago, E. B. & Carvalho, S.: *Inorg. Chim. Acta* **136**, 159 (1987).
87Pr1   Prasad, K., Rao, A. K., & Mohan, M. S.: *J. Coord. Chem.* **16**, 251 (1987).
87Pr2   Prasad, K. & Mohan, M. S: *J. Coord. Chem.* **16**, 1 (1987).
87Sh   Shuaib, N. M., Marafie, H. M., Hassan, M. M., & El-Ezaby, M. S.: *J. Inorg. Biochem.* **31**, 171 (1987).
87Si   Sigel, H. ed.: *Metal Ions in Biological Systems*, Vol 22, Marcel Dekker, New York (1987).

87So1 Sodijev, U. M., Musajev, Z. M., Hodzajev, O. F., Uschmanhodzajeva, J.
       S., & Parnijev, N. A.: *Koord. Khim.* **13**, 179 (1987)
87So2 Sóvágó, I. & Petőcz, Gy.: *J. Chem. Soc., Dalton Trans.* 1717 (1987).
87To   Tomassetti, M., Cardarelli, E., Curini, R., & D'Ascenzo, G.: *Thermo-chim. Acta* **113**, 243 (1987).
88Be   Beck, W.: *Pure and Appl. Chem.* **60**, 1357 (1988).
88Bu   Burger, K., Sipos, P., Véber, M., Horváth, I., Noszál, B., & Löw, M.:
       Inorg. Chim. Acta **152**, 233 (1988)
88Ku   Kurzak, B., Kroczewska, D., Jezierska, J., & Huza-Koralewicz, M.:
       Transition Met. Chem. **13**, 297 (1988).
88Le   Leporati, E.: *J. Chem. Soc., Dalton Trans.* 421 (1988).
88Li   Lin, H-K., Gu, Z-X., & Chen, X-M.: *Thermochim. Acta* **123**, 201 (1988).
88Ma1 Martin, R. B.: *Met. Ions Biol. Syst.* **23**, 123 (1988).
88Ma2 Martin, R. B. & Sigel, H.: *Comments Inorg. Chem.* **6**, 285 (1988).
88Ma3 Masoud, M. S., Abdel-Nably, B. A., Soliman, E. M., & Abdel-Hamid, O.
       H.: *Spectrochim. Acta* **128**, 75 (1988).
88Pa1 Pavone, V., Lombardi, A., di Blasio, D., Bennedetti, E., & Pedone, C.:
       Inorg. Chim. Acta **153**, 171 (1988).
88Pa2 Palade, D. M., Ozherelev, I. D., & Linkova, V. S.: *Zh. Neorg. Khim.* **32**, 1140 (1988).
88Pe   Pessca, J. C., Boas, L. F. V., Gillard, R. D., & Lancashire, R. J.:
       Polyhedron **7**, 1245 (1988).
88Pr   Prasad, K., Mohan, M. S., & Bathina, H. B.: *J. Coord. Chem.* **17**, 63 (1988).
88Ra   Raymond, K. N. & Garrett, T. M.: *Pure and Appl. Chem.* **60**, 1807 (1988)
       and references therein
88Sz   Szabó-Plánka, T., Rockenbauer, A., Györ, M., & Gaizer, F.: *Coord. Chem.* **17**, 69 (1988).
88Ta   Tabata, M. & Tanaka, M.: *Inorg. Chem.* **27**, 3190 (1988).
88Tu   Turowski, P. N., Rodgers, S. J., Scarrow, R. C., & Raymond, K. N.:
       Inorg. Chem. **27**, 474 (1988) and references therein
88Va   Várnagy, K., Sóvágó, I., & Kozlowski, H.: *Inorg. Chim. Acta* **151**, 117 (1988).
88Ze   Zelano, V., Zerbinati, O., & Ostacoli, G.: *Ann. Chim. (Rome)* **78**, 273 (1988).
89Fa   Farkas, E., Szőke, J., Kiss, T., Kozlowski, H., & Bal, W.: *J. Chem. Soc., Dalton Trans.* 2247 (1989).
89Ki1 Kiss, T., Sóvágó, I., & Martin, R. B.: *J. Am. Chem. Soc.* **111**, 3611 (1989).
89Ki2 Kiss, T., Farkas, E., & Kozlowski, H.: *Inorg. Chim. Acta* **155**, 281 (1989).
89Le   Leporati, E.: *J. Chem. Soc., Dalton Trans.* 1299 (1989).
89Ma   Matzak-Jon, E. & Wojciehowski, W.: *Inorg. Chim. Acta* (in the press).
89No1 Noszál, B. & Sándor, P.: *Anal. Chem.* (submitted for publication).
89No2 Noszál, B. & Tánczos, R.: *Int. J. Pept. Prot. Res.* (in preparation).
90Ba   Balla, J., Kiss, T., Jezowska-Bojczuk, M., Kozlowski, H., & Kafarski, P.:
       J. Chem. Soc., Dalton Trans. (submitted for publication).
90Fa1 Farkas, E. & Kiss, T.: *J. Chem. Soc., Perkin 2 Trans.* 1549, 1990).
90Fa2 Farkas, E. & Buglyó, P.: *J. Chem. Soc., Dalton Trans.* 1549, (1990).

90Ki   Kiss, T., Sóvágó, I., & Gergely, A.: *Pure and Appl. Chem.* (submitted for publication).

IV

Metal complexes of peptides and their derivatives

Imre Sóvágó
Department of Inorganic and Analytical Chemistry, L. Kossuth University,
4010-Debrecen, Hungary

1. INTRODUCTION

Various metal ions play an important role in the binding and transport of organic molecules and in the catalysis of different acid–base and redox processes in biological systems. Proteins comprise one of the major groups of organic substances that bind metal ions. This is demonstrated by the existence and role of metalloenzymes, in which metal ions are bonded to specific amino acid residues of the polypeptide chain. The discovery of the biuret reaction more than a hundred years ago revealed the possibility of strong interactions between metal ions and proteins, but even now very little information is available on the structure and binding mode of metalloproteins. Investigations of the metal complexes of various oligopeptides have made a significant contribution to the understanding of metal ion–protein interactions. The most important results achieved in this field will be discussed in this review.

As a result of the significant developments in coordination chemistry during the past 30 years, a considerable amount of information is now available on metal ion–peptide interactions. The results also permit conclusions as to some general features of the coordination ability of oligopeptides. The most important results and conclusions are already to be found in both textbooks and review papers [61Ma, 67Fr, 69Sh, 73Fr, 74Ma1, 82Si, 87Ha]. Among them, the review by *Sigel* and *Martin* [82Si] provides a general picture of the coordination chemistry of amides and related ligands. Consequently, only the recent results and the general features of the coordination ability of oligopeptides will be covered by the present chapter.

2. ACID–BASE PROPERTIES AND COORDINATION ABILITY OF PEPTIDES

2.1 General

It is well known that amino acids can form stable bis or tris complexes with various transition metal ions, where the coordination takes place via the α-amino and carboxylate groups so as to form 5-membered chelate rings (Ia). This type of coordination is greatly affected by the other donors atoms present in the side chain (R), as discussed in the previous chapter of this book. The amino and carboxylate groups are in terminal position in a dipeptide (Ib), and thus the steric requirements exclude the formation of 5-membered chelate rings. On the other hand, it is obvious that at least four donor atoms (amino-N, carbonyl-O, amide-N, and carboxylate-O) are present in a dipeptide, all of which, in principle, are capable of metal ion coordination. Other donor groups may additionally be present in the side chains (R_1 and R_2) of the molecule, which results in a great variety of metal ion–peptide interactions. The basicity of the various donor groups in the peptide molecules is one of the most important parameters influencing the coordination chemistry of oligopeptides.

I.

2.2 Acid–base properties

Scheme I illustrates the fact that all dipeptides contain three separated functional groups: terminal–amino group($-NH_2$), carboxylate group ($-COO^-$), and amide group ($-CONH-$) which is referred to as peptide linkage. This also holds for the longer oligopeptides, where the presence of a new amino acid results in an increase in the distance between the terminal groups and in repetition of the amide groups.

The acid–base properties of various di- and tripeptides have already been studied extensively, and a survey of the literature clearly demonstrates that if the side chains do not contain any additional functional groups (e.g. for glycine peptides) only two pK values are characteristic of oligopeptides in the measurable pH range ($0 < pH < 14$). The pK values of glycine and glycine oligopeptides are listed in Table 4.1.

Table 4.1 — pK values of glycine and its oligopeptides ($T = 298$ K; $I = 0.1$ mol. dm^{-3})

Ligand	$pK_2(NH_3^+)$	Ref.	$pK_1(COOH)$	Ref.
glycine	9.60	90Ki	2.37	90Ki
diglycine	8.13	72Si	3.21	64Ki
triglycine	7.96	72Si	3.27	66Ki
tetraglycine	7.97	72Si	3.24	66Ki

Deprotonation of ammonium and carboxylic groups are well separated in both amino acids and peptides, and pK_1 and pK_2 relate to carboxylate and amino groups, respectively. Table 4.1 shows that the extent of this separation is the most significant in the amino acid form, and that the presence of the peptide linkage decreases the basicity of the amino group and the acidity of the carboxylic group. The similarity of the pK values of triglycine and tetraglycine, however, suggests that further increase of the peptide chain does not have a significant effect on the basicity of peptides.

The amide group offers two potential donor atoms (the carbonyl)O and amide-N) for the binding of protons or metal ions. However, the tetrahedral amino nitrogen which possesses a lone pair of electrons in the amino acid, loses its basicity when it reacts to give an amide group, because amide groups are planar with a double-bond character of the carbon–nitrogen bond. Consequently, a peptide linkage is neutral throughout the whole pH range, though it can behave as a very weak acid or base in very basic or acidic solutions. Scheme II can be used to summarize the proton-binding ability of amide groups.

II.

Protonation studies on *N*-methylacetamide gave a pK value of $pK = -0.7$ [69Li] which was initially interpreted in terms of the proton-binding ability of the amide-*N*. NMR studies, however, definitely proved that the carbonyl-O is the main protonation site throughout the entire acidic pH range [72Ma, 73Ma, 78Ma]. The ratio of O- to N-protonated amide cations was estimated as 10^7, which supports the extremely weak basic character of the amide group in peptides.

Proton loss from the $-CONH-$ group takes place in highly basic solution, and this makes equilibrium measurements rather difficult. A value of $pK \sim 15.2$ was measured in the case of GlyTyr [79Ap], and similar high values have been obtained for other simple amides. Further details of the acid–base properties of the amide group were discussed in the previous review [82Si].

The acid–base properties of peptides are significantly modified by the occurrence of functional groups in the side chains. Deprotonation of these groups can overlap with the deprotonation of the terminal ammonium and carboxylic groups, which results in the existence of microscopic protonation processes. Details of these studies are discussed in Chapter 2, and in the following sections of the present chapter.

2.3 Possible modes of coordination in peptides

Because of the 'neutrality' of the amide group the terminal amino and carboxylate groups are the most effective binding sites for metal ions in peptides. As mentioned in section 2.1 steric requirements preclude the simultaneous coordination of the terminal groups to the same metal ion. The amino and carboxylate groups are consequently independent primary ligating groups (or 'anchors') for metal ion binding, and their interaction will largely depend on the nature of the metal ion. Glycineamide (IIIa) and N-acetylglycine (IIIb) are the simplest ligands to model the amino and carboxylate terminals of peptides. A literature survey reveals that N-acetylglycine (and other N-acetyl amino acids without chelatable side chains) can coordinate to metal ions as simple carboxylates [78Ud, 82Si]. This interaction results in the formation of parent complexes in which monodentate binding occurs via the negatively charged carboxylate groups; this is not able to prevent metal hydroxide precipitation in neutral aqueous solution. On the other hand, the amide group is in a chelatable position relative to the carboxylate moiety, but its coordination can occur only after deprotonation, because of the very low basicity of the amide group. To date there has been no report of amide deprotonation and coordination in simple N-acetyl amino acids. (Coordination of N-acetyl amino acids can be different if another functional group is present in the side chain, as will be discussed in sections 6 and 7.)

$$CH_2-C \overset{O}{\underset{NH_2}{\diagup}} \quad\quad\quad CH_2-C \overset{O}{\underset{NH}{\diagup}} OH$$

(a)   (b)

III.

Owing to the monodentate binding, terminal carboxylate groups of peptides are weak binding sites, and the presence of terminal groups will be the governing factor during complex formation with peptides. This is demonstrated by the complex-forming ability of glycineamide, represented in Scheme IV, which shows that glycineamide forms 5-membered chelate rings via the coordination of amino-N and carbonyl-O donors, ([CuA_2] complexes). This type of binding is reminiscent of that of amino acids, except for the charge neutralization. Consequently, the stability constants of amino acid complexes are higher than those of peptides or glycineamide, but higher for the latter than for carboxylates (e.g. $\log \beta_1$ values for [CuA] complexes of glycine, diglycine, glycineamide, and acetate are 8.07, 5.56, 5.30, and 1.76, respectively [76So, 77Ge, 86So1, 70Bu]. The increase in stability relative to the carboxylate complexes makes it possible to avoid metal hydroxide precipitation, and the deprotonation and coordination of the amide group of glycineamide with copper(II) and with some other metal ions take place in slightly basic solution. Accordingly two terminal amino groups and two deprotonated amide nitrogens can saturate the coordination sphere of copper(II) in the species [CuA_2H_{-2}]. Throughout this chapter the stoichiometric compositions and stability constants of the complexes will be referred by the following equation:

$$pM + qA + rH \rightleftharpoons M_p A_q H_r$$

$$\beta_{pqr} = \frac{[M_p A_q H_r]}{[M]^p [A]^q [H]^r} \qquad (1)$$

The pK values for amide deprotonation in the copper(II) complexes of glycineamide are 7.07 and 8.33, which is a decrease of at least seven orders of magnitude as compared with the free ligand.

IV.

In the case of dipeptides the same amide (or peptide) deprotonation and coordination are accompanied by the possibility of the formation of a new 5-membered chelate ring, with the participation of terminal carboxylate groups, as depicted in Scheme V. The $[\text{CuAH}_{-1}]$ species of diglycine contains two fused 5-membered chelate rings, which results in an enhanced stability of the complex and in a low pK of amide group (p$K = 4.23$ [77Ge]).

V.

To summarize the general features of the coordination ability of simple oligopeptides, it can be stated that the terminal amino group is the primary ligating group for various transition metal ions. Via the coordination of the amino-N and neighbouring carbonyl-O a five-membered chelate ring with moderate stability can be formed (Va). The formation of such complexes can prevent metal ion hydrolysis in alkaline solution only after amide deprotonation and binding (Vb). The occurrence of this process depends largely on the nature of the metal ions, and only a few of them are able to promote amide binding, as will be discussed in the next section. Amide deprotonation and coordination are influenced by the side chain donor groups, too. Almost any amino acid residue has some effect on the stability of the complexes and on the pK value of the amide groups, and some functional groups can induce or prevent the binding of amide as will be shown in sections 4 to 7.

3. GLYCINE OLIGOPEPTIDE COMPLEXES

3.1 Copper(II) complexes

Diglycine is the simplest model for peptide coordination, and its complexes with various metal ions have been thoroughly studied. The general mode of this coordination has already been mentioned in section 2.3. The formation of [MA] (Va) is characteristic for all metal ions, but its stability is much lower than that of the amino acid complexes. Consequently, the complexes of metal ions which are able to promote amide deprotonation and coordination are the most interesting in this field. A literature survey shows that palladium(II), copper(II), and nickel(II) are the most effective in this respect, and peptide–NH ionization can occur below or around the physiological pH range. The copper(II) complexes are the best studied, and details of the copper(II)-diglycine interaction are given in Scheme VI [55Dol, 64Ki, 68Br, 77Ge, 89Fa1, 89Sh, and references therein].

VI.

Scheme VI reveals that complex-formation processes between copper(II) and diglycine are quite simple in the acidic pH range because only the species [CuA]$^+$ and [CuAH$_{-1}$] can form. The possible binding modes in these species were shown in

Scheme V, while the binding modes in the complexes present in alkaline solution are depicted in Scheme VII. The predominating species $[CuAH_{-1}]$ has one free coordination site, thus other species can be formed in basic solution, depending on the metal ion to ligand ratio and the pH [75Ka, 77Ge, 89Fa1].

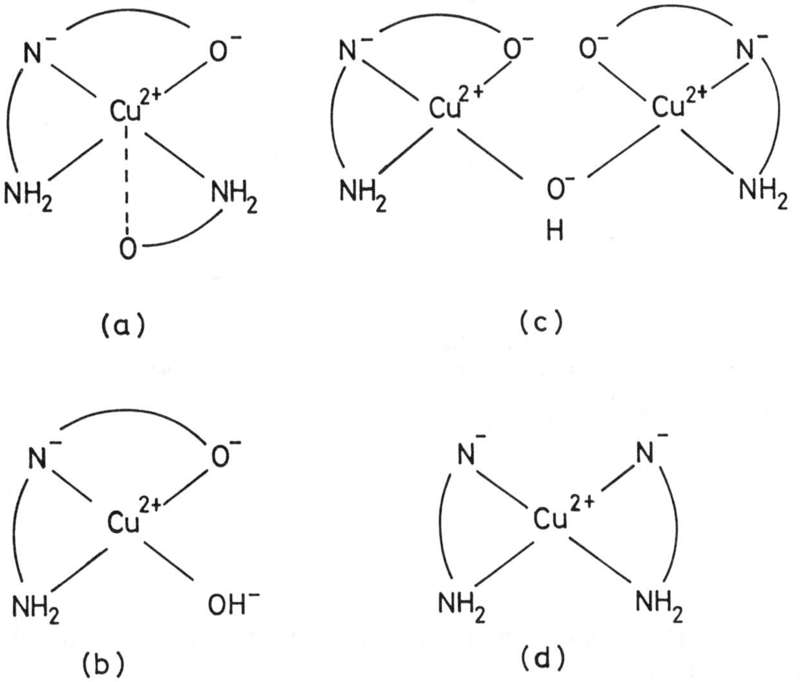

$$
\begin{array}{cccc}
\text{(a)} & & \text{(c)} &
\end{array}
$$

(a) — Cu^{2+} coordinated by N$^-$, O$^-$, NH$_2$, and O

(c) — two Cu^{2+} centres: one coordinated by N$^-$, O$^-$, NH$_2$, O$^-$H; the other by O$^-$, N$^-$, NH$_2$

(b) — Cu^{2+} coordinated by N$^-$, O$^-$, NH$_2$, OH$^-$

(d) — Cu^{2+} coordinated by N$^-$, N$^-$, NH$_2$, NH$_2$

VII.

The bis complex $[CuA_2H_{-1}]^-$ contains the second ligand in axial–equatorial position via the coordination of terminal amino and neighbouring carbonyl groups without amide binding and deprotonation (VIIa). This fact is reflected in the low values of the corresponding stability constants collected in Table 4.2.

The low stability of the species $[CuA_2H_{-1}]^-$ cannot prevent the hydrolysis of $[CuAH_{-1}]$, which results in the formation of various mixed hydroxo complexes above pH ~ 9 (VIIb, VIIc). The formation of the hydroxo complexes is generally accompanied by a small red shift in the absorption spectra, and by a negative shift in the redox potential [86Sa]. In very basic solutions, the biuret reaction, characteristic of 4N-coordination, can be observed in the presence of a very high (100–1000-fold) ligand excess [67Sh, 77Ko1, 89Fa1]. This process has been interpreted in terms of an increase in the number of coordinated amide groups (VIId). The formation of this species is not characteristic for all dipeptides [89Fa1], and the factors influencing the existence of the species $[Cu(AH_{-1})_2]^{2-}$ will be discussed later.

The copper(II) complexes of oligopeptides containing more than two glycyl residues (tri-and tetraglycine) have also been extensively studied [55Do2, 66Ki,

Table 4.2 — Stability constants of copper(II)–oligoglycine complexes

Species	diglycine		triglycine[§]	tetraglycine[§]
$[CuA]^+$	$5.56^\dagger$	$5.55^\ddagger$	5.24	5.08
$[CuAH_{-1}]$	1.33	1.56	0.02	−0.42
$[CuAH_{-2}]^-$	−8.04	—	−6.58	−7.31
$[CuAH_{-3}]^{2-}$	—	—	—	−16.60
$[CuA_2H_{-1}]^-$	4.46	—	—	—
$[Cu_2A_2H_{-3}]^-$	−4.51	—	—	—
$pK_1^\P$	4.23	3.99	5.22	5.50
$pK_2^\P$	$(9.37)^{\dagger\dagger}$	—	6.60	6.89
$pK_3^\P$	—	—	$(11.9)^{\dagger\dagger}$	9.29
$\log K^{\ddagger\ddagger}$	3.13	—	—	—

†Ref. [77Ge] ($T = 298$ K; $I = 0.2$ mol.dm^{-3} KCl)
‡Ref [75Si] ($T = 298$K; $I = 0.1$ mol.dm^{-3} NaClO$_4$)
§Ref [72Si] ($T = 298$K; $I = 0.1$ mol.dm^{-3} NaClO$_4$)
$^\P pK_i$ belong to the deprotonation of the first, second, and third amide groups.
††hydroxo complex formation.
‡‡equilibrium constant for the process:

$$[CuAH_{-1}] + [A^-] \overset{K}{\rightleftharpoons} [CuA_2H_{-1}]^-$$

69Ki, 72Si]. The stability constants of the various species present in solution are collected in Table 4.2.

Table 4.2 reveals that the formation of various deprotonated species is characteristic of tri- and tetraglycine. This means that the amide deprotonation process continues with the complexation of copper(II). Namely, the binding in the $[CuAH_{-1}]$ complex of triglycine occurs via amino-N, amide-N, and carbonyl-O. The corresponding protonation constant is $pK = 5.24$, which is larger than that of diglycine ($pK = 3.99$) but lower than that of glycineamide ($pK = 7.07$). On increase of the pH the second (for triglycine) or the second and third protonated peptide-N atoms will saturate the coordination sphere of copper(II). The binding modes of these species in alkaline solution are depicted in Scheme VIII. VIIIa relates to the $[CuAH_{-2}]^-$ complex of triglycine, where the amino-N, two protonated peptide-N, and the C-terminal carboxylate-O donor atoms will occupy the coordination sites of copper(II). This type of coordination prevents the formation of bis complexes and hinders hydrolysis, which occurs only in highly basic solution ($pK \sim 11.9$). The terminal amino-N and three deprotonated peptide-N atoms will occupy the coordination sphere of copper(II) in the $[CuAH_{-3}]^{2-}$ species of tetraglycine (VIIIb). The pK values for successive amide deprotonation increase (Table 4.2), but they are quite low as compared with that for the free ligand. The results presented above were mainly obtained from potentiometric studies, but the results of other calorimetric and spectroscopic studies on solutions [68Br, 69Ki] and crystal structure studies on solid forms [61St, 65Fr1, 65Fr3, 67Fr] provide unambiguous proof of the existence of amide binding.

$$[CuAH_{-2}]^{-}$$
(a)

$$[CuAH_{-3}]^{2-}$$
(b)

VIII.

Electronic absorption spectroscopy and CD are especially useful tools for the study of metal ion–peptide interactions [69Ts, 74Ma1, 75Ma]. The applicability of these methods and the most important results obtained by spectroscopy are discussed in detail in the reviews by Martin [74Ma1, 82Si]. As regards the absorption spectra of copper(II) complexes the d-d bands are situated from 520 to 650 nm; an increase in the number of coordinated nitrogen donor atoms significantly increases the energy of absorption. Absorption maxima (and molar absorptivities) of d-d bands were reported to appear at 635 nm ($\varepsilon = 84$), 555 nm ($\varepsilon = 149$), and 520 nm ($\varepsilon = 145$) for the fully deprotonated copper(II) complexes of diglycine, triglycine, and tetraglycine, respectively. The blue shift is associated with the increasing tetragonality, which is almost complete in the fully coordinated complexes of tetra- or pentaglycine. An intense charge transfer band at around 250 nm (220 nm for the free ligand) is also characteristic of the transition metal complexes of peptides coordinated via deprotonated amide groups [69Ts].

3.2   Complexes of nickel(II) and palladium(II)

Many studies have dealt with the complexation of oligopeptides with nickel(II), because this metal ion is also able to promote amide deprotonation and coordination in slightly basic solution [60Ma1, 67Ki, 69Da, 75Do]. The stoichiometric compositions and structures of nickel(II) complexes, however, are significantly different from those of copper(II). Namely, copper(II) forms 1:1 tetragonal complexes with all oligoglycines, while nickel(II) forms complexes with two different stereochemistries. Both potentiometric and spectroscopic studies show that bis complexes are formed in the nickel(II)-diglycine system and that both peptide-N atoms are deprotonated and coordinated by pH ~ 10. The corresponding pK values for formation of the species $[NiA_2H_{-1}]^{-}$ and $[NiA_2H_{-2}]^{2-}$ are 9.35 and 9.96, respectively, and both species are paramagnetic and octahedral. X-ray studies [78Fr1] quite clearly prove the existence of hexacoordinated nickel(II) ions in the solid state. The binding mode of the species $[NiA_2H_{-2}]^{2-}$ is illustrated in Scheme IX, which shows the tridentate chelation of both peptide molecules.

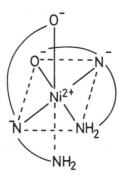

IX.

In contrast with diglycine, the formation of yellow coloured, diamagnetic complexes is characteristic for the interaction of nickel(II) with triglycine or tetraglycine. Potentiometric titrations show the liberation of two (for triglycine) or three (for tetraglycine) protons at around pH 8 to 9. The pK values for amide deprotonation cannot be easily determined, however, because deprotonation takes place in a cooperative manner. p$K_1 = 8.8$ and p$K_2 = 7.7$ have been calculated for the nickel(II)–triglycine system [70Bi], which means that deprotonation of the second peptide linkage occurs with a lower pK than the first one. This contradiction can be solved only by the assumption of cooperative deprotonation, which is forced by the formation of the square–planar diamagnetic species. On the other hand, it means that the concentration of [NiAH$_{-1}$] is negligible at any metal ion to ligand ratio and any pH, and the individual pK values have no real meaning. The same is true for tetraglycine, where the cooperative deprotonation of three peptide groups results in the formation of [NiAH$_{-3}$]$^{2-}$ with square–planar geometry. This means that the binding modes in the nickel(II) complexes of triglycine and tetraglycine are similar to those of the copper(II) complexes (VIIIa,b) without any metal ion–ligand interaction in the apical positions. The crystal structures of the complexes [78Fr2] and the absorption spectra and NMR studies in solution [67Ki, 69Ki, 75Do] support the formation of amide-bonded, square–planar diamagnetic complexes.

Another important feature of nickel(II) complexes with oligopeptides is the kinetically slow formation of the various deprotonated species. The inertness of the complexes increases with increase in the number of deprotonated amide groups, which makes equilibrium studies of the various tri- and tetrapeptides rather difficult. Details on the kinetic studies of peptide complexes are available from the review by Margerum & Dukes [74Ma2].

Palladium(II) is the most effective metal ion as concerns promotion of peptide-NH ionization and coordination. Substitution of the amide-H by palladium(II) generally occurs at such low pH values (pH ~ 2) that it is difficult to obtain reliable stability constants. On the other hand, the complexation with palladium(II) is characterized by slow formation kinetics, which also hampers equilibrium investigations. Potentiometric and spectroscopic studies, however, have definitely proved that palladium(II) forms diamagnetic, planar complexes with oligopeptides [70Wi, 72Pi, 77Li, 85Pe1]. This type of coordination is similar to that of copper(II) with diglycine (Vb), or that in the yellow, square–planar complexes of nickel(II) with tri- and tetraglycine (VIIIa,b). The pK of amide deprotonation is at around 2.0 for diglycine, and the cooperative deprotonation of two and three amide groups (at around pH ~ 4) is characteristic of tri- and tetraglycine, respectively, in a kinetically slow process. On the other hand, evidence has been presented that coordinated palladium(II) promotes the hydrolysis of peptide esters [84Ha, 86Ha].

3.3   Complexes of other metal ions
The complex-formation processes of glycine oligopeptides have been studied with various transition and non-transition elements. The cobalt(II), cobalt(III), and zinc(II) complexes have been investigated the most extensively, and many contra-dictory findings have been published concerning the possibility of amide binding [82Si].

From a literature survey, it can be concluded that at around pH ~ 10 cobalt(II) ions are able to substitute the peptide-NH hydrogens, which results in the formation of a hexacoordinated cobalt(II) bis complex ($[Co(AH_{-1})_2]^{2-}$), similar to the corresponding nickel(II) complexes shown in Scheme IX [69Mi, 71Mo1]. Cobalt(II) ions exhibit unusual absorption spectra in the dipeptide complexes type $[Co(AH_{-1})_2]^{2-}$. Three absorption bands appear in the d-d range, owing to a high spin–low spin equilibrium in the octahedral complex. In the presence of oxygen, the light blue bis complexes of cobalt(II) are readily oxidized to a wine-red cobalt(III) complex, which has the same binding mode as that in the cobalt(II) and nickel(II) complexes [66Gi, 70Ba]. A peroxo-bridged binuclear complex appears to be an intermediate during oxidation. Details on the oxygen-binding ability of cobalt(II) peptide complexes are discussed by Sigel & Martin [82Si]. It has already been mentioned that X-ray structural studies [66GHi, 70Ba] definitely proved the exis-tence of cobalt(III)-promoted amide deprotonation and binding. Cobalt(III) com-plexes are kinetically inert, however, thus the corresponding pK values cannot be determined, but cobalt(III) ion seems to be as effective as palladium(II).

In connection with the complexation of triglycine the formation of both octa-hedral bis-complexes of cobalt(II) and cobalt(III) [71Gi] (via the coordination of terminal amino, one deprotonated amide, and carbonyl-O) and octahedral cobalt(III) mixed complexes with two deprotonated amide nitrogens in the X-Y plane (similarly to copper(II)-triglycine) and with two additional ligands in apical positions has been reported [80Ev, 86Ge].

In acidic and slightly basic solutions, the complex formation processes between zinc(II) and glycyl peptides are similar to those of cobalt(II): at pH >5 chelation occurs via the terminal amino and neighbouring carbonyl groups, and the stability constant for the $[ZnA]^+$ species of diglycine is log $\beta_1 = 3.13$ [72Ra], which is very close to that of the $[CoA]^+$ species, log $\beta_1 = 3.23$ [56Da]. In spite of the similar stabilities of the complexes of the two metal ions, there is no real evidence to support amide deprotonation and binding in basic solutions of zinc(II) complexes. This can be attributed to the lack of ligand field stabilization in zinc(II) complexes, and to the higher tendency for the formation of hydroxo complexes than in the case of cobalt(II). Thus, one may conclude that the zinc(II) ion can coordinate peptide molecules, but it is unable to substitute amide H-atoms of simple dipeptides at any pH. The presence of side-chain donor groups, however, can promote this process in zinc(II) complexes, as will be discussed for histidine-containing peptides in section 6.1.

Complex-formation processes with oligopeptides have also been studied for manganese(II) [56Da, 75Ba], iron(II) and (III) [70Mo, 71Bo], cadmium(II) [72Ra, 85Wa, 86So2], chromium(III) [86Mu], chromium(II) [86Mi], lead(II) [72Ra], vanadyl(IV) [88Re], and platinum(II) [70Fr1]. From these studies, it is evident that manganese(II), cadmium(II), chromium(II), vanadyl(IV), and lead(II) are not able to promote amide coordination, although these metal ions form complexes with oligopeptides via the coordination of terminal amino and neighbouring carbonyl groups, similarly as with zinc(II). There are reports of the iron(II) and iron(III)-induced deprotonation of the amide group [70Mo, 71Bo]. These suggestions have not been proved, however, and they contradict many other findings [82Si].

Crystal structure study of the chromium(III)-diglycine complex with composition of $[Cr(AH_{-1})_2]^-$ nevertheless definitely shows the participation of deprotonated peptide-N in coordination [86Mu]. The inertness of Cr(III) complexes does not permit the acquisition of reliable equilibrium data on the chromium(III)–promoted amide deprotonation. A similar phenomenon has been observed for the complexes of platinum(II), where the existence of Pt(II)-peptide-N^- bonding was proved in an X-ray study of platinum(II) complex of GlyMet [70Fr1], in which the coordination occurs as in the palladium(II) complexes.

Another important feature of the coordination chemistry of oligopeptides is that they are able to stabilize the trivalent oxidation states of copper and nickel. Copper(III) and nickel(III) complexes of tri- and tetrapeptides have already been reviewed by Margerum and co-workers [75Bu, 81Ma, 83Ma1]. The copper(III) complex of tetraglycine with composition $[Cu(AH_{-3})]^-$ can be obtained from the corresponding copper(II) complex with oxidizing agents such as $[IrCl_6]^{2-}$ or by electrolysis. The resulting complex (similarly to other d^8 systems) undergoes very slow substitution reactions, it has different absorption spectra, and it is EPR-silent because of the square–planar geometry [83Di]. The nickel(III) complexes are six-coordinated with tetragonal distortion. Axial substitutions are very labile, whereas equatorial substitutions are sluggish. The existence of silver(III) complexes of tri- and tetraglycine have also been reported [84Ki3].

The complexes of trivalent metal ions are strong oxidizing agents. The redox potentials of the copper complexes largely depend on the nature of the peptides; $\varepsilon_0 = 0.92$ V for the triglycine complex and 0.63 V for the tetraglycine complex. These potentials are accessible *in vivo*, which renders the trivalent metal complexes biologically important.

To summarize the effects of metal ions on the promotion of amide deprotonation in peptides, the following series of metal ions can be given (pK values in parentheses):

$$Pd^{2+}(2) > Cu^{2+}(4) > Ni^{2+}(8) > Co^{2+}(10) \ .$$

Three inert metal ions (cobalt(III), platinum(II), and chromium(III)) have also been reported to substitute amide protons, but equilibrium data and pK values are not available, for kinetic reasons.

From this series of metal ions, conclusions can be drawn concerning some general features of metal ions in the promotion of amide coordination. The possibility of square–planar (e.g. Pd(II), Pt(II), Ni(II), etc.) or distorted octahedral (e.g. Cu(II), Co(II), Co(III), Ni(II), etc.) geometry seems to be one of the most important factors influencing the interactions with peptides. On the basis of the geometry of the complexes, of course, certain other metal ions should interact in the same way (e.g. Mn(II), Fe(II,III), Zn(II), Cd(II), etc.). The lack of this interaction proves that some other factors must play an important role in this process. The high affinity of palladium(II) or platinum(II) for peptides is indicative of the softness of amide-N^-. On the other hand, the fact that zinc(II) and cadmium(II) are not able to promote amide binding points to the complex nature of metal ion–peptide interaction. At the same time, it should be borne in mind that the series given above is valid only for glycine-type oligopeptides. Strongly coordinated donor groups in the side chain can prevent or induce amide deprotonation relative to that for oligoglycines.

The results presented in this section demonstrate that even a simple tetrapeptide can saturate the coordination sphere of copper(II) or nickel(II). Consequently, further extension of the peptide chain will not result in the formation of new metal ion-amide-N bonds. This conclusion is supported by crystal structure study of the copper(II) complex of pentaglycine [70Bal]. In this case a species with composition $[Cu(AH_{-3})]^{-2}$ can be obtained (similarly as with tetraglycine) via the coordination of terminal amino and three subsequent deprotonated amide groups. The fourth peptide group and the terminal carboxylate do not participate in direct interaction with the metal ion. This result cannot be generalized to all longer peptides or proteins. In this case, the conformation of the macromolecule and the other donor atoms present in the side chains provides specific binding sites for the metal ions, as discussed in the next chapter.

4.   COMPLEXES OF PEPTIDES WITH NON-COORDINATING SIDE CHAINS

As emphasized in the previous section, the complex-formation processes of peptides depend on the following three factors:

- the nature of metal ion,
- the 'size' of the peptide (the number of peptide linkages),
- the presence of donor atoms in the side chains.

The roles of the metal ions and the number of peptide linkages were illustrated via the complex-forming capability of oligoglycines. The presence of donor atoms in the side chain and the side-chain disposition, however, can result in new types of coordination, which also depend on the nature of the metal ions. If there are no donor atoms in the side chains, new binding modes cannot be achieved, but a bulky side chain can influence the stability and structure of complexes for steric reasons. Such bulky groups have almost the same effect on all metal ion complexes. Consequently, most of the results are available for the copper(II) complexes, and this section summarizes the results obtained for oligopeptides containing aliphatic or aromatic side chains.

4.1   Stability and structure of complexes
The complexes of peptides consisting of aliphatic or aromatic amino acids (α- and β-alanine, valine, leucine, *iso*-leucine, tryptophane, and phenylalanine) will be

discussed in this section. The guanidyl residue of arginine can also be treated as a non-coordinating side chain, because significant interactions between metal ions and guanidino nitrogens have not been reported.

These amino acids can appear in peptides of type Gly−X, X−Gly, and X−X (or X−Y) which allows conclusions on the role of the location of a side chain. Stability constants for selected copper(II) complexes are listed in Table 4.3. From a literature

Table 4.3 — Equilibrium data for copper(II) complexes of dipeptides ($T = 298K$; $I = 0.1$ mol.dm^{-3})

Dipeptide	$pK_2(-NH_3^+)$	$\log\beta_{110}$	pK(amide)	$pK_2-\log\beta_{110}$	Ref.
GlyGly	8.15	5.55	3.99	2.60	75Si
GlyAla	8.25	5.79	4.04	2.46	75Si
GlyLeu	8.28	5.89	4.76	2.39	75Si
GlyIleu	8.26	5.83	4.71	2.43	75Si
AlaGly	8.17	5.26	3.64	2.91	75Si
LeuGly	8.10	4.75	3.26	3.35	75Si
IleuGly	8.07	4.75	3.26	3.32	75Si
GlyPhe	8.08	5.59	3.86	2.49	86Ki[†]
PheGly	7.46	4.93	3.67	2. 53	86Ki[†]
AlaAla	8.17	5.54	3.72	2.63	74Na
LeuLeu	7.91	5.21	3.88	2.70	74Na
Gly-β-Ala	8.16	5.74	4.67	2.42	81Si
β-Ala-Gly	9.58	6.15	4.75	3.43	81Si
β-Ala-β-Ala	9.54	5.76	6.82[‡]	3.78	81Si

All amino acids of L-configuration.
[†] $I = 0.2$ mol.dm^{-3} (KCl)
[‡] hydrolysis

survey and Table 4.3 it can be concluded that the complex-formation processes of aliphatic dipeptides are very similar to those of GlyGly. The formation of [CuA]$^+$ is characteristic at around pH 3 to 4; in this, coordination occurs via the terminal amino group and neighbouring carbonyl oxygen atom. This is followed by amide deprotonation and coordination, which results in the formation of a species [CuAH$_{-1}$] having the same structure as that shown in Scheme V.

The first part of Table 4.3 gives data on peptides of types Gly−X and X−Gly. It can be seen that alkyl or aryl substitution in C-terminal amino acids does not significantly influence the protonation constants of the amino group (pK_2), while for the case of peptides X−Gly pK_2 follows the trend observed for the free amino acids. The stability of the complexes [CuA]$_+$ can be assessed via the pK_2-$\log\beta_{110}$ values. There is a small increase in stability for the peptides Gly−X, and a measurable decrease for X−Gly (X=Leu, Ileu, etc.). The relative stabilities of the copper(II) complexes of the peptides X−X are therefore very close to that of GlyGly. The pK values for the process [CuA]$^+ \rightleftharpoons$ [CuAH$_{-1}$]$+$[H$^+$] provide information on the stability of the species [CuAH$_{-1}$]. Naturally, the overall stability constants include the pK values of the amino and amide groups, the latter not being known for dipeptides. It can be seen from Table 4.3 that the presence of an alkyl side chain promotes amide deprotonation in the peptides Gly$-$X and hinders this process in the peptides X$-$Gly. Complex formation processes of aliphatic dipeptides have been studied with various other metal ions [60Mal, 75Brl, 82Kil, 83Ki, 82Ra2]. The same effect of the side chains was observed in nickel(II)-complexes [75Brl], while coordination via oxygen donors was proposed for the complexation of calcium(II) [82Ra3, 83Pr].

Another striking difference between the complexes of peptides containing N- or C-terminal side chains is the formation of [Cu(AH$_{-1}$)$_2$]$^{2-}$ which is observed only for X$-$Gly peptides. This seems to be due to the steric hindrance caused by the side chain. It has been supported by potentiometric [89Fa1] and EPR measurements [89Sz].

There is direct evidence for a metal ion–aromatic ring interaction in the case of dipeptides containing aryl side chains. The data for PheGly and GlyPhe in Table 4.3 show that a phenylalanine residue does not decrease the relative stability of [CuA]$^+$ complexes and to a small extent it promotes ionization of the amide groups. In the case of the square–planar diamagnetic palladium(II) or nickel(II) for (tripeptides) complexes there is a chance for the detection of metal–aromatic ring interaction via the measurement of rotamer populations by NMR spectroscopy [77Ko4, 80Ve]. The data reveal that the mole percentage of the rotamer that directs the aromatic side chains toward the metal ion rises to above 60% from about 20% in the unbound ligand. The metal ion–aromatic ring interaction also affects the stability of mixed ligand complexes in which the B-ligands possess hydrophobic side chains [82Sh, 84Ki2]. Copper(II)–aromatic ring interaction was proposed for tryptophan-containing peptides on the basis of EPR measurements [77Sp].

Stability constants for the copper(II) complexes of β-alanine containing dipeptides appear in the last three lines of Table 4.3. It is well known that β-alanine can form a six-membered chelate ring, thus the stabilities of β-alanine complexes are lower than those of complexes of α-alanine or other amino acids. On the other hand, the results obtained for β-alanine-containing peptides make possible to conclude on the role of the size of the chelate rings in peptides.

The complex [CuA]$^+$ of Gly$-$β$-$Ala has almost the same stability as that of GlyGly, suggesting the same type of coordination. In contrast, there is a characteristic stability decrease for the peptides β$-$Ala$-$X, because of the six-membered chelate ring with the participation of amino and carbonyl groups. The pK of amide deprotonation varies in the sequence GlyGly<Gly$-$β$-$Ala<β$-$AlaGly<β$-$Ala$-$β$-$Ala, which is in correlation with the sizes of the fused chelate rings: (5,5)>(5,6)> (6,5)>(6,6). The pK values for β$-$Ala$-$β$-$Ala probably correspond to hydrolysis of the species [CuA]$^+$. This means that the formation of two fused 6-membered chelate rings with the participation of amide-N donors is especially unfavoured [81Si].

The effects of side chains have also been studied in the complexes of tri- and tetrapeptides. The formation of [MAH$_{-2}$]$^-$ was characteristic in copper(II) and nickel(II) complexes of leucine-containing tripeptides [78Bi]. The stability constants of the corresponding copper(II) complexes are given in Table 4.4. The data in Table 4.4 reveal that substituents on the α-carbon atom decrease the stability of the chelate

Table 4.4 — Stability constants of the copper(II) complexes of leucine-containing tripeptides. ($T = 298K$; $I = 0.1$ mol.dm^{-3} NaClO$_4$)

Ligand	logβ_{110}	pK$_1$[†]	pK$_2$[‡]	Ref.
GlyGlyGly	5.12	5.11	6.68	71Ha
	5.24	5.22	6.60	72Si
LeuGlyGly	4.10	4.17	6.72	78Bi
GlyLeuGly	5.37	5.87	6.43	78Bi
GlyGlyLeu	5.11	5.03	7.20	78Bi

[†]pK$_1$ for first amide deprotonation.
[‡]pK$_2$ for second amide deprotonation.

ring in which it is located (see log β_{110} for LeuGlyGly). In contrast with this destabilizing effect, a C-substituent promotes the deprotonation of a peptide group located in the same chelate ring (see pK$_1$ for LeuGlyGly, or pK$_2$ for GlyLeuGly).

In the course of the study of copper(III) complexes [83Di] stability constants of the copper(II) complexes of some di-, tri-, and tetrapeptides containing α-amino-isobutyric acid have been determined [83Ha2].

In this case the relative stability of the peptide complexes is influenced by the combined effect of the enhanced stability owing to the inductive effects and destabilization due to the steric hindrance caused by the bulky side chain in α-amino-isobutyric acid. The complexes of glycine-containing tri- or tetrapeptides of α-aminoisobutyric acid are generally slightly more stable than the complexes of oligoglycines, but coordination of the third peptide-N is hindered in the copper(II) complex of the tetrapeptide consisting only of α-aminoisobutyric acid residues.

4.2 Stereoselectivity in the metal complexes of peptides

Stereoselectivity in the amino acid complexes requires the binding of three donor groups. In the non-glycyl aliphatic dipeptides, however, there are two optically active centres, which results in the existence of diastereomers; the stereoselectivity observed in their complexes was first reviewed by Pettit & Hefford [79Pe].

A dipeptide obtained from optically pure amino acids can appear in the forms of four isomers. The LL and DD isomers are optically active or 'pure' isomers, whereas the LD and DL isomers are meso or 'mixed' isomers. The complexes of enantiomeric dipeptides have the same stability, but the diastereomers can form complexes with different stability. This is illustrated in Table 4.5, where the pK values and stability constants of the copper(II) complexes of several dipeptides are listed. It can be seen from Table 4.5 (and from other literature data [79Pe] that stereoselectivity can be observed in the complexes of AlaAla which has only methyl groups in the side chains. The terminal ammonium is less acidic and the carboxylic group is more acidic in the LD-dipeptides than in the pure LL-isomers. In the copper(II) complexes this results in the enhanced stability of the species [CuAH$_{-1}$], which is reflected in the lower pK of amide deprotonation in the pure forms than in the LD or DL-dipeptides. Table 4.5 also reveals that the stereoselectivity is the higher the bulkier the side chains of the peptides. Consequently, the changes in the stabilities of the complexes of the

Table 4.5 — pK values and stability constants of the copper(II) complexes of various diastereomers of dipeptides

Dipeptide	pK(COOH)	pK(NH$_3^+$)	logβ_{110}	pK(amide)	Ref.
					84Im
L-Ala-L-Ala	3.31	8.19	5.54	3,72	86Bo
					84Im
L-Ala-D-Ala	3.19	8.34	5.71	3.96	86Bo
L-Leu-L-Leu	3.45	7.91	—	3.88	74Na
L-Leu-D-Leu	3.05	8.20	—	4.88	74Na
L-Ala-L-Phe	3.25	7.87	5.20	3.44	74Na
L-Ala-D-Phe	3.02	8.08	5.42	3.93	74Na
L-Met-L-Met	3.23	7.39	5.07	3.37	87Bo
D-Met-L-Met	2.91	7.54	5.12	3.69	87Bo

diastereomers are related to the intramolecular interactions of the hydrophobic side chains of dipeptides. This 'noncovalent bonding' is possible only for LL-diastereomers in which the substituents on the α-carbon atoms are on the same side of the coordination plane. On the basis of thermodynamic data, Bonomo *et al.* [86Bo] concluded that the complexes of LL-diastereomers are enthalpically favoured. Only a few data are available on complexation with other metal ions, but the same trends can be observed for the nickel(II) and zinc(II) complexes [79Pe].

Data reveal that the stereoselectivity in dipeptides does not depend on the presence of coordinating side chains, but it is due to the noncovalent hydrophobic and electrostatic interactions between the substituents [88Ia]. The role of the hydrophobic interactions between the side chains is strengthened by the data obtained for tripeptide complexes containing two leucyl residues. The copper(II) complexes of LeuLeuGly of the same chirality were more stable than those of different chirality, but the stereoselectivity was not observed when glycine was in intermediate position (LeuGlyLeu) [88Ha]. Only a few data are available on the stereoselectivity of peptides with donor atoms in lateral positions. Pettit *et al.* [87Li] studied the copper(II) complexes of diastereomers of ProHis and HisHis, and the stereoselectivity of the ligands and copper(II) complexes was explained by the preferred trans conformation of the peptide chains. On the other hand, a significant stereoselectivity was not observed in the pK values and copper(II) complexes of diastereomeric tripeptides (ArgLysAsp) containing mainly hydrophylic side chains [89So, 89No].

4.3 L-proline as a 'breakpoint' in the metal ion–peptide systems

L-proline is a naturally occurring amino acid which has no additional donor group in the side chain. Thus, the complex-formation processes of oligopeptides containing a prolyl residue in the N-terminal position are very similar to those of glycyl peptides [75Si, 87Li]. However, proline has a secondary nitrogen atom which cannot be protonated when incorporated in a peptide bond. Therefore the amide–nitrogen is not a metal-binding site, and it works as a 'breakpoint' during complex formation when proline is in an intermediate or C-terminal position in the peptide chain. The same phenomenon was described for the peptides of sarcosine (*N*-methylglycine) where complexes $[CuAH_{-1}]$ were formed with sarcosylglycine, but not with glycyl-sarcosine [60Ko].

On the other hand, proline is an important amino acid residue in many neuropeptides and hormones, and various proline-containing oligopeptides were recently studied by Kozlowski *et al.*

Melanostatin (L-prolyl-L-leucylglycineamide) is a hypothalamic hormone and a therapeutic agent for Parkinson's disease. Spectroscopic and potentiometric studies [81Ko1] revealed the successive deprotonation and coordination of the amide groups and the formation of $[CuAH_{-3}]^{2-}$ as the final species via the coordination of 4N-donor atoms. Nickel(II) complexes are formed in very slow reactions, because of the cooperative deprotonation of the amide groups, but $[NiAH_{-3}]^{2-}$ can readily be characterized as the usual square–planar diamagnetic 4N-species. The same type of coordination has been reported for the tetrapeptide, ProAlaAlaAla [81Ko2]. Transition metal complexes of L-prolyl-L-leucylglycinehydroxamic acid (an effective inhibitor of human collagenase) have been recently studied [90Fa]. In the copper(II) systems complex formation starts via the hydroxamate residue, but the usual tripeptide-like coordination is characteristic of both copper(II) and nickel(II) complexes in slightly basic solution. These results prove that N-terminal prolyl residues (in di-, tri-, and tetrapeptides alike) can be directly coordinated to metal ions.

Spectroscopic and thermodynamic data on copper(II) complexes with tetrapeptides containing prolyl residues in different positions of a peptide sequence suggest the very critical role of the proline location. The main reason for this specific behaviour stems from the fact that proline breaks the metal ion coordination to the subsequent amide-N atoms, but it does not necessarily stop further metal-peptide binding [84Ba, 84Be]. It means that the insertion of a prolyl residue into the peptide sequence divides the ligand molecule into two fragments which are potentially able to interact with the same metal ion to form 'large chelate rings' or 'loop structures' [85Pe2, 85Pe3, 88Li]. The latter case is especially favoured when other donor groups are also present in the side chain [85Pe3, 83Ko1]. For example, large chelate rings can be present in the copper(II) complexes of proline-containing tetrapeptides [85Pe2]. Stepwise protonation constants are given in Table 4.6, which shows that the pK values of amide deprotonation of tetraglycine and ProGlyGlyGly are very similar, suggesting the same type of coordination. It can also be seen from Table 4.6 that amide deprotonation occurs in the other Pro-containing peptides, too, but the structures of the species are totally different from those of tetraglycine. The binding modes of the species $[CuAH_{-1}]$ and $[CuAH_{-2}]^-$ of GlyProGlyGly are shown in Scheme X, which shows that an unusually large (11-membered) chelate ring is present in the complex $[CuAH_{-1}]$ of GlyProGlyGly, and thus deprotonation of the amide group is hindered relative to tetraglycine. The coordination of the C-terminal amide group now makes it possible to bind another amide nitrogen which results in the formation of a species $[CuAH_{-2}]^-$ which contains an 8-membered chelate ring (Xb). The hindrance of amide binding elevates the concentration of the species $[CuA]^+$ and leads to the formation of the bis-complexes $[CuA_2]$, too. In the case of

Table 4.6 — pK values for the amide deprotonation in the copper(II) complexes of proline-containing peptides [85Pe2]

Ligand	pK_1	pK_2	pK_3
Triglycine [72Si]	5.22	6.60	11.9[†]
Tetraglycine [72Si]	5.50	6.89	9.29
ProGlyGlyGly	5.42	6.95	9.32
GlyProGlyGly	7.92	9.40	9.48[†]
GlyGlyProGly	5.01	9.40[†]	10.01[†]
GlyGlyGlyPro	5.64	7.02	10.08[†]

[†]formation of hydroxo complexes.

GlyGlyProGly diglycine-like, while for GlyGlyGlyPro triglycine-like coordination is suggested. Analogous observations have been made on sarcosyl (X−Sar−X−X, etc.) and alanyl peptides (AlaProAlaAla, etc.) [87Pe].

(a)   X.   (b)

Copper(II) complexes of peptides containing more than one proline residue have also been studied, and $[CuA]^+$ and $[CuAH_{-1}]$ were found to be the main species in equimolar solutions of copper(II) and ProProGly. The pK for amide deprotonation is around 7.7; the formation of an 8-membered chelate ring was proposed with participation of the secondary amine of the N-terminal proline and the deprotonated amide-N of the C-terminal glycine [87Pe]. The bis complex is also an important species in the presence of ligand excess. Potentiometric and spectroscopic studies have been demonstrated that increase of the peptide chain length to $(ProProGly)_{2 \, or \, 3}$ does not significantly influence the binding mode suggested for ProProGly.

The formation of large chelate rings or loop structures is especially favoured when other donor groups are also present in the peptide chain. For example, the existence of a 15-membered loop was presumed in the copper(II) complex of PheGlyProTyr, via the coordination of amino-N, subsequent amide-N and phenolate-O of tyrosine [88Li]. Similar results were obtained for the copper(II) complexes of GlyGlyProLys, where the ε-amino group of the lysyl residue is also involved in the coordination [85Ba]. This peptide is a model for the octapeptide canine tuftsinyl–tuftsin (ThrLysProLysThrLysProLys), in which the prolyl residues likewise leads to very unusual binding properties [83Ko4]. The formation of a 3N-species was described for the octapeptide, with the participation of N-terminal NH_2 and amide-N) and C-terminal (amide-N and COO^-) donors of the ligand. This means that the complex contains a loop consisting of five unbound amino acid residues.

Proline residues occur widely among the biologically important or natural polypeptides. Arginine[8]-vasopressin (AVP) and arginine[8]-vasotocin are neuro-physical hormones possessing a 20-membered ring joined by a disulphide bridge and having a tripeptide side chain: X-ProArgGlyNH$_2$. Copper(II) and nickel(II) complexes of five derivatives of AVP and AVT have been studied by potentiometric and spectroscopic methods [89Ko]. The data confirm the dominant species are the 4N-complexes. The marked stabilities of the complexes are due to the conformation of the binding sites within the ring formed by the disulphide bridge.

Copper(II) complexes of several pentapeptides containing two prolyl residues (and modelling substance P, which is a peptide of eleven amino acids) have recently been studied [89Ba1]. The formation of large chelate rings and a bent conformation was again suggested, with participation of the terminal amino-N, peptide-N, and ε-amino group of lysine.

5. COMPLEXES OF PEPTIDES CONTAINING OXYGEN OR NITROGEN DONORS IN THE SIDE CHAIN

The complex-forming capabilities of peptides containing alcoholic-OH groups (serine and threonine peptides), phenolic-OH groups (tyrosine peptides) carboxylate groups (peptides of aspartic and glutamic acids), acid amide groups (peptides of asparagine and glutamine), or amino groups (lysine peptides) will be discussed in this section. Histidine also contains an N-donor in the side chain, but with regard to the considerable biological importance of the imidazole moiety its peptides are dealt with in the following section.

5.1 Peptides of serine and threonine

The alcoholic-OH groups of the amino acids serine and threonine do not generally bind directly to metal ions, but to some extent they enhance the stability of the corresponding transition metal complexes. Thus, the alcoholic-OH groups can be classified as a weakly coordinating side chain, and the complex-formation processes of serine or threonine-containing peptides are quite similar to those of oligoglycines or aliphatic dipeptides. The stability constants of the copper(II) complexes of some seryl or theonyl peptides are collected in Table 4.7, from which it can be seen that there is a stability increase for the complexes $[CuA]^+$ of the dipeptides containing an alcoholic-OH group in the N-terminal side chain. This contradicts the results obtained for aliphatic dipeptides, and supports an axial interaction between the metal ion and the OH-group. This interaction promotes amide binding, too, which is reflected in the low pK value of amide deprotonation. Other potentiometric and spectroscopic studies led to the same conclusion, and the role of alcoholic-OH groups was detected at high pH values, too [82Ge, 86Fa, 89Fa1]. The formation of $[Cu(AH_{-1})_2]^{2-}$ bis-complexes (coordinated bidentately via amino and amide groups) was found to be favoured in the case of SerGly and ThrGly, while these

Table 4.7 — Stability conbstants of the copper(II) complexes of some serine and threonine-containing peptides ($T = 298K$, $I = 0.1$ mol.dm^{-3} NaClO$_4$)

Peptide	$pK_2(NH_3^+)$	pK_1 (COOH)	$\log\beta_{110}$	$\log\beta_{11-1}$	pK_2-log β_{110}	pK(amid)	Ref.
GlyGly	8.15	3.11	5.55	1.56	2.60	3.99	75Si
Gly-L-Ser	8.14	2.99	5.66	1,89	2.48	3.77	77Si
Gly-L-Thr	8.14	3.00	5.57	1.43	2.57	4.14	77Si
L-SerGly	7.33	3.21	4.96	1.36	2.37	3.60	77Si
L-ThrGly	7.34	3.14	5.06	1.51	2.28	3.55	77Si

species were not formed with the C-terminal peptides. On the basis of spectral studies and the low pK values of second amide protonation (10.40 and 10.41 for SerGly and ThrGly, respectively, as compared with 11.60 for GlyGly), the enhanced stability of the bis-complexes were interpreted as due to hydrogen-bond formation between the noncoordinated carboxylate and alcoholic-OH groups [89Fa1].

On the other hand, a direct interaction between copper(II) and the deprotonated alcoholate group was proposed above pH ~ 12 in the case of GlySer and GlyThr [89Fa1]. The spectral parameters of the copper(II) and palladium(II) complexes of AlaSer were discussed in the same way, and the populations of the various rotamers were obtained by NMR spectroscopy both in the free ligand and palladium(II) complexes [78Ko4]. The high ratio of gauche rotamer wass explained by the hydrogen bonding in the free ligand and by the metal ion–alcoholic-OH interaction in the palladium(II) complex. Involvement of the deprotonated alcoholate of GlySer has been proposed in various mixed complexes of copper(II), too [77Sp, 82Ge, 83Sh].

5.2   Tyrosine-containing peptides

The phenolic-OH group is generally not a binding site in the transition metal complexes of L-tyrosine. This is probably due to the steric requirements; for in other compounds the deprotonated phenolate group readily interacts with metal ions as demonstrated by the complexes of *ortho*-tyrosine or catecholates [79Gel, 84Ki1].

Accordingly, phenolate cannot be classified as a 'weakly coordinating' side chain, but its involvement in the coordination in peptides will largely depend on the location of the tyrosyl residue. Some conclusions can be drawn from the stability constants of copper(II) complexes listed in Table 4.8.

For the tyrosine peptides, pK_2 cannot be unambiguously described by deprotonation of the ammonium group, because it is slightly overlapped by the deprotonation of phenolic-OH, and microconstants should be used to interpret the process [82Ki2]. From Table 4.8 it can be concluded that in neutral solutions the complex-formation processes of Tyr peptides are very similar to those of Phe: a stability increase can be observed in the [CuAH$_{-1}$] complexes, due to the interaction between the aromatic ring and the d-orbitals of the metal ion. This is also characteristic for the nickel(II) and palladium(II) complexes, as proved by NMR, EPR, and crystal structural studies [70Fr2, 77Ko3, 78Ko1, 78Ko2, 80Ve]. Significant differences have been

Table 4.8 — Stability constants of copper(II) complexes of dipeptides containing phenylalanine or tyrosine. ($I = 0.2$ mol.dm^{-3}(KCl); $T = 298$ K)

Dipeptide	pK_1	pK_2	pK_3	$\log\beta_{110}$	pK_2-$\log\beta_{110}$	pK(amid)	$\log K_{\frac{CuAH_{-1}}{CuA_2H_{-1}}}$	Ref.
GlyAla	3.17	8.20	—	5.76	2.44	4.22	3.08	77Ge
GlyPhe	2.99	8.08	—	5.59	2.49	3.86	3.14	86Ki
GlyTyr	3.03	8.10	9.96	5.66	2.44	3.96	2.96	86Ki
AlaGly	3.15	8.19	—	5.25	2.74	3.91	2.60	77Ge
PheGly	3.12	7.46	—	4.93	2.53	3.67	2.78	86Ki
TyrGly	3.13	7.54	9.86	4.86	2.68	3.57	2.75	86Ki

observed by some authors, however, between the complexes of GlyTyr and TyrGly, in slightly basic solutions [81He, 83Bo, 86Ki]. A dimeric species [(CuAH$_{-1}$)$_2$] was detected in the complexes of peptides containing an N-terminal tyrosyl residue (TyrGly or TyrLeu), the (N,N,O)-coordinated monomer units being linked via the other phenolate groups of the dimeric complex. These findings have been confirmed by the detailed equilibrium and structural studies of Yamauchi *et al.* [85Ya]. No dimeric species were found, however, when tyrosine was present in the C-terminal position. The important role of N-terminal tyrosine was confirmed by recent results for other tyrosyl peptides containing phenylalanine or lysine as second amino acid [90Ra]. The dimerization constant of TyrPhe was reported to be higher than that of TyrGly (2.41 and 2.12, respectively), because of the stacking interaction between the aromatic rings.

The coordination sphere of the metal ions is saturated in the fully deprotonated complexes of tripeptides ([MAH$_{-2}$]$^-$), and there is no chance for the formation of phenolate-bridged polynuclear complexes. Thus, the results for tyrosine-containing tripeptides demonstrate a simple triglycine-like coordination. The existence of a metal ion–phenolate interaction was excluded, regardless of the location of the tyrosyl residue [87Ki]. The copper(II) complex of GlyGlyTyr$-N-$methylamide (a model for dog serum albumin) was studied by Sarkar *et al.* [84Mü]. The formation of [CuAH$_{-3}$]$^{2-}$ and [Cu(AH$_{-1}$)$_2$]$^{2-}$ were detected in highly basic solution, with the participation of 4N-coordination. The absence of any charge-transfer band around 400 nm (characteristic of copper(II)-phenolate bond) strongly indicates that copper(II) does not bind to the phenolate group. This is in accord with the low copper(II) binding affinity of dog serum albumin, as compared with that of human serum albumin which has GlyGlyHis as the N-terminal amino acid sequence.

The results on the complexation between cobalt(II) or zinc(II) and GlyTyr revealed the formation of bis complexes via coordination of the N-terminal amino and carbonyl groups. No evidence could be found for direct metal ion-promoted phenolic-OH dissociation, but cobalt(II)-promoted amide deprotonation was reported to occur in basic solution [79Ap].

The tyrosyl residue can play an important role in metal binding if other donor groups are also present in the oligopeptides. This is illustrated by the Pro- and Tyr-containing peptides, where the presence of the prolyl residue makes the copper(II)-phenolate interaction possible [83Ko1, 85Pe2, 88Li] (for details see previous section). Copper(II)-phenolate interactions have been reported to occur during complexation with poly-(L−Lys, L−Tyr) and poly-(L−Glu, L−Tyr) [78To, 79Pa].

N-terminal tyrosyl residues are present in the methionine and leucine enkephalins (TyrGlyGlyPheMet and TyrGlyGlyPheLeu). Because of the saturation of the coordination sphere, however, the copper(II) complexes of these ligands are very similar to those of oligoglycines, and the occurrence of copper(II)-phenolate or copper(II)-thioether interactions has been excluded [85Ko2].

5.3 Peptides with carboxylate or acid amide groups in the side chain

There are two carboxylate groups in aspartic and glutamic acids in α-, β-, and α, γ-positions, respectively. In the transition metal complexes of these amino acids the carboxylate groups enhance the stability of the complexes; this is especially so for aspartic acid, where 5- and 6-membered chelate rings can be formed with the participation of amino and carboxylate groups. Similar observations have been made for asparagine and glutamine, where the metal ion-promoted amide deprotonation of asparagine is also possible in basic solution. (For details, see the previous chapter).

The copper(II), nickel(II), and zinc(II) complexes of GlyAsp, GlyGly, and GlyAsn were studied by Gergely *et al.* [80Fa, 82Ge]. The existence of similar species as for GlyGly was detected in neutral or slightly basic solution, but the equilibrium data showed the enhanced stability of the complexes. The following stability sequence was established for the complexation with copper(II) in [CuA] complexes:

$$GlyAla < GlyAsn < GlyGlu < GlyAsp \ .$$

This sequence suggests that the most significant interaction can be achieved with the β-carboxylate groups of the aspartyl residue. At the same time, this interaction hinders the amide binding which is reflected in the pK of amide deprotonation (p$K = 4.82$ for GlyAsp [82Ge]). The copper(II) complexes of all dipeptides undergo hydrolysis in basic solution, but in the case of GlyAsn a new base-consuming process was observed. This was identified as deprotonation of the amide group of asparagine, which leads to coordination of the deprotonated amide-N instead of carboxylate-O in the equatorial position. The same conclusion was reached by Martin *et al.* [69Ts], but amide deprotonation of the asparagine residue during complexation with nickel(II) or zinc(II) was ruled out [80Fa]. The formation of $[Cu(AH_{-1})_2]^{2-}$ was also excluded, similarly to the copper(II) complexes of other Gly−X peptides [89Fal], but a 4N-species was presumed to exist during complexation with palladium(II) [76Ko].

The important role of C-terminal and intermediate aspartyl residues was demonstrated for some oligopeptides of biological significance. Thymopoietin is a 49 amino acid polypeptide hormone of the thymus. The synthetic segments corresponding to positions 32–34 to 32–36 in the amino acid sequence contain aspartic acid and are biologically active (TP3=ArgLysAsp, TP4=ArgLysAspVal, and TP5= ArgLysAspValTyr). The lanthanride(III) complexes of TP5 were first studied, and NMR evidence was found for Ln(III)-carboxylate interactions [81Va]. The formation of 1:1 complexes between zinc(II) and TP4 or TP5 was proved by polarographic studies [85Ve]. Copper(II) and nickel(II) complexes of TP3, TP4, and TP5 were identified by potentiometric and spectroscopic studies [84So, 90So]. Tripeptide-like coordination was reported for all the tri-tetra-, and pentapeptides, via the coordination of terminal amino, and two subsequent amide groups and the β-carboxylate of the aspartyl residue. It means that the β-carboxylate group of aspartic acid can block the deprotonation of the subsequent amide group in tetra- or pentapeptides. This observation was strengthened by the data on the copper(II)–AlaAlaAspAla system, where formation of the same (N,N,N,O)-coordinated species was found [86De2].

AlaAspSerGly is an N-terminal tetrapeptide segment of fibrinopeptide A, which is a polypeptide of 16 amino acid with outstanding biological significance. Its complexation with copper(II) and nickel(II) again revealed that the β-carboxylate of the aspartyl residue increases the stability of the species [CuAH$_{-1}$], but it significantly hinders the deprotonation of the second and third peptide groups [86De1].

AspArgValTyr is a tetrapeptide segment of angiotension II, containing N-terminal aspartic acid residue. The copper(II) and nickel(II) complexes of the tetrapeptide were studied by Kozlowski *et al.* [88De, 89Pe]. Tetraalanine-like coordination was proposed, but with a different concentration distribution, because of the enhanced stability of the species [CuA] with β-alanine-like coordination. The coordination of the β-carboxylate also makes bis complex ([CuA$_2$]$^{2-}$) formation possible, which results in an increase in the pK of amide deprotonation. The same conclusion was reached in connection with the copper(II) and nickel(II) complexes of AspPheNH$_2$ [88Va1].

AspAlaHis−N−methylamide (a model for human serum albumin) is another example of the N-terminal aspartyl peptides. During complexation with copper(II) the formation of a 4N-species was proposed (similarly as for GlyGlyHis) without involvement of the aspartic acid residue [81Ra]. The free β-carboxylate group of the species [CuAH$_{-2}$]$^-$, however, can bind another metal ion (cadmium(II) or zinc(II)) which results in the formation of mixed metal complexes [83Da].

5.4 Lysine-containing peptides

The primary amino group is an effective binding site for transition metal ions. The steric requirements generally exclude the direct interaction of the ε-amino group in the metal complexes of lysine. However, the side-chain amino nitrogen can serve as a site for polynuclear complex formation. This type of interaction is possible in the lysyl-containing peptides, too, although only a few data are available on complexation with simple lysine-containing peptides.

Copper(II) complexes of LysTyr and TyrLys are the only examples of complexation of lysyl dipeptides containing N- and C-terminal lysine, respectively [90Ra]. The results of these interactions cannot be generalized, however, because the phenolate group of the tyrosyl residue may bind metal ions, especially in the N-terminal position [86Ki]. Consequently, the formation of binuclear species has been suggested for both peptides. The dimerization constants of the copper(II) complexes of LysTyr and TyrLys were higher than those of other Tyr-containing dipeptides, suggesting that both the phenolate and amino groups can act as bridging ligands. In the case of TyrLys it is the phenolate, while in the case of LysTyr it is the ε-amino group which is more involved in the coordination. This is supported by the intensity of the Cu(II)−O charge-transfer band and by the EPR results. The higher affinity of the N-terminal lysyl residue to take part in coordination is in accord with the results of Rainer & Rode [85Ra2], who found that the C-terminal lysyl residue plays only a minor role in the stabilization of copper complexes of GlyLys, ValLys, and HisLys.

The solid state structure of the 1:1 complex of copper(II)–LysTyr was determined by X-ray studies, and a polymeric structure was found in which the metal ions are coordinated in a similar way to simple dipeptides, and the ε-amino groups act as bridging ligands [89Ra1].

There are examples of complexes of lysine-containing oligopeptides in which the ε-amino group does not take part in the coordination. In these cases, the stable amide binding or the saturation of the coordination sphere rules out interaction with the ε-amino group of lysine. GlyHisLys (growth-stimulating factor) forms 1:1 complexes with copper(II) that involves the same type of coordination as for GlyHis [83Ma2, 84Pe, 85Ra2]. Interactions between copper(II) or nickel(II) and the ε-amino group of the lysine residue were also ruled out in the oligopeptide segments of thymopoietin [84So, 90So] and for the tetrapeptide ThrLysAlaAla [77Ko5].

6. COMPLEXES OF HISTIDINE-CONTAINING PEPTIDES

The imidazole–N donor atom is one of the major binding sites in metalloenzymes. The results on the transition metal complexes of histidine show that the imidazole–N can compete with the amino acid binding sites, and it significantly enhances the stabilities of the corresponding complexes. Consequently, imidazolyl side chains play an important role during complex formation with oligopeptides. The effect, however, depends greatly on the location of the histidyl residue. There are many publications which deal with the complexes of histidine-containing peptides, and some of them have already been reviewed by Sigel & Martin [82Si]. In this section we focus on the most recent results obtained with histidine peptides and draw some general conclusions about the role of imidazolyl moieties.

6.1 The dipeptides His-X and X-His

The literature data demonstrate that complex formation processes of N- and C-terminal histidyl-containing peptides are completely different. Scheme XI gives a general picture of the complexation of copper(II), while the binding modes of the various species present in solution are depicted in Scheme XII [82So].

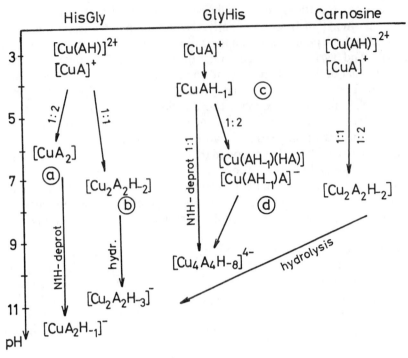

XI.

(a)   (b)

(c)   (d)

XII.

On the basis of Scheme XI, it can be stated that the complex-formation processes of HisGly depend on the metal ion to ligand ratio. In the N-terminal histidyl peptides the amino–N and imidazole–N are in suitable positions to form a stable 6-membered chelate ring. In the presence of excess ligand amide deprotonation can be excluded and stable bis-complexes $[CuA_2]$ with histamine-like coordination can be formed (XIIa). In equimolar solutions the coordination sphere of copper(II) is unsaturated, however, which makes amide deprotonation and coordination possible. Consequently, (NH_2, N^-, O^-)-coordination (as for GlyGly) is characteristic for the species above $pH \sim 6$. The fourth binding site of copper(II) is occupied by the imidazolyl side chain of another $[Cu(AH_{-1})]$ unit which results in the formation of a dimeric species (XIIb). The formation of a deprotonated complex can be interpreted via hydroxo bridges too [75Ag], but other potentiometric [74Ail, 75Br2, 88Da] and spectroscopic studies [75Bo] definitely prove that amide deprotonation takes place. In the copper(II)–HisVal system, glycine-like coordination was proposed on the basis of EPR measurements [84Hu2], but the same bis complex and binuclear complex formation were reported for HisPhe, HisTyr, HisMet, HisAla, HisVal, and HisLys [82En, 85Ra2, 87Ra, 87So]. On the other hand, the existence of a monomer-–dimer equilibrium was reported for the copper(II) complexes and His-X having a bulky non-coordinating side chain on the amino acid in C-terminal position [87Ra].

For GlyHis, steric requirements rule out the possibility of histamine-like coordination. However, there is a chance for the formation of a 3N-species with the involvement of terminal-amino deprotonated amide, and imidazole–N3 in the species $[CuAH_{-1}]$ (XIIc). This type of binding was first proposed by Martin & Edsall [60Ma2] and proved by many other potentiometric and spectroscopic studies [74Ai2, 75Ag, 75Br2, 82So, 84Hu2, 82En]. The same type of coordination was reported for the solid complex from X-ray structural investigations [67Bl]. The species $[CuAH_{-1}]$ has one free coordination site which makes the binding of another GlyHis possible. The second ligand in the species $[CuA_2]$ and $[CuA_2H_{-1}]^-$ is bonded monodentately via the terminal amino or imidazole-N3 moeties (XIId).

The same type of coordination is observed for other dipeptides containing C-terminal L-histidine [82En, 85Ra2, 87Li], except for carnosine (β-alanylhistidine), which is a naturally occurring dipeptide. In this case the 3N-coordination would result in the formation of two 6-membered chelate rings, which are not stable enough, as was discussed in section 3.1. Thus, L-carnosine readily forms a dimeric species (XIIb), in which coordination takes place via the terminal amino, deprotonated amide, and terminal carboxylate groups (as for β-alanylglycine), while the imidazole-N3 in the side chain acts as a bridging ligand. This type of coordination was unambiguously proved in the solid phase by crystal structure studies [65Fr2], but very contradictory results were reported for the solution equilibria between copper(II) and L-carnosine. Visible spectroscopic, calorimetric, NMR-relaxation [82So], and EPR [82Br] data, however, prove the predominance of the same dimeric species $[Cu_2A_2H_{-2}]$, in solution. At the same time, there is a possibility of the existence of monomeric complexes when carnosine is present in very large excess [80Br].

Another important feature of histidine-containing peptides arises from the presence of the pyrrolic–N1H group in imidazole. In the free ligand this is a very weak acid ($pK > 14$), and it generally does not take part in the coordination in the complexes of histidine [74Su]. The interaction of imidazole–N3 with metal ions, however, promotes the ionization of pyrrolic-N1H [78So]. This process also occurs in the copper(II) complexes of histidine-containing dipeptides, but the pK of N1H depends considerably on the binding mode of the complexes. The lowest value (p$K \sim 10.5$) was obtained for the copper(II)–GlyHis system. In this case, the deprotonation is accompanied by a significant blue shift of the d-d band, suggesting the coordination of the deprotonated pyrrolic-N atom. Morris & Martin [71Mo2] proposed that a tetrameric complex is formed ($[Cu_4A_4H_{-8}]^{4-}$) from the $[CuAH_{-1}]$ units via imidazole–N1 bridges. This type of coordination was not reported for the complexes of ProHis, however, probably because of the steric requirements [87Li]. A p$K > 11$ can be obtained for imidazole–N1H ionization in the bis complexes $[CuA_2]$ of HisGly, where the imidazole takes part in the chelate ring, while hydrolysis is characteristic for the dimeric species of HisGly and carnosine, where the imidazole is only a bridging ligand [82So].

Histidine-containing peptides easily form complexes with other transition metal ions. The composition of these complexes depends on the location of the imidazolyl side chain. Amide deprotonation and coordination do not take place at any metal ion to ligand ratios in the complexes of HisGly (and other His-X peptides), with cobalt(II), nickel(II), zinc(II), and cadmium(II). This is due to the stable histamine-like coordination, which leads to the formation of bis complexes $[MA_2]$ (XIIa) [83Fa]. On the other hand, cobalt(II), zinc(II), nickel(II), and palladium(II) form complexes $[MAH_{-1}]$ with GlyHis, with the same type of coordination as described for copper(II) (XIIc) [60Ma2, 71Mo2, 83Fa]. pK values of 6.03, 6.73, and 7.40 were reported for the nickel(II), zinc(II), and cobalt(II) promoted ionization of amide group in GlyHis [83Fa]. The peptides X-His are so far the only examples of such low pK values for complexation with zinc(II) or cobalt(II). NMR studies, however, definitely prove the occurrence of amide binding in the zinc(II) complexes of GlyHis and AlaHis [85Ra1]. A base-consuming process at around pH ~ 9 has also been described during complexation between cadmium(II) and GlyHis, but amide coordination was not proved beyond doubt [86So2].

L-carnosine is also a dipeptide of type X-His, with β-alanine in the N-terminal position. This excludes the complexes $[MAH_{-1}]$ with cobalt(II) and zinc(II), and amide deprotonation and coordination take place only in the copper(II) and nickel(II) complexes of L-carnosine [83Fa].

Cobalt complexes of histidyl peptides are especially interesting becasue of their ability to participate in reversible O_2-binding [69Mi, 71Mo1]. It was found that the cobalt(II) complexes of all three dipeptides (HisGly, GlyHis, and carnosine) are able to take up molecular oxygen which is subsequently released partially, or almost completely. For GlyHis, it is likely that $[Co(AH_{-1})]$ is the active species, while the presence of bis complexes is presumed to be necessary for oxygen uptake for HisGly and carnosine [83Fa].

6.2 Tripeptides of histidine

The amino acid sequences X-Y-His, X-His-Y, and His-X-Y indicate the possible modes of occurrence of histidine in tripeptides. The transition metal complexes of GlyGlyHis have been most widely studied, because this tripeptide serves as a model for human serum albumin.

It has been quite clearly proved by several authors and via various methods [74La, 77Ag, 80Sa, 81La2, 83De, 84Fa] that copper(II) and nickel(II) form complexes $[MAH_{-2}]^-$ with GlyGlyHis through the coordination of terminal amino, two deprotonated amide nitrogens, and the imidazole–N3 donors (Scheme XIII). The C-terminal carboxylate group does not take part in coordination. This is supported by the results of an X-ray study of the copper(II) complex of GlyGlyHis-N-methylamide, where the same type of binding was identified [76Ca]. Formation of the same chelate rings was proposed for copper(II) complexes of AspAlaHis-N-methylamide, which is another model for the metal-binding sites of human serum albumin [81Ra]. The nickel(II) and copper(II) complexes of GlyGlyHis readily undergo oxidative decarboxylation [76Me, 79Sa], and the end product has the same binding mode, as proved by X-ray structural studies [78Ma].

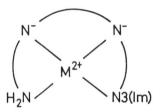

XIII.

In the presence of a ligand excess the existence of bis complexes in solution has been proposed by several authors [77Ag, 81La2, 83De]. This finding was not strengthened by other studies, however, where the exclusive formation of $[MAH_{-2}]^-$ was suggested [74Ai1, 84Fa]. pK values for successive amide deprotonation cannot be obtained (the concentration of $[CuAH_{-1}]$ is negligible), suggesting the cooperative deprotonation of two peptide groups.

A cobalt(III) complex of GlyGlyHis has been prepared, and the same donor atoms were suggested in the equatorial plane, with two monofunctional ligands in the axial positions. The inertness of the cobalt(III) complex made it possible to determine the pK values of the uncoordinated carboxylate and N1H-groups; values of 4.06 and 9.81, respectively, were obtained [83Ha1]. GlyGlyHis complexes of other metal ions have not been widely studied. Zinc(II) was reported to promote amide ionization, but polynuclear complexes are also formed [77Ag].

The results obtained for the copper(II) complexes of GlyHisGly show the same binding mode as reported for GlyHis (XIIc). Namely, $[CuAH_{-1}]$ is the main species, with 3N-coordination and the third amino acid moiety does not take part in the coordination [74Ai2], 84Fa]. X-ray structural studies in solid state led to the same conclusion [72Ös]. Palladium(II) complexes of GlyHisGly were interpreted as involving the same donor groups [82Ra1].

Two biologically important tripeptides, thyrotropin releasing factor (TRF= L-pyroglutamyl-L-histidyl-L-prolineamide) and growth stimulating factor GHL= Glycycl-L-histidyl-L-lysine) contain a histidyl residue in an intermediate position. Consequently, their complex-formation processes are very similar to those of GlyHisGly, which supports the finding that the histidyl residue in peptides X-Y-His or X-His-Y can block the coordination of the amino acid residues following L-histidine. This means that the proline residue is not involved in coordination in the copper(II) and nickel(II) complexes of TRF [80Ko1, 79Ko, 83Ko2, 84Fa], and cobalt(II) is able to induce amide deprotonation in TRF, as for GlyHis [81Ko3].

Metal complexes of GHL and related dipeptides have been studied by various authors [84Ra, 85Ra2, 83Ma2, 84Pe]. Binding via three nitrogen donor atoms was reported both in solution and in the solid state in the copper(II) complexes. Coordination of the ε-amino groups of the lysyl–residue during the complexation with copper(II) and zinc(II) was ruled out. Amide deprotonation was presumed in the zinc(II) complexes of GHL and related dipeptides, such as AlaHis and LeuHis [84Ra]. It was also reported that palladium(II) forms a very stable complex with GHL, via 3N-coordination, and pK < 2 was proposed for amide ionization [87La].

Far fewer data are available for the complexation tripeptides containing N-terminal histidine. Copper(II) complexes of HisGlyGly were studied by Aiba *et al.* [74Ai1] and Demaret *et al.* [83De]. The formation of a bis complex (XIIa) or dimeric species (XIIb) was reported in neutral solution, as for the dipeptide, HisGly. The dimer is decomposed in alkaline solution, however, and the deprotonation of the second amide group was presumed, with triglycine-like coordination.

6.3   Other histidine-containing peptides

The results outlined in the previous two sections show that the histidyl residue is a major binding site in peptides. In the N-terminal position it hinders the deprotonation of amide groups, while in X-His or X-Y-His peptides it enhances the stability of the complexes, but hinders binding of the amino acid residues following histidine.

To understand the role of histidine in more complicated peptides, the anchoring capability of the imidazolyl moiety should be considered. N-Acetylhistamine and N-acetylhistidine are the simplest models for this purpose. Amide deprotonation and coordination have been ruled out in the copper(I) and (II) complexes of both ligands [60Ma2, 75Te]. On the other hand, these processes have been suggested to occur in certain mixed complexes with copper(II). Values of pK=7.24 and 7.88 were reported for amide deprotonation in the 2,2'-bipyridyl mixed complexes of *N*-acetylhistamine and *N*-acetyl-L-histidine, respectively [86So1]. Nickel(II) and copper(II) complexes of N-protected histidine peptides yield harder evidence of this anchoring effect. In the copper(II)-*N*-acetyl-GlyGlyGlyHis system coordination starts at the C-terminal imidazole locus, followed by the successive deprotonation of adjacent peptide bonds (pK = 6.50, 7.35, and 8.80) resulting in a 4N-coordinated complex [65Br, 66Br].

Further conclusions can be drawn from the complex-formation processes of peptides containing two histidyl residues. Copper(II) complexes of HisHis have been studied by several authors [60Ma2, 64Do, 76Ag, 87Li]. Pettit and co-workers [87Li] concluded that coordination starts with the participation of the terminal amino-N and imidazole-N donors (as in HisGly) which is followed by 3N-coordination (as in GlyHis), but the N-terminal imidazole–N3 can behave as a bridging ligand, which leads to the formation of dimeric species in slightly basic solution (Scheme XIV). Zinc(II)-promoted amide deprotonation was not reported to occur in the zinc(II)-HisHis system [76Ag], suggesting that histidine-like coordination is more favoured than the amide binding for zinc(II).

XIV.

The copper(II) complexes of the tetrapeptides, HisGlyHisGly and HisHis GlyGly, have been studied by EPR and CD spectroscopy [87Ue1]. The coordination of HisGlyHisGly was found to be similar to that of GlyGlyHis (XIII), while the existence of dimeric species (similar to those of HisHis (XIV)) was presumed for HisHisGlyGly.

Luteinizing hormone-releasing hormone (LHRH=pGluHisTrpSerTyrGly LeuArgProGlyNH$_2$) is a decapeptide-containing histidine as the second amino acid. Its copper(II) complex proved to be similar to that of TRF (3N-coordination via pyroglutamyl, amide, and imidazole-N atoms) and no significant effect of the further side chain was reported [88Ge]. Amide deprotonation was ruled out in the zinc(II) complexes of LHRH, however, and the imidazole-N3 was suggested as the main binding site [89Ba2].

L-Histidine frequently occurs among the cyclic dipeptides which serve as model compounds for metalloenzymes. The complex-forming capabilities of cyclic peptides have already been reviewed by Sigel & Martin [82Si]. It can be stated that substitution of amide hydrogens with metal ions is much less favourable than in the linear counterparts, because of the absence of an anchoring primary amino group. The imidazole–N3 takes part in the coordination only in the copper(II) complexes of cyclo(GlyHis) and cyclo(MetHis), and no evidence of amide deprotonation was obtained [85Ko1]. On the other hand, there is evidence of amide binding in the copper(II) complex of cyclo(HisHis) both in the solid state and in solution [77Ko2, 87Ar].

7. COMPLEXES OF SULPHUR-CONTAINING PEPTIDES

There are several possibilities for the occurrence of sulphur donors in natural peptides:

- *thioether* group is present in methionine and its derivatives,
- *thiol* group is the main binding site in L-cysteine,

- *disulphide* group can be obtained by the oxidation of thiols, and it occurs in many polypeptides,
- *thioamides* are ligands in which the sulphur donor appears in the peptide linkage.

7.1 Peptides containing the thioether group

Transition metal complexes of thioether-containing ligands have already been studied by many authors, and the results have been treated in review papers [82Si, 74Mc, 85De]. From these studies it is obvious that the thioether sulphur is a soft base. Thus, methionine of S-methyl-L-cysteine can readily coordinate to silver(I), palladium(II), or platinum(II) via the thioether group, whereas the main binding site for 3d-transition metal ions is the amino acid locus.

As concerns the interactions of copper(II) with thioether-containing peptides, it seems to be clear that the effect of the thioether moieties is significant only in S-methyl-L-cysteine [77Si, 85De, 86Pe]. The increase in stability of the complex $[CuA]^+$ and the decrease in the value of pK for amide deprotonation, as compared with those in GlyGly, were explained by the coordination of sulphur atoms in the apical positions [77Si].

In the case of methionine, the corresponding interaction is negligible for steric reasons, but a significant stereoselectivity was found in the copper(II) complexes of Met-peptides, owing to hydrophobic interactions [79Pe, 87Bo]. On the other hand, it has been shown that methionine can bind to copper(II) via the sulphur donor atom under certain experimental conditions (very low pH and temperature, or non-aqueous solvents) [78Ko3]. An increase of pH destroys this interaction, however, and the characteristic (N,N,O)-chelation exists in aqueous solutions.

Copper(II) and nickel(II) complexes of GlyMet, MetGly, HisMet, and MetHis were studied recently [87So]. It was demonstrated again that the presence of the thioether moiety does not influence the binding properties of the ligands, but in the case of an N-terminal L-methionine residue (MetGly and MetHis) the weak axial interaction of the thioether group inhibits the formation of bis complexes $[Ma_2H_{-1}]^-$.

In the complex $[CuAH_{-1}]$ of methionyl-S-methyl-L-cysteine a direct copper-(II)–thioether interaction in solid state was ruled out by X-ray study [86Pe]. The unbound thioether side chains can easily bind another metal ion, however, resulting in the formation of mixed metal complexes. A very dramatic stereoselectivity was reported in the stability constants of the complexes $[AgCuAH_{-1}]^+$ of various dipeptides containing two thioether groups in the side chains. The ternary complexes of dipeptides of the same chirality were much more stable than those of the opposite chirality.

Lyons & Pettit [84Ly] studied the silver(I) complexes of various thioether-containing dipeptides, and the data were compared with those for ValVal. It was found that silver(I) coordinates to simple dipeptides only via the terminal amino group, and it does not promote amide ionization. The complexes of GlyMet and MetGly are significantly more stable than those of ValVal, while the complexes of ligands containing two sulphur donors (MetMet) are significantly more stable still. The stability increase and the existence of the protonated species $[AgHA]^+$ prove the occurrence of silver(I)-thioether interaction, which was found to be more enhanced with C-terminal dipeptides.

The complexation of palladium(II) or platinum(II) with thioether-containing peptides is especially interesting, because the soft metal ions readily bind the S-atom; at the same time palladium(II) is the most effective metal ion for the promotion of amide binding. Thus, the compositions of palladium(II) complexes of methionine-containing peptides depend on the location of the thioether side chains and on the metal ion to ligand ratio. The binding modes of the various species present in solution are depicted in Scheme XV. It can be seen that in equimolar solutions palladium(II) is bonded to the dipeptide, GlyMet, via (NH_2, N^-, S)-donor set [78Ir, 85De]. This means that the C-terminal carboxylate does not take part in the coordination (XVa). At higher pH values, and in the presence of excess ligand, (NH_2,N)-coordinated bis complexes are favoured (XVc). The N-terminal thioether side chain of S-methylcys-teinylglycine can form a stable 5-membered chelate ring with participation of the amino group, which results in the formation of stable bis complexes without amide binding (XVb). At high pH, the 4N-species will again predominate [82De, 83Ko3]. The results obtained for the palladium(II) complexes of thioether ligands are similar to those obtained for copper(II) complexes of histidyl peptides, where amide deprotonation and binding were prevented by an N-terminal side chain and promoted by a C-terminal side chain.

XV.

7.2 Thiol-containing peptides

The deprotonated thiol group is a strong soft base and a very effective metal-binding site for common transition metal ions. The main properties of this binding site have already been reviewed by several authors [72Mc, 79Ge2], and the results are also discussed in the previous chapter.

However, very few data are available on simple L-cysteine-containing dipeptides because of the experimental difficulties in the preparation of these compounds in reduced form. Another important feature of thiol compounds is that they can easily reduce various metal ions. Consequently, the copper(II) complexes of thiol-containing peptides have scarcely been studied, and most of the results are available for complexation with nickel(II), palladium(II), or zinc(II).

N-Acetyl-L-cysteine (NAC) and α-mercaptopropionylglycine (MPG) are the simplest ligands that contain both amide and thiol groups, and their complex-formation processes are summarized in Scheme XVI. The formation of bis complexes via (S,O)-coordination is characteristic for the reactions of transition metal ions with NAC [81So2, 83Ha3], which supports the conclusion that the thiol group is the major binding site, but it does not promote amide binding if there is another group in a chelatable position (XVIa).

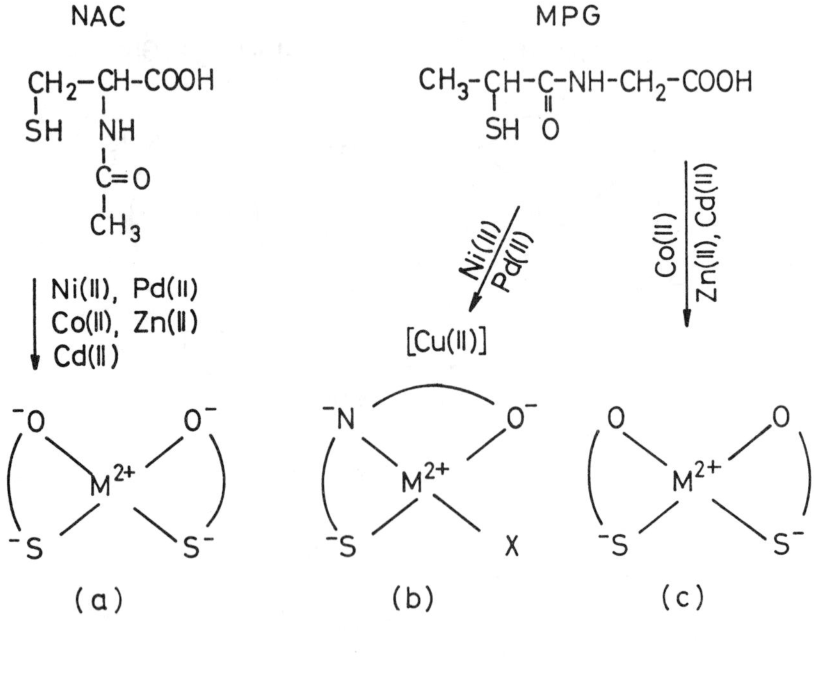

NAC

$$CH_2-CH-COOH$$
$$|\qquad|$$
$$SH\quad NH$$
$$\qquad\quad|$$
$$\qquad\quad C=O$$
$$\qquad\quad|$$
$$\qquad\quad CH_3$$

Ni(II), Pd(II)
Co(II), Zn(II)
Cd(II)

MPG

$$CH_3-CH-C-NH-CH_2-COOH$$
$$\qquad\;|\quad\|$$
$$\qquad\;SH\;\;O$$

Ni(II) / Pd(II)

Co(II) | Zn(II), Cd(II)

[Cu(II)]

(a)   (b)   (c)

XVI.

MPG is the best model for study of the ability of the thiol group to serve as an anchor for amide binding. It has been proved by a number of authors that nickel(II) and palladium(II) are able to induce the deprotonation and binding of the amide group in MPG [75Su1, 81So2, 85Fi]. The (S,N,O)-chelation (XVIb), however, will not saturate the coordination sphere of the metal ions in square–planar diamagnetic complexes, therefore there is a possibility of polynuclear complex formation, too, via sulphur bridges [81So2]. This is supported by recent structural studies, where a trimeric nickel(II) complex of MPG was isolated [89Ba3]. On the other hand cobalt(II), zinc(II), and cadmium(II) ions are not able to induce amide deprotonation, and (S,O)-coordinated bis complexes (XVIc) and polynuclear species are present in solution [83Ha, 85Fi, 90Ko].

Another interesting feature of MPG is that it can interact with copper(II) without inducing redox reactions [75Su2]. The mixing of copper(II) and MPG at neutral pH results in the formation of a green coloured copper(II) complex coordinated via thiol, amide, and carboxylate groups, as for nickel(II). Its spectral parameters are reminiscent of those of blue copper proteins. Complexes of the same type were obtained with several derivatives of MPG, too [76Su]. Especially stable copper(II) and nickel(II) complexes can be obtained if another thiol or imidazole-N donors are present in the ligand (α-mercaptopropionyl-L-cysteine and N-mercaptoacetyl-L-histidine) [77Su]. The deprotonation and binding of two amide groups are characteristic for the copper(II) complexes of N-mercaptoacetylglycyl-L-histidine [78Su].

L-cysteine-containing dipeptides (GlyCys and CysGly) have recently been prepared, and the complexation was studied with nickel(II), palladium(II), cobalt(II), zinc(II), and cadmium(II) [87Ko1, 88So, 90Ch, 90Ko]. Complex formation processes of the two dipeptides are shown in Scheme XVII.

XVII.

In the case of CysGly there is a possibility of stable cysteine-like (S,N)-coordination, which is favourable for all the metal ions studied. This means that the N-terminal cysteine residue prevents amide deprotonation and coordination, and bis complexes $[MA_2]^{2-}$ are formed (XVIIa). Steric reasons rule out this type of coordination for GlyCys, but bis complexes can be present via the coordination of thiol and carboxylate groups (XVIIb). Cobalt(II), zinc(II), and cadmium(II) prefer this type of binding which results in the formation of various protonated species, too, because of the presence of the free terminal amino group.

The soft palladium(II) and nickel(II) ions do not prefer (S,O)-coordination, but they promote the amide deprotonation and coordination. This results in the formation of (N,N,S)-coordinated species, which can readily be dimerized via sulphydryl bridges (XVIIc). Spectrophotometric and NMR studies yield an unambiguous proof of the existence of a dimer $\rightleftharpoons$ monomer equilibrium. In the case of the GlyCys the dimeric species predominates, while the presence of a bulky side chain in the N-terminal position (e.g. PheCys) increases the concentration of the monomer [90Ch]. An excess of ligand destroys the dimeric species, and complexes $[MA_2H_{-1}]^{3-}$ are formed in which the second peptide molecule is bonded monodentately via the thiol group (XVIId).

Transition metal complexes of C-terminal L-cysteine-containing tripeptides (GlyGlyCys and AlaAlaCys) have also been studied [89Ch]. The coordination abilities of these tripeptides are very similar to those of GlyCys. Cobalt(II), zinc(II), and cadmium(II) prefer (S,O)-coordination of the C-terminal part of the molecule, without amide binding (XVIIb), while palladium(II) and nickel(II) promote the ionization of two peptide groups, which results in the formation of the species $[MAH_{-2}]^{2-}$ with (NH$_2$,N$^-$,N$^-$,S$^-$)-coordination (Scheme XVIII), which is very similar to the stable coordination of GlyGlyHis.

XVIII.

Metal complexes of tri- and tetrapeptides containing two L-cysteine residues are under investigation in our laboratories. The peptide sequences Cys-X-Cys and Cys-X-X-Cys are very important from a biological aspect because they mimic the binding sites of thioneins. On the other hand, these peptides contain both N- and C-terminal cysteinyl residues, which have different effects on metal binding. As concerns the results obtained with CysAlaCys, it can be stated that the N-terminal L-cysteine is the main binding site, and (S,N)-coordinated bis complexes are the predominant species with cobalt(II), zinc(II), and cadmium(II). The presence of the other thiol group can lead to the formation of polynuclear complexes in equimolar solutions. Amide deprotonation and coordination occur in the mixture of nickel(II) and CysAlaCys, which can be interpreted by (N,N,N,S)-coordination, as for AlaAlaCys. In CysAlaAlaCys the N- and C-terminal thiol groups are too far from each other to coordinate to the same metal ion. Thus, none of the metal ions studied (cobalt(II), nickel(II), and zinc(II)) promote amide ionization, and the formation of bis and polynuclear complexes is favoured for this tetrapeptide.

Palladium(II) and iron(III) complexes of the same tetrapeptide sequence (Cys-X-Y-Cys), but in N-protected forms, have also been studied, and coordination of the thiol groups without amide binding was suggested [81Ue, 83Na, 87Ue2].

7.3 Coordination chemistry of glutathione

Glutathione (γ-L-glutamyl-L-cysteinylglycine) is one of the most important naturally occurring tripeptides (Scheme XIX). As the structural formula shows, it contains altogether eight donor atoms, which gives rise to the possibility of various types of coordination:

- at the N-terminal part of the molecule there is a possibility of amino acid-like binding.
- the thiol group is a very effective binding site and can interact with metal ions independently, or with (S,O)-coordination, or with the incorporation of deprotonated amide-N,
- glutathione is a tripeptide, thus the coordination of two amide groups is also possible, at least in principle.

$$^-OOC-\underset{\underset{NH_3^+}{|}}{C}H-(CH_2)_2-\underset{\underset{O}{\parallel}}{C}-NH-\underset{\underset{\underset{SH}{|}}{CH_2}}{C}H-\underset{\underset{O}{\parallel}}{C}-NH-CH_2-COOH$$

XIX.

Because of the steric requirements, all the binding sites cannot coordinate to the same metal ion. Various protonated and polynuclear complexes can therefore be formed with several metal ions. Accordingly, the coordination chemistry of glutathione is rather complicated, and in spite of the numerous investigations the results are very contradictory. We shall not discuss all aspects of metal ion glutathione interactions here because review papers are available on this topic [79Ra, 81So1].

One of the most intriguing questions concerning the coordination chemistry of glutathione is the uncertainty of the coordination of the amide groups. Palladium(II), nickel(II), copper(II), cobalt(II), and zinc(II) are the potential metal ions for the promotion of this interaction. It has been proved that glutathione reduces copper(II) to give copper(I) complexes and disulphide. This is a difference from MPG which was able to keep copper in divalent oxidation states in its complexes [75Su2]. Although a similar (S,N,O) donor set is available in glutathione, the different environments of the donor atoms do not allow the formation of stable copper(II) complexes. In connection with the other redox reactions of glutathione, a relatively long-lived chromium(V) species should be mentioned because of its potential biological hazard [86Go].

Almost all authors agree that the interactions of nickel(II), cobalt(II), and zinc(II) with glutathione start at the amino acid locus in slightly acidic or medium pH-range [59Ma, 80Ko2, 81So2]. In the case of nickel(II) this is followed by the formation of square–planar complexes due to sulphur or amide coordination. Williams *et al.* [80Ko2] concluded that deprotonation of the amide in glutathione occurs at around pH ~ 8, while the formation of S-coordinated polynuclear species was suggested by other studies. Nickel(II) and palladium(II)-promoted amide binding was presumed to occur above pH ~ 12 and 9, respectively [81So2].

On the basis of NMR measurements, zinc(II)-induced amide deprotonation was suggested above pH 10.5 [73Fu]. The same type of interaction was ruled out in the reaction of cobalt(II) with glutathione [83Ha3]. The thiol and carbonyl group of L-cysteine were proposed as the main binding sites in the tetrahedral bis complexes of cobalt(II). The existence of polynuclear species was also postulated with the participation of glutamic acid in the coordination.

7.4 Complexes of disulphides

Disulphide linkage occurs frequently in proteins and polypeptides, but only a few studies have been performed on the role of disulphide moieties in peptide coordination. D-Penicillamine disulphide is the most simple and well-studied ligand-containing disulphide linkage. Because of the presence of the two amino acid residues, it forms stable dimeric complexes $[M_2A_2]$ with copper(II), nickel(II), zinc(II), and cobalt(II), in which the disulphide is not involved in the coordination. The existence of such species was proved by solution equilibria [81La1, 88Va2] and by X-ray structural studies [74Ti].

Oxidized glutathione is the most easily available compound for studies of the effect of the disulphide moieties on peptide coordination. This ligand, however, also contains amino acid residues, which are the main binding sites in neutral or slightly basic solutions. This means that copper(II), nickel(II), cobalt(II), and zinc(II) form stable $[MA]^{2-}$ complexes with oxidized glutathione (Scheme XX), and the disulphide and peptide moieties do not take part in the coordination. The 'loop' around the central metal ion contains 19 atoms, which is large enough to give enhanced stability to the species $[MA]^{2-}$ [82Bl, 84Hul, 85Po, 88Va2] (Scheme XX).

XX.

Above pH ~ 10, there is a significant spectral shift toward lower wavelengths in the copper(II)-oxidized glutathione system, which is attributed to amide deprotonation and coordination [79Ra]. This process was interpreted in terms of the formation of $[Cu_2AH_{-4}]^{2-}$ species where the copper(II) is coordinated via the deprotonated amide groups, and the disulphide linkage bridges the copper(II) units [75Kr]. An exact evaluation of the copper(II)-oxidized glutathione system is very difficult, however, because the amide binding is accompanied by copper(II)-catalysed disproportionation of disulphide bonds, as follows [80Os]:

$$2\ RSSR + 2\ OH^- \rightarrow 2\ RS^- + RSH + RSO_2H\ .$$

There are no amino acid-binding sites in L-cysteinylglycine disulphide, consequently its complex-formation processes exhibit a strong metal ion dependence, similarly to those of other oligopeptides [88Va2]. The ligand behaves as a doubled dipeptide, the amino and carbonyl groups being the main binding sites in acidic solution. In the cases of copper(II) and nickel(II), this is followed by amide deprotonation and coordination, as with aliphatic dipeptides. Because of the two separated dipeptide linkages, polynuclear species are also formed, but their concentration depends on the metal ion to ligand ratio. In the absence of structural studies, a weak disulphide–copper(II) interaction cannot be ruled out, but equilibrium studies do not provide any support for such an interaction [88Va2].

7.5 Complexes of thioamides
Sections 7.1 to 7.4 have dealt with complexes of peptides in which the sulphur atom is in the 'side chain'. In the thioamides the sulphur is present in the amide linkage itself,

$$-C\diagup^{S}_{\diagdown NH-}$$
. It is well known from other studies (see thiourea and derivatives) that
the thiocarbonyl group is quite a strong donor as compared with carbonyl groups, and its introduction into a peptide linkage may have a significant effect on the metal binding capability of peptides. Recent studies on the copper(II) and nickel(II) complexes of various thioamides support this statement [87Ko2, 88Va1].

Complexes of methionine-N-methylamide and leucine-N-methylamide and the corresponding thio analogues were studied by potentiometric and spectroscopic methods [87Ko2]. It was found that the complex-formation processes of the amino acid amides are very similar to those of glycineamide. The coordination starts with the participation of amino-N and carbonyl-O donors, while at above pH 7 copper(II) and nickel(II) induce amide deprotonation, which results in the formation of the species $[MA_2H_{-2}]$ with 4N-coordination.

Similar species were detected in the case of the thio-analogues, but with much higher stability. Complexes formed with (NH_2,S)-chelation are 3 to 4 orders of magnitude more stable than the respective complexes involving carbonyl binding. This is demonstrated by the fact that complex formation with amino acid amides starts at around pH ~ 5, whereas all the metal ions are involved in (N,S)-coordinated bis complexes by that pH in the case of thioamides. It can also be concluded that thiocarbonyl coordination does not prevent, but to some extent hinders, deprotonation of the amide group in basic solution. pK values of 7.04 and 8.16 were reported for the deprotonation of the first amide group in the copper(II) complexes of methionine-N-methylamide and the thioanalogues, respectively [87Ko2].

The effect of the thiocarbonyl group is even more dramatic if it is present in the C-terminal residue of a dipeptide containing side-chain donor groups. This is best exemplified by the copper(II) and nickel(II) complexes of AspPhe-NH$_2$ and thio-AspPhe-NH$_2$ [88Va1]. Concentration distribution curves for the copper(II) complexes of these two ligands are depicted in Fig. 4.1. It can be seen that the complexes of AspPhe-NH$_2$ are very similar to those of GlyGly-NH$_2$ [75Do], but the presence of the N-terminal aspartic acid residue will increase the concentration of the species $[MA]^+$ and $[MA_2]$ and the deprotonation constant of the amide group. In the case of thio$-$AspPhe$-$NH$_2$ the thiocarbonyl group is already bonded by pH ~ 3, which prevents β-alanine-like coordination, but promotes the deprotonation and coordination of the peptide linkage.

8.   CONCLUSIONS

The results outlined in the previous sections permit conclusions as to some general features of the coordination chemistry of oligopeptides. It is obvious that, with the participation of the terminal amino and neighbouring carbonyl or C-terminal carboxylate groups, the peptide molecules can interact with almost all metal ions. These interactions are rather weak, however, and generally do not compete with the hydrolysis of the metal ions in alkaline solution. More stable complexes can be obtained if the metal ion is able to substitute the amide-H atoms, resulting in the formation of fused 5-membered chelate rings. The deprotonation and coordination of the peptide linkage is a function of the nature of metal ions. The large number of data available clearly demonstrate that among the labile metal ions palladium(II), copper(II), and nickel(II) are the most effective in that respect. Cobalt(II) and zinc(II)-induced amide deprotonations have also been reported, this mainly being due to the effect of the side-chain donor groups. The involvement of amide nitrogen in the coordination was reported for the inert complexes of platinum(II), chromium (III), and cobalt(III), too. Although all the metal ions have not been studied from this aspect, it is not probable that other (biologically important) metal ions will provide further examples of metal ion-promoted peptide-NH ionization.

It should also be noted that the formation of amide-bounded metal complexes requires the presence of an 'anchoring' donor group in a chelatable position relative to the amide bond. The terminal amino group is the most common anchor. Terminal carboxylate does not generally serve as an anchor, but phosphate–oxygen in the phosphate derivatives of dipeptides has been reported to behave as an anchor [89Ki] (for details see the previous chapter). Examples of imidazole or thiol-promoted amide binding have also been published.

The presence of various side-chain (or lateral) donor groups has the most pronounced effect on the coordination chemistry of peptides. Weakly coordinating donor groups (alcoholic-OH, phenolic-OH, lysil-amino, etc.) are generally more effective in the N-terminal position, and they can form dimeric or polynuclear species in many cases. The β-carboxylate group of aspartyl residues increases the

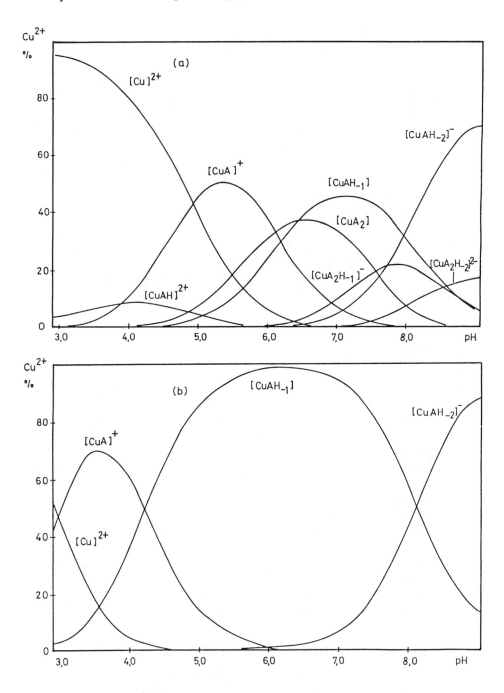

Fig. 4.1 — Concentration distribution of the complexes formed in copper(II)-AspPhe-NH$_2$ and copper(II)-thio-AspPhe-NH$_2$ systems.
(a) A = AspPhe-NH$_2$
     $C_{Cu}2+ = 6.9.10^{-4}$; $C_A = 1.452.10^{-3}$ mol.dm^{-3}
(b) A = thio-AspPhe-NH$_2$
     $C_{Cu}2+ = 6.66.10^{-4}$; $C_A = 1.618.10^{-3}$ mol.dm^{-3}.

stability of peptide complexes. In the N-terminal position the β-alanine-like coordination stabilizes the species $[MA]^+$ and $[MA_2]$, while in the C-terminal position it enhances the coordination of the intermediate and precludes the deprotonation of subsequent peptide linkages.The non N-terminal prolyl residue is unique because it cannot bind metal ions. Prolyl generally works as a 'breakpoint' but it permits the formation of unusually large chelate rings or loop structures.

The imidazole moiety of histidine and the thiol group of cysteine have the most significant effects on the coordination ability of oligopeptides. In the N-terminal position these side chains generally prevent amide coordination, but stable bis complexes can be formed, with histamine- or cysteine-like coordination. C-terminal histidyl or thiol residues promote the binding of intermediate peptide groups, but preclude the coordination of the subsequent amino acids.

The coordination chemistry of peptides can be even more complex if two or more strongly-coordinated side chains are present in the peptide sequence. The data available in the literature, however, do not allow a summary of the effects of all possible amino acid arrangements. Consequently, subsequent investigations of the metal complexes of peptides should be directed toward the attainment of a deeper understanding of the side-chain interactions.

REFERENCES

55Do1   Dobbie, H. & Kermack, W. O.: *Biochem. J.*, **59**, 246 (1955).
55Do2   Dobbie, H. & Kermack, W. O.: *Biochem. J.*, **59**, 257 (1955).
56Da   Datta, S. P. & Rabin, D. R.: *Trans. Faraday Soc.*, **52**, 1117 (1956).
59Ma   Martin, R. B. & Edsall, J. T.: *J. Am. Chem. Soc.*, **81**, 4044 (1959).
60Ko   Koltun, W. L., Fried, M., & Gurd, F. R. N.: *J. Am. Chem. Soc.*, **82**, 233 (1960).
60Ma1   Martin, R. B., Chamberlin, M., & Edsall, J. T.: *J. Am. Chem. Soc.*, **82**, 495 (1960).
60Ma2   Martin, R. B. & Edsall, J. T.: *J. Am. Chem. Soc.*, **82**, 1107 (1960).
61Ma   Martin, R. B., *Fed. Proc. (USA)*, **20**, 54 (1961).
61St   Strandberg, B., Lindqvist, I., & Rosenstein, R. *Z. Kristallogr.* **116**, 266 (1961).
64Do   Doran, M. A., Chaberek, S., & Martell, A. E.: *J. Am. Chem. Soc.*, **86**, 2129 (1964).
64Ki   Kim, M. K. & Martell, A. E.: *Biochemistry,* **3**, 1169 (1964).
65Br   Bryce, G. F., Roeske, R. W., & Gurd, F. R. N.: *J. Biol. Chem.*, **240**, 3837 (1965).
65Fr1   Freeman, H. C., Schoone, J. C., & Sime, J. G.: *AScta Cryst.,* **18**, 381 (1965).
65Fr2   Freeman, H. C. & Szymanski, J. T.: *Chem. Commun.*, 344, (1965).
65Fr3   Freeman, H. C. & Taylor, M. R.: *Acta Cryst.*, **18**, 939 (1965).
66Br   Bryce, G. F., Roeske, R. W., & Gurd, F. R. N.: *J. Biol. Chem.*, **241**, 1072 (1966).
66Gi   Gillard, R. D., McKenzie, E. D., Mason, R., & Robertson, G. B.: *Nature,* **209**, 1347 (1966).

66Ki   Kim, M. K. & Martell, A. E.: *J. Am. Chem. Soc.,* **88**, 914 (1966).

67Bl   Blount, J. F., Fraser, K. A., Freeman, H. C., Szymanski, J. T., & Wang, C. H.: *Acta Cyst.,* **22**, 396 (1967).

67Fr   Freeman, H. C.: Crystal structures of metal peptide complexes; In: *Advances in Protein Chemistry,* **22**, Academic press, New York (1967).

67Ki   Kim, M. K. & Martell, A. E.: *J. Am. Chem. Soc.,* **89**, 5138 (1967).

67Sh   Sheinblatt, M. & Becker, E. D.: *J. Biol. Chem.,* **242**, 3159 (1967).

68Br   Brunetti, A. P., Lim, M. C., & Nancollas, G. H.: *J. Am. Chem. Soc.,* **90**, 5120 (1968).

69Da   Davies, G., Kustin, K., & Pasternack, R. F.: *Inorg. Chem.,* **8**, 1535 (1969).

69Ki   Kim, M. K. & Martell, A. E.: *J. Am. Chem. Soc.,* **91**, 872 (1969).

69Li   Liler, M.: *J. Chem. Soc. (B.)* 385 (1969).

69Mi   Michailidis, M. S. & Martin, R. B.: *J. Am. Chem. Soc.,* **91**, 4683 (1969).

69Sh   Amino acids, peptides and proteins, ed. Sheppard R., *Royal Society of Chemistry Specialist Periodical Reports.,* London, Vol. 1. (1969), Vol. 18. (1987).

69Ts   Tsangaris, J. M., Chang, J. W., & Martin, R. B.: *J. Am. Chem. Soc.,* **91**, 726 (1969).

70Ba   Barnet, M. T., Freeman, H. C., Buckingham, D. A., Nan Hsu, I., & van der Helm, D.: *J. Chem. Soc., Chem. Commun.,* 367 (1970).

70Bi   Billo, E. J. & Margerum, D. W.: *J. Am. Chem. Soc.,* **92**, 6811 (1970).

70Bl   Blount, J. F., Freeman, H. C., Holland, R. V., & Milburn, G. H. W.: *J. Biol. Chem.,* **245**, 5177 (1970).

70Bu   Bunting, J. W. & Thong, K. M.: *Can. J. Chem.,* **48**, 1654 (1970).

70Fr1   Freeman, H. C. & Golomb, M. L.: *J. Chem. Soc., Chem. Commun.* 1523 (1970).

70Fr2   Franks, W. A. & van der Helm, D.: *Acta Cryst.,* **B27**, 1299 (1970).

70Mo   Motekaitis, R. J. & Martell, A. E.: *J. Am. Chem. Soc.,* **92**, 4223 (1970).

70Wi   Wilson, E. W. & Martin, R. B.: *Inorg. Chem.,* **9**, 528 (1970).

71Bo   Bowles, A. M., Szarek, W. A., & Baird, M. C.: *Inorg. Nucl. Chem. Letters,* **7**, 25 (1971).

71Gi   Gillard, R. D. & Phipps, D. A.: *J. Chem. Soc. (A),* 1074 (1971).

71Ha   Hauer, H., Billo, E. J., & Margerum, D. W.: *J. Am. Chem. Soc.,* **93**, 4173 (1971).

71Mo1   Morris, P. J. & Martin, R. B.: *Inorg. Chem.,* **10**, 964 (1971).

71Mo2   Morris, P. J. & Martin, R. B.: *J. Inorg. Nucl. Chem.,* **33**, 2913 (1971).

72Ma   Martin, R. B.: *J. Chem. Soc., Chem. Commun.,* 793 (1972).

72Mc   McAuliffe, C. A. & Murray, S. G.: *Inorg. Chim. Acta Rev.,* **6**, 103 (1972).

72Ös   Österberg, R., Sjöberg, B., & Söderquist, R.: *Acta Chem. Scand.,* **26**, 4184 (1972).

72Pi   Pitner, T. P., Wilson, E. W., & Martin, R. B.: *Inorg. Chem.,* **11**, 738 (1972).

72Ra   Rabenstein, D. L. & Libich, S.: *Inorg. Chem.,* **11**, 2960 (1972).

72Si   Sigel, H., Griesser, R., & Prijs, B.: *Z. Naturforsch. (B),* **27**, 353 (1972).

73Fr   Freeman, H. C.: Metal complexes of amino acids and peptides. In: *Inorganic biochemistry,* Vol. 1. Chap. 4. Elsevier, Amsterdam (1973).

73Fu   Fuhr, B. J. & Rabenstein, D. L.: *J. Am. Chem. Soc.,* **95**, 6944 (1973).

73Ma   Martin, R. B. & Hutton, W. C.: *J. Am. Chem. Soc.*, **95**, 4752 (1973).
74Ai1   Aiba, H., Yokoyama, A., & Tanaka, H.: *Bull. Chem. Soc. Japan*, **47**, 136 (1974).
74Ai2   Aiba, H., Yokoyama, A., & Tanaka, H.: *Bull. Chem. Soc. Japan*, **47**, 1437 (1974).
74La   Lau, S. J., Kruck, T. P. A., & Sarkar, B.: *J. Biol. Chem.*, **249**, 5878 (1974).
74Ma1   Martin, R. B.: *Metal Ions in Biological Systems*, **1**, 129 (1974).
74Ma2   Margerum, D. W. & Dukes, G. R.: *Metal Ions in Biological Systems*, **1**, 158 (1974).
74Mc   McCormick, D. B., Greisser, R., & Sigel, H.: *Metal Ions in Biological Systems*, **1**, 213 (1974).
74Na   Nakon, R., & Angelici, R. J.: *J. Am. Chem. Soc.*, **96**, 4178 (1974).
74Su   Sundberg, R. J. & Martin, R. B.: *Chem. Rev.*, **74**, 471 (1974).
74Ti   Tich, J. A., Mastropaolo, P., Potenza, J., & Schugar, H. J.: *J. Am. Chem. Soc.*, **96**, (1974).
75Ag   Agarwal, R. P. & Perrin, D. D.: *J. Chem. Soc. Dalton Trans.* 268 (1975).
75Ba   Basosi, R., Tiezzi, E., & Valensin, G.: *J. Phys. Chem.*, **79**, 1725 (1975).
75Bo   Boggess, R. K. & Martin, R. B.: *J. Inorg. Nucl. Chem.*, **37**, 1097 (1975).
75Br1   Brookes, G. & Pettit, L. D.: *J. Chem. Soc., Dalton Trans.*, 2106 (1975).
75Br2   Brookes, G. & Pettit, L. D.: *J. Chem. Soc., Dalton Trans.*, 2112 (1975).
75Bu   Burce, G. L., Paniago, E. B., & Margerum, D. W.: *J. Chem. Soc., Chem. Commun.*, 261, (1975).
75Do   Dorigatti, T. F. & Billo, E. J.: *J. Inorg. Nucl. Chem.*, **37**, 1515 (1975).
75Ka   Kaneda, A. & Martell, A. E.: *J. Coord. Chem.*, **4**, 137 (1975).
75Kr   Kroneck, P.: *J. Am. Chem. Soc.*, **97**, 3839 (1975).
75Ma   Martell, A. E., Kim, M. K., & Kaneda, A.: *J. Coord. Chem.*, **4**, 159 (1975).
75Si   Sigel, H.: *Inorg. Chem.*, **14**, 1535 (1975).
75Su1   Sugiura, Y., Hirayama, Y., Tanaka, H., & Ishizu, K.: *J. Inorg. Nucl. Chem.*, **37**, 2367 (1975).
75Su2   Sugiura, Y., Hirayama, Y., Tanaka, H., & Ishizu, K.: *J. Am. Chem. Soc.*, **97**, 5577 (1975).
75Te   Temussi, P. A. & Vitagliano, A.: *J. Amer. Chem. Soc.*, **97**, 1572 (1975).
76Ag   Agarwal, R. P. & Perrin, D. D.: *J. Chem. Soc., Dalton Trans.*, 89 (1976).
76Ca   Camerman, N., Camerman, A., & Sarkar, B.: *Can. J. Chem.*, **54**, 1309 (1976).
76Ko   Kozlowski, H. & Trzebiatowska, B. J.: *Chem. Phys. Letters*, **42**, 246 (1976).
76Me   de Meester, P. & Hodgson, D. J.: *J. Am. Chem. Soc.*, **98**, 7086 (1976).
76So   Sóvágó, I. & Gergely, A.: *Inorg. Chim. Acta*, **20**, 27 (1976).
76Su   Sugiura, Y. & Hirayama, Y.: *Inorg. Chem.*, **15**, 679 (1976).
77Ag   Agarwal, R. P. & Perrin, D. D.: *J. Chem. Soc., Dalton Trans.*, 53 (1977).
77Ge   Gergely, A. & Nagypál, I.: *J. Chem. Soc., Dalton Trans.*, 1104 (1977).
77Ko1   Kozlowski, H.: *Chem. Phys. Letters*, **46**, 519 (1977).
77Ko2   Kojima, Y., Hirotsu, K., & Matsumoto, K.: *Bull. Chem. Soc. Japan*, **50**, 3222 (1977).

77Ko3   Kozlowski, H. & Jezowska, M.: *Chem. Phys. Letters*, **47**, 452 (1977).
77Ko4   Kozlowski, H., Kozlowska, G. F., & Trzebiatowska, B. J.: *Org. Magn. Res.*, **10**, 146 (1977).
77Ko5   Kozlowska, G. F., Kozlowski, H., & Trzebiatowska, B. J.: *Inorg. Chim. Acta*, **25**, 1 (1977).
77Li   Lim, M. C.: *J. Chem. Soc., Dalton Trans.*, 15 (1977).
77Si   Sigel, H., Naumann, C. F., Prijs, B., McCormick, D. B., & Falk, M. C.: *Inorg. Chem.*, **16**, 790 (1977).
77Sp   Sportelli, L., Neubacher, H., & Lohmann, W.: *Z. Naturforsch.*, **32c**, 643 (1977).
77Su   Sugiura, Y., & Hirayama, X.: *J. Am. Chem. Soc.*, **99**, 1581 (1977).
78Bi   Billo, E. J.: *J. Inorg. Nucl. Chem.*, **40**, 1971 (1978).
78Fr1   Freeman, H. C. & Guss, J. M.: *Acta Cryst.*, **B34**. 2451 (1978).
78Fr2   Freeman, H. C., Guss, J. M., & Sinclair, R. L.: *Acta Cryst.*, **B34**, 2459 (1978).
78Ko1   Kozlowski, H.: *Inorg. Chim. Acta*, **31**, 135 (1978).
78Ko2   Kozlowski, H. & Baranowski, J.: *Inorg. Nucl. Chem. Letters*, **14**, 315 (1978).
78Ko3   Kozlowski, H. & Kovalik, T.: *Inorg. Nucl. Chem. Letters*, **14**, 201 (1978).
78Ko4   Kozlowski, H. & Siatecki, Z.: *Chem. Phys. Letters*, **54**, 498 (1978).
78Ma   Martin, R. B.: *Nature*, **271**, 94 (1978).
78Me   de Meester, P. & Hodgson, D. J.: *Inorg. Chem.*, **17**, 440 (1978).
78So   Sóvágó, I., Kiss, T., & Gergely, A.: *J. Chem. Soc., Dalton Trans.*, 964 (1978).
78Su   Sugiura, Y.: *Inorg. Chem.*, **17**, 2176 (1978).
78To   Tosi, L. & Garnier, A.: *Inorg. Chim. Acta*, **29**, L261 (1978).
78TR   Trzebiatowska, B. J., Kowalik, T., & Kozlowski, H.: *Bull. Acad. Pol. Sci.*, **26**, 223, (1978).
78Ud   Udupa, M. R. & Krebs, B.: *Inorg. Chim. Acta*, **31**, 251 (1978).
79Ap   Appleton, D. W., Kruck, T. P. A., & Sarkar, B.: *J. Inorg. Biochem.*, **10**, 1 (1979).
79Ge1   Gergely, A. & Kiss, T.: *Metal Ions in Biological Systems*, **9**, 143 (1979).
79Ge2   Gergely, A. & Sóvágó, I.: *Metal Ions in Biological Systems*, **9**, 77 (1979).
79Ko   Kozlowska, G. F., Kozlowski, H., Trzebiatowska, B. J., Kupryszewski, G., & Przybylski, J.: *Inorg. Nucl. Chem. Letters*, **115**, 387 (1979).
79Pa   Pastor, J. M., Garnier, A., & Tosi, L.: *Inorg. Chim. Acta*, **37**, L549 (1979).
79Pe   Pettit, L. D. & Hefford, J. W.: *Metal Ions in Biological Systems*, **9**, 173 (1979).
79Ra   Rabenstein, D. L., Guevremont, R. & Evans, C. A.: *Metal Ions in Biological Systems*, **9**, 103 (1979).
79Sa   Sakurai, T. & Nakahara, A.: *Inorg. Chim. Acta*, **34**, L243 (1979).
80Br   Brown, C. E., Antholine, W. E., & Froncisz, W.: *J. Chem. Soc., Dalton Trans.*, **590**, (1980).
80Ev   Evans,, E. J., Grice, J. E., Hawkins, C. J., & Heard, M. R.: *Inorg. Chem.*, **19**, 3496 (1980).
8oFa   Farkas, E., Beke, B., & Gergely, A.: *Magyar Kém. Foly.*, **86**, 345 (1980).

80Ko1   Kozlowska, G. F., Kozlowski, H., & Kupryszewski, G.: *Inorg. Chim. Acta*, **46**, 29 (1980).

80Ko2   Kozlowska, G. F., May, R. P., & Williams, D. R.: *Inorg. Chim. Acta*, **46**, L51 (1980).

80Os   Ostern, M., Pelczar, J., Kozlowski, H., & Trzebiatowska, B. J.: *Inorg. Nucl. Chem. Letters*, **16**, 251 (1980).

80Sa   Sakurai, T. & Nakahara, A.: *Inorg. Chem.*, **19**, 847 (1980).

80Ve   Vestues, P. I. & Martin, R. B.: *J. Am. Chem. Soc.*, **102**, 7906 (1980).

81He   Hefford, R. J. W. & Pettit, L. D.: *J. Chem. Soc., Dalton Trans.*, 1331 (1981).

81Ko1   Kozlowska, G. & Kozlowski, H., Bezer, M., Pettit, L. D., Kupryszewski, G., & Przybylski, J.: *Inorg. Chim. Acta*, **56**, 79 (1981).

81Ko2   Kozlowska, G. F., Kozlowski, H., Siemion, I. Z., Sobczyk, K., & Nawrocka, E.: *J. Inorg. Biochem.*, **15**, 201 (1981).

81Ko3   Kozlowska, G. F., Kozlowski, H., & Kupryszewski, G.: In: *Structure and activity of natural peptides*, p. 287, Walter de Gruyter, Berlin–New York, (1981).

81La1   Laurie, S. H., Mohammed, E. S., & Prime, D. H.: *Inorg. Chim. Acta*, **56**, 135 (1981).

81La2   Lau, S. J. & Sarkar, B.: *J. Chem. Soc., Dalton Trans.*, 491 (1981).

81Ma   Margerum, D. W. & Owens, G. D.: *Metal Ions in Biological systems*, **12**, 75 (1981).

81Ra   Rakhit, G. & Sarkar, B.: *J. Inorg. Biochem.*, **15**, 233 (1981).

81Si   Sigel, H., Prijs, B., & Martin, R. B.: *Inorg. Chim. Acta*, **56**, 45 (1981).

81So1   Sóvágó, I. & Gergely, A.: *Actions and Agents, Supplements*, **8**, 291 (1981).

81So2   Sóvágó, I. & Martin, R. B.: *J. Inorg. Nucl. Chem.*, **43**, 425 (1981).

81Ue   Ueyama, N., Nakata, M., & Nakamura, A.: *Inorg. Chim. Acta*, **55**, L61 (1981).

81Va   Vaughn, J. B., Stephens, R. L., Lenkinski, R. E., Krishna, N. R., Heavner, G. A., & Goldstein, G.: *Biochim. Biophys. Acta*, **671**, 50 (1981).

82Bl   Blais, M. & Berthon, G.: *J. Chem. Soc., Dalton Trans.*, 1803 (1982).

82Br   Brown, C. E., Vidrine, D. W., Julian, R. L., & Froncisz, W.: *J. Chem. Soc., Dalton Trans.*, 2371 (1982).

82De   Decock-Le Reverend, B., Loucheux, C., Kowalik, T., & Kozlowski, H.: *Inorg. Chim. Acta*, **66**, 205 (1982).

82En   Ensuque, A., Demaret, A., Abello, L., & Lapluye, G.: *J. Chim. Phys.* **79**, 185 (1982).

82Ge   Gergely, A. & Farkas, E.: *J. Chem. Soc., Dalton Trans.*, 381 (1982).

82Ki1   Kittl, W., S. & Rode, B. M.: *Inorg. Chim. Acta*, **66**, 105 (1982).

82Ki2   Kiss, T. & Tóth, B.: *Talanta*, **29**, 539 (1982).

82Ra1   Rabenstein, D. L., Isab, A. A., & Shoukry, M. M.: *Inorg. Chem.*, **21**, 3234 (1982).

82Ra2   Rainer, M. J. A. & Rode, B. M.: *Inorg. Chim. Acta*, **58**, 59 (1982).

82Ra3   Rainer, M. J. A. & Rode, B. M.: *Monatsh. Chem.*, **113**, 399 (1982).

82Sh   Shelke, D. N.: *J. Coord. Chem.*, **12**, 35 (1982).

82Si   Sigel, H. & Martin, R. B.: *Chem. Rev.*, **82**, 385 (1982).

82So   Sóvágó, I., Farkas, E., & Gergely, A.: *J. Chem. Soc., Dalton Trans.*, 2159 (1982).

83Bo   Bojczuk, M. J., Baranowski, J., & Kozlowski, H.: *Pol. J. Chem.*, 685 (1983).

83Da   Daniele, P. G., Amico, P., Ostacoli, G., & Marzona, M.: *Annali Chim.*, **73**, 299 (1983).

83De   Demaret, A., Ensuque, A., & Lapluye, G.: *J. Chim. Phys.*, **80**, 475 (1983).

83Di   Diaddario, L. L., Robinson, W. R., & Margerum, D. W.: *Inorg. Chem.*, **22**, 1021 (1983).

83Fa   Farkas, E., Sóvágó, I., & Gergely, A.: *J. Chem. Soc., Dalton Trans.*, 1545 (1983).

83Ha1   Hawkins, C. J. & Martin, J.: *Inorg. Chem.*, **22**, 3879 (1983).

83Ha2   Hamburg, A. W., Németh, M. T., & Margerum, D. W.: *Inorg. Chem.*, **22**, 3535 (1983).

83Ha3   Harman, B. & Sóvágó, I.: *Inorg. Chim. Acta*, **80**, 75 (1983).

83Ki   Kittl, W. S. & Rode, B. M.: *J. Chem. Soc., Dalton Trans.*, 409 (1983).

83Ko1   Kozlowski, H., Bezer, M., Pettit, L. D., Bataille, M., & Hecquet, B.: *J. Inorg. Biochem.*, **18**, 231 (1983).

83Ko2   Kozlowska, G. F., Bezer, M., & Pettit, L. D.: *J. Inorg. Biochem.*, **18**, 335 (1983).

83Ko3   Kozlowski, H., Decock-Le Reverend, B., Delaruelle, J., Loucheux, C., & Ancian, B.: *Inorg. Chim. Acta*, **78**, 31 (1983).

83Ko4   Kozlowska, G. F., Konopinska, D., Kozlowski, H., & Decock-Le Reverend, B.: *Inorg. Chim. Acta*, **78**, L47 (1983).

83Ma1   Margerum, D. W., *Pure and Appl. Chem.*, **55**, 23 (1983).

83Ma2   May, P. M., Whittaker, J., & Williams, D. R.: *Inorg. Chim. Acta*, **80**, L5 (1983).

83Na   Nakata, M., Ueyama, N., Terakawa,, T., & Nakamura, A.: *Bull. Chem. Soc. Japan*, **56**, 3647 (1983).

83Pr   Probst, M. M. & Rode, B. M.: *Inorg. Chim. Acta*, **78**, 135 (1983).

83Sh   Shelke, D. N.: *Inorg. Chim. Acta*, **80**, 255 (1983).

84Ba   Bataille, M., Kozlowska, G. F., Kozlowski, H., Pettit, L. D., & Steel, I.: *J. Chem. Soc., Chem. Commun.*, 231, (1984).

84Be   Bezer, M., Pettit, L. D., Steel, I., Bataille, M., Djemil, S., & Kozlowski, H.: *J. Inorg. Biochem.* **20** 13 (1984).

84Fa   Farkas, E., Sóvágó, I., Kiss, T., & Gergely, A.: *J. Chem. Soc., Dalton Trans.*, 611 (1984).

84Ha   Hay, R. W. & Pujari, M. P.: *J. Chem. Soc., Dalton Trans.*, 1083 (1984).

84Hu1   Huet, J., Juoni, M., Abello, L., & Lapluye, G.: *J. Chim. Phys.*, **81**, 505 (1984).

84Hu2   Heut, J. & Vilkas, E.: *Inorg. Chim. Acta*, **91**, 43 (1984).

84Im   Impellizeri, G., Bonomo, R. P., Cali, R., Cucinotta, V., & Rizzarelli, E.: *Thermochim. Acta*, **72**, 2631 (1984).

84Ki1   Kiss, T. & Gergely, A.: *J. Chem. Soc., Dalton Trans.*, 1951 (1984).

84Ki2   Kim, S. H. & Martin, R. B.: *J. Am. Chem. Soc.*, **106**, 1707 (1984).
84Ki3   Kirschenbaum, L. J. & Rush, J. D.: *J. Am. Chem. Soc.*, **106**, 1003 (1984).
84Ly   Lyons, A. Q. & Pettit, L. D.: *J. Chem. Soc., Dalton Trans.*, 2305 (1984).
84Mü   Müller, D., Decock-Le Reverend, B., & Sarkar, B.: *J. Inorg. Biochem.*, **21**, 215 (1984).
84Pe   Perkins, C. M., Rose, N. J., Weinstein, B., Stenkamp, R. E., Jensen, L. H., & Pickart, L.: *Inorg. Chim Acta*, **82**, 93 (1984).
84Ra   Rainer, M. J. A. & Rode, B. M.: *Inorg. Chim. Acta*, **93**, 109 (1984).
84So   Sóvágó, I., Kiss, T., & Gergely, A.: *Inorg. Chim. Acta*, **93**, L53 (1984).
85Ba   Bataille, M., Pettit, L. D., Steel, I., Kozlowski, H., & Tatarowski, T.: *J. Inorg. Biochem.*, **24**, 211 (1985).
85De   Decock-Le Reverend, B. & Kozlowski, H.: *J. Chim. Phys.*, **82**, 883 (1985).
85Fi   Filella, M. & Williams, D. R.: *Inorg. Chim. Acta*, **106**, 49 (1985).
85Ko1   Kojima, Y., Ishio, N., & Yamashita, T.: *Bull. Chem. Soc. Japan*, **58**, 759 (1985).
85Ko2   Kozlowska, G. F., Pettit, L. D., Steel, I., Livera, C. E., Kupryszewski, G., & Rolka, K.: *J. Inorg. Biochem.*, **24**, 299 (1985).
85Pe1   Pettit, L. D. & Bezer, M.: *Coord. Chem. Rev.*, **61**, 97 (1985).
85Pe2   Pettit, L. D., Steel, I., Kozlowska, G. F., Tatarowski, T., & Bataille, M.: *J. Chem. Soc., Dalton Trans.*, 535 (1985).
85Pe3   Pettit, L. D., Steel, I., Kowalik, T., Kozlowski, H., & Bataille, M.: *J. Chem. Soc., Dalton Trans.*, 1201 (1985).
85Po   Postal, V. S., Vogel, E. J., Young, C. M., & Greenaway, F. T.: *J. Inorg. Biochem.*, **25**, 25 (1985).
85Ra1   Rabenstein, D. L., Daigenault, S. A., Isab, A. A., Arnold, A. P., & Shoukry, M. M.: *J. Am. Chem. Soc.*, **107**, 6435 (1985).
85Ra2   Rainer, M. J. A. & Rode, B. M.: *Inorg. Chim. Acta*, **107**, 127 (1985).
85Ve   Véber, M., Horváth, I., & Burger, K.: *Inorg. Chim. Acta*, **108**, L21 (1985).
85Wa   Wang, S. M. & Gilpin, R. K.: *Talanta*, **32**, 329 (1985).
85Ya   Yamauchi, O., Tsujide, K., & Odani, A.: *J. Am. Chem. Soc.*, **107**, 659 (1985).
86Bo   Bonomo, R. P., Cali, R., Cucinotta, V., Impellizzeri, G., & Rizzarelli, E.: *Inorg. Chem.*, **25**, 1641 (1986).
86De1   Decock-Le Reverend, B., Andrianariajona, L., Livera, C., Pettit, L. D., Steel, I., & Kozlowski, H.: *J. Chem. Soc., Dalton Trans.*, 2221 (1986).
86De2   Decock-Le Reverend, B., Lebriki, A., Livera, D., & Pettit, L. D.: *Inorg. Chim. Acta*, **124**, L19 (1986).
86Fa   Farkas, E., Tözsér, J., & Gergely, A.: *Magyar Kém. Foly.*, **92**, 49 (1986).
86Ge   Gerega, K., Kozlowska, G. F., Trzebiastowska, B. J., & Decock-Le Reverend, B.: *Inorg. Chim. Acta*, **125**, 183 (1986).
86Go   Goodgame, D. M. L. & Joy, A. M.: *J. Inorg. Biochem.*, **26**, 219 (1986).
86Ha   Hay, R. W. & Pujari, M. P.: *Inorg. Chim. Acta*, **123**, 47 (1986).
86Ki   Kiss, T. & Szücs, Z.: *J. Chem. Soc., Dalton Trans.*, 2443 (1986).
86Mi   Micskei, K. & Nagypál, I.: *J. Chem. Soc., Dalton Trans.*, 2721 (1986).
86Mu   Murdock, C. M., Cooper, M. K., Hambley, T. W., Hunter, W. N., & Freeman, H. C.: *J. Chem. Soc., Chem. Commun.*, 1329 (1986).

86Pe   Pettit, L. D. & Lyons, A. Q.: *J. Chem. Soc., Dalton Trans.*, 499 (1986).

86Sa   Saato, M., Matsuki, S., & Ikeda, M., Nakaya, J.: *Inorg. Chim. Acta*, **125**, 49 (1986).

86So1   Sóvágó, I., Harman, B., Gergely, A., & Radomska, B.: *J. Chem. Soc., Dalton Trans.*, 235 (1986).

86So2   Sóvágó, I., Várnagy, K., & Bényei, A.: *Magyar Kém. Foly.*, **92**, 114 (1986).

87Ar   Arena, G., Bonomo, R. P., Impellizzeri, G., Izatt, R. M., Lamb, J. D., & Rizzarelli, E. : *Inorg. Chem.*, **26**, 795 (1987).

87Bo   Bonomo, R. P., Maccarone, G., Rizzarelli, E., & Vidali, M.: *Inorg. Chem.*, **26**, 2893 (1987).

87Ha   Hay, R. W. & Nolan, K. B.: In *Specialist Periodical Report*, **19**, 290 (1987) (and previous volumes).

87Ki   Kiss, T.: *J. Chem. Soc., Dalton Trans.*, 1263 (1987).

87Ko1   Kozlowski, H., Decock-Le Reverend, B., Ficheux, D., Loucheux, C., & Sóvágó, I.: *J. Inorg. Biochem.*, **29**, 187 (1987).

87Ko2   Kowalik, T., Kozlowski, H., Sóvágó, I., Várnagy, K., Kupryszewski, G., & Rolka, K.: *J. Chem. Soc., Dalton Trans.*, 1 (1987).

87La   Laussac, J. P., Pasdeloup, M., & Hadjiliadis, N.: *J. Inorg. Biochem.*, **30**, 227 (1987).

87Li   Livera, C. E., Pettit, L. D., Bataille, M., Perly, B., Kozlowski, H., & Radomska, B.: *J. Chem. Soc., Dalton Trans.*, 661 (1987).

87Pe   Pettit, L. D., Livera, C., Steel, I., Bataille, M., Cardon, C., & Kozlowska, G. F.: *Polyhedron*, **6**, 45 (1987).

87Ra   Radomska, B., Kiss, T., & Sóvágó, I.: *J. Chem. Soc., Chem. Res. (S)*, 156 (1987).

87So   Sóvágó, I. & Petöcz, Gy.: *J. Chem. Soc., Dalton Trans.*, 1717 (1987).

87Ue1   Ueda, J., Ikota, N., Hanaki, A., & Koga, K.: *Inorg. Chim. Acta*, **135**, 43 (1987).

87Ue2   Ueyama, N., Ueno, S., & Nakamura, A.: *Bull. Chem. Soc. Japan*, **60**, 283 (1987).

88An   Anliker, S. L., Beach, M. W., Lee, H. D., & Margerum, D. W.: *Inorg. Chem.*, **27**, 3809 (1988).

88Da   Daniele, P. G., Zerbinati, O., Aruga, R., & Ostacoli, G.: *J. Chem. Soc., Dalton Trans.*, 1115 (1988).

88De   Decock-Le Reverend, B., Liman, F., Livera, C., Pettit, L. D., Pyburn, S., & Kozlowski, H.: *J. Chem. Soc., Dalton Trans.*, 887 (1988).

88Ge   Gerega, K., Kozlowski, H., Masiukiewicz, E., Pettit, L. D., Pyburn, S., & Rzeszotarska, B.: *J. Inorg. Biochem.*, **33**, 11 (1988).

88Ha   Hanaki, A., Ikota, N., Motono, K., & Yamauchi, O.: *Nippon Kagaku Kaishi*, 578 (1988).

88Li   Livera, C., Pettit, L. D., Bataille, M., Krembel, J., Bal, W., & Kozlowski, H.: *J. Chem. Soc., Dalton Trans.*, 1357 (1988).

88Re   Rehder, D.: *Inorg. Chem.*, **27**, 4312 (1988).

88So   Sóvágó, I., Kiss, T., Várnagy, K., & Decock-Le Reverend, B.: *Polyhedron*, **7**, 1089 (1988).

88Ta   Tabata, M. & Tanaka, M.: *Inorg. Chem.*, **27**, 3190 (1988).
88Va1   Várnagy, K., Kozlowski, H., Sóvágó, I., Jankowska, T. K., Kruszynski, M., & Zboinska, J.: *J. Inorg. Biochem.*, **34**, 83 (1988).
88Va2   Várnagy, K., Sóvágó, I., & Kozlowski, H.: *Inorg. Chim. Acta*, **151**, 117 (1988).
89Ba1   Bal, W., Bataille, M., Kozlowski, H., Delstanche, M. L., Pettit, L. D., Pyburn, S., & Scozzafava, A.: *J. Chęm. Soc., Dalton Trans.*, (in press).
89Ba2   Bal, W., Kozlowski, H., Masiukiewicz, E., Rzeszotarska, B., & Sóvágó, I.: *J. Inorg. Biochem.*, **36** (1989).
89Ba3   Baidya, N., Olmstead, M. M., & Mascharak, K. P.: *Inorg. Chem.*, **28**, 3426 (1989).
89Fa1   Farkas, E. & Kiss, T.: *Polyhedron* **8**, 2463 (1989).
89Ki   Kiss, T., Farkas, E., Kozlowski, H., Siatecki, Z., Kafarski, P., & Lejczak, B.: *J. Chem. Soc., Dalton Trans.*, 1053 (1989).
89Ko   Kozlowski, H., Radomska, B., Kupryszewski, G., Lammek, B., Livera, C., Pettit, L. D., & Pyburn, S.: *J. Chem., Soc., Dalton Trans.*, 173 (1989).
89No   Noszál, B., Schön, I., Nyéki, O., Tánczos, R., & Nyiri, J.: *Int. J. Pept. Prot. Res.* (in press).
89Pe   Pettit, L. D., Pyburn, S., Kozlowski, H., Decock-Le Reverend, B., & Liman, F.: *J. Chem. Soc., Dalton Trans.*, 1471 (1989).
89Ra1   Radomska, B., Kubiak, M., Glowiak, T., Kozlowski, H., & Kiss, T.: *Inorg. Chim. Acta*, **159**, 111 (1989).
89Sh   Shtyrlin, V. G., Gogolashvili, E. L., & Zakharov, A. V.: *J. Chem. Soc., Dalton Trans.*, 1293 (1989).
89Sz   Szabó-Planka, T., Peintler, G., Rockenbauer, A., Györ, M., Varga-Fábián, M., Institórisz, L., & Balázspiri, L.: *J. Chem. Soc., Dalton Trans.*, 1925 (1989).
90Ch   Cherifi, K., Decock-Le Reverend, B., Várnagy, K., Kiss, T., Sóvágó, I., & Kozlowski, H.: *J. Inorg. Biochem.* **38**, 69 (1990).
90Fa   Farkas, E. & Kiss, I.: *J. Chem. Soc., Dalton Trans.* 749 (1990).
90Ra   Radomska, B., Sóvágó, I., & Kiss, T.: *J. Chem. Soc., Dalton Trans.* 289 (1990).
90So   Sóvágó, I., Radomska, B., Schön, I., & Nyéki, O.: *Polyhedron* **9**, 825 (1990).
90Ki   Kiss, T., Sóvágo, I., & Gergely, A.: *J. Pure Appl. Chem.* (in press).
90Ko   Kozlowski, H., Sóvágó, I., Spychala, J., Urbanska, J., Várnagy, K., Kiss, A., & Cherifi, K.: *Polyhedron* **9**, 831 (1990).

V

Thermodynamic and kinetic aspects of metalloenzymes and metalloproteins

Junzo Hirose and Yoshinori Kidani
Faculty of Pharmaceutical Sciences, Nagoya City University, Japan

INTRODUCTION

Metal ions are present in cells, where they play important physiological roles. The metal content of a cell is much smaller than generally that of metal organic compounds. In spite of the low content of transition metal ions, they play important roles relating to the catalysis of the reactions of the substrates at the active sites of the enzymes, and in the regulation of the phsiological reactions in a cell. The metalloenzymes and metal-requiring enzymes comprise about 30% of the total enzymes known at present. It is not known how metal ions were selectively incorporated into the metalloenzymes after the synthesis of the protein on the ribosomes, and how the metal ions required for enzyme activity are selectively located near the metal-requiring enzymes in a cell. This is a very large question in the metal metabolism of the cell, and it is still an unresolved problem, because the amount of transition metal ions in the cell is too small for their natural state in the cell to be established, and a complicated equilibrium state involving metal-binding proteins and small ligands may be set up in a cell. The detailed physiological metabolism of the metal ions in a cell is not known.

The study of metalloenzymes further our understanding of metal binding in the protein, because the metal ions are strongly bound at the metal-binding sites of the protein. The thermodynamic and metal-binding characters of metalloenzymes will provide much information towards an understanding of the basic character of the interactions between metal ions and proteins, which are very different from metal–small ligand interactions. Metal-binding constants and binding-rate constants for a metalloenzyme will give basic information concerning metal selection in metalloenzymes. Free metal ions may be coordinated to a small molecule in a cell, hence the rate of metal binding and the rate of dissociation from the metal-binding sites of metalloenzymes may be influenced by small ligands. This behaviour may also be correlated with the metal-binding selectivity or metal metabolism in a cell. In this chapter we describe the metal-binding and dissociation mechanism of metallo-enzymes and proteins.

In metalloenzymes, the metal ion plays an important role in the catalysis of the reactions of substrates. To understand this role it is very important to study the character of the metal ion in the intermediate state of the reaction product of the substrate with the metalloenzyme.

We focus mainly on the intermediate state of the catalytic step of reactions of the substrate. The binding mode and thermodynamic character of the inhibitor likewise yield much information about the catalytic mechanism of the reactions of metallo-enzymes. We also describe the binding and dissociation mechanisms of small ligands (inhibitors and substrates) in reactions involving metalloenzymes and metalloproteins.

1. METAL-BINDING AND DISSOCIATION MECHANISM FOR METALLOENZYMES AND PROTEINS

1.1 Introduction

Various chelating agents remove the metal ion from carboxypeptidase A and carbonic anhydrase according to the following equation system [86Ki, 77Kil, 76Ki, 78Rol, 79Bi, 80Bi].

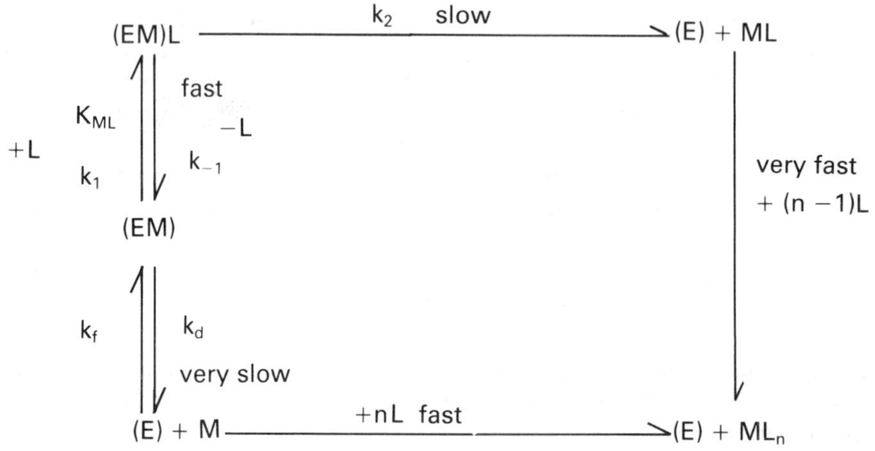

$$
\begin{array}{c}
\text{(EM)L} \xrightarrow{\quad k_2 \quad \text{slow} \quad} \text{(E)} + \text{ML} \\[2mm]
\end{array}
$$

$$
+L \quad K_{ML} \; k_1 \;\Big\Updownarrow\; \begin{array}{l}\text{fast} \\ -L \\ k_{-1}\end{array} \qquad\qquad\qquad \begin{array}{l}\text{very fast} \\ + (n-1)L \end{array} \quad (1)
$$

$$
\text{(EM)}
$$

$$
k_f \;\Big\Updownarrow\; \begin{array}{l} k_d \\ \text{very slow}\end{array}
$$

$$
\text{(E)} + \text{M} \xrightarrow{\quad +nL \;\; \text{fast} \quad} \text{(E)} + \text{ML}_n
$$

A small chelating agent (L) forms a ternary complex ((EM)L) with the metal-loenzyme and removes the metal ion from the enzyme ((EM)L → (E) + ML). The rate of formation of the ternary complex is very high, while the rate of dissociation of the ternary complex to apoenzyme ((E)) and metal complex (ML) is low [86Ki]. Polydentate ligands (EDTA, NTA, etc.) cannot enter the cavity of the enzyme where the active site is located, and thus they cannot form ternary complexes. Polydentate ligands can react only with the metal ion dissociated from the metallo-enzyme, thereby behaving as scavengers $((E)M \xrightarrow{L} (E) + M \rightarrow ML_n + (E))$. Detailed data and results on such reactions were published in a previous review [86Ki].

A metal bound to a polymer such as a protein has a different character from a metal bound to a small ligand. In this chapter, therefore, the metal-binding constants and metal-binding and dissociation mechanism for metalloenzymes and proteins will be described and compared with those for small ligands.

1.2 Carboxypeptidase A

Carboxypeptidase A (peptidyl-L-amino acid hydrolase, EC 3.4.17.1: Zn(CPA)) catalyses the hydrolysis of the C-terminal peptide bond of proteins and polypeptides. It consists of a single polypeptide chain with a molecular weight of 35 000 and has a zinc ion at the active site. The zinc ion is at the bottom of the deep crevice of the enzyme [89Ch]. X-ray analysis has shown that the zinc ion at the active site possesses five-coordinate geometry; it is coordinated by the two oxygens of the carboxylate group of Glu-72, two nitrogens of His-69 and 196, and the oxygen of a water molecule (Fig. 5.1) [86Re].

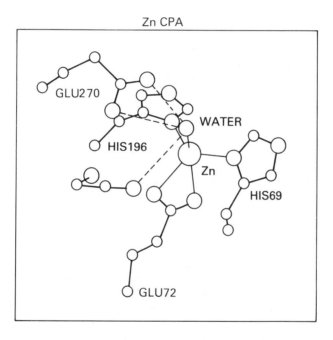

Fig. 5.1 — The zinc-binding mode in Zn(CPA). Thin solid lines represent metal–ligand interactions, while dashed lines indicate hydrogen bonds.

Zinc can be removed from the enzyme by a chelating agent and replaced by other metal ions. The metallo derivatives of the enzyme and the corresponding activities are shown in Table 5.1 [86Au1].

The binding constants of some metal ions for apocarboxypeptidase A were measured by Vallee *et al.* by means of the equlibrium dialysis method, and are shown in Table 5.1 [61Co].

Table 5.1 — Activities and metal-binding constants of metallocarboxypeptidases
(Reproduced from refs. 86Au1 and 61Co)

Metal	% Activity ($v/v_{zinc} \times 100$)		Binding constant[§] (M^{-1})
	Peptidase[†]	Esterase[‡]	
Apo	0	0	
Zinc	100	100	$10^{10.5}$
Cobalt	200	110	$10^{7.0}$
Nickel	50	40	$10^{8.2}$
Manganese	30	160	$10^{5.6}$
Cadmium	5	140	$10^{10.8}$
Mercury	0	90	10^{21}
Rhodium	0	70	
Lead	0	60	
Copper	0	0	$10^{10.6}$

[†] 0.02 M benzyloxycarbonylglycyl-L-phenylalanine, pH 7.5, 0°C;
[‡] benzoylglycyl-D,L-phenylacetate, pH 7.5 25°C;
[§] Stability constant corrected for competition by 1 M Cl$^-$ and 0.05 M tris buffer, pH 8.0.

The binding constant of cobalt-carboxypeptidase A (Co(CPA) was measured directly by determining the metal ion content released from the metal–enzyme via equilibrium dialysis. Other binding constants of metallo derivatives of carboxypeptidase A were measured by competitive reaction between cobalt and other metal ions for the apoenzyme [61Co].

$$Co(CPA) + M^{2+} \rightleftharpoons M(CPA) + Co^{2+} \qquad (2)$$

The sequence of binding constants in various metallo carboxypeptidase A complexes is Mn(CPA) < Co(CPA) < Ni(CPA) < Cu(CPA) > Zn(CPA) and is consistent with the Irving–Williams series. X-ray analysis data on various metallo derivatives indicate that the Co and Ni derivatives also involve five-coordination geometry [86Re]. On the basis of Vis-absorption, and MCD spectra and NMR data, it is proposed that Co(CPA) exhibits five-coordination geometry or distorted tetrahedral geometry in the solution state [83Be1], but that Ni(CPA) has octahedral geometry [75Ro].

Wilkins et al. [78Bi] measured the rate constants for the formation (k_f) and dissociation (k_d) of metallocarboxypeptidase A, as given by the following equations:

$$M^{2+} + CPA \underset{k_d}{\overset{k_f}{\rightleftharpoons}} M(CPA) \qquad (3)$$

$$K = \frac{k_f}{k_d} . \qquad (4)$$

Proton release occurs on the uptake of a metal ion by apocarboxypeptide A. The formation rate could be determined by monitoring this small pH change by using indicators in a stopped-flow apparatus. The change in the pH level reflects the rate of metal ion incorporation into the apoenzyme [78Bi]. The observed first order rate constant (k_{obs}) for metal ion entry was proportional to the metal ion concentration used, because the metal concentration was in excess over that of the apoenzyme [78Bi]. From these data, the second order rate constant was determined [78Bi]. The second order formation rate constants (k_f) for the binding of some metal ions of apocarboxypeptidase A at various pH values between 5.5 and 8.5 are shown in Fig. 5.2 [78Bi]. k_f increases with increasing pH and the k_f values of metallocarboxypepti-

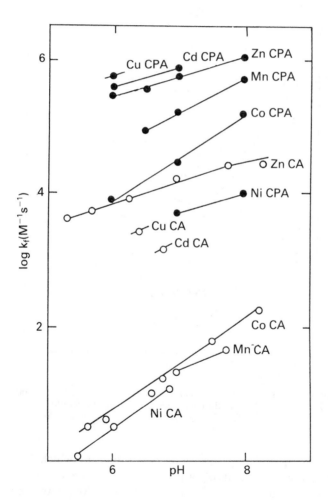

Fig. 5.2 — Plot of log formation rate constants ($M^{-1}s^{-1}$) as a function of pH for the reaction of apocarboxypeptidase A (CPA) and apocarbonic anhydrase (CA) with various metal ions at 25°C and $I = 1$ M NaCl. Reproduced from ref. 78Bi.

dase A are larger than those of the corresponding metallocarbonic anhydrase. The rate of dissociation of metal ions from metallocarboxypeptidase A was measured by following the loss in enzyme activity when bulky chelating agents (EDTA, NTA) were used as scavengers (the polyaminocarboxylates remove metal ions from metalloenzymes by metal self dissociation pathway in Eq. (1) [86Ki, 78Hi]. The pH-dependence of the rate constants of metal ion dissociation from metallocarboxypeptidase A are shown in Fig. 5.3 [78Bi]. k_d is independent of the concentration or

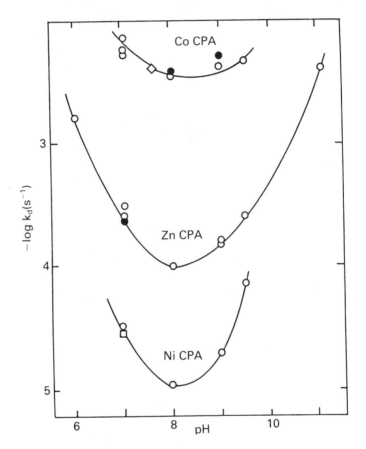

Fig. 5.3 — Rate constants as a function of pH for the dissociation of Co(CPA), Zn(CPA), and Ni(CPA) at 25°C and $I = 1$ M NaCl. Buffers: 50 nM acetate (pH 6); 50 mM Tris (pH = 7–11). Scavengers: ○, EDTA; ●, NTA; □, CyDTA; ◇, Zn^{2+}. Reproduced from ref. 78Bi.

nature of the scavenger. Under the same conditions for which k_f and k_d were determined, the binding constants of various metal ions for apocarboxypeptidase A were measured via the competition reaction for the metal ions between apocarboxypeptidase A and a chelating agent [78Bi]. The binding constants calculated from the ratio of the rate constants for the formation and dissociation of these metallocarboxypeptidase A complexes were in good agreement with the values obtained via the competitive reaction with chelating agents or by the dilution method [78Bi] (Table 5.2). The zinc binding constants ($\log K$) for apo-carboxypeptidase A increased with increasing pH and the relationship between $\log K$ and pH from 5.0 to 8.0 was given by the equation [78Bi] (Fig. 5.4):

$$\log K = 2.5 + 1.0\,\text{pH} \ . \tag{5}$$

The K value becomes constant at a value of $10^{11}\,\text{M}^{-1}$ above pH 8.5. The same behaviour is found for bovine and human carbonic anhydrase (see the following section) [62Li, 64Li]. This value might be a real stability constant because of the pH-independence. Below pH 8.0, the slope of this line was about 1.0. This behaviour implied that apparently one proton competes with the metal ion for the binding site in the enzyme [78Bi] (Fig. 5.4).

Table 5.3 gives the rate constants ($k_0^{(\text{CPA})}$) for the reactions of apocarboxypeptidase A with various metal ions. Rate constants for the reactions of a neutral 1,10-phenanthroline molecule ($k_f^{(\text{phen})}$) with metal ions and the rate constants ($k_{\text{ex}}^{\text{H}_2\text{O}}$) of water exchange (an inner-sphere water molecule ligated to metal ions is exchanged with a water molecule in the outer sphere) [78Bi] are also shown.

$$\text{M(H}_2\text{O)}_n^{2+} + \text{H}_2\text{O*} \rightleftharpoons \text{M(H}_2\text{O)}_{n-1}^{2+}(\text{H}_2\text{O*}) + \text{H}_2\text{O} \ \ k_{\text{ex}}^{\text{H}_2\text{O}} \tag{6}$$

The rate constants for the information of the complexes of a neutral ligand such as 1,10-phenanthroline with metal ions are about one order of magnitude lower than the corresponding water exchange rate constants [78Bi, 74Wil]. The rate constants for the formation of various metal complexes for apocarboxypeptidase A ($k_0^{(\text{CPA})}$) are almost the same as those of the reactions of various metal ions with 1,10-phenanthroline (k_f) [78Bi]. Wilkins *et al.* interpreted this behaviour as reflecting the magnitude of K_{as}, the outer-sphere association constant between M^{2+} and the neutral ligand, on the basis of the following mechanism [78Bi]:

$$\text{M}^{2+} + \text{L} \underset{}{\overset{K_{\text{as}}}{\rightleftharpoons}} \text{M}^{2+}\text{L} \overset{k_{\text{ex}}}{\rightarrow} \text{ML} \tag{7}$$

$$\text{outer-sphere} \qquad \text{inner-sphere}$$
$$\text{complex} \qquad \qquad \text{complex}$$

The rate constants of formation of the complexes of a small ligand with various metal ions are very similar to those for apocarboxypeptidase A. This behaviour indicates that the metal-binding site in apocarboxypeptidase A is very flexible, despite the metal-binding site being in the deep crevice of the protein [78Bi].

The rates of reaction of metal complexes of various chelating agents with apoarsanilazo-Tyr-248 carboxypeptidase A (a derivative with the selectivity modified chromophoric arsanilazo-Tyr-248 residue introduced without significantly affecting the activity of the enzyme (Azo-CPA)) were measured by Hirose & Wilkins [84Hi1]. Only the species with a 1:1 metal:ligand (bidentate or tridentate ligand) ratio can react with apo-Azo-CPA and not via dissociation of the metal ions. The species with 1:2 and 1:3 metal:ligand ratios were not active species for the metal transfer reaction. The rate constants of the reactions of the cobalt complex with various chelating agents are shown in Table 5.4 [84Hi1]. The rate constants of the reactions

Table 5.2 — Binding constants of metal ions by kinetic and equilibrium dialysis method in carbopeptidase A and carbonic anhydrase

	k_f (M^{-1}s^{-1})	k_d (s^{-1})	k_f/k_d (M^{-1})	Equilibrium dialysis (K) (M^{-1})	Conditions	Source	Ref.
Metallocarboxypeptidase A							
Zn^{2+}	6×10^5	3×10^{-4}	2×10^9	3.0×10^{10}	pH7	bovine	78Bi
				3.0×10^{11}	pH8		86Ki
Co^{2+}	3×10^4	6×10^{-3}	5×10^6	1.0×10^7	pH7	bovine	78Bi
Ni^{2+}	5×10^3	3×10^{-5}	3×10^8	1.5×10^8	pH7	bovine	78Bi
Mn^{2+}	1×10^5			0.4×10^5	pH7	bovine	78Bi
Cd^{2+}	1×10^6				pH7	bovine	78Bi
Cu^{2+}	1×10^6				pH6	bovine	78Bi
Metallocarbonic anhydrase							
Zn^{2+}	3.0×10^3	1.0×10^{-6}	3.0×10^9	2.5×10^9	pH5	bovine II	68He, 77Ki1,
		8.7×10^{-7}					77Ki1
Cu^{2+}				3.2×10^{10}	pH5.5	human II	64Li
Ni^{2+}				4.0×10^{11}	pH5.5	human II	64Li
				3.2×10^9	pH5.5	human II	64Li
Co^{2+}	0.72	1.2×10^{-6}	6.0×10^5	1.5×10^6	pH5	bovine II	77Ki2
	1.5×10^2				pH8.2	bovine II	78Bi
				1.6×10^7	pH5.5	human II	64Li
Mn^{2+}	56	3.5×10^{-4}	8×10^4		pH7.75	bovine II	74Wi2
	9.0	1.1×10^{-3}			pH6.25	bovine II	74Wi2
Cd^{2+}				6.3×10^3	pH5.5	human II	64Li
				1.6×10^9	pH5.5	human II	64Li

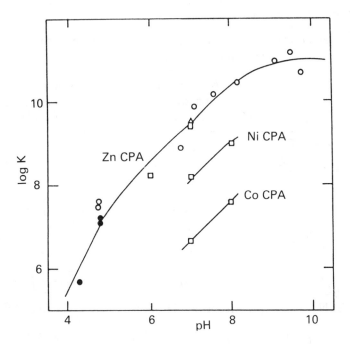

Fig. 5.4 — Conditional formation constants of Co(CPA), Zn(CPA) and Ni(CPA) as a function of pH at 25°C and $I = 1$ M NaCl. 50 mM acetate (pH 4–5); 50 mM Tris (pH = 6–10). $\bigcirc$, 1,10-phenanthroline competition; $\bullet$, proton competition; $\triangle$, activity of dilute solution (100 mM Tris); $\square$, k_f/k_d (k_f determined in 25 μM Tris). Reproduced from ref. 78Bi.

Table 5.3 — Comparison of rate constants for water exchange and formation of metallocarboxypeptidase and 1,10 phenanthroline. (Reproduced from ref. 78Bi)

Me^{2+}	$k_{ex}^{H_2O\dagger}$, s^{-1}	$k_f^{(phen)}$, M^{-1}s^{-1}	$k_O^{(CPA)}$, M^{-1}s^{-1}
Zn^{2+}	3.2×10^7	6.3×10^6	1×10^6
Co^{2+}	2.5×10^6	2×10^5	3×10^5
Ni^{2+}	2.5×10^4	2.5×10^3	1.5×10^4
Mn^{2+}	3×10^7	2.5×10^6	7×10^5
Cu^{2+}	2×10^9	7.6×10^7	10^6
Cd^{2+}	2×10^8	$> 10^7$	1×10^6

[†] Ref. 74Wil.

of the cobalt complex vary with the nature of the ligands. In comparison with the hexa-aquocobalt(II)ion, the coordination of some ligands (1,10-phenanthroline, terpyridine, and 1,10-phenanthroline-5-sulfonate) accelerates the uptake of cobalt

Table 5.4 — Interaction of Co(II) complexes (CoL^{n+}) with apocarbonic anhydrase[†] and apocarboxypeptidase A[‡]

L	Apocarbonic anhydrase[†] (M^{-1}s^{-1})	Apoazocarboxy-peptidase[‡] (M^{-1}s^{-1})
(8-hydroxyquinoline, N, O$^-$)	$\sim 4 \times 10^3$	
(quinoline with SO$_3^-$, N, O$^-$)	1.9×10^2	9.5×10^4 (7.0×10^5)[¶]
H$_2$O	81	7.7×10^5 (5.0×10^6)[¶]
SCN$^-$	Strong acceleration[§]	No effect
($^-$OOC–N–COO$^-$)	71	9×10^4
(phenanthroline, N N)	30	3×10^6 (4.0×10^6)[¶]
(phenanthroline, SO$_3^-$, N N)		2.4×10^6
(terpyridine, N N N)	3	1.2×10^6
(terpyridine, SO$_3$, N N N)		3.2×10^5
(O$_3$S–, $^-$O$_3$S, HO OH, SO$_3^-$, N N)		$< 1.0 \times 10^5$

† pH 7.5, $I = 0.1$ M [75Ge];
‡ pH 8.2, $I = 0.1$ M [84Hi1], [80Ha];
§ For zinc complexes.

ions by apo-Azo-CPA. Only monophasic kinetics were observed in the reaction of cobalt complexes with apo-Azo-CPA, and this may imply a simple bimolecular attack of CoL on the apoprotein. Nevertheless, the reaction may proceed in two steps as in Eq. (7), because the rate constants for the reactions of the cobalt complex are influenced by the nature of the ligands [84Hi1]. There is a possibility that the first step is the formation of the outer-sphere complex and the metal–ligand complex subsequently binds to the metal-binding site in the enzyme to form the ternary complex (LCo(Azo-CPA)) [84Hi1].

In the metal ion transfer from a small ligand to the positive charged polyamine-macrocyclic ligand, ligands with a negative charge accelerated the metal transfer from the metal ligand complex to the polyamine-macrocycle [85Wu]. This reaction system is a very good model for the reaction between the apoenzyme and the metal complex. The active species in this reaction was also a 1:1 metal:ligand complex [85Wu] (Table 5.5). In the copper complexes with bi- and tridentate aminocarboxy-

Table 5.5 — Interaction of Cu(II) complexes (CuL^{n+}) with a macrocyclic ligand (1,4,8,11-tetraazacyclotetradecane). $I = 0.5\,M$ (KNO_3) (Reproduced from ref. [85Wu])

L	Rate constants[†] $(M^{-1}s^{-1})$
Succinate^{2-} (Suc^{2-})	$56(8) \times 10^6$
Glycolate$^-$ (Glol$^-$)	$18(2) \times 10^6$
Picolinate$^-$ (Pic$^-$)	$7.1(2) \times 10^6$
Malonate^{2-} (Mal^{2-})	$5.3(3) \times 10^6$
Glycine$^-$ (Gly$^-$)	$5.2(2) \times 10^6$
Iminodiacetate (IDA^{2-})	$3.7(1) \times 10^6$
H_2O	2.4×10^6
Nitrilotriacetate (NTA^{3-}	$8.5(1) \times 10^6$
N-(2-hydroxyethyl)- -ethylenediaminetriacetate (HEDTA^{3-})	4.3
Ethylenediamine (en)	$7.1(1) \times 10^6$
1,3-Diaminopropane (pn)	$3.9(1) \times 10^6$
Diethylenetriamine (dien)	$4.5(2) \times 10^2$
bis(3-aminopropyl)amine (dipn)	$9.1(5) \times 10$

† Standard errors in brackets.

lates (glycine and iminodiacetic acid) no change in reactivity is found, but the rates are decreased for the copper complex of a tetradentate aminocarboxylate (nitrotriacetic acid) [85Wu].

	$Cu(H_2O)_6^{2+}$	$Cu(Gly)^+$	$Cu(IDA)$	$Cu(NTA)$
$k[M^{-1}s^{-1}]$	2.4×10^6	5.2×10^6	3.7×10^6	8.5×10^5

The presence of a water molecule ligated to the copper ion may be very important in making the ternary complex an intermediate species, because the macrocyclic ligand can easily attack a 1:1 copper:ligand complex by replacing the water molecule bound to the copper. Some ligands with a negative charge neutralize the positive charge of metal ions, so that the outer-sphere complex formed between the metal complex and the macrocyclic compound may be more stable than that formed between an aquometal complex ion and a macrocyclic ligand [85Wu]. The coordination of a negatively charged ligand to the copper ion will therefore accelerate complexation of the copper ion with a macrocyclic ligand.

In the reaction of a cobalt complex with apo-Azo-CPD, the 1:1 ligand:metal complex was again the active species. Since the active site in the enzyme is present in the deep crevice of the protein, 1:2 or 1:3 species may not be able to enter the active site because of the steric hindrance of the protein, and the water molecules ligated to the cobalt ion in the 1:1 species may be very convenient as concerns formation of the ternary complex between the apoenzyme and the 1:1 cobalt complex (LCo(II)(CPA)) [84Hi1].

Why can terpyridine, 1,10-phenantroline, and 1,10-phenanthroline 5-sulpho-nate, but apparently no other ligands, accelerate the metal transfer from the 1:1 metal-ligand complex to apo-Azo-CPA? As regards the active site of carboxypepti-dase A, there is a hydrophobic site near the metal-binding site, to which the hydrophobic phenyl group of the substrates can bind [89Ch]. This hydrophobic site may play an important role in formation of the outer-sphere complex between the 1:1 metal:ligand complex and apo-Azo-CPD. However, the cobalt complex containing a very bulky ligand such as 1,10-phenanthroline-4,7-diparabenzenesulphonate has a small rate constant because of the steric hindrance [84Hi1].

1.3 Carbonic anhydrase
Carbonic anhydrase (carbonate hydrase, EC 4.2.1.1, Zn(CA)) is a metalloenzyme containing 1 Zn atom per molecular weight 30 000. Carbonic anhydrase catalyses the hydration of carbon dioxide. Three genetically and immunologically distinct but structurally homogeneous cytosolic isozymes are known, which have different activities and small ligand-binding properties [88Si]. Isozyme II has the highest specific activity and is present together with isozyme I, which has a less efficient enzyme activity, in the red blood cells. Isozyme III is present in the muscles and has lower activity [88Si]. Nevertheless, the basis features of the catalytic mechanisms for these isozymes are very similar to one another [88Si].

$$CO_2 + H_2O \rightleftharpoons HCO_3^- + H^+ \tag{8}$$

The results of an X-ray analysis of this enzyme by a Swedish group [80Ka, 72Li, 86Er] are shown in Fig. 5.5. The catalytic zinc ion is present at the active site at the bottom of the 15 Å† deep narrow cavity [80Ka]. One part of the cavity is dominated by hydrophobic amino acid side chains, whereas the other part has mostly hydrophilic residues. The zinc ion in human carbonic anhydrase II is coordinated by three

† 1Å = 0.1 nm

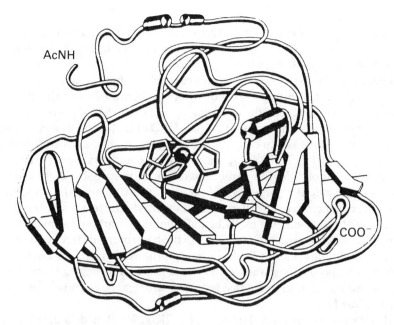

Fig. 5.5 — A simplified drawing of the backbone structure of human carbonic anhydrase II. The zinc ion is coordinated by three His residues. The arrows represent pleated sheet strands pointing from the amino toward the carboxyl end, and the cylinders represent helical regions. Reprinted with permission from ref. 72Li.

imidazole groups from His-94, His-96, and His-119 [80Ka]. The fourth ligand is a water molecule, and the coordination geometry of the zinc ion is tetrahedral [80Ka].

A residue with a pK_a of 7.0 was found to be very important for the catalysis of CO_2 hydration. The chemical nature of this activity-linked group was for a long time unknown and was the subject of research [73Li, 82Li]. Recently, a number of investigations have indicated that this important group is the water molecule coordinated to the zinc ion [88Si]. A simplified model of the catalytic mechanism is shown by Eqs. (9) and (10). This mechanism involves a direct nucleophilic attack of the zinc-bound OH^- on CO_2 [88Si].

$$EZnOH^- + CO_2 \rightleftharpoons EZn(OH^-)CO_2 \rightleftharpoons EZnHCO_3^- \rightleftharpoons$$

$$\rightleftharpoons EZnH_2O + HCO_3^- \qquad (9)$$

$$EZnH_2O \rightleftharpoons EZnOH^- + H^+ \qquad (10)$$

Zinc ions can be removed by dialysis against chelating agents (2,6-pyridinedicarboxylate or 1,10-phenanthroline) and the enzyme activity disappears on removal of the zinc ions [86Ki, 77Kil, 76Ki]. However, the activity is recovered completely when zinc ions are added to apo-(CA). Lindskog and Malmström determined the metal-binding constant for apo-human CA using a chelating ligand as metal competitive agent [62Li, 64Li]. The results are shown in Table 5.6.

$$M(CA) + L_n \rightleftharpoons LM(CA) \rightleftharpoons ML_n + (CA) .$$

The sequence of binding constants for various metallocarbonic anhydrases was Zn^{2+} < Cu^{2+} > Ni^+ > Co^2 > Mn^{2+}. This is consistent with the Irving–Williams series for small ligand complexes [62Li, 64Li].

Wilkins *et al.* [84Hi1, 75Ge] found that only cobalt complexes with a 1:1 ligand:metal ratio react with apo-bovine carbonic anhydrase II, and that the reaction rates depend on the nature of the chelating agent [75Ge, 84Hi1]. This behaviour is completely consistent with that of apo-(Azo-CPD) for the cobalt-ligand complex. The cavity of the active site in carbonic anhydrase should be narrower than that of carboxypeptidase A [89Ch, 80Ka], because carbonic anhydrase catalyzes the reactions of only small molecules like carbon dioxide or carbonate, but carboxypeptidase A catalyses the reactions of large peptides at its active site. The cavity of carboxypeptidase A is much larger than that of carbonic anhydrase. However, since carboxypeptidase A cannot react with 1:2 metal:ligand complexes carbonic anhydrase is even less capable of reacting with such species. On the basis of the different rate constants with various chelating agents in Table 5.4, it was proposed that the reactions of cobalt complexes with apo-bovine carbonic anhydrase proceed through a ternary complex [75Ge]. Henkens *et al.* also found that bidentate azo ligands can bind to the zinc ions at the active site [72He].

On the basis of kinetic data, Hirose *et al.* [77Ki1, 76Ki] made a detailed study of the reaction of metal removal from bovine carbonic anhydrase II by chelating agents, and found that the chelating agents form ternary complexes with metallocarbonic anhydrase (LM(CA)). The formation constants (K_{EML}) of the ternary complex are relatively small (Table 5.11 in section 2.3), thus the formation of the ternary complex has no influence on the measurement of the stability constant of the metal apo-enzyme complex, which was done by using a low chelating agent concentration in equilibrium dialysis [86Ki]. The formation of a ternary complex with metallocarbonic anhydrase was also confirmed by Roman *et al.* [78Ro1] and Hendy *et al.* [82He].

Bulky ligands such as EDTA and NTA cannot form a ternary complex with metallocarbonic anhydrase because the cavity is too narrow for them, so that the metal removal reaction proceeds by the reaction of self-dissociation of the metal ion from the metalloenzyme according to Eq. (1) [86Ki, 77Ki2]. This reaction was used to measure the self-dissociation constant of metal ions from metalloenzymes.

The rate constants of formation of the complex between zinc ions and apocarbonic anhydrase was first measured by Henkens & Startevant [68He] and was shown in Table 5.2. The rate constants for formation of the complexes of various metals and carbonic anhydrase and the pH-dependence were measured by Wilkins *et al.* [78Bi] and were also given in Table 5.2 and Fig. 5.2. The rate constants of the binding of zinc ions and other metal ions with apocarbonic anhydrase were much smaller than those for carboxypeptidase A [78Bi]. The pH-dependence (the slopes of the lines in the plots of $\log k_f$ versus pH) of the formation rate constants of various metal ions (Fig. 5.2) is very similar to that for carboxypeptidase A [78Bi]. For zinc-, cobalt-, and manganese-carbonic anhydrase, the formation (k_f) and dissociation (k_d) rate constants have been measured (Table 5.2) and the binding constants of the metal ions have been calculated in terms of the ratio k_f/k_d. These values are very similar to those obtained by equilibrium dialysis.

The enthalpy changes accompanying the binding of Zn^{2+}, Co^{2+}, Cu^{2+}, Cd^{2+}, and Ni^{2+} to apocarbonic anhydrase I and II were determined through reaction calorimetry by Henkins & Startevant [69He] and are shown in Table 5.6. Small

Table 5.6 — Thermodynamic parameters for the specific binding of metal ions to apocarbonic anhydrase at 25°C. (Reproduced from ref. 69He)

Metal ions	pH	$\Delta H_{(kcal\,mol^{-1})}$	$\Delta S'_{(cal\,deg^{-1}\,mol^{-1})}$	$\log K$
Bovine apocarbonic anhydrase II(A)				
Zn^{2+}	7.0	8.49		
	5.5	5.65		
Co^{2+}	7.0	8.83		
	5.5	4.33		
Cu^{2+}	7.0	3.4		
Cd^{2+}	7.0	4.4		
Ni^{2+}	7.0	3.2		
Bovine apocarbonic anhydrase II(B)				
Zn^{2+}	7.0	9.84	88.0	12.0
	5.5	3.86	60.9	10.5
Co^{2+}	7.0	8.13		

increases in enthalpy, ranging from 3 to $10\,kcal\,mol^{-1}$, were observed for the reactions of all metals with apocarbonic anhydrase. On the basis of the binding constants of the zinc ion for apocarbonic anhydrase, the thermodynamic parameters for the binding process were calculated as follows: $\Delta G = -16.4\,kcal\,mole^{-1}$, $\Delta S = +88\,cal\,deg_{-1}\,mole^{-1}$ and $\Delta H - 9.8\,kcal\,mole^{-1}$ at pH 7.0 [69He]. This indicates that the enzyme cavity in which the zinc ion is bound is polar in character in the apoenzyme but becomes relatively nonpolar on the addition of zinc ions [69He]. The thermodynamic parameters may be interpreted by the protein conformation change involved in the metal-binding process [69He].

1.4   Cu, Zn-superoxide dismutase

This enzyme was first discovered as bovine cuprozinc protein in bovine erythrocytes by Mann & Kelein [38Ma]. Until 1969, the enzyme activity of the cupro-zinc protein was unknown and it was thought that the protein might be connected with metal transport. In 1969, McCord & Fridovich [69Mc] found that the cupro-zinc protein had superoxide dismutase activity *in vitro*:

$$2O_2^- + 2H^+ \rightleftharpoons H_2O_2 + O_2 \ . \tag{11}$$

Accordingly, this enzyme was named superoxide dismutase.

Bovine erythrocyte superoxide dismutase (Cu_2Zn_2SOD) has two identical sub-units, with a zinc or a copper ion at the active site [81Va]. The X-ray structure has been determined by Tainer *et al.* [82Ta] at 2.0Å resolution. The main feature of the metal-binding region is that copper and zinc ions coordinate to the same imidazole ring of the His residue. This is shown in Fig. 5.6 [87Ba].

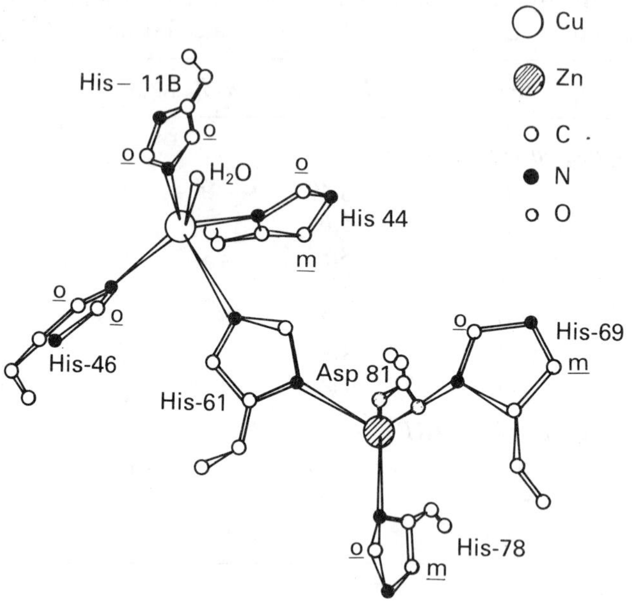

Fig. 5.6 — Scheme of the copper and zinc binding modes in bovine superoxide dismutase (Cu₂Zn₂SOD). Reproduced from ref. 87Ba1.

The copper ion is coordinated by four His and one water molecule in five-coordination geometry. The zinc ion is coordinated by three His and one Asp residues, and its coordination geometry is tetrahedral. The copper ion is located at the bottom of the crevice, which has three width stages as shown in Fig. 5.7 [89Be], and is exposed to the solvent. The distance Cu-O(water) is 2.8 Å and the water molecule semicoordinates to the copper ion in the enzyme [83Ta]. The zinc ion is completely surrounded by amino acid residues and is not exposed to the solvent. There are Arg and Thr residues as a gate in front of the narrowest cavity in which the copper ion is located [89Be, 83Ta]. The Arg residue is very important for fixing O_2^- near the copper ion in the active site [87Be]. The Thr residue is also important for the hydrophilic character of the cavity [89Be]. The bridging His residue likewise has an important role, because the bond between the coppper ion and the bridging imidazole is disrupted when the copper(II) ion is reduced to copper(I) ion by the O_2^- ion [81Va, 83Ta]. A proton then binds to the nitrogen atom of the disrupted imidazole in the His residue, and this proton on the imidazole may be supplied to the O_2^{2-} in the substrate catalytic cycle [81Va, 83Ta].

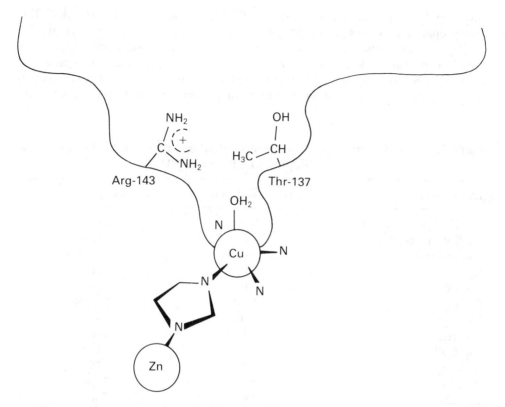

Fig. 5.7 — Schematic of the active site cavity, illustrating the positions of Arg-143 and Thr-137. Reproduced from ref. 89Be.

$$(His)_3Cu(I) \; HN \!\!-\!\!-\!\! N - Zn \, (Glu) \, (His)_2 \xrightleftharpoons[+H^+, \, +e^-]{-H^+, \, -e^-}$$

$$(His)_3Cu(II) \; -N \cdots\cdots N - Zn(Glu) \, (His)_2$$

$$(12)$$

The large number of metallo derivatives of Cu_2Zn_2SOD are prepared by selectively removing the copper or zinc from the enzyme. Apoenzyme (E_2E_2SOD: E = empty) in which both the zinc and the copper are removed was prepared by dialyzing the enzyme with EDTA at pH 3.8 [69Mc]. The copper-free enzyme (E_2Zn_2SOD) was prepared by dialyzing the parent enzyme with CN^- at pH 7.0 after reducing the copper ion with $Fe(CN)_6^{4-}$ [77Ri1]. The zinc-free enzyme (Cu_2E_2SOD) was prepared by dialysing the parent enzyme against buffer solution at pH 3.8 [79Pa]. Because of the selective metal-removing method, various metallo derivatives could be prepared. Many metal-substituted derivatives have been reported and characterized. The superoxide dismutase activity of metallo derivatives is shown in Table 5.7. The copper ion at the copper-binding site is absolutely necessary for the

Table 5.7 — SOD activities of metal-substituted derivatives of bovine cuprozinc protein. (Reproduced from ref. [81Va])

Derivatives	Activity (% of native)	References
$Cu_2^{II}Zn_2^{II}$ (as isolated)	100	69Mc
$Cu_2^{II}Zn_2^{II}$ (reconstituted)	110	73Be
E_2E_2 (Apo)	0, $<0.1\%$	69Mc, 73Be
$E_2Co_2^{II}$	0	73Fe
$E_2Zn_2^{II}$	0	72Ro
$Cu_2^{II}E_2$	20–80	73Be, 75Fe
$Cu_2^{II}Co_2^{II}$	90	73Be
$Cu_2^{II}Hg_2^{II}$	90	73Be
$Cu_2^{II}Cd_2^{II}$	70	73Be
$Ag_2^{I}Cu_2^{II}$	5	77Be
$Ag_2^{I}Co_2^{II}$	Trace	77Be
$Cu_2^{II}Cu_2^{II}$	100	75Fe

enzyme activity, but any metal may occupy the zinc-binding site without influencing the enzyme activity. Valentine *et al.* [88Ro] proposed that the zinc ion is important for stabilizing the conformation of the enzyme. The UV spectrum of the apoenzyme is very different from that of the native enzyme, indicating a considerable conformational difference between the two. On the other hand, metal coordination at the zinc site would be expected to facilitate the deprotonation from the disrupted His residue according to Eq. (12) [88Oz]. However, the role of the zinc ion is not yet clear.

Hirose *et al.* [81Hi, 82Hi] measured the copper-binding constants for apo-SOD(E_2E_2SOD) by means of equilibrium dialysis by using 2-pyridinecarboxylate as competitive chelating agent. The binding constants of the copper for the copper and the zinc sites were completely pH-dependent; they are shown in Fig. 5.8. The binding constants of the copper ion for the native zinc sites were very small at low pH (4.0–3.2). Therefore, the copper ion is easily dissociated from the native zinc sites at low pH (Fig. 5.8), whereas the copper ion is not dissociated from the native copper sites at lower pH because of the high binding constants of the copper ion relative to the native copper sites at lower pH [82Hi]. At a high pH, the binding constants of the copper ion for the native zinc sites are almost the same as those for the native copper site [82Hi].

Valentine *et al.* [79Va] showed that there was a marked decrease in EPR signal

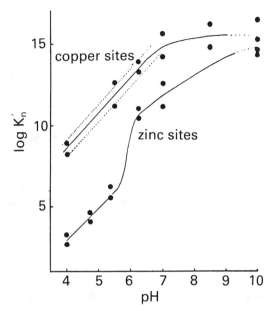

Fig. 5.8 — Relationship between the logarithm of the copper binding constants to aposuperoxide dismutase and the pH. Reproduced from ref. 82Hi.

intensity for Zn-free SOD (Cu_2E_2SOD) at 30°C when the pH was increased as shown in Fig. 5.9. This reaction was reversible. The EPR spectrum of Cu_2Cu_2SOD showed

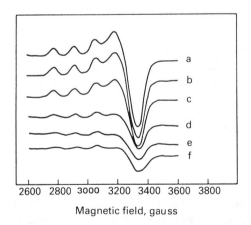

Magnetic field, gauss

Fig. 5.9 — pH-dependence of the ESR spectrum of Zn-free superoxide dismutase (Cu_2E_2SOD) at 30 ± 2°C. a, pH 7.4; b, pH 8.0; c, pH 8.4; d, pH 9.0; e, pH 9.4; f, pH 9.8. Reproduced from ref. 79Va.

no signal at 30°C, because of antiferromagmetic interaction between the copper ions [79Va]. On the basis of this behaviour, Valentine *et al.* proposed that a reversible, pH-dependent migration of a copper(II) ion occurred from the native copper sites to the empty zinc sites [79Va]. This migration would be interpreted by the pH-dependence of the apparent binding constants of the copper ion for the enzyme, because the difference in the apparent binding constants of the copper ion between the native copper sites and the native zinc sites is fairly large below pH 7.0, but small above pH 10.0 (Fig. 5.8) [82Hi, 79Va]. Therefore, migration of the copper ion from the native copper sites to the native zinc sites in Cu_2E_2SOD occurs between pH 7.0 and 10.0. The EPR-detectable copper contents were calculated on the basis of the apparent binding constants of the copper ion at pH 7.0 and 10.0. These were 88% and 35& at pH 7.0 and 10.0, respectively [82Hi]. These contents were completely consistent with those obtained from the decreasing EPR intensity [82Hi, 79Va].

How can Apo-SOD(E_2E_2SOD) selectively bind the copper and zinc ions in each native site? To study this, Hirose *et al.* [84Hi2, 85Hi] measured the apparent binding constant of the zinc ion for each native site by means of equilibrium dialysis methods; the results are shown in Table 5.8 [84Hi2, 85Hi].

Table 5.8 — Stepwise apparent binding constants of various metalloderivatives of bovine Cu, Zn-superoxide dismutase. (Reproduced from refs. [84Hi2, 85Hi])

	$\log K_1'$	$\log K_2'$	$\log K_3'$	$\log K_4'$	Conditions
Zn_2Zn_2SOD	10.9	11.1	7.8	6.5	pH 6.25
	zinc sites (Zn^{2+})		copper sites (Zn^{2+})		
E_2Zn_2SOD	11.1	10.9	—	—	pH 6.25
	zinc sites (Zn^{2+})				
Cu_2Cu_2SOD	13.9	13.4	11.1	10.6	pH 6.25
	copper sites (Cu^{2+})		zinc sites (Cu^{2+})		
Co_2Zn_2SOD	6.7	5.2	—	—	pH 7.4
	copper sites (Co^{2+})				

When the apparent binding constants for Cu_2Cu_2SOD and Zn_2Zn_2SOD at pH 6.25 are simply compared, the copper sites have a 10^6 times larger binding constant for the copper ion than for the zinc ion, thus the native copper sites selectively bind to the copper(II) ion [84Hi2]. On the other hand, the zinc sites have almost the same binding constants for the zinc and copper ions [84Hi2]. To confirm the selectivity of the metal binding for the zinc and copper sites, the competitive reaction constants for the copper and zinc ions with the native copper sites in the following equations were determined via a metal-competitive reaction.

$$Zn_2Zn_2SOD + Cu^{2+} \overset{K_{1s}'}{\rightleftharpoons} CuZnZn_2SOD + Zn^{2+}$$

$$CuZnZn_2SOD + Cu^{2+} \overset{K'_{2s}}{\rightleftharpoons} Cu_2Zn_2SOD + Zn^{2+}$$

$$K'_{1s} = \frac{[CuZnZn_2SOD][Zn^{2+}]}{[Zn_2Zn_2SOD][Cu^{2+}]} \quad K'_{2s} = \frac{[Cu_2Zn_2SOD][Zn^{2+}]}{[CuZnZn_2SOD][Cu^{2+}]} \tag{13}$$

The competitive equilibrium constants K_{1s}, and K_{2s}, were $10^{7.1}$ and $10^{6.2}$ at pH 6.25, respectively, and accordingly the copper ion can bind very selectively to the native copper site site in a competitive reaction at pH 6.25 [84Hi2]. K'_{1s} and K'_{2s} are reasonably consistent with the difference (in Table 5.8) between the binding constants of the zinc and copper ions for the native copper sites. This indicates the validity of the assumption that K'_1 and K'_2 for Zn_2Zn_2SOD in Table 5.8 are the apparent binding constants of the zinc ion for the native zinc sites, while K'_3 and K'_4 for Zn_2Zn_2SOD are those for the native copper sites [84Hi2].

The reaction of metal removal from yeast Cu_2Zn_2SOD by chelating agents was investigated by Dunbar et al. [82Du]. These reactions were biphasic. The reaction kinetics are influenced by two rate constants, corresponding to two equivalent parts of Cu_2Zn_2SOD per dimer, strongly suggesting asymmetric binding sites for both copper and zinc [82Du]. However, the metal-binding constants of bovine Cu_2Zn_2SOD in Table 5.8 indicate symmetric binding sites for both copper and zinc.

Rigo et al. [78Ri] measured the copper(II)-binding rate for copper-free superoxide dismutase (E_2Zn_2SOD). The copper ion-binding reaction has two phases (a fast and a slow reaction). UV absorption and solvent proton relaxation rate measurement show that the fast binding of Cu^{2+} occurs partly at the native active site. The recovery of the enzyme activity, the UV spectrum, and the ESR data indicated a slow rearrangement of the site around the bound metal [78Ri].

1.5   Human transferrin and ovotransferrin

The iron-binding proteins in extracellular biological fluids were extracted from various sources. Sero-, ovo-, and lactotransferrin were isolated from human serum, hen egg white, and milk, respectively [87Cr]. Schade et al. [49Sc, 87Cr] first established that serotransferrin binds two atoms of iron per molecule and that the iron binding is accompanied by the concomitant binding of a bicarbonate ion for each iron atom bound. Transferrin is very important as a protein transporting iron from biological fluids into the cells. Transferrin is a glycoprotein with a molecular weight of about 80 000; it can bind two iron(III) ions together with the binding of an anion, usually bicarbonate [87Cr, 73Ai]. It was presumed that transferrin contains two homologous domains, because some peptidases cleaved the polypeptide chain into two halves, each containing one iron, and the amino acid sequence shows a strong two-fold internal homology [87An]. For lactotransferrin there is a 40% sequence identity between its N- and C-terminal halves. Anderson et al. recently succeeded in the X-ray analysis of lactotransferrin at 3.2 Å resolution [87An]. The X-ray analysis data clearly indicate that this protein has two globular lobes (N and C) with one iron(III) ion binding site in each (Fig. 5.10), the lobes being connected by a short α-helix [87An]. Each lobe has the same protein folding, which consists of two domains with similar supersecondary structure. The iron-binding site is between the two domains in each lobe [87An]. At each iron-binding site, where the iron is buried

A   B

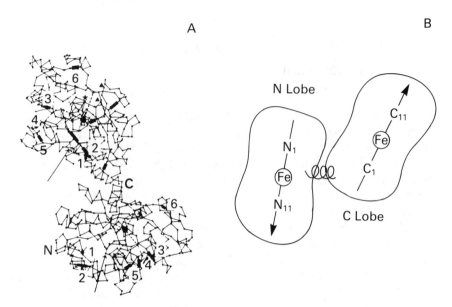

Fig. 5.10 — Structure of human lactotransferrin. A: Stereo view of the whole lactotransferrin molecule. N and C termini are labelled. The iron atoms, 42 Å apart, are shown as ○; disulphide bridge, as ━■■━. B. Schematic model of lactotransferrin. Reprinted with permission from ref. 87An.

in hydrophobic residues, two Tyr, one His, and one Asp coordinate the iron(III) ion (Fig. 5.11). The iron-binding sites are about 42 Å apart (Fig. 5.10) [87An]. The four ligands bound to iron are the same in lactotransferrin, ovotransferrin, and serotransferrin. The four ligands occupy four sites of the iron(III) ion in octahedral coordination geometry. Two ligands in *cis* positions are not clear on 3.2 Å resolution X-ray analysis [87An]. However, additional electron density is observed near the iron. The Arg residue with a positive charge is present near the iron-binding site. It is known that an anion such as bicarbonate or carbonate binds to a protein simultaneously with the binding of iron, and the Arg residue provides the anion to bind to the protein [78Ro2]. Therefore, the high electron density near the Arg residue should be the anion. It is proposed that a water molecule also coordinates to the iron(III) [81Pe, 83Ko] so that one more high electron density site might be water [87An]. The coordination geometry around the iron(III) ion is schematized in Fig. 5.11.

Structurally, the region in the N lobe resembles that in the C lobe, but is not completely consistent with it in terms of the molecule [87An]. The anion site in the C lobe appears more restricted than that in the equivalent site in the N lobe because the access 'channel' is close to the interface between the lobes [87An]. Further, there is a small structure difference between the C and N lobes with variations in some residues around the iron sites [87An].

The binding constant of a metal ion in apotransferrin is shown by the following equations; HCO_3^- also binds, and several protons are released on the binding of metal ions to apotransferrin.

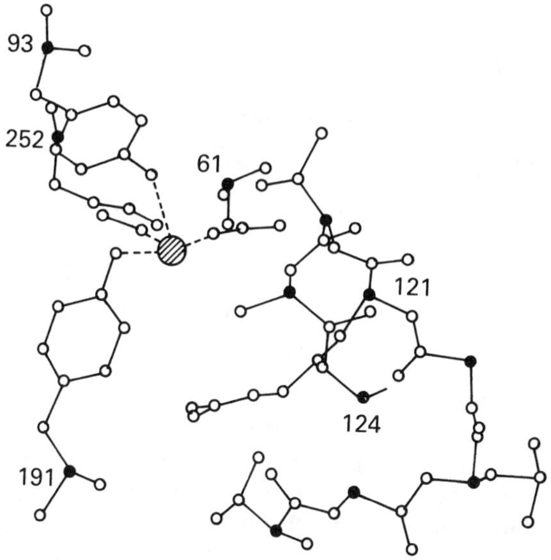

Fig. 5.11 — Stereo view of the iron coordination in lactotransferrin. Only the protein ligands are shown. Density that may be attributable to the CO_3^{2-} anion, and possibly a water molecule as well, spans the two remaining octahedral coordination positions. Reprinted with permissions from ref. 87An.

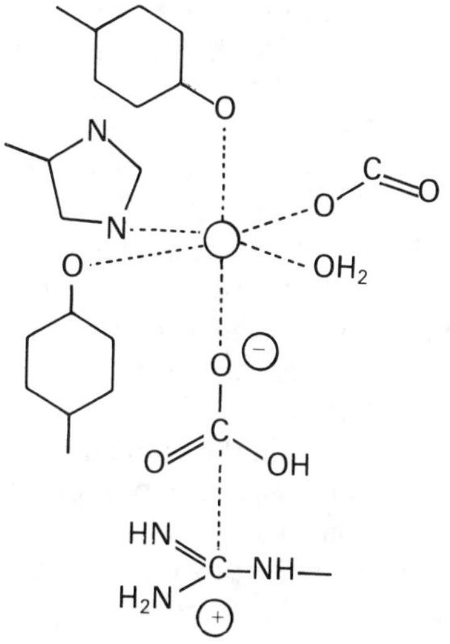

Fig. 5.12 — Schematic picture of the iron coordination in lactotransferrin. CO_3^{2-} may be located between the Arg residue and the iron ion. The interaction between the positive charge of Arg and the negative charge of CO_3^{2-} may be important for binding of the carbonate ion.

$$M^{n+} + \text{apotransferrin} + HCO_3^- \rightleftharpoons$$
$$\rightleftharpoons (M^{n+}\text{transferrin } HCO_3^-) + mH^+ \tag{14}$$

$$K_1 = \frac{[(M^{n+} \text{ transferrin } HCO_3^-)][H^+]^m}{[M^{n+}][\text{apotransferrin}][HCO_3^-]} \tag{15}$$

$$M^{n+} + (M^{n+} \text{ transferrin } HCO_3^-) + HCO_3^- \rightleftharpoons$$
$$\rightleftharpoons (M_1^{n+} \text{ transferrin } (HCO_3^-)_2 +)mH^+ \tag{16}$$

$$K_2 = \frac{\cdot\ [((M^{n+})_2 \text{ transferrin } (HCO_3^-)_2)][H^+]^m}{[M^{n+}][(M^{n+} \text{ transferrin } HCO_3^-)][HCO_3^-]} \tag{17}$$

The number of protons released in these reactions depends on the nature of the metal ions [78Ai, 83Ha1]. It is difficult to determine K_1 and K_2, thus conditional stability constants (K_1' and K_2') were calculated at a constants HCO_3^- concentration and a constant pH. The relationships between the conditional stability constants and the stability constants are shown by the following equations.

$$K_1' = \frac{[HCO_3^-]}{[H^+]^m} K_1 \tag{18}$$

$$K_2' = \frac{[HCO_3^-]}{[H^+]^m} K_2 \tag{19}$$

The conditional stability constants of the complexes of various metal ions with apotransferrin at various pH values and HCO_3^- concentrations are shown in Table 5.9. Without metal ions, bicarbonate can bind to apotransferrin through the positive charge of Arg at the active site [83Ha2]. The stepwise apparent binding constants of bicarbonate to apotransferrin at pH 7.4, as determined by difference UV spectroscopy, are $10^{2.74}$ and $10^{2.13}$ [83Ha2]. Therefore the binding constants of metal ions are governed by the concentration of the bicarbonate ion on the basis of Eqs. (15) and (17). Thus, in Table 5.9, the concentration of HCO_3^- is shown. The derivatives which have the iron only in the N or the C site can be prepared by using the difference between the iron-binding characters for the N and C site [86Kr].

Transferrin has a high stability for the iron(III) ion, and a low stability for the iron(II) ion (Table 5.9). This character may be very important from a physiological point of view, because the iron(III) ion, which binds to apotransferrin in the serum, may be easily reduced to the iron(II) in the cell and released from the protein. The ionic radius and charge of the metal may be important as concerns the selectivity of this protein for various metals. The possibility has been pointed out that some small chelating ligands may participate in the process of iron removal from transferrin in the cell. Studies on the reaction of iron removal from transferrin by chelating agents are also very important as regards the search for effective drugs to excrete iron from iron-overloaded blood in certain blood disorders. Therefore the mechanism of the reaction of iron removal from transferrin by chelating agents has been studied.

Carrano & Raymond [79Ca] measured the rate constant of iron removal from transferrin by the chelating agent LICAMS. The constants exhibited saturation with increase of the LICAMS concentration. This behaviour was completely consistent

Table 5.9 — Conditional stability constants of metal binding to apotransferrin

Metal	Ionic radius (Å)	Stability constants (log K)	Conditions	Source	Method	References
Ni^{2+}	0.69	4.39 ± 0.21 (N site), 4.11 ± 0.14 (C site)	pH 7.4, 0.01 M HEPES, 5 mM HCO_3^-	human	DF	86Ha1
Cu^{2+}	0.72	10.1 ± 0.06, 10.2 ± 0.2	pH 7.6, 0.03 M HEPES, 15 mM HCO_3^-	ovo	EL	86Sy
Zn^{2+}	0.74	7.8 ± 0.2 (C site), 6.4 ± 0.4 (N site)	pH 7.4, 0.01 M HEPES, 15 mM HCO_3^-	human	DF	83Ha
Cd^{2+}	0.95	5.95 ± 0.10 (C site), 4.86 ± 0.13 (N site)	pH 7.4, 0.01 M HEPES, 5 mM HCO_3^-	human	DF	88Ha
Fe^{2+}	0.92	6.7 (C site), 3.2, 5.4 (N site), 2.5	pH 7.4, 0.01 M HEPES, 15 mM HCO_3^-; pH 7.4, 0.01 M HEPES, 5 mM HCO_3^-	human	LI, LI	83Ha1, 86Ha
Fe^{3+}	0.64	20.7 (C site), 19.4 (N site)	pH 7.4, 0.01 M HEPES, 14 mM HCO_3^-	human	VI	78Ai
Nd^{3+}	0.983	6.09 ± 0.15 (C site), 5.04 ± 0.46 (N site)	pH 7.4, 0.01 M HEPES, 0.2 mM HCO_3^-	human	DF	86Ha2
Sm^{3+}	0.958	7.13 ± 0.24 (C site), 5.39 ± 0.32 (N site)	pH 7.4, 0.01 M HEPES, 0.2 mM HCO_3^-	human	DF	86Ha2
Ga^{3+}	0.62	19.53 ± 0.25, 18.58 ± 0.25	pH 7.4, 0.01 M Tris-HCl, 5 mM HCO_3^-	human	DF	83Ha3
VO_3^-	0.59	~ 6.5, 6.5 ± 0.2	pH 6.26–9.5, 0.1 M HEPES	human	DF	84Ha1

DF, Difference absorption spectra; EL, Cu^{2+} electrode; VI, Visible absorption; LI, These values were calculated from the zinc binding constants via the linear free energy correlation.

with that of carbonic anhydrase, and it was proposed that the reaction proceeds through a ternary complex, as shown in Eq. (20).

$$\text{FeTf} + \text{L} \overset{K_{eq}}{\underset{}{\rightleftharpoons}} \text{LFeTf} \underset{k_{-2}}{\overset{k_2}{\rightleftharpoons}} \text{FeL} + \text{Tf} \ . \tag{20}$$

In the metal removal reaction, k_{-2} can be ignored. The observed metal removal rate constant (k_{obs}) is represented by the following equation.

$$k_{obs} = \frac{k_2 K_{eq}[\text{L}]}{1 + K_{eq}[\text{L}]} \ . \tag{21}$$

The iron removal reaction is biphasic because the rates of iron removal by chelating agents were different in the N and C lobes [79Ca]. On the basis of Eq. (21) the relationship between 1/[LICAMS] and $1/k_{obs}$ in each phase should give a linear plot, and this relationship was satisfied for various chelating agents [79Ca, 84Ha2, 87Ch]. If the metal removal reaction (k_{-2} is ignored) by chelating agents proceeds as in Eq. (20), the presence of a ternary complex in the metal removal reaction should be detected by a spectroscopic or some other method [86Ki]. However, the presence of a ternary complex has not been detected for various agents.

Bates *et al.* [82Co] demonstrated the formation of a ternary complex by a spectrophotometric method in the reaction between the iron–ligand complex and apotransferrin ($\text{Tr} + \text{LFe} \rightleftharpoons \text{LFeTr}$), and found that the decomposition of the ternary complex to transferrin and ligand was very slow (this behaviour is inconsistent with Eq. (20), because the rate of the decomposition from LFeTf to FeTf + L must be very high in Eq. (20) [82Co, 86Co].

On the basis of these results and those of Raymond, Bates *et al.* [82Co] proposed the following equation for the reaction of iron removal by a chelating agent:

$$\text{FeTf} \underset{k_{-1}}{\overset{k_1}{\rightleftharpoons}} \text{FeTf}^* \underset{k_{-2}-\text{L}}{\overset{k_2+\text{L}}{\rightleftharpoons}} \text{LFeTf}^* \underset{k_{-3}}{\overset{k_3}{\rightleftharpoons}} \text{LFe} + \text{Tf}^* \tag{22}$$

$$k_{obs} = \frac{k_1 k_2 k_3 [\text{L}]}{k_1(k_{-2} + k_3) + k_2 k_3 \approx \text{L}]} \tag{23}$$

where * means 'open conformation', Tf is apotransferrin, and L is the chelating agent. The rate-determining step of Eq. (22) is the conformational change of the protein between open transferrin (FeTf^*) and closed transferrin (FeTf). According to Eq. (23), the relationship between $1/k_{obs}$ and $1[\text{L}]$ gives a linear plot for the reaction of iron removal by the chelating agent (k_{-3} can be ignored). If the reaction of metal removal from transferrin with a chelating agent proceeds according to Eq. (22), then the maximum velocity obtained in the presence of excess chelating agent should be independent of the nature of the chelating agent. However the maximum velocity for various chelating agents reveals various rate constants $(0.001–0.125\,min^{-1})$ [84Ha2].

It is still unclear which iron-removal mechanism is correct, because the metal-removal rate constant is influenced by various factors such as the ionic strength and the anions present [81Ba], hence the reaction is very complicated. The ClO_4^- ion accelerates the reaction of iron removal from the C site in iron transferrin, but suppresses the removal from the N site [88Be1,88Be2]. In contrast, the ClO_4^- ion accelerates the reaction of copper removal from both sites in copper-transferrin [88Be2].

2. MECHANISM OF BINDING AND DISSOCIATION OF SMALL LIGANDS (INHIBITORS AND SUBSTANCES SUCH AS SUBSTRATES) IN METALLOENZYMES AND METALLOPROTEINS

2.1 Introduction

It is very difficult, by physical methods, to follow how the residues at the active site of metalloenzymes function in catalysing the reaction and keeping the substrate complex of the metalloenzyme in the solution state, because the substrate is rapidly catalysed by the enzyme and released from the coordination sphere of the metal ion at the active site. Therefore it is very important to study how an inhibitor interacts with the residues and the metal ions at the active site. In this section, investigations of the interactions between inhibitors or substrates and the enzyme by physicochemical methods are described.

2.2 Carboxypeptidase A

Arsanilazo-Tyr-248-carboxypeptidase [86Au2, 75Ha] in which Tyr-248 is modified by the arsanilazo group, has an appropriate character for investigation of the interactions between substrates or inhibitors and the enzyme [75Ha]. The arsanilazo-Tyr residue coordinated to the zinc ion at the active site has a characteristic red colour, and the dissociation of this residue from the zinc ion at the active site gives a colour change from red to yellow [75Ha]. The rate constant of the relaxation of the arsanilazo-Tyr residue for the zinc ion at the active site was found to be $64\,000\,s^{-1}$, and this process is faster than the catalytic step involving the enzyme (Fig. 5.13) [75Ha]. This derivative was therefore used to follow the substrate binding with the enzyme by the temperature-jump method (Fig. 5.14) [86Au2, 75Ha, 78Ha]. When Gly-L-Phe, Gly-L-Tyr, Gly-L-Ile or Phe interacted with the enzyme, two reaction steps were observed: the rapid substrate or product binding process was followed by the slower interconversion of at least two conformational forms of the enzyme-ligand complex [78Ha]. Since Gly-L-Phe, Gly-L-Try and Gly-L-Ile are bad substrates which are catalysed by the enzyme very slowly, it is possible to follow the process by the temperature-jump method.

$$E + S \underset{}{\overset{K_1}{\rightleftharpoons}} ES_1 \underset{k_{-2}}{\overset{k_2}{\rightleftharpoons}} ES_2 \ . \tag{24}$$

The N-dansylated peptides are excellent substrates for zinc and cobalt carboxy-

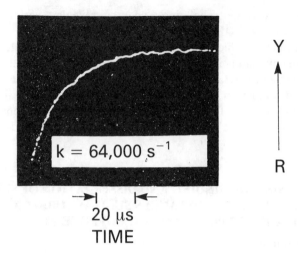

Fig. 5.13 — Temperature jump of azo-Tyr-248-carboxypeptidase A and concomitant relaxation effect. 0.1 M Tris-HCl, pH 8.5, 1 M NaCl. Initial temperature 21°C, pulsd by 4 degrees to result in a final temperature of 25°C. Reproduced from ref. 75Ha.

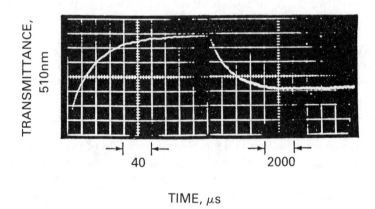

Fig. 5.14 — Temperature jump of azo-Tyr-248-carboxypeptidase A in the presence of 4 mM Gly-L-Phe. 0.1 M Tris-HCl, pH 8.0, 1 M NaCl. Temperature conditions as in Fig. 5.13. Reproduced from ref. 78Ha.

peptidase A. The binding and hydrolysis of Dns-Ala-Ala-Phe by carboxypeptidase A was followed by means of a stopped flow-technique (Fig. 5.15) [86Au1, 83Ga, 83Ge]. At 20°C, the initial and rapid increase in dansyl fluorescence indicates that the formation of the steady-state complex of ES is too rapid to follow with the stopped-flow apparatus, as shown in Fig. 5.15A [86Au1, 83Ga]. The reaction, in which the ES

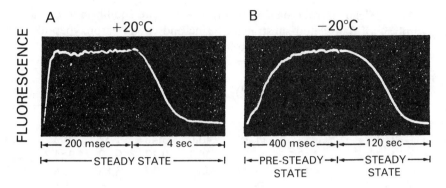

Fig. 5.15 — Stopped-flow measurement of the binding and hydrolysis of 50 μM Dns-Ala-Ala-Phe by 1 μM carboxypeptidase A in 10 mM Hepes, 4.5 M NaCl, pH 7.5, at (A) + 20°C and (B) − 20°C. Reproduced from ref. 83Ge.

complex was converted to product and the enzyme (E + P), was followed by the reduction of the fluorescence. Fortunately, carboxypeptidas A can be dissolved in a high concentration of NaCl buffer solution (4.5 M) and the values of K_m and k_{cat} in 4.5 M NaCl buffer deviate by less then 2-fold from those obtained in 1 M NaCl buffer, a normal component of the assay mixture [86Au1, 83Ga]. Therefore at low temperature (− 20°C) it became possible to measure the intermediate steps of substrate catalysis involving the enzyme with the stopped-flow method [86Au, 83Ga]. The stopped-flow apparatus for use at low temperature was devised by Vallee *et al.* [86Au1, 83Ga, 83Ge]. At − 20°C two pre-steady states are observed for Dns-(Ala)$_2$-Phe. When the substrate is mixed with the enzyme, an initial very rapid increase (< 15 ms) of danyl fluorescence occurs within the dead-time of the instrument (Fig. 5.15B), corresponding to formation of the Michaelis complex (ES$_1$). A slower exponential rise thereafter is due to the formation of a second intermediate (ES$_2$), with a first order rate constant. This is followed by the breakdown of the intermediate state (ES$_2$), which is very slow. The following reaction mechanism is proposed [86Au1, 83Ga]:

$$E + S \overset{K_s}{\rightleftharpoons} ES_1 \underset{k_{-2}}{\overset{k_2}{\rightleftharpoons}} ES_2 \overset{k_3}{\to} E + P \ . \tag{25}$$

The equilibrium and rate constants of each step are shown in Table 5.10 [86Au1, 83Ga, 84Au]. The presence of two intermediate states is completely consistent with the results obtained by the temperature-jump method on azo-Tyr-248-carboxypeptidase A.

To understand more about the role of the metal ion in the catalysis of substrates, Vallee *et al.* investigated the interactions between substrates and cobalt–carboxypeptidase A with a rapid-scanning stopped-flow method at subzero temperature [86Au1, 82Ge, 84Au]. The spectrum of the cobalt enzyme during catalysis at − 17°C directly visualized the formation and breakdown of the intermediate state. The d–d* transition spectrum of cobalt in the cobalt enzyme from 600 to 500 nm changed with the formation and breakdown of the intermediate states. The formation of ES$_1$ did

Table 5.10 — Michaelis–Menten parameters and rate and equilibrium constants for the hydrolysis of dansylated substrates by zinc and cobalt carboxypeptidase A[†]. (Reproduced from ref. 86Aul)

Substrate	K_m (μM)	k_{cat} (s^{-1})	k_2 (s^{-1})	k_{-2} (s^{-1})	k_3 (s^{-1})	K_s (μM)	K'_m (μM)
Peptides							
Dns-Ala-Ala-Phe	13.5	1.18	40.0	3.48	1.32	102	10.9
Dns-Ala-Ala-Phe[§]	2.77	0.57	36.1	0.17	0.59	154	3.2
Dns-Gly-Ala-Phe	12.9	3.72	228	6.12	3.88	418	17.6

† Conditions of assay: $-20°C$, in 10 mM Hepes, 4.5 M NaCl, pH 7.5.
‡ $K'_m = K_s(k_{-2} + k_3)/(k_{-2} + k_2 + k_3)$ for Eq. (25).
§ Data for cobalt enzyme [84Au].

not influence the spectrum of the cobalt enzyme because the first rapid scan spectra obtained at 16.8 ms (ES$_1$ is formed in this stage) is completely consistent with that of the cobalt enzyme alone. This implies that the first substrate binding does not cause a change in the coordination sphere of the cobalt enzyme. The formation of ES$_2$ is accompanied by a spectral change ($<$ 200 ms) [84Au]. This reaction is followed by the breakdown of the intermediate state (ES$_2 \rightleftharpoons E + P$), and finally the original spectrum of the cobalt enzyme is obtained. The spectrum shape for ES$_2$ is completely different from that for the cobalt enzyme in the resting state. This behaviour indicates that the coordination of substrate occurs in the ES$_2$ stage and the coordination geometry of the cobalt ion in the enzyme is significantly changed after the substrate binding [86Au1, 84Au].

Recently, Christianson & Lipscomb [89Ch, 86Ch] made a close study of the binding mode of a substrate analogue to the active site of carboxypeptidase A. $(-)$-3-(p-methoxybenzoyl)-2-benzylpropanoic acid (BMBP) is a competitive inhibitor of carboxypeptidase A.

The X-ray analysis data on the BMBP carboxypeptidase A complex demonstrated a unique binding mode for the ketone to the active site of the enzyme [89Ch, 86Ch]. The benzyl group is located in the hydrophobic pocket of the enzyme (this is the reason why a substrate which has hydrophobic amino acids in its C-terminal part can easily be hydrolyzed by carboxypeptidase A), and the carboxy group forms a salt link with the guanidinium residues of Arg-145 (this is the reason why carboxypepti-

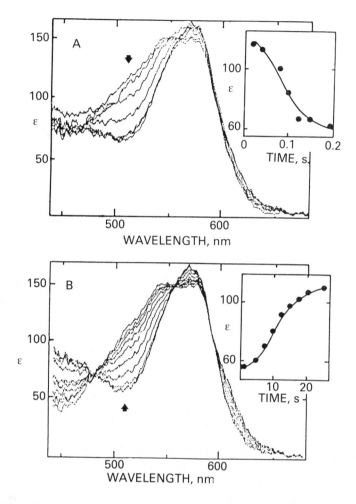

Fig. 5.16 — Absorption spectra of the formation (A) and breakdown (B) of the ES_2 intermediate formed in the reaction of cobalt carboxypeptidase A, 0.1 nM, with Dns-Ala-Ala-Phe, 0.2 mM, at −17°C. Individual spectra were collected in 16.48 ms. Insets: absorbance change at 510 nm, indicated by the arrows, as a function of the time after mixing. Reproduced from ref. 84Au.

dase A recognizes the C-terminal carboxylate of the substrate [89Ch]). Tyr-248 is in the 'down' position, and its phenolic oxygen forms a hydrogen bond with the carboxylate oxygen of BMBP. (Recently, the site-direct mutagenic method [86Ga, 86Hi] showed that Tyr-248 is not essential for hydrolysis of the substrate, but it is important to bind the substrate to the enzyme.) These interactions were observed in many substrate analogue adducts of carboxypeptidase A. In the BMBP adduct, however, the carbonyl oxygen of the p-methoxybenzoyl group of BMBP does not coordinate to the zinc ion (the carbonyl oxygen of Gly in Gly-Tyr — a slowly hydrolyzed CPA substrate — is coordinated to the zinc ion at the active site [89Ch])

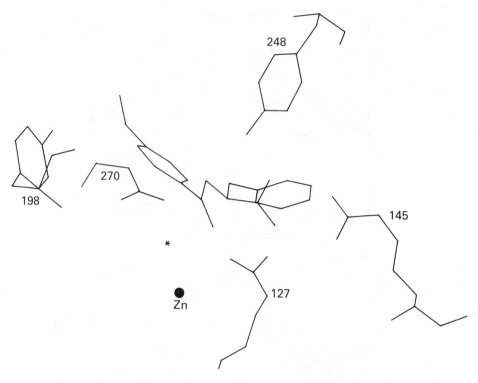

Fig. 5.17 — BMBP bound to the active site of CPA. Enzyme residues, in a clockwise direction starting with Zn, are Tyr-198, Glu-270, Tyr-248, Arg-145, and Arg-127. The zinc-bound water molecule is depicted as a star. Reprinted with permission from ref. 86Ch.

but interestingly, it is hydrogen-bonded to Arg-127 [89Ch, 86Ch]. Surprisingly, in the BMBP adduct, the water molecule ligated to the zinc ion at the active site still remains at the same coordination site as in the native enzyme, and the coordination geometry of the residues (His-69, His-196, and Glu-72) ligated to the zinc ion is likewise unchanged in the BMBP adduct. This adduct-binding mode closely resembles that in the ES_1 intermediate state observed by the rapid-scan stopped-flow method in cobalt–carboxypeptidase A [84Au]. (Substrate is not coordinated to the cobalt ion of the enzyme in the ES_1 intermediate state [89Ch, 86Ch].)

Christianson & Lipscomb [89Ch, 86Ch] also investigated a possible transition state analogue (2-benzyl-3-formylpropionic acid: BFP) adduct of carboxypeptidase A by X-ray analysis (Fig. 5.18).

BFP: $\text{H} \underset{\underset{\text{O}}{\|}}{\overset{}{\text{C}}} \text{CH}_2 \underset{\text{H}}{\overset{\text{CH}_2\phi}{\text{C}}} \text{CO}_2^-$

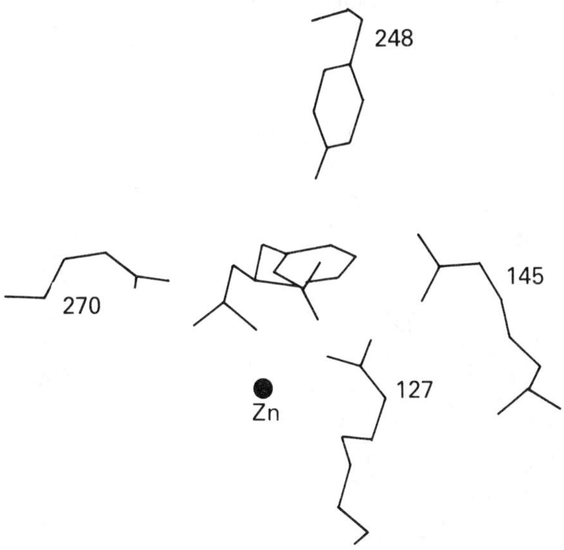

Fig. 5.18 — BFP bound to the active site of CPA. Enzyme residues, in a clockwise direction, starting with Zn, are Glu-270, Tyr-248, Arg-145 and Arg-127. Reprinted with permission from ref. 86Ch.

The X-ray analysis data showed that the aldehyde substrate is bound to the active site in hydrated form. The benzyl group is located at the hydrophobic site, the terminal carboxylate interacts with Arg-145 through electrostatic forces, and the phenolic oxygen of Tyr-248 is located 2.5 Å away from one of the carboxylate oxygens of BFP. The two hydrate oxygens of BFP are 2.5 Å and 2.7 Å from the zinc ion at the active site [89Ch, 86Ch]. The water molecule ligated to the zinc ions is removed by the substrate, and the coordination geometry of the residues (His-69, His-196, and Glu-72) ligated to the zinc ion in the enzyme is also changed by the binding of the substrate. This intermediate state, detected by X-ray analysis, may correspond to the ES_2 state found by the kinetic method [84Au], because the substitution of a ligand for a water molecule occurs in the coordination sphere of the metal ion in the enzyme [89Ch]. On the basis of these results proposed, Christianson & Lipscomb the catalytic mechanism for carboxypeptidase A [89Ch].

2.3 Carbonic anhydrase

Carbonic anhydrase is inhibited by sulphanilamide derivatives [68Li]. The binding constants of these derivatives are very large (Table 5.11), and sulphanilamide derivatives can bind strongly to the active site of the enzyme. X-ray analyses of the adducts btween sulphanilamide derivatives (acetazolamide, sulphanilamide, and acetoxy-mercuri-benzeneselphanilamide) and human CA were performed by a Swedish group [80Ka, 86Er]. The nitrogen of the sulphanilamide group can bind to the zinc ion at the active site and hydrogen bond to oxygen of Thr-199. The acetamide group interacts with the sidechains of residues Glu-92, Val-121, and Phe-131 in HCA II (Fig. 5.19).

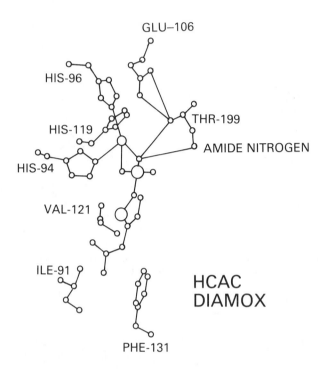

Fig. 5.19 — Acetazolamide (Diamox) binding to human carbonic anhydrase II. Reprinted with permission form ref. 80Ka.

The heterocyclic moiety of acetazolamide is present between hydrophylic and hydrophobic residues at the enzymes active site and interacts with the hydrophobic residues (Val-121, Phe-131) [80Ka]. This may be a force driving sulphanilamide derivatives to bind to the enzyme. The binding rate constants of sulphanilamide derivatives are shown in Table 5.11. The binding-rate constants for sulphanilamide derivatives are very large despite the narrow crevice of the enzyme.

The polydentate chelating agent can remove the metal ions from CA. On removal of the metal from the enzyme, the ternary complex consisting of a metal, a chelating agent, and apo-CA is formed. The formation and dissociation rate constants for the ternary complex (k_f and k_d) between 2,6-pyridinedicarboxylate (L) and cobalt bovine CA II (ECo) were determined by Wilkins $et\ al.$ with the stopped-flow method [80Ha].

$$\text{ECo} + \text{L} \underset{k_d}{\overset{k_f}{\rightleftharpoons}} \text{ECoL} \rightarrow \text{E} + \text{CoL} \tag{26}$$

Table 5.11 — Binding equilibrium (K) and rate (k) constants for small ligands with metallo-carbonic anhydrase

Enzymes	Ligand	k_f (M⁻¹s⁻¹)	k_d (s⁻¹)	k_f/k_d (M⁻¹)	K_{EML} (M⁻¹)	Ref.
(1) Chelating						
CoBCA II	2,6-pyridinedicarboxylate	1.3×10^2	1.1	1.2×10^2	1.3×10^2	80Ha
	5-methyl-1,10-phenanthroline				3.9×10^2	78Hi
ZnBCA II	2,6-pyridinedicarboxylate				2.3×10^2	78Hi
	1,10-phenanthroline				3.8×10^2	78Hi
	5-methyl-1,10-phenanthroline				7.8×10^2	78Hi
(2) Sulphanilamides						
ZnHCA II	(azo-naphthalene-disulfonate with –SO₂NH₂, structure)	5.81×10^5	0.075	7.74×10^6	6.9×10^6	70Ta1
	(thiadiazole CH₂CONH–, –SO₂NH₂; $O_2NC_6H_4SO_2NH_{rp}$)	4.83×10^6	0.068		7.1×10^7	70Ta1
		7.37×10^5	0.048	1.53×10^7	1.60×10^7	70Ta1
	((CH₃)₂N–naphthalene –SO₂NH₂, structure)	2.4×10^5	0.390	6.15×10^5	5.80×10^5	70Ta1

Table 5.11 — *continued*

Enzymes	Ligand	k_f $(M^{-1}s^{-1})$	k_d (s^{-1})	k_f/k_d (M^{-1})	K_{EML} (M^{-1})	Ref.
ZnHCA I	(azo-naphthalene sulfonamide structure: OH, N=N–C6H4–SO2NH2, O3S, SO3)	9.7×10^5	0.018	5.39×10^7		70Ta1
	$O_2NC_6H_4SO_2NH_{rp}$	2.86×10^5	0.049	5.84×10^6		70Ta1
ZnBCA II	((CH$_3$)$_2$N–naphthalene–SO$_3$NH$_2$)	1.34×10^5	0.030	4.46×10^6	4.0×10^6	70Ta1
CoHCA II	((CH$_3$)$_2$N–naphthalene–SO$_3$NH$_2$)	1.43×10^5				69Ch
(3) Anions	$O_2NC_6H_4SO_2NH_{rp}$	1.77×10^6	0.082	2.16×10^7		70Ta2
ZnHCA I	Cl^- (pH 6.46)	5×10^6	1×10^6			72Wa
	Cl^- (pH 8.59)	3.7×10^6	7.3×10^6			72Wa
ZnHCA II	CN^-	$> 10^9$	$> 10^2$			71Ta1
CoBCA II	CNO^-	$\sim 8 \times 10^6$	$\sim 4 \times 10^2$			74Ge
CoHCA II	$HCOO^-$	3.9×10^8	6.0×10^4		6.5×10^3	71Ta2
	CH_2FCOO^-	2.2×10^8	2.0×10^5		1.1×10^3	71Ta2
	CHF_2COO^-	1.9×10^8	1.0×10^5		1.9×10^3	71Ta2
	CF_3COO^-	2.0×10^8	1.5×10^5		1.3×10^3	71Ta2

These values are about 100–1000 times smaller than those obtained for sulphanilamide derivatives that can bind to the metal ion as monodentate ligands. Formation of the ternary complex requires that the polydentate ligands bind to the metal ion through multisites in the narrow crevice of the enzyme, hence the formation rate constant may be much smaller than those obtained for sulphanilamide derivatives (Table 5.11).

It has been proposed that the coordination geometry of the ternary complex between bi- or tridentate ligands (1,10 phenanthroline and 2,6-pyridinedicarboxylate) and cobalt(II)-bovine CA is five or six-coordinate as indicated by the low molecular absorption coefficients of the ternary complexes in the visible region [80Ha, 80Hi, 81Hi, 78Ki]. The MCD spectrum of the ternary complex between a bidentate ligand and cobalt(II)-bovine CA II has a low molar absorption coefficient in the visible region; it also has two clear negative MCD bands [84Hi3]. These MCD spectral shapes resembles those of the model cobalt complexes which have five-coordination geometry. The water proton relaxation, MCD, and visible spectral data indicate the possibility of five-coordination geometry in the ternary complexes of bidentate ligands with cobalt-CA [80Hi, 81Hi, 78Ki, 84Hi3]. The MCD and visible spectra of the ternary complexes between some bidentate ligands and cobalt-BCA II are shown Fig. 5.20.

The formation constants of the ternary complexes between these bidentate ligands and native CA or cobalt-CA are several orders of magnitude lower than the analogous stability constants obtained for the metal ion in aqueous solution (Table 5.12) [78Hi]. The thermodynamic parameters for formation of the ternary complex are shown in Table 5.12 [78Hi]. For the formation of ternary complexes of zinc- and cobalt-enzyme with 5-methyl-1,10-phenanthroline, the values of ΔH and ΔS were almost the same. When the thermodynamic parameters of the ternary complex were compared with those of 1,10-phenanthroline metal complexes, ΔH for the ternary complex was almost the same as that for 1,10-phenanthroline, but ΔS had a large negative entropy value (positive entropy for 1,10-phenanthroline metal complexes in aqueous systems) [78Hi, 81Hi2]. The large enthalpy (ΔH) drives the reaction, which is opposed by a large negative entropy change. This is why the formation constant of the ternary complex is much smaller than the stability constants of 1,10-phenanthroline metal complexes. The relatively large net enthalpy is more in character with a complex stabilized by a dominant ligand–metal bond rather than hydrophobic interaction [78Hi]. However, it is more difficult to interpret the large negative entropy.

Anions such as N_3^-, NCO^-, and I^- also inhibit the CO_2 hydration activity of carbonic anhydrase. The binding mode of the anions to cobalt-CA was investigated in detail by the group of Bertini [78Be]. On the basis of the visible and near-infrared spectra (Fig. 5.21) and the water proton relaxation data on the anion adducts of Co(II)-CA, it is proposed that the adducts of NCO^-, CN^-, HS^-, etc., with the Co-enzyme have typical tetrahedral coordination geometry; the CH_3COO^-, NO_3^-, SCN^-, etc., adducts have five-coordination geometry; and the Br^-, N_3^-, Cl^-, I^-, benzoate, etc., adducts are in the equilibrium between tetrahedral and five-coordination geometry, as shown by the following equation [78Be]:

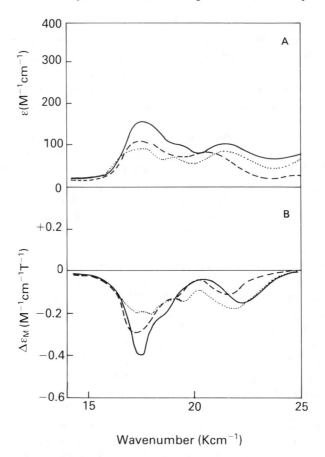

Fig. 5.20 — Absorption (A) and MCD (B) spectra of Co(II)-bovine carbonic anhydrase adducts. ——— oxalate; – – – – acetate; · · · · · 8-quinolinecarboxylate. 0.1 M Hepes buffer, pH 8.0. Reproduced from ref. 84Hi3.

Table 5.12 — Thermodynamic parameters for association of bovine carbonic anhydrase II (BCA) and small chelating agents at 25°C (Reproduced from ref. [78Hi])

Ion	Chelating agents	ΔH (kcal mol^{-1})	ΔS (cal deg^{-1} mol^{-1})	$\log K$
Zn^{2+} ZnBCA II	1,10-phenanthroline	-7.5	4.4	6.55
	5-methyl-1,10-phenanthroline	~ -9	~ -19	2.5
Co^{2+} CoBCA II	1,10-phenanthroline	-9.1	2.1	7.3
	5-methyl-1,10-phenanthroline	~ -8	~ -17	2.2

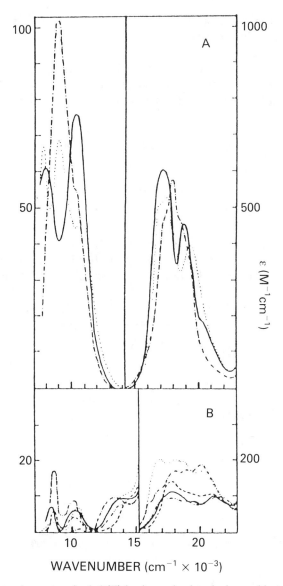

Fig. 5.21 — Electronic spectra of cobalt(II)-bovine carbonic anhydrase adducts with various ligands. (A) –.–.–.–, aniline; · · · · ·; acetazol-amid; ———, cyanate; (B) · · · · ·, bromide; –.–.–.–, iodide; ———, acetate; – – – –; nitrate; ..–..–.., thiocyanate. Reproduced from ref. 78Be.

$$
\begin{array}{ccc}
\text{(His)N} & \diagdown & \text{OH}_2 \\
\text{(His)N} & \!\!-\!\text{Co} & \\
\text{(His)N} & \diagup & \text{anion}
\end{array}
\quad
\begin{array}{c}
+\text{H}_2\text{O} \\
\rightleftharpoons \\
-\text{H}_2\text{O}
\end{array}
\quad
\begin{array}{ccc}
\text{(His)N} & \diagdown & \\
\text{(His)N} & \!\!-\!\text{Co}\!-\!\!-\!\! & \text{anion (2.2-3)} \\
\text{(His)N} & \diagup &
\end{array}
\qquad (27)
$$

The MCD spectrum of the acetate adduct of Co(II)-BCA [84Hi3] is also shown in Fig. 5.20 and clearly indicates that the acetate adduct of Co(II)-CA has five-coordination geometry.

The affinity of anions for carbonic anhydrase in its high pH form is so weak that the pH-dependence of anion binding can be interpreted as the competition between the anion and the hydroxide ion for a coordination site on the metal [88Si].

$$
\text{EM} - \text{OH} \underset{-\text{H}^+}{\overset{+\text{H}^+}{\rightleftharpoons}} \text{EM} - \text{OH}_2 \underset{-\text{ anion}}{\overset{+\text{ anion}}{\rightleftharpoons}} \text{EM} \bigg\langle {}^{\text{OH}_2}_{\text{anion}} \quad \text{or EM-anion (2.3-3)}
$$

$$(28)$$

Lindskog *et al.* [88Re] investigated the interaction of N_3^- with Co-CA III by kinetic study of the enzyme activity and by spectrophotometric examinations of the visible spectrum of Co-CA III. In Fig. 5.22 the visible d-d* transition spectrum of

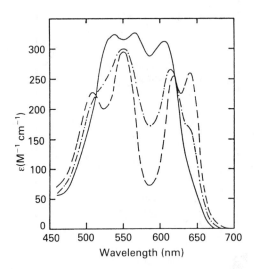

Fig. 5.22 — Titration spectra of Co(II)-bovine carbonic anhydrase III with N_3^-. pH 8.9, 25°C; N_3^- concentration: ———, 52.6 M; –.–.–.–, 4.85 mM; ––––, 0 mM. Reproduced from ref. 88Be.

Co^{+2}-CA III is changed on binding of the N_3^- ion. The pH-dependence of the N_3^- dissociation constant (K_i) of above association obtained by spectrophotometric titration (Fig. 5.23) [88Re] showed that the pKa of deprotonation of the water ligated to the metal ion at the active site is about 5.0.

The N_3^- anion inhibits the CO_2 hydration of the enzyme. The inhibition dissociation constants (K_i'), obtained from the maximum velocity inhibition of CO_2

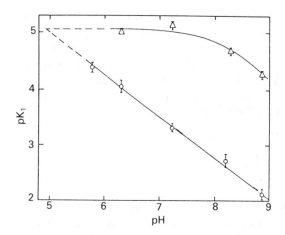

Fig. 5.23 — Ph-dependence of the interaction of N_3^- with Co(II)-bovine carbonic anhydrase III. $\triangle$: The inhibition constant obtained from inhibition of the maximum velocity of CO_2 hydration. $\bigcirc$: The dissociation constant from spectrophotometric titration with N_3^-. Reproduced from ref. 88Re.

hydration kinetics by N_3^-, were much smaller than the dissociation constants obtained by the spectrophotometric titration method. The pH-dependences of K_i and K_i' between pH 8.9 and 5.8 are also shown in Fig. 5.23. To study the N_3^- adduct formed in the steady state in the presence of a high substrate (CO_2) concentration, time-resolved spectra were measured in the rapid-scan stopped-flow instrument. The rapid-scan spectra are shown in Fig. 5.24. In the absence of CO_2 (Fig. 5.22), about

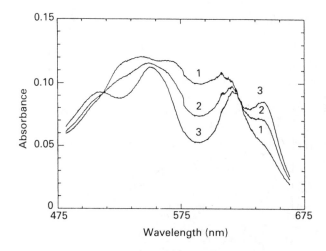

Fig. 5.24 — Rapid-scan stopped-flow spectra of (Co(II)-carbonic anhydrase III during CO_2 hydration in the presence of NaN_3 (1 mM). Spectrum 1 was recorded 10–15 ms, spectrum 2 30–35 ms, and spectrum 3 70–75 ms after stopping of the flow. Reproduced from ref. 88Re.

10% N_3^- adduct with Co-CA III is formed at 1 mM N_3^- concentration [88Re]. In the presence of saturated CO_2, the shape of spectrum 1 (<15 ms) in Fig. 5.24 is completely similar to that obtained for the N_3^- adduct of CO-CA III by spectrophotometric titration (N_3^-: 52.6 mM) [88Re]. This behaviour indicates that N_3^- can bind strongly to the cobalt ion at the active site in the presence of CO_2. Why does N_3^- have a smaller dissociation constant in the presence of substrate (CO_2)? Lindskog *et al.* [88Re] interpreted this behavior via the following eqaution:

$$E^- + CO_2 \rightleftharpoons X \rightleftharpoons EH + HCO_3^- \tag{29}$$

$$EH \rightleftharpoons E^- + H^+ \tag{30}$$

$$EH + I^- \rightleftharpoons EHI^- \tag{31}$$

where E^- and EH are the enzyme with metal-bound OH^- and H_2O, respectively, and X is the Michaelis complex. The anion has a high affinity for the EH form, because the anion competes with OH^- for a coordination site on the metal ion (Eq. 28) [88Re]. In the catalysis of CO_2, the rate-determining step is Eq. (30), thus the high-affinity N_3^- binding observed at high CO_2 concentrations results from an accumulation of the EHI^- form in the steady-state condition [88Re].

Bertini *et al.* [78Be] proposed that the coordination geometry of the azide adduct form of Co-BCA (EHI^-) is in equilibrium between tetrahedral and five-coordination geometry (Eq. 27).

The absorption spetrum of the Co-enzyme in the steady-state at a very high CO_2 concentration in the absence of inhibitory anions was also measured with a rapid-scan stopped-flow apparatus and is shown in Fig. 5.25 [88Re]. The spectrum of the

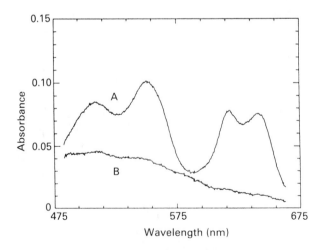

Fig. 5.25 — Rapid-scan stopped-flow spectrum of activated Co(II)-carbonic anhydrase III during CO_2 hydration. Spectra were recorded 6–7 ms after stopping of the flow. (A) No CO_2; (B) enzyme mixed with the buffer, which was saturated with CO_2 at 25°C. Reproduced from ref. 88Re.

intermediate state, recorded in the early stage of the reaction, has a lower intensity than that of the resting state of the enzyme, and the shape of the spectrum for the intermediate state is very similar to that obtained for the anion adduct of Co-CA II which has five-coordination geometry (Figs 5.20, 5.21) [88Re, 86Tu]. Therefore there is the possibility that an intermediate state of the CO_2 hydration reaction in CA has five-coordination geometry.

2.4 Superoxide dismutase

Cu,Zn-superoxide dismutase is inhibited by various anions in a competitive manner. The K_i values were obtained by Rotilio *et al.* [77Ri2]. The modes of anion binding to Cu,Zn-SOD were investigated by the group of Bertini, using cobalt derivatives of superoxide dismutase (Cu_2Co_2SOD) with almost the same activity as that of the native enzyme [87Ba1, 97Ba2, 87Fr, 88Ba, 88Mi, 83Be2]. The antiferromagnetic interaction between Cu(II) and Co(II) through a histidinato bridge permitted detection of the proton NMR peaks of the residues near the metal ions at the active site [87Ba2, 87Fr, 88Ba, 88Mi, 83Be2]. Bertini *et al.* [87Ba1, 87Ba2, 87Fr, 88Ba, 88Mi, 83Be2] have studied the interactions between anions and Cu_2Co_2SOD by NMR, EPR, and absorption spectrophotometry. They proposed three anion-binding modes: (i) the anions replace a His, with the release of water and the formation of square–planar coordination geometry (CN^- and N_3^-); (ii) the anions replace a His without the release of water, and five-coordination is produced (NCS^- and NCO^-); (iii) the anions do not bind directly to the copper ion (phosphate), but interact with the positive charge of Arg-143 at the active site (Arg-143 is present at the gate of the copper binding site (Fig. 5.7)). The binding modes of anions are shown in Fig. 5.26 [83Be2].

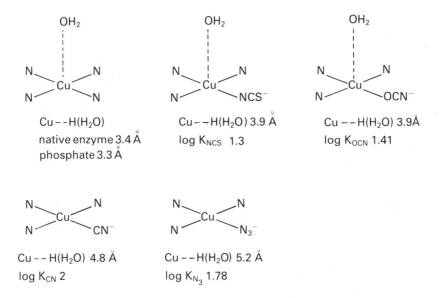

Fig. 5.26 — Stereochemistry around copper(II) in some SOD derivatives. Reproduced from refs. 88Ba and 83Be2.

Bertini *et al.* measured the distances between Cu^{2+} and the proton of water in the presence of various anions by means of a ^1H nuclear magnetic relaxation dispersion method [88Ba]. These results are also shown in Fig. 5.26, with the relevant affinity constants. These results indicate that there is no water in the vicinity of the copper ion in the case of the cyanide and azide adducts, whereas there is some, though at a non-bonding distance, in the case of the cyanate derivative [88Ba]. Thiocyanate displays intermediate behaviour. In the case of phosphate, the distance between the water proton and a copper ion is almost the same as that observed in the native enzyme [88Ba].

Hirose *et al.* [88Oz, 83Hi] investigated the electron transfer between the copper ion of Cu_2Zn_2SOD and $Fe(CN)_6^{3-}/Fe(CN)_6^{4-}$ in the presence of anions. In the native enzyme the electron transfer from $Fe(CN)_6^{4-}$ to the Cu^{2+} ion occurs at pH 6.0, and the Cu^{2+} ion is completely reduced to the Cu^+ state. In the presence of N_3^- and CN^-, the electron transfer was not observed from $Fe(CN)_6^{4-}$ to the anion adduct of $Cu_2^{2+}Zn_2SOD$, whereas the OCN^- and SCN^- adducts of $Cu_2^{2+}Zn_2SOD$ were able to accept electron from $Fe(CN)_6^{4-}$. This behaviour is interpreted in terms of the redox potential of the anion adduct of Cu_2Zn_2SOD: the redox potentials of the N_3^- and CN^- adducts of Cu_2Zn_2SOD [79La] were too low to be reduced by $Fe(CN)_6^{4-}$, whereas those of OCN^- and SCN^- were almost the same as that of the native enzyme. The N_3^- and CN^- adducts have square–planar coordination geometry, but OCN^- and SCN^- give rise to five-coordinate geometry [88Oz]. These results therefore indicate that the coordination geometry of the anion adduct is directly correlated with the redox potential of the copper ion in the enzyme [88Oz].

Hirose *et al.* [85Hi] and Rotilio *et al.* [84De] proposed that phosphate is bound directly to the cobalt ion in Co_2Zn_2SOD, with a change in the coordination geometry from five-coordinate to tetrahedral, as detected by MCD, the ^{32}P-NMR relaxation time, and the near-infrared spectra. The MCD spectral changes in the presence of various concentrations of phosphate ion are shown in Fig. 5.27. The MCD spectrum of Co_2Zn_2SOD in the absence of phosphate ion mainly displays two negative MCD bands in the visible region, and is very similar to that obtained for chelating agent adduct of CoBCA (Fig. 5.20). Via the electronic and ^1H-NMR spectra, Bertini *et al.* [87Ba2] showed that His-61, which bridges the zinc and cobalt ions, is detached upon the addition of phosphate. The phosphate ion probably binds both to Arg-143 (Fig. 5.7) and the cobalt ion [87Ba2, 88De].

Dooley & McGuirl [86Do] have measured the thermodynamic parameters (K, ΔH, ΔS) for N_3^- and SCN^- binding to Cu_2Zn_2SOD through the difference UV spectra. Comparative data have also been obtained for the anation reactions of Cu^{2+}-diethylenetriamine (dien). These results are shown in Table 5.13. Anion binding to Cu(dien) is accompanied by a large negative ΔH and a small negative ΔS, thus the driving force of anion binding to Cu(dien) is the large negative ΔH. On the other hand, anion binding to Cu_2Zn_2SOD gives a small negative ΔH or a large positive ΔH with a large positive ΔS. The large positive entropy may be interpreted in terms of the dissociation of water molecules around the anion or the displacing of ordered water molecules at the active site [86Do]. The thermodynamic parameters of N_3^- and SCN^- binding to Cu_2Zn_2SOD are qualitatively different, whereas those of N_3^- and SCN^- binding to Cu(dien) are very similar. This behaviour may be correlated to the difference in the binding modes of these anions, because N_3^- and

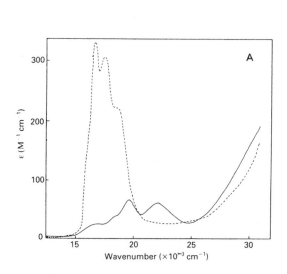

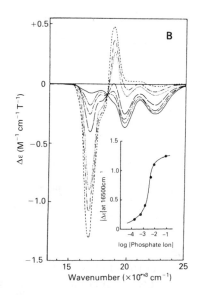

Fig. 5.27 — Electronic and magnetic circular dichroic (MCD) spectra of Co$_2$Zn$_2$SOD. (A) Electronic spectra of phosphate and water form of Co$_2$Zn$_2$SOD at pH 7.4. ———, 0.01 M Hepes buffer ----, 8.4 × 10^{-2} M phosphate ion in 0.01 M Hepes buffer. (B) Magnetic circular dichroic spectra of Co$_2$Zn$_2$SOD with increase of the phosphate ion concentration. Phosphate ions: ———, 0 M; ···–···, 2 × 10^{-4} M; –·–·–, 5.3 × 10^{-4} M; ----, 1.52 × 10^{-3} M; ······, 3.52 × 10^{-3} M; –·–·–·–, 7.52 × 10^{-3} M. ----, 8.40 × 10^{-2} M. Reproduced from ref. 85Hi.

Table 5.13 — Thermodynamic parameters for ligand substitution reactions (Reproduced form ref. [86Do])

Reactions	Binding constant	ΔH (kJ/mol)	ΔS (J/mol K)	ΔG (kJ/mol)
Cu$_2$Zn$_2$SOD + N$_3^-$	88 ± 2	− 4.0 ± 1.0	28 ± 2	− 11.0 ± 1.0
Cu$_2$Zn$_2$SOD + SCN$^-$	22 ± 3	11.7 ± 1.2	64 ± 4	− 7.4 ± 2.4
Cu(dien)$^{2+}$ + N$_3^-$	75 ± 2	− 10.6 ± 0.7		
			− 2 ± 2	− 10.0 ± 1.0
Cu(dien)$^{2+}$ + SCN$^-$	36 ± 2	− 9.5 ± 1.0	− 3 ± 3	− 8.6 ± 1.0

ΔH and ΔS were calculated from the temperature-dependence of K via RT ln K.

SCN$^-$ adducts exhibit square–planar and five-coordination geometry respectively [86Do].

The affinity constants of N$_3^-$ for various mutants in position 143 (an Arg residue in the native enzyme) parallel the decrease in the specific activity of the various mutants (Table 5.14) [87Be1, 90Be]. The positive charge of the Arg residue is very

Table 5.14 — Comparison between catalytic activity and azide binding constant for SOD (Arg-143) and mutants at position 143. (Reproduced from refs. [87Be, 90Be])

Derivative	Activity	N_3^- binding constant (%)
Arg-143	100%	154 (100%)
Lys-143	43%	63 (41%)
Ile-143	11%	16 (10%)
Glu-133	<1%	6 (4%)

important for the enzyme activity. It is suggested that O_2^- is fixed into the cavity with the same affinity pattern as N_3^- by the Arg residue located at the gate of the active site cavity [83Ta, 87Be].

2.5 Transferrin

The coordination sphere of iron in transferrin is in a closed state, therefore various spectroscopic methods cannot detect the ternary complex when various anions and small chelating agents are added to iron-transferrin. However, Swope *et al.* [88Sw] recently showed that the cyanide ion slowly forms a low-spin adduct at only the C-terminal iron(III) binding site of transferrin. The stoichiometry of the reaction reveals that a water molecule, the bicarbonate ion, and the carboxylate of Asp, which coordinate to the iron(III) at the C site of transferrin, are replaced by the cyanide ion. This result indicates that the iron(III) character is different at the C and N site and the presence of bicarbonate is not essential for the formation of a stable iron adduct with apotransferrin [88Sw].

ACKNOWLEDGEMENTS

We would like to thank Mr Maarten Alink for many comments. Thanks are also expressed to Prof. Ivano Bertini of Firenze University in Italy, and to Prof. Mamoru Nakanishi of Nagoya City University, for their encouragement.

REFERENCES

38Ma   Mann, T. & Keilin, D.: *Proc. Roy. Soc.* (*London*) *Ser. B.* **126**, 303 (1938).

49Sc   Schade, A. L., Reinhart, R. W., & Lery, H.: *Arch. Biochem. Biophys.* **20**, 170 (1949).

61Co   Coleman, J. E. & Vallee, B. L.: *J. Biol. Chem.* **236**, 2244 (1961).

62Li   Lindskog, S. & Malmstrom, B. G.: *J. Biol. Chem.* **237**, 1129 (1962).

64Li   Lindskog, S. & Nyman, P. O.: *Biochem. Biophys. Acta* **85**, 462 (1954).

68He   Henkens, R. W. & Sturtevant, J. M.: *J. Am. Chem. Soc.* **90**, 2669 (1968).

68Li   Lindskog, S. & Thorsland, A.: *Eur. J. Biochem.* **3**, 453 (1968).

69Ch   Chen, R. F., Schechter, A. N., & Berger, R. L.: *Anal. Biochem.*, **29** 68 (1969).

69He   Henkens, R. W., Watt, G. D., & Sturtevant, J. M.: *Biochemistry* **8**, 1874 (1969).

69Mc   McCord, J. M. & Fridvich, I.: *Biol. Chem.* **244**, 6049 (1969).

70Ta1   Taylor, P. W., King, R. W., & Burgen, A. S. V.: *Biochemistry* **9**, 2638 (1970).

70Ta2   Taylor, P. W., King, R. W., & Burgen, A. S. V.: *Biochemistry* **9**, 3894 (1970).

71Ta1   Taylor, P. W. & Burgen, A. S. V.: *Biochemistry* **10**, 3859 (1971).

71Ta2   Taylor, P. W., Feeney, J., & Burgen, A. S. V.: *Biochemistry* **10**, 3866 (1971).

72He   Henkins, R. W. & Sturtevant, J. M.: *Biochemistry* **11**, 206 (1972).

72Li   Liljas, A., Kannan, K. K., Bergsten, P. C., Fridborg, K., Strandberg, B., Carbom, V., Jarup, L., Lovgren, S., and Petef, M.: *Nature New Biol.* **235**, 131 (1972).

72Ro   Rotilio, G., Calabrese, L., Bossa, F., Barra, D., Finazzi, A., & Mondovi, B.: *Biochemistry* **11**, 2182 (1972).

72Wa   Ward, R. L. & Cvull, M. D.: *Arch. Biochem. Biophys.* **150**, 436 (1972).

73Ai   Aisen, P.: *Inorganic biochemistry*; Elsevier Sci. Pub. Inc., Netherland, (1973) p. 280.

73Be   Beem, K. M., Rich, W. E., & Rajagopalan, K. V., *J. Biol. Chem.* **248**, 4299 (1973).

73Fe   Fee, J. A.: *J. Biol. Chem.* **248**, 4229 (1973).

73Li   Lindskog, S. & Coleman, J. E.: *Proc. Natl. Acad. Sci. USA* **70**, 2505 (1973).

74Ge   Gerber, K., Ng, F. T. T., & Wilkins, R. G.: *Biochemistry* **13**, 2663 (1974).

74Wi1   Wilkins, R. G.: *The study and mechanism of reactions of transition metal complexes*; Allyn and Bacon, Boston (1974) p. 219.

74Wi2   Wilkins, R. C. & Williams, K. R.: *J. Am. Chem. Soc.* **96**, 2441 (1974).

75Fe   Fee, J. A. & Briggs, R. G.: *Biochem. Biophys. Acta* **400**, 439 (1975).

75Ge   Gerber, K., Ng, F. T. T., & Wilkins, R. G.: *Bioinorg. Chem.* **4**, 153 (1975).

75Ha   Harrison, L. W., Auld, D. S., & Vallee, B. L.: *Proc. Natl. Acad. Sci. USA* **72**, 3930 (1975).

75Ro   Rosenberg, R. C., Root, C. A., & Gray, H. B.: *J. Am. Chem. Soc.* **97**, 21 (1975).

76Ki   Kidani, Y., Hirose, J., & Koike, H.: *J. Biochem.* **79**, 43 (1976).

77Be   Beem, K. M., Richardson, D. C., & Rajagopalan, K. V.: *Biochemistry* **16**, 1930 (1977).

77Ki1   Kidani, Y. & Hirose, J.: *J. Biochem.* **81**, 1383 (1977).

77Ki2   Kidani. Y. & Hirose, J.: *Chemistry Letters* 475 (1977).

77Ri1   Rigo, A., Viglino, P., Calabrese, L., Cocco, D., & Rotilio, G.: *Biochem. J.* **161**, 27 (1977).

77Ri2   Rigo, A., Stevanato, R., Viglino, P., & Rotilio, G.: *Biochem. Biophys. Res. Commun.* **79**, 776 (1977).

78Ai   Aisen, P., Leibman, A., & Zweier, J.: *J. Biol. Chem.* **253**, 1930 (1978).

78Be   Bertini, I., Canti, G., Luchinat, C., & Scozzafava, A.: *J. Am. Chem. Soc.* **100**, 4873 (1978).

78Bi   Billo, E. J., Brito, K. K., & Wilkins, R. G.: *Bioinorganic Chemistry* **8**, 461 (1978).

78Ha   Harrison, L. W. & Vallee, B. L.: *Biochemistry* **17**, 4359 (1978).

78Hi   Hirose, J. & Kidani, Y.: *Chem. Pharm. Bull.* **26**, 1768 (1978).

78Ki   Kidani, Y. & Hirose, J.: *Biochem. Biophys. Res. Commun.* **82**, 506 (1978).

78Ri   Rigo, A., Viglino, P., Bonori, M., Cocco, D., Calabrese, L., & Rotilio, G.: *Biochem. J.* **169,** 277 (1976).

78Ro1   Roman, A., Graihen, M. E., Lochmuller, & Henkens, R.: *Bioinorganic Chemistry* **9**, 217 (1978).

78Ro2   Roger, T. B., Borrensen, T., & Feeney, R. E.: *Biochemistry* **17**, 1105 (1978).

79Bi   Billo, E. J.: *J. Inorg. Biochem.* **11**, 339 (1979).

79Ca   Carrano, C. J. & Raymond, K. N.: *J. Am. Chem. Soc.* **101**, 5401 (1979).

79La   Laurence, G. D. & Sawyer, D. T.: *Biochemistry* **18**, 3047 (1979).

79Pa   Pantoliano, M. W., McDonnell, P. J., & Valentine, J. S.: *J. Am. Chem. Soc.* **101**, 6454 (1979).

79Va   Valentine, J. S., Pantoliano, M. W., McDonell, P. J., Burger, A. R., & Lippard, S. J.: *Proc. Natl. Acad. Sci. USA* **76**, 4245 (1979).

80Bi   Billo, E. J.: *J. Inorg. Biochem.* **12,** 335 (1980).

80Ha   Harrington, P. C. & Wilkins, R. G.: *J. Inorg. Biochem.* **12**, 107 (1980).

80Hi   Hirose, J. & Kidani, Y.: *Biochem. Biophys. Acta* **622**, 71 (1980).

80Ka   Kannan, K. K.: *Biophysics and physiology of carbon dioxide*; Springer-Verlag, Heidelberg, (1980) p. 184.

80Va   Vallee, B. L. & Holmquist, B.: *Advances in Inorganic Chemistry*; **2** (1980) p. 27.

81Ba   Baldwin, D. A. & Sousa, M. R. D.: *Biochem. Biophys. Res. Commun.* **99** 1101 (1981).

81Hi1   Hirose, J., Iwatzuka, K., & Kidani, Y.: *Biochem. Biophys. Res. Commun.* **98**, 58 (1981).

81Hi2   Hirose, J. & Kidani, Y.: *J. Inorg. Biochem.* **14**, 313 (1981).

81Pe   Pecarora, V. L., Harris, R. W., Carrano, C. J., & Raymond, K. N.: *Biochemistry* **20**, 7033 (1981).

81Va   Valentine, J. S. & Pantoliano, M. W.: *Copper protein*; Wiley, New York, (1981) p. 291.

82Co   Cowart, R. E., Kojima, N., & Bates, G. W.: *J. Biol. Chem.* **257**, 7560 (1982).

82Du   Dunbar, J. C., Johansen, J. T., & Uchida, T.: *Calsberg Res. Commun.* **47**, 163 (1982).

82He   Hendy, D., Lindoy, L. F., & Yelloless, D.: *Inorg. Chim. Acta* **65**, L237 (1982).

82Hi   Hirose, J. Ohhira, T., Hirata, H., & Kidani, Y.: *Arch. Biochem. Biophys.* **218**, 179 (1982).

82Li   Lindskog, S.: *Advance Inorg. Biochem.* **4**, 115 (1982).
82Ta   Tainer, J. A., Getzoff, E. D., Beem, K. M., Richardson, J. S., & Richardson, D. C.: *J. Mol. Biol.* **160**, 181 (1982).
83Be1   Bertini, I.: *The coordination chemistry of metalloenzymes*; D. Reidel Publishing Co. Holland, (1983) p. 1.
83Be2   Bertini, I., Luchinat, C., & Scozzafava, A.: *The coordination chemistry of metalloenzymes*; D. Reidel Publishing Co. Holland, (1983) p. 155.
83Ga   Galdes, A., Auld, D. S., & Vallee, B. L.: *Biochemistry* **22**, 1888 (1983).
83Ge   Geoghegan, K. F., Galdes, A., Martinelli, R. A., Auld, D. S., & Vallee, B. L.: *Biochemistry* **22**, 2255 (1983).
83Ha1   Harris, R. W.: *Biochemistry* **22**, 3920 (1983).
83Ha2   Harris, R. W. & Stenback, J. Z.: *J. Inorg. Biochem.* **33**, 211 (1983).
83Ha3   Harris, R. W. & Pecoraro, V. L.: *Biochemistry* **22**, 292 (1983).
83Hi   Hirose, J., Ueoka, M., Tsuchiya, T., Nakagawa, M., Noji, M., & Kidani, Y.: *Chem. Lett.* 1983, 1429 (1983).
83Ko   Koenig, S. H. & Brown, R. D.: *The coordination chemistry of metalloenzymes*; D. Reidel Publishing Co. Holland, (1983) p. 19.
83Ta   Tainer, J. A., Getzoff, E. D., Richardson, J. S., & Richardson, D. C.: *Nature* **306**, 284 (1983).
84Au   Auld, D. S., Galdes, A., Geoghegam, K. F., Holmquist, B., Martinelli, R. A., & Vallee, B. L.: *Proc. Natl. Acad. Sci. USA* **81**, 5041 (1984).
84De   Desideri, A., Cocco, D., Calabrese, L., & Rotilio, G.: *Biochem. Biophys. Acta* **785**, 111 (1984).
84Ha1   Harris, R. W. & Carrano, C. J.: *J. Inorg. Biochem.* **22**, 201 (1984).
84Ha2   Harris, R. W.: *J. Inorg. Chem.* **21**, 263 (1984).
84Hi1   Hirose, J. & Wilkins, R. G.: *Biochemistry* **23**, 3149 (1984).
84Hi2   Hirose, J., Yamada, M., Hayakawa, C., Nagao, H., Noji, M., & Kidani, Y.: *Biochem. Int.* **8**, 401 (1984).
84Hi3   Hirose, J., Noji, M., & Kidani, Y.; *Chem. Pharm. Bull.* **32**, 2481 (1984).
85Hi   Hirose, J., Hayakawa, C., Noji, M., & Kidani, Y.: *Inorg. Chim. Acta* **107**, L7 (1985).
85Wu   Wu, Yi-he & Kaden, T. A.: *Helv. Chim. Acta* **68**, 1611 (1985).
86Au1   Auld, D. S. & Vallee, B. L.: *Zinc enzymes*; Birkhauser Boston Inc. Birkhauser (1986), p. 167.
86Au2   Auld, D. S., Larson, K., & Vallee, B. L.: *Zinc enzymes*; Birkhauser Boston Inc. Birkhauser (1986), p. 133.
86Ch   Christianson, D. W. & Lipsomb, W. N.: *Zinc enzymes*; Birkhauser Boston Inc. Birkhauser (1986) p. 121.
86Co   Cowart, R. E., Swope, S., Loh, T. T., Chasteen, N. D., & Bates, G. W.: *J. Biol. Chem.* **261**, 4607 (1986).
86Co   Dooley, D. M. & McGuirl, M. A.: *Inorg. Chem.* **25**, 1261 (1986).
86Er   Eriksson, E. A., Jones, T. A., & Liljas, A.: *Zinc enzymes*; Birkenhauser, Boston Inc. Birkenhauser, (1986) p. 317.
86Ga   Gardell, S. J., Craik, C. S., Hilvert, D., Urdea, M. S., & Rutter, W. J.: *Zinc enzymes*; Birkhauser, Boston Inc. Birkhauser (1986) p. 191.
86Ha1   Harris, R. W.: *J. Inorg. Biochem.* **27**, 41 (1986).
86Ha2   Harris, R. W.: *Biochemistry* **25**, 2041 (1986).

86Hi   Hilvert, D., Gardell, S. J., Rutter, W. J., & Kaiser, E. T.: *J. Am. Chem. Soc.* **108**, 5298 (1986).

86Ki   Kidani, Y. & Hirose, J.: *Zinc enzymes*; Birkhauser Boston Inc. Birkhauser (1986) p. 59.

86Kr   Kretchmar, S. A. & Raymond, K. N.: *J. Am. Chem. Soc.* **108**, 6212 (1986).

86Re   Ree, D. C., Howard, J. B., Chakrabarti, P., Yeates, T., Hsu, B. T., Hardman, K. D., & Lipscomb, W. N.: *Zinc enzymes*; Birkhauser, Boston Inc. Birkhauser (1986) p. 158.

86Sy   Syvertsen, C., Gaustad, R., Schroder, K., & Ljones, T.: *J. Inorg. Biochem.* **26**, 63 (1986).

86Tu   Tu, C. K. & Silverman, D. N.: *J. Am. Chem. Soc.* **108**, 6065 (1986).

87An   Anderson, B. F., Baker, H. M., Dodson, E. J., Norris, G. E., Rumball, S. V., Waters, J. M., & Baker, E. N.: *Proc. Natl. Acad. Sci. USA* **84**, 1769 (1987).

87Ba1   Banci, L., Bertini, I., Luchinat, C., & Scozzafava, A.: *J. Am. Chem. Soc.* **109**, 2328 (1987).

87Ba2   Banci, L., Bertini, I., Monnani, R., & Scozzafava, A.: *Inorg. Chem.* **26**, 153 (1987).

87Be1   Beyer, W. F., Fridvich, I., Mullenbach, G., & Hallewell, R.: *J. Biol. Chem.* **262**, 1182 (1987).

87Ch   Cheuk, M. S., Loh, T. T., Hui, Y. V., & Keung, W. M.: *J. Inorg. Biochem.* **29**, 301 (1987).

87Cr   Crichton, R. R. & Wauters, C. M.: *Eur. J. Biochem.* **164**, 485 (1987).

87Fr   Mota de Freitas, D., Luchinat, C., Banci, L., Bertini, I., & Valentine, J. S.: *Inorg. Biochem.* **26**, 2788 (1987).

88Ba   Banci, L., Bertini, I., Luchinat, C., Monnanni, R., & Scozzafava, A.: *Inorg. Chem.* **27**, 107 (1988).

88Be1   Bertini, I., Hirose, J., Luchinato, C., Messori, L., Piccioli, M., & Scozzafava, A.: *Inorg. Chem.* **27**, 2405 (1988).

88Be2   Bertini, I., Hirose, J. Kozlowski, H., Messori, L., & Scozzafava, A.: *Inorg. Chem.* **27**, 1081 (1988).

88De   Desideri, A., Paci, M., & Rotilio, C.: *J. Inorg. Biochem.* **33**, 91 (1998).

88Ha   Harris, R. W. & Madsen, L. J.: *Biochemistry* **27**, 284 (1988).

88Mi   Ming, L. J., Banci, L., Luchinat, C., Bertini, I., & Valentine, J. S.: *Inorg. Chem.* **27**, 728 (1988).

88Oz   Ozaki, S., Hirose, J., & Kidani, Y.: *Inorg. Chem.* **27**, 3746 (1988).

88Re   Ren, X., Sandstrom, A., & Lindskog, S.: *Eur. J. Biochem.* **173**, 73 (1988).

88Ro   Roe, J. A., Buttler, A., Scholler, D. M., Valentine, J. S., Marky, L., & Breslauer, K. J.: *Biochemistry* **27**, 950 (1988).

88Si   Silverman, D. N. & Lindskog, S.: *Acc. Chem. Res.* **21**, 30 (1988).

88Sw   Swope, S. K., Chasteen, D. N., Weber, K. T., & Harris, D. C.: *J. Am. Chem. Soc.* **110**, 3835 (1988).

89Be   Bertini, I., Banci, L., Luchinat, C., Bielski, B. H. J., Cabelli, D. E., Mullenbach, G. T., & Hallewell, R. A.: *J. Am. Chem. Soc.* **111**, 714 (1989).

89Ch   Christianson, D. W. & Lipscomb: *Acc. Chem. Res.* **22**, 62 (1989).
90Be   Bertini, I., Banci, L., Luchinat, C., & Picciolo, M.: *Coord. Chem. Rev.* (in press).

VI

Metal complexes of carbohydrates and sugar-type ligands

Kálmán Burger and László Nagy
Department of Inorganic and Analytical Chemistry, A. József University,
6701 Szeged, Hungary

1. GENERAL PROPERTIES

Carbohydrates and related compounds have long been known to form complexes with metal ions. In spite of this, the field of carbohydrate complexes continued to remain largely unexplored. The complexing of metal ions with carbohydrates and their derivatives has received considerable interest only in the past two decades, mainly because of the possible importance of such interactions in a variety of biological processes, for example, in the binding of metal ions to cell walls, etc.

Attention has been paid to the application of carbohydrate complexes in analytical chemistry and in chemical technology. Differences in solubility or stability of complexes permit the large-scale separation of polyhydroxy compounds. Differences in electrophoretic mobility of carbohydrates in the presence of a metal salt can be used to separate mixtures on a small scale and to assist in identifying carbohydrates. The ability of sugar derivatives to sequester metal ions can be utilized to develop new classes of metal-based affinity catalysts, of metal-chelators for clinical use, and of models for biologically important chelates. Further interests include NMR studies of the interactions of metal ions with sugars. Some sugar complexes are used in analytical chemistry.

The quantitative characterization of the metal ion coordination equilibria of polyalcohols and other sugar-type ligands not containing donor atoms apart from alcoholic and aldehyde (or ketone) oxygens is made difficult by the low stabilities of the complexes in neutral or acidic aqueous solutions. This is not only due to the low electron densities on the donor oxygens of the sugars, which accordingly do not readily substitute the water molecules bound in the first coordination sphere of metal ions. In solutions of sugar-type ligands, the species are in anomeric and conformational equilibria, and only isomers which have suitably positioned sequences of alcoholic hydroxy groups can interact specifically with metal ions. The fraction of such suitable isomers in the total concentration of the ligand, and any shifts due to metal coordination in the anomeric and conformational equilibria, thereby resulting in changes in the above fraction, will also influence the complex stabilities. Thus, complex stability constants determined by the techniques used in equilibrium chemistry (potentiometric, spectrophotometric, calorimetric studies, etc.) must be regarded as overall values reflecting the association of all forms of the ligand. To specify clearly which processes are involved in the metal-binding equilibria is rather difficult (if not impossible) by the conventional methods.

In the literature, a few surveys of carbohydrate complexes have been published. Von Lippmann reviewed much of the work published before 1904 [04Li], while in 1964 Sawyer [64Sa2] cited 140 references. In 1966 Rendleman summarized the results obtained on the carbohydrate complexes of alkali metals and alkaline earth metals [66Re]. Weigel reviewed the anionic complexes formed between a variety of oxyacids and polyhydroxy compounds [63We]. Cook & Bugg [77Co] dealt with the crystal structures of calcium–carbohydrate complexes. Critical reviews were published by Angyal [80An1, 89An] on sugar-cation complexes in neutral solution, and by Yano [88Ya] on the coordination compounds of transition metal ions with amino sugars and related compounds. A comprehensive review of carbohydrate complexes is not available. The aim of this chapter is to summarize the results obtained by the different methods (electrochemical, spectroscopic, magnetic measurements, X-ray diffraction, etc.) on the complexes formed between mono- and polysaccharides and some of their derivatives with metal (especially transition metal) ions.

As concerns the findings on carbohydrate complexes, some generally accepted rules can be stated. In aqueous solution, coordination complexes are formed by the displacement of water molecules in the solvation sphere of cations by hydroxy groups of polyols. Since water molecules solvate cations much better than monohydric alcohols or diols, the latter will not displace water from the coordination sphere. Thus, simple diols or alcohols do not form stable complexes with cations in neutral aqueous solution. It seems to be generally true that at least three hydroxy groups in a favourable steric arrangement are required for complex formation in such systems. The general rule is that cyclitols and sugars containing an axial–equatorial–axial sequence of three hydroxy groups in a six-membered ring (pyranose form), or a *cis-cis* sequence in a five-membered ring (furanose form), form 1:1 complexes with metal cations in hydroxylic solvents [73An1, 87Al, 80An1]. The formation of such complexes is direct proof of the coordination of oxygen atoms of non-deprotonated alcoholic hydroxy groups. The low stability of such complexes reflects (among others) the low donor ability of such oxygens. In alkaline solution, the hydroxy groups of sugars can, after deprotonation, form much stronger complexes with cations than in neutral solution.

On introduction of a carboxy or amino group (or both) into a sugar molecule, the complex-forming ability is enhanced by several orders of magnitude, even in acidic or neutral solution [65Ta1, 90Na1, 65Ta2, 85Mi, 65Ta3, 64Sa2]. In strongly alkaline solution, the complexes formed with transition metal ions have an anionic character. In some cases, the formation of various amounts of dimeric complexes is also detectable. The complex formation sometimes changes the conformational equilibrium of sugars, or may even cause epimerization [73An1, 88Ya].

Besides the conventional electrochemical and spectrochemical equilibrium methods, which reflect the metal or ligand activity (concentration) decrease caused by complex formation, the metal ion coordination of carbohydrate ligands can also be characterized by structural chemical methods which reflect the conformation or/ and configurational changes caused by such processes.

NMR spectroscopy has been found to be especially suitable for the study of processes in which the coordination of the metal ion forces the transformation of the organic ligand into a form in which the alcoholic hydroxy groups are in positions favourable for coordination of the metal. In aqueous solution, D-allose is present in an equilibrium mixture of α- and β-pyranose forms. Since only the α form contains three hydroxy groups in axial-equatorial-axial positions favourable for metal ion binding, complex formation will alter the α:β ratio by complexing the α form of D-allose. The H-NMR spectrum clearly reflects this change, the extent of which characterizes the strength of the coordination interaction [72An].

X-ray crystal structure analysis of crystalline compounds containing a carbo-hydrate and a metal salt in stoichiometric proportions does not consitute evidence of complex formation in solution. The well-defined crystal structure merely indicates that in the solid state the carbohydrate molecule with the metal ion and the anion can fill the space in a regular packing, usually held together by coordinate and H bonds. When the crystals are dissolved in polar solvents (e.g. water), these hydrogen bonds may be broken and the solvent water may displace the coordination hydroxy groups of the carbohydrate from the coordination sphere of the metal ion. Thus, on the basis of crystal structures, one cannot predict complex formation in solution. On the other hand, when complex formation is known to occur in solution from independent equilibrium measurements, it is very probable that the main binding sites are the same in the crystal and in solution [80An1]. In the crystal, some additional weak binding sites may also be present, but the sugar hydroxy oxygens (usually three per metal ion) coordinating the metal ion are the same in the solid state and solution.

2.   COORDINATION EQUILIBRIA

2.1   Sugar (small molecular carbohydrate) complexes

Carbohydrates are very weak acids; the first acidity constants fall in the range $10^{-14} - 10^{-12}$ mol dm^{-3} [73Re]. A number of investigations have been published on the ionization of free sugars (among them pentoses, hexoses, and sugar derivatives) [13Mi, 29Hi, 66Iz, 70Ch1, 71De] and acyclic polyhydroxy alcohols [13Mi, 29Hi, 50So, 52Th, 64Mu, 66Ry, 81Vi]. In exceptional cases, such investigations have made possible the assignment of the hydroxy group which deprotonates in the system. It has been shown, for instance, that the 2-OH group is the most acidic in glycopyrano-sides [60Cr, 60Le, 71Ro, 80Ro]. The acidities of methyl arabino-, ribo-, xylo-, and lyxofuranosides comprise two groups: the *trans* 1,2-glycosides have acidity constants 2 to 4 times larger than those of their *cis* counterparts [83Ve2]. A possible explanation [65El] for this acidity difference is the stronger solvation of the 2-oxyanion than that of the neutral hydroxy group, and a neighbouring *cis*-methoxy group impedes solvation more effectively than does the *trans* group. Therefore, in each pair, the anomer having the *cis*-1,2-configuration is the weaker acid. Because of the low acidity of sugar molecules, the metal complexes formed by mono- and disaccharides are usually weak in acidic and slightly alkaline solutions. This is the reason for the lack of equilibrium data for such systems. Just the opposite can be said about complexes formed with sugar acids or other sugar derivatives containing stronger (more basic) donor groups than the alcoholic oxygens.

One of the first convincing pieces of evidence on the metal ion coordination of sugar molecules stems from paper electrophoretic measurements [61Mi]: in a study of the acidity of sugar molecules by paper electrophoresis, Mills [61Mi] found that in the presence of metal salts some sugar molecules migrate even at pH ~ 7, toward the cathode indicating the positive charge of the sugar-containing species in the system. This behaviour could be understood only by assuming that the neutral sugars coordinate metal ions, resulting in metal complex cations. Since the extent of migration could be considered to be a rough measure of the complexing ability (other reasons for mobility changes being neglected), paper electrophoretic measurements were used to acquire information on the conditions needed for metal binding by sugars. Inositol, which contains three syn-axial hydroxy groups in each of its two equivalent chair forms (Fig. 6.1) exhibited the greatest mobility in a series of sugars.

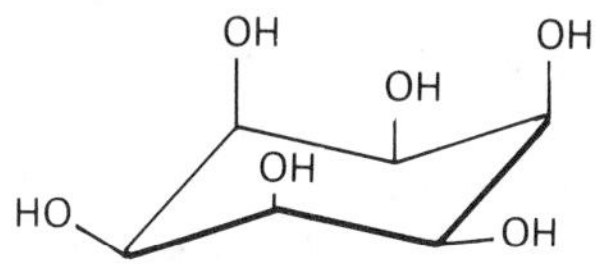

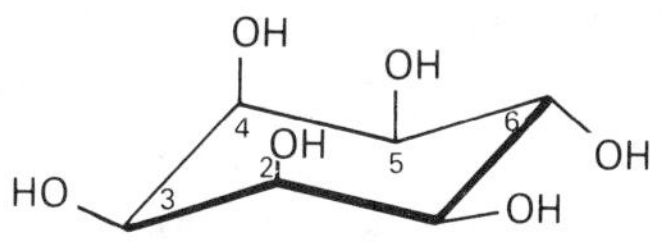

Fig. 6.1 — Structure of *epi*-inositol (1) and *allo*-inositol (2).

It was a reasonable assumption to correlate this feature with the complexing ability of inositol. Later, NMR investigations are carried out by Angyal & Hickman [75An2] on epi-inositol, which does not contain three *syn*-axial hydroxy groups in its stable chair form, but would provide this structure by flipping into the less stable chair form. Their results demonstated that *epi*-inositol undergoes the above transformation in the presence of calcium ions.

On the basis of the latter results, the electrophoretic migration and the NMR spectra of a number of sugars were investigated in the presence of calcium chloride. The results indicated a clear-out correlation between the migration and the NMR shifts due to the presence of calcium ions in the case of sugars carrying hydroxy oxygens in the axial-equatorial-axial sequence. Some other, less pronounced but still significant complexing groupings were also identified as a consequence of these investigations. These results, which were later confirmed by a large number of independent studies, showed that the metal ion-binding ability of sugar molecules is determined by the steric situation of their hydroxy groups.

The complexing sites of monosaccharides, in descending sequence of their strengths, are as follows: axial-equatorial-axial triol grouping on a six-membered ring > *cis-cis* triol grouping on a five-membered ring > acyclic *threo-threo* triol grouping > acyclic *threo* pair adjacent to a primary hydroxy group > acyclic *erythro-threo* triol grouping > acyclic *erythro* pair adjacent to a primary hydroxy group > acyclic *erythro-erythro* triol grouping > *cis* diol grouping on a five-membered ring > *cis* diol grouping on a six-membered ring > *trans* diol grouping on a six-membered ring [89An].

In cyclic monosaccharides, no more than three hydroxy oxygens can coordinate to the same metal ion, but in molecules containing more than one monosaccharide moiety, complexing can occur at more than three oxygen atoms [86Ca]. In each of its two monosaccharide moieties, α-D-allopyranosyl α-D-allopyranose, for example, has an axial-equatorial-axial-sequence of hydroxy groups and it forms a pentadentate complex with calcium ions [78Ol].

In oligo- or polysaccharides, the extent of complexing usually increases with increasing chain length, partly because of the increasing number of complexing sites on the molecule and partly because of interchain crosslinking resulting in favourable binding environments.

Symons *et al.* [76Ha] showed that separate water and hydroxy proton resonances can be resolved for sugars in dilute aqueous solution. This demonstrated the possibility of using H-NMR for the detection of complex formation by sugars in aqueous solution. The interactions of D-ribose with calcium chloride or perchlorate served as the first examples [82Sy] of this technique. These results strongly supported the results of others on these systems [72An, 76Le2]. The very detailed NMR study by Symons *et al.* [84Sy] on the interactions between calcium ions and monosaccharides (D-xylose, D-glucose, 2-deoxy-D-glucose, D-mannose, D-galactose, D-fucose, L-sorbose, and D-fructose) directly proved the applicability of this method [84Sy]. The shifts in the hydroxy resonances on the addition of calcium chloride to dilute aqueous solutions of the monosaccharides were followed. The spectra reflected three types of changes: (1) normal, linear, upfield shifts, comparable with that of water protons due to the calcium ion–water interaction, indicating no preferential interaction of the metal with the sugar; (2) reduced upfield or weak downfield shifts, which could be assigned to the weak binding of sugar hydroxy groups to calcium ions, and (3) downfield shifts becoming normal upfield shifts at high salt concentrations, indicating the strong binding of calcium ions to three adjacent hydroxy groups in an axial-equatorial-axial configuration, resulting in a conformational change in the sugar. Earlier, Angyal *et al.* [76An1, 76An2, 71An] had already shown that the addition of calcium chloride to solutions of cyclitols in deuterium oxide causes downfield shifts of some protons because of complex formation. This shift was also explained by the cation binding to three oxygen atoms which have the same arrangement as discussed above.

The cation-induced conformational changes in the sugars due to strong complexation [86Fr] were also revealed by Raman studies on α and β-glucose. Calcium-binding was shown to favour the α-anomer. The effect of complex formation on the anomeric equilibria, resulting in a favourable steric arrangement of the carbohydrate compound, was successfully used in the preparation of a large series of carbohydrates [75An1, 80An2, 77An, 75Pa] in given anomeric form. The concentrations of the desired compounds were followed by NMR spectrocopy.

The complex formation of ten alditols with praseodymium(III) (isosteric with calcium ion) in aqueous solution was studied by ^{1}H NMR [75Ki]. The following stability sequence of the configuration of ligands was established: *xylo > threo > arabino(lyxo) >* glycerol *> erythro > ribo.* Secondary hydroxy groups appear to undergo better complexation than primary ones. The structure of the sorbitol–ytterbium(III) complex in water was found to be the same as that of the sorbitol–praseodymium(III) complex [77Ki]: tridentate complexation of ytterbium(III) with 0-2, 0-3, and 0-4 was found, with a mean Yb . . . O distance of 230 pm [73Au, 75Gr]. The complex has axial symmetry. Complexing of the europium(III) ion occurs at three consecutive oxygen atoms of alditols which are in a gauche-gauche arrangement [74An2].

Paramagnetic shift reagents, such as europium(III), gadolinium(III), thulium(III), and praseodymium(III), have been used to elucidate of the NMR spectra of carbohydrate derivatives in chloroform, carbon tetrachloride, and aqueous solution [71Ho, 71Ar, 72Yo, 71Bu, 72Ar, 73An2, 75Ha]. It seems to be generally true for the lanthanide ion-induced shifts in aqueous solution that some of the signals are shifted upfield and others downfield [73An2]. A quantitative treatment of the lanthanide-induced shifts for some carbohydrate systems is to be found in ref. [81Mc].

Dill *et al.* [84Da1] performed ^{13}C-NMR spectroscopy on the interaction of *epi*-inositol with gadolinium(III) demonstrated that this carbohydrate molecule includes also the binding sites consisting of three vicinal hydroxy groups having the axial-equatorial-axial arrangement. This is apparently a strong metal-ion binding arrangement for metal ions with an ionic radius of 100 pm [74An1, 74An2]. On the other hand, the binding sites of *cis*-inositol for manganese(II) and calcium(II) involve the three axial hydroxy groups. The binding of these metal ions to *epi*-inositol and *cis*-inositol may distort the chair conformation of the carbohydrate slightly [85Ca].

Not only electrophoretic and NMR studies, but other equilibrium methods too, reflect the metal ion-binding of sugar molecules. All these methods of investigation have shown that the metal complexes of neutral sugar molecules in aqueous solution display low equilibrium stability. As in most complex systems, the stability increases with increasing charge of the metal ion and with decreasing relative permittivity of the solvent.

Metal ions with a positive charge of 3 or 4 show a greater affinity toward donor atoms than metals with lower charges. This is probably the reason why, in aqueous solution of pH 2.5, the iron(III) ion forms complexes with simple sugars, e.g. fructose, containing one sugar molecule per iron with a stability constant of 2.0 [64Sa1].

Iron(III) ions may coordinate to the oxygen donor atoms of sugar-type ligands in pH-dependent and pH-independent processes. The coordination of iron(III) results in the deprotonation of the carboxy groups of sugar acids, for example, and sometimes even in that of alcoholic hydroxy groups of polyalcohols or sugars in slightly acidic solution. The formation of mixed hydroxo and polynuclear complexes in solution makes the equilibrium study of such systems even more complicated. This is almost certainly the reason why so surprisingly few data have been reported in the literature on the stability of this type of iron complexes.

Similar difficulties accompany the exact evaluation of equilibrium data in systems containing central metal atoms in a high (+ V, + VI) oxidation state.

A potentiometric study revealed that D-lyxose, D-mannose, and L-rhamnose form dinuclear anionic complexes with tungstate or molybdate ions in acidic solution. The complexes contain the ligands in pyranose form [89Ve]. In each case, the tungstate(VI) complexes are more stable than their molybdate(VI) homologues. The differences between the $\log K$ values determined for tungstate(VI) and molybdate(VI) appear to be almost independent of the nature of the sugar, which is a strong indication that they may be due to intrinsic properties of the inorganic elements.

The stability of tungstate complexes of pentoses has been shown to decrease in the following sequence: D-lyxose > D-ribose > D-arabinose > D-xylose [88Ve]; the same sequence seems to hold for nolybdate complexes [78Al]. It appears that a C-2, C-3 diol group with *erythro* configuration is a much more favourable site for complexation in such systems than a *threo*-diol group.

Calorimetric studies [77Ev] have shown that the calcium ion forms complexes of 1:1 stoichiometry with galactose, lactose, and *myo*-inositol. The processes are mainly enthalpy-controlled. This observation was strongly supported by calorimetric studies of the calcium(II)-D-ribose and -D-arabinose systems [85Mo, 86Mo]. The low stability constant ($K = 0.93$) was shown to be due to the large unfavourable entropic term ($\Delta S = -83 \, \text{J K}^{-1}$), which overcompensated the large negative enthalpic contribution ($\Delta H = -24 \, \text{kJ mol}^{-1}$) favourable for complex formation [86Mo]. The stability constants that were determined calorimetrically [87Al] and electrochemically [85Mo] for the same systems were found to be in excellent agreement, but different from the value determined via NMR spectroscopy [76Le]. The reason for such differences is most probably that the different concentration ranges were studied by different methods. For reliable NMR equilibrium studies, much more concentrated solutions are needed than for the other two methods.

An analysis of calorimetric results showed that the interaction of calcium with D-fructose was much stronger [84Gr] than that with D-ribose [87Al]. Solubility measurements indicated that the stability constants of the calcium complexes with *epi*-inositol, methyl-α-D-ribofuranoside, D-glucitol, and D-mannitol decrease in this sequence [79Ki].

2.2 Glycofuranoside complexes

Most of the publications cited so far deal with the complexation of carbohydrates having six-membered rings. Lönnberg *et al.* [80Lö1, 80Lö2, 83Ve1, 81Ve, 82Ve] filled a gap in the coordination chemistry of carbohydrates by performing equilibrium studies of the complex formation of carbohydrates with five-membered rings. Using potentiometric and calorimetric measurements, they not only determined the compositions and stabilities of the complexes formed, but also pointed out the correlation between the complexing ability and the reaction rate of carbohydrates.

Glycofuranosides served as their favourite model systems, and the complex formation of a series of methylglycofuranosides with alkali metal, alkaline earth metal, and lanthanide ions was investigated. In all these systems, the equilibrium studies reflected the formation of complexes with 1:1 stoichiometry and rather low stabilities. To ensure reliable stability constants, the equilibrium study was performed in both aqueous and methanolic solutions. In methanol, the stability constants proved to be higher by about two orders of magnitude. Calorimetric studies revealed the different contributions of reaction enthalpy and entropy to the complex stabilities in different systems. The acid-catalysed solvolysis of glucofuranosides was shown to depend on the stabilities of the metal complexes formed in the system.

2.3 Sugar acid complexes

The carboxy group is completely deprotonated at physiological pH (~ 7) in aqueous solution. Sugar acids therefore have a much stronger complexing ability than sugars.

Iron complexes with sugar acids are potential pharmaceutical preparations for iron therapy. Their stability is high enough to prevent the hydrolysis of iron(III) in biological systems. The formation of iron(III) and iron(II) complexes with these ligands aroused, therefore, much interest. The iron(III) ion and gluconic acid form a series of complexes as the pH is increased. An early study by means of polarography, spectrophotometry, pH titration, and potentiometry [55Pe1] reflected only the formation of mononuclear species. The sequence of the relative stabilities of 1:1 iron(III) complexes with polyhydroxy acids in weakly acidic (pH 6) and alkaline (pH 12.5) solutions was later found to be gluconic > galactonic > galacturonic > glucuronic acids [740d, 74Ch, 77Pa].

The interactions of lactobinionic acid, gluconic acid, and lactose with iron(III) were investigated [83Za] by spectrophotometry. The stabilities of the complexes were shown to decrease in the sequence gluconate > lactobionate > lactose. For the coordination of two iron(III) ions by one ligand, a deprotonated carboxy or hydroxy group is needed in these systems. Lactobionate and gluconate ions can both coordinate two iron ions, in spite of the lower chain length of the latter [83Za]. The same sequence of stability constants was observed by means of polarography in 0.1 M acetate buffer solution at pH 4.2 [89Na1].

The formation constants of iron(III) complexes of different organic acids (among them galacturonic and gluconic) have been determined, and the concentration distribution of these complexes over a range of physiological pH values have been computed [74Ca]. The results showed that galacturonate and malate are promising ligands in oral iron therepy.

The composition and stability constants of D-(+)-saccharic acid complexes with manganese(II), iron(III), cobalt(II), nickel(II), copper(II), and cadmium(II) ions have been determined [75Ve, 76Ve1, 76Ve2]. The results showed that two carboxy groups are coordinated in each 1:3 iron complex and also in the 1:2 copper complexes; the latter form a polymer chain. The stability constants of the various complexes followed the Irving–Williams stability series. The results of polarographic studies revealed that one hydroxide ion also participates in the formation of copper(II)–saccharic acid compounds, resulting in mixed ligand complexes [77Ba].

The first report on the copper(II)-gluconate chelates was published by Pecsok *et al.* [55Pe2]. They demonstrated the existence of five species: with metal:ligand ratios of 2:4, 2:2, and 2:1 in strongly basic solution, and two other different 2:2 complex species in less basic solution. Via potentiometric measurements in the copper(II)–gluconate system in the pH range 3.5–6.5, Mario [83Ma] has shown the formation of the mononuclear complexes CuL^+, CuL_2, $CuH_{-1}L$, $CuH_{-2}L$, and $CuH_{-1}L_2$. He also found less convincing evidence for the existence of the polynuclear deprotonated complex $Cu_2H_{-3}L$. The formation of dimeric species in copper(II) chelate-containing solutions of polyhydroxy carboxylates was supported by ESR studies [71To1]. The potentiometric results published on the copper(II)–gluconate system in [81Fr, 76Pa] are inconsistent with the result in [71To], as they did not reflect the existence of dimeric complexes in solution. The measured equilibrium constants of analogous complexes were also found to be different.

Fridman *et al.* [85Fr] established that copper(II) forms mixed ligand complexes with gluconic acid and amino acids in aqueous solution. The stability constants of CuA(HL) (HA = amino acid; H_2L = gluconic acid) decrease with increasing dissociation constant of HA, but the formation constant for $CuAl^-$ is independent of the dissociation constant of HA. These trends were explained in terms of the Pearson-type hardness of HA.

Glucuronic acid is a major constituent of several important polysaccharidic articular cartilage tissue components [78Co1]. Accordingly, several papers have dealt with the interactions of copper(II) with glucuronic acid. The methods used so far have been polarography [82Pa], potentiometry [77Ma1], calorimetry [81Ar], circular dichroism [77Fi], and spectrophotometry [78Fi]. The copper(II)–oxygen bond distance has also been measured [83Sa]. The binding sites could not be established from equilibrium measurements, although several models have been proposed. In [77Ma1, 77Fi, 78Fi] the coordination of copper(II) in a simple monodentate fashion to the carboxylate group was assumed, although [77Fi, 78Fi] mentioned that a multiple equilibrium was present at pH 5. On the basis of comparative thermodynamic data, Aruga [81Ar] postulated that at pH 4.3 copper was bound in a bidentate manner: directly to the carboxylate group and, through an outer sphere electrostatic bond, to the endocyclic oxygen of the sugar ring. A very surprising model was proposed by Rao *et al.* [83Sa]. They assumed that copper(II) is coordinated to the ligands through the ether oxygen of C-1 and through the aldehyde oxygen, and the carboxylate group does not participate in the coordination of copper(II).

One of the most detailed studies on the coordination mode of copper(II) with glucuronic acid was preformed by Ternai *et al.*, using NMR spectroscopy [86Co]. The model is presented in Fig. 6.2. The initial step in the mechanism involves the complexation of copper(II) to the carboxylate group and the endocyclic oxygen. The next step is a conformational change of the ring to a boat form through complexation with 0–3, which places a labile proton in the correct orientation to effect proton transfer and ring opening. At this stage copper(II) is coordinated to the carboxy group and 0–3 and 0–5. The rate-determining step is the ring closure to the mutarotated sugar. Thus, the time-averaged complex appears to have structure IV, as reflected by NMR.

Less attention has been paid to carbohydrate complexes of nickel(II) and cobalt(II) ions than to the corresponding copper or iron complexes. An early study [63Zo] showed that in acidic solution the stabilities of the nickel(II) and cobalt(II)

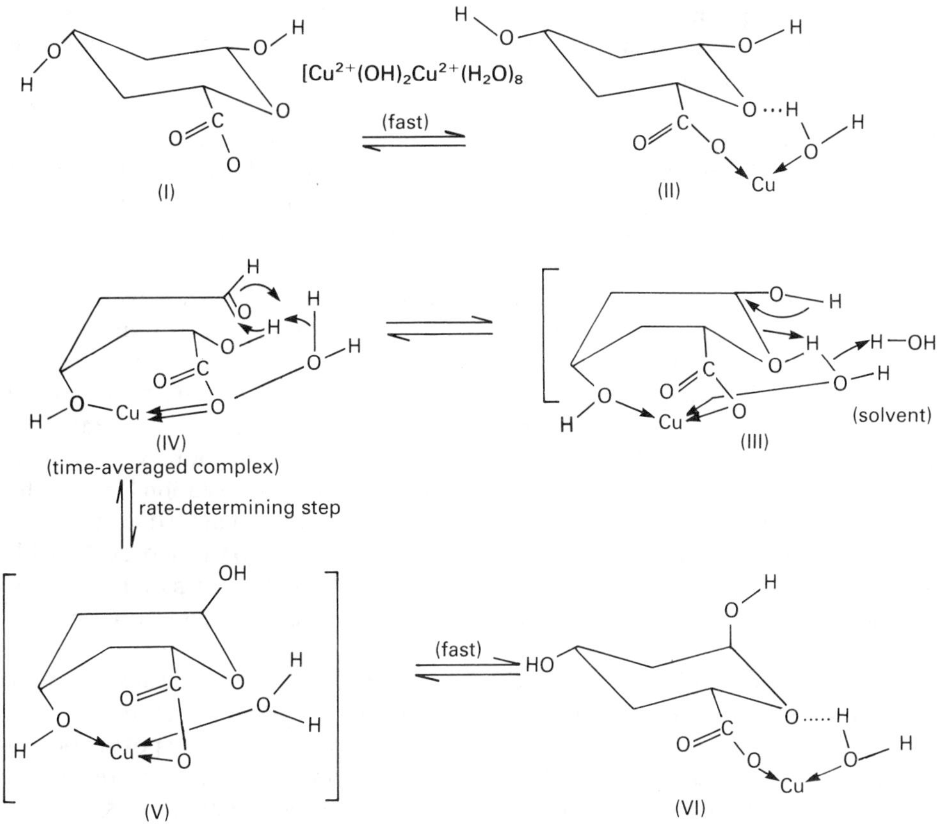

Fig. 6.2 — Model for the interaction of copper(II) with glucuronic acid [86Co].

complexes decrease in the following sequence of ligands: maleic acid > trihydroxy-glutaric acid > gluconic acid [63Zo]. The stability sequence for the cobalt(II) complexes in alkaline solution was found to be the same, but for nickel(II) complexes trihydroxyglutaric acid proved to have the highest stability. Several techniques were used to examine the nickel(II)–gluconate system [65Jo]. Three complex species were identified; NiG^+, in a solution of pH 7; an anionic species $[Ni_2(OH)_4G]^-$ in a solution of pH > 9; and an insoluble product, $Ni_2(OH)_3G$, precipitated from solutions of pH 7–9. The formation of polynuclear nickel(II) and cobalt(II) complexes of gluconate [78Ar] was concluded from electrophoretic studies.

A polarographic study of the zinc(II)–gluconate system reflected the formation of five different species [78Pa]. Recent critical studies [87Ve] did not support these data. They revealed the formation of only two different zinc(II)–gluconate complexes in either neutral or alkaline media, with 1:1 and 1:2 metal:ligand ratios. In the analogous lactobionate-containing system, only one complex was observed, with a 1:1 metal:ligand ratio.

Potentiometric measurements reflected the formation of cadmium(II)-gluconate complexes in aqueous acidic solution [81Ta1]. A polarographic study [80Ag] indicated that cadmium(II) forms mixed ligand species with gluconate and itaconate or with pyridine and gluconate.

The formation of lead(II)–gluconate complexes was first reported by Pecsok *et al.* [56Pe] on the basis of polarographic and optical rotary measurements. They demonstrated the presence of a 1:1 species with low stability ($\log K = 2.6$) in the pH range 1 to 6, and one of high stability ($\log K = 16.2$) in the pH range 10.5 to 12.5. The formation of a negatively charged 3:2 complex was also demonstrated in the latter system.

Coccioli & Vicedomini [78Co2, 78Co3] reinvestigated this system by means of potentiometric measurements and proved the formation of hydroxo mixed complexes, but they did not find any evidence of the formation of polynuclear species.

Potentiometric and conductometric measurements reflected the formation of only 1:1 complexes of lead(II) with either saccharic or mucic acid [72Pa]. The formation of similar simple species was reported for the lead complexes of *meso*-eritrite, mannitol, dulcitol, and sorbitol in $1 \, \text{mol dm}^{-3}$ alkaline solution [68Vi]. The formation of only 1:1 parent complexes in the latter systems is mainly due to the low affinity of polyalcohols for the lead ion. Because of their carboxy group, sugar acids are much stronger donor molecules than are polyalcohols. The formation of higher complexes in former systems was prevented by the low ligand concentrations used in [72Pa].

The reactions of aluminium(III) with different hydroxy carboxylic acids (among them gluconic and saccharic acid) were studied by Martell [84Mo]. The coordination reactions of aluminium(III) with gluconic acid were explained by the postulation of the formation of three species, each with 1:1 composition, but containing the ligand in different protonation states. Aluminium(III) has a strong tendency to displace the protons of the hydroxy groups of hydroxy acids, a property that has previously not been taken into consideration. The increase in pH of such solution may also lead to the formation of hydroxo mixed complexes, due to the hydrolysis of aluminium(III). Saccharic acid as ligand displayed similar behaviour towards the aluminium(III) ion, but with the formation of only two differently protonated 1:1 complexes.

An electrophoretic study of the saccharic acid complexes with aluminium(III), gallium(III), chromium(III), and molybdenum(VI) ions reflected olation equilibria in the systems [79Go]. The validity of the reported stability constants is limited by the nature of the experimental procedure.

Anionic complexes are formed by reactions between a variety of inorganic oxyacids and polyhydroxy compounds [63We]. Molybdenum(VI) complexes of different sugars [60Bo, 64An, 66Ve, 66Ba1, 66Ba2, 68Vo] have been investigated by optical rotatory dispersion (ORD) and circular dichroism (CD). One of the reasons for these investigations was that sugars are possible complexing sites for molybdenum in certain enzymes [68Vo].

The stability constants of the adducts formed by molybdenum(VI) with mannitol and with citric, gluconic and saccharic acids are of the order of 10^8 [72Fe]. Increase of the number of carboxy groups in such molecules increases the stability of the molybdenum complexes.

Potentiometric studies reflected the complex formation of molybdenum(VI) with sorbitol, and with lactic, citric, and gluconic acids [75Fe]. A spectroscopic study of the tungsten(VI)–gluconic acid system at a ligand excess has shown the presence of two mononuclear and two dinuclear species, with 1:2 and 2:2 compositions, respectively [85Ll]. The coordination site of the organic ligand to the tungsten(VI) centre was studied by means of ^{13}C NMR [86Ll]. The impregnation of cellulose with tungstate influences the chromatographic migration rates of a number of poly-hydrodroxy compounds. From the migration rates, the pseudostability constants of the complexes have been determined [78An].

Spectrophotometric, conductometric, and salt-cryoscopic equilibrium measurements have shown that, if metavanadate solutions are acidified with hydroxy-carboxylic acids, or with perchloric acid in the presence of polyalcohols (sorbitol, mannitol, dulcitol, arabitol, adonitol, or xylitol), coloured (from yellow to deep red) vanadium(V)-hydroxy-carboxylate (or poly-alcohol) complexes are formed besides the decavanadate ions. The results show that primarily tetravanadato complexes are present in the system, which are partly converted into monovanadate complexes by an excess of the ligand [72Pr]. Spectrophotometric, CD and polarographic measurements in the pH range 5–14 have revealed that the stoichiometry of the vanadium(IV)-gluconate complexes varies from a metal to ligand ratio of 1:1 (at pH 6.0) to 1:2 (above pH 12) [76Go].

Anionic complexes are formed between germanic acid or germanate ion and sugar acids or disaccharides [86Pe]. Both 1:1 and 1:2 complexes have been shown to exist in solution by means of potentiometry. In the case of arsenious and telluric acids, only 1:1 complexes are formed with the same ligands [87Se].

Potentiometric equilibrium studies have demonstrated [74Ko] the formation of hydroxo mixed complexes of hafnium(V) and zirconium(V) with saccharic acid. Both complexes have a composition $M(OH)_3L$ (where H_2L is saccharic acid), with nearly equal stabilities.

Sawyer & Ambrose [62Sa1] reported the formation of extremely stable cerium(IV) complexes of gluconate ion in basic solution. Among the several complexes formed, the most stable was found to contain two cerium(IV) ions per three gluconate ions [62Sa2]. Gluconate ion forms a stable complex of uranium(VI) with composition $[(UO_2)(GH_4)(OH)_2]^-$ (where GH_4^- denotes gluconate ion) [62Sa1].

The stability constants of the mixed-ligand complexes of lanthanum(III), cerium(III), praseodymium(III), neodymium(III), and samarium(III) with EDTA as primary ligand and malic, lactic, glycolic, or gluconic acid as secondary ligand have been determined. The stability constant values decrease in the above sequence of ligands in the complexes with each metal ion and an increase in the order of metal ion sequence La < Ce < Pr < Nd < Sm [84Li].

Carbohydrates (e.g. sugar acid, lactone) bound to the amino group of amino acids (or to the N-terminal amino group of peptides) serve as models of biologically important molecules. The preparation and complex formation of 2-(polyhydroxyal-kyl)-thiazolidine-4(R)-carboxylic acid were investigated for this reason in [59We]. It was found that the pK_1 and pK_2 values are 2.15 and 5.5, respectively, and depend slightly on the nature of the sugar bound to L-cysteine. These compounds form metal complexes with 1:1, and 1:2 metal:ligand molar ratios. The values of the stability constants depend on the nature of the sugars; they are in the pK range 5–6.6 for the zinc(II) complexes and 5.8–6.75 for the cobalt(II) complexes. These systems have recently been reinvestigated [90Ga1]. The carboxylate group was found to be much more acidic ($pK_1 \sim 1.60$) than reported in [59We]. On the other hand, the formation of parent and of hydroxo mixed ligand complexes of zinc(II) was demonstrated in slightly alkaline solution. The protonation and complex stability constants obtained were explained in terms of the structure of the sugar moiety. The managanese(II) complexes were shown to dimerize almost completely in slightly alkaline solution [90Ga2].

2.4 Amino sugar complexes

Amino sugars, which are among the most abundant natural organic compounds, are known to bind metal ions.

A number of papers have been published on the metal complexes of D-glucosamine and related compounds [65Ta1, 90Na1, 65Ta2, 85Mi, 65Ta3]. It has been shown that the copper(II) ion forms the strongest complexes with these ligands. Besides mononuclear complex formation the dimerization of copper(II) complexes must also be considered [65Ta1, 90Na1]. A recently reported EPR, CD, and UV-visible spectroscopic study [85Mi] showed that D-glucosamine is bound to the copper(II) ion through its amino-nitrogen and a deprotonated hydroxy group in solutions at pH = 6–9.

The complexes are either enthalpy or entropy-stabilized. The thermodynamic functions ΔG_1^0, ΔH_1^0 and ΔS_1^0 have been determined for the copper(II) and iron(III) complexes of glucosamine, glucosoxime, and galactosoxime. The entropy values were large and positive, owing to the liberation of water molecules from the coordination sphere of the central ions [86Ma].

The stability constants of the copper(II), nickel(II), and cobalt(II) – D-glucosamine complexes followed the Irving–Williams stability series [86Le1], similarly to those of the analogoues metoxy-D-glucosamine complexes [87Pu]. In the latter case the basic binding site is the $\{NH_2, C4\text{-}0^-\}$ donor set. The formation of ternary complexes was also shown in the copper(II)–prometone–D-glucosamine system [86Le2].

Equilibrium studies reflected the steric effects of the substituents on the complex stabilities. The copper(II) complex of methyl-β-D-glucosamine has the same stability ($\log K_1 = 4.5$) as that of the 3,4,6-tri-O-methyl-D-glucosamine complex, but both of them have slightly smaller stabilities than that of the D-glucosamine complex (5.0). This seems to indicate that the methoxy groups at C-1 and C-3 in glucosamine hinder the complex formation by interfering with the mutual approach of the copper(II) ion and the ligand molecule [65Ta3]. No complex formation was observed in the copper(II)-N-acetyl-D-glucosamine system [85Mi].

The complex formation of different metal ions with N-methyl-D-glucosamine, D-glucosamine, D-galactosamine, D-mannosamine, and D-talosamine was investigated by means of pH-metric titrations [79Mi]. The stability constants ($\log K_1$) of the metal complexes decreased in the sequence $Cu^{2+} > Pb^{2+} > Zn^{2+} > Ni^{2+} > Cd^{2+} > Ca^{2+} \sim Mg^{2+}$. The stabilities of the copper(II) complexes increased in the above sequence of ligands. Recent potentiometric and spectroscopic study on the coordination ability of D-mannosamine [89Ko] have shown that this amino sugar is more effective ligand than e.g. D-glucosamine or D-galactosamine [88Ur, 88Ra]. All these monomeric amino sugars act as bidentate ligands with their amino group as a main donor toward metal ions. The second donor centre derives from one of the amino-sugar hydroxyl groups. The polarographic technique was shown to detect amino-sugar complexes in minor concentration not seen by the potentiometric titrations [90Ur]. 2-Amino-2-deoxy-D-gluconic acid forms complexes with higher stability than those of D-glucosamine, owing to the presence of the carboxylate group [66Mi] indicating the coordination of the latter oxygen in the system.

The interactions of sodium(I), magnesium(II), aluminium(III), calcium(II), chromium(III), iron(III), nickel(II), copper(II), zinc(II), strontium(II), cadmium(II), and barium(II) as central atoms with a large number (46) of different kind of sugar-type compounds have been investigated by thin-layer ligand-exchange chromatography [81Br]. All of the compounds were shown to form complexes with the copper(II) ion. In general, alditols form stronger complexes than do mono-saccharides; the amino sugar complexes of copper(II) usually have outstanding stability. Sugars containing three hydroxy groups in the axial-equatorial-axial arrangement were found to be most suitable for metal coordination, as stated earlier by Angyal [74Au1, 80An1].

2.5   Ion-binding properties of macromolecular carbohydrates

Macromolecular (polymeric) carbohydrates are of vital importance in life processes. Hyaluronic acid, for instance, is a major component of connective tissues and body fluids in vertebrates. Its local structure [62St] consists of repeating disaccharide units composed of D-glucuronic acid and N-acetyl-D-glucosamine residues. Other impor-tant glycosaminoglycans (GAGs) are heparin, chondroitin sulphate, keratin sulphate, etc. The complexation of metal ions by GAGs and other polysaccharides is of considerable interest in view of their biological role [70Ve, 73Al, 73Mu, 74Al], particularly in the availability of nutrients to plants [71Le, 77Ra], and in the storage of accumulated metals in plant cell walls [69Pe, 77Fa].

Interactions between charged polymers (e.g. macromolecular carbohydrates) and their counterions are far more complicated than those occurring in solutions of simple salts and ligands, because the charge density on the chain may reach much higher values. This is the reason why the equilibrium constants of such simple processes as the dissociation of the carboxylic groups of the macromolecule or the metal ion coordination of the latter show a polymer concentration-dependence. This explains the use of such terms as 'ionatmosphere' or 'counterion condensation' to describe the accumulation of counterions in the polyelectrolyte domain.

Among molecular polyelectrolyte theories, the most attractive from the formal aspect of the analytical equations are those proposed by Oosava [71Oo] and Manning [69Ma, 78Ma2]. The dependences of the apparent dissociation constant and the enthalpy of dissociation on the degree of dissociation of some derivatives obtained by selective oxidation of amylose, scleroglucon, and cellulose are presented in [87Ce]. The results are discussed and compared with those obtained via a simple model based on the polyelectrolyte theory.

The value of pK derived for hyaluronic acid (2.9 ± 0.1) is 0.3–0.4 pK unit lower than that for the monomeric analogue D-glucuronic acid (pK = 3.23 ± 0.02) [82Cl]. This decrease may be due largely to the effect of the hexosamine residue substituting the C-4 hydroxy group of glucuronic acid in the polymer. This explanation is supported by Kohn's observation of a 0,3 unit decrease in the pK of D-galacturonic acid when a methoxy group substitutes the C-4 hydroxy group. Substitution of the C-2, C-3, or C-4 hydroxy groups by methoxy groups led to a decrease of 0.38 and 0.25 pK unit for D-galacturonic acid and D-glucuronic acid, respectively [78Ko].

The $^{23}Na^+$ NMR method is a very powerful tool for the establishment of changes occurring in counterion binding as a result of both the changed degree of ionization and the changed chemical structure of GAGs [78Gu, 83De]. The binding of calcium(II), magnesium(II), and sodium ions to hyaluronic acid at neutral pH can be explained by electrostatic coulombic interactions, without the assumption of any more specific binding [84Pi].

The factors influencing the hydration of hyaluronic acid have been studied via compressibility and density measurements [83Da1]. It was found that the change of the counterions of this polyelectrolyte from univalent to bivalent, e.g. sodium to calcium(II), leads to a small though significant increase in the total hydration sheath surrounding the polymer. In X-ray studies of sodium hyaluronate, it has been shown that the sodium ion plays an important role in the polyanion structure [75Gu]. Three non-coaxial helical chains of hyaluronate are stabilized by a network of hydrogen-bonds and $O \cdots Na^+ \cdots O$ bridges. These bridges are formed by the carboxylate and acetylamide oxygens of different polysaccharide chains. Intramolecular hydrogen bonds occur in the trigonal lattice structures reported for stretched hyaluronate films [75Wi] and one is situated in the rectangular lattices as well [75Gu]. In solution, the situation can be very different. The local conformation varies in the presence of different counterions (Na^+, Li^+, Ca^{2+} and Mg^{2+}) [77Ch]. Evidence for the existence of intramolecular hydrogen bonds in solution is not yet definite, although the effect of pH in alkaline solution on the limiting viscosity of hyaluronate strongly suggests that hydrogen bonds are broken under alkaline conditions [77Cl, 77Ma2].

The enthalpy of mixing with sodium chloride solution per mole of polymer charge equivalent has been determined for the sodium salts of hyaluronate, chondroitin 4-sulphate, and chondroitin [79Cl]. The results have been explained by the extension of the infinite line charge theory [70Cl].

The binding of divalent cations to proteoglycans and their constituent GAGs has been demonstrated in numerous studies [68Ur, 71Mc, 77Fi, 78Mu, 83Ba, 88Hu]. The infrared spectroscopy of calcium–proteoglycan complexes indicated that sulphate groups, as well as carboxylates, interact with calcium at least at high calcium ion concentrations [71Mc]. Balt *et al.* [83Ba] suggested that cations interact with carboxylate groups, as specific binding sites, but only with 'atmospheric' (electro-static) binding to sulphates [83Ba1]. Recent results [88Hu] have shown that both sulphate and carboxylate groups are involved in cation binding. The order of calcium-binding affinities is heparin > chrondroitin sulphate > keratan sulphate > hyaluronate, and is critically dependent upon the charge density. For example, heparin has approximately 1.5 times more anionic groups than condroitin has, but it binds calcium with 10 times higher affinity.

A spectrophotometric method has been applied to characterize the binding of magnesium(II), calcium(II), [81Jo], and zinc(II) and manganese(II) [81Ma] to the polyelectrolyte dextran sulphate (DS), a highly sulphated polymer of α-D-glucose. No significant difference was found between the bindings of these cations, support-ing the idea of a non-specific electrostatic interaction between these divalent ions and the sulphate groups of DS. Surprisingly, in [82Ma] the same authors suggested some specific univalent counterion binding.

Parrish and Fair [81Pa1] concluded that the zinc ion seems to interact in a specific way with GAGs. It is bound by heparin, but not by other members of this group of polyelectrolytes.

There are only a few reports in the literature on the binding of trivalent cations to polyelectrolytes. Mathews used a spectrophotometric technique to study the binding of hexamminecobalt(III) ions to heparin, its derivatives, and sulphated chitosans [64Ma]. The $[Co(NH_3)_6]^{3+}$ cation was shown to form outer-sphere ion pairs with the anionic groups of the polyanions. The interaction of La^{3+} with chondroitin sulphate [79Ma] appeared to obey the charge fraction rule of Manning's theory of counterion condensation [69Ma]. From binding studies [84Ma1] in the system La^{3+} + DS + Cl^-, it is apparent that the two-variable theory of Manning [69Ma] gives a quite accurate quantitative description of the binding process.

The binding of copper(II) to the carboxylate groups of hyaluronic acid has been demonstrated by spectrophotometry [77Fi]. The results indicate a 2:1 HA:Cu^{2+} ratio in the complex formed, with an equilibrium constant of 3×10^3 (HA representing a fragment containing one carboxylate group of the macromolecule). It was reported later that different types of complexes are formed, depending on the pH of the solution and the concentration of the polymer [78Fi]. The composition of the HA-Cu^{2+} complex observed in [77Fi] was supported by the results of polarographic measurements [85Ko]. Analysis of the ^{13}C and ^1H relaxation data for the HA-Cu^{2+} complexes indicates binding sites involving the carboxy group and O–1 of the glucuronic acid moiety [85St]. In spite of this, Chakrabarti *et al.* favoured the acetamido group as the second binding site of the ligand [77Ch, 77Fi, 84Ma1].

The NH group of HA can probably be ruled out as playing a role in copper(II) complexation in the above systems. Hard-pulse Redfield sequence measurements [81Wr] on the system in aqueous solutions did not reflect the influence of copper on the NH relaxation time. The tetragonal geometry of HA-Cu^{2+} complex was determined in [84Co].

Ligand-field data and ESR measurements on metal complexes formed with chondroitin sulphate indicated that the metal ion is surrounded only by oxygen donor atoms [83Ba2].

An investigation of the interactions between polygalacturonic acid and copper(II), vanadate(II), manganese(II), nickel(II), and cobalt(II) suggested that in the fully hydrated gels copper(II) and vanadate(II) ions form inner-sphere carboxylate complexes, whereas $[Mn(OH_2)_6]^{2+}$ retains its inner hydration sphere [80De]. The binding of iron(II) was shown to depend on the water content of the samples. The Mössbauer spectra reflected the presence of different hexacoordinated iron(II) species in the fully hydrated, air-dried, and anhydrous samples. On the other hand, the iron(III) ion give rise to polynuclear structures, which are stable over a wide pH range [81Mi].

Studies of interactions between the copper(II) ion and ionic polysaccharides of natural origin have shown that the association processes are strongly influenced by many independent variables, e.g. by the chemical nature, composition, and charge density of the polysaccharide, the copper:polymer ratio, the polymer concentration, and the ionic strength [84Ma2]. Polarographic investigations reflected the following sequence of copper-binding affinity for a series of polysaccharides: polyacrylate > pectate > bacterial alginate $\simeq$ algal alginate > glucuronate-rich oligomer > alternating man-glu oligomers > mannurate-rich oligomers [84Re].

Among the cell surface carbohydrates, the most significant is probably sialic acid (N-acetyl neuraminic acid). It is usually the terminal monosaccharide in oligosaccharide chains attached to membrane glycoproteins. As a medium strong acid (pK ~ 2.6), it is completely deprotonated at physiological pH and therefore provides negative charges on biomembrane surfaces. ^{13}C and ^{1}H NMR spectra have been used to study the interactions of sialic acid with metal ions. It was demonstrated that sialic acid strongly and preferentially complexes the calcium ion at pH 7 [77Ja]. Only a 1:1 complex was shown to be formed, with a stability constant of 121 mol^{-1}dm^{3}. The calcium(II) binding occurs with greater strength with the β-anomeric form of sialic acid then with the biologically more significant α-anomer [77Cz]. It was suggested that the main binding sites for calcium(II) are glycerol oxygens; the carboxylate oxygen participate in the interaction only indirectly, through a hydrogen-bonded water molecule. An investigation of the interactions of gadolinium(III) and manganese(II) with the α-anomer of sialic acid and its derivatives indicated the participation of the glycerol side chain and the acetamido group in the metal-binding process [82Da].

In some cell membranes (e.g in red blood cells), glycophorin A is the major sialoglycoprotein. It is composed of 131 amino acids and contains about 60% (by weight) carbohydrates, half of which is α-sialic acid. The oligosaccharides are situated near the N-terminal part of the molecule and are linked to it by a tetrasaccharide unit. ^{13}C NMR studies have shown that gadolinium is bound not only by the α-sialic moiety of glycophorin but also by the tetrasaccharide unit acting as link in the macromolecule [83Da2].

3.   STRUCTURE AND BONDING

3.1   Sugar (mono-, di- and oligosaccharide) complexes

The acquisition of reliable structural information on bond distances and bond angles for transition metal–sugar complexes is hampered by difficulties in the preparation of single crystals in these systems. On the other hand, several papers have been published on the structures of carbohydrate complexes formed with alkali metal and alkaine earth metal ions.

Solution and structural studies show that simple carbohydrates bind calcium ions only if they can provide three or more hydroxy groups in geometrical arrangements permitting calcium ion coordination.

The crystal structures of many of these carbohydrate complexes have been reviewed [77Co, 79Po1, 83Dh]. In most of the crystal structures, the sugars are positioned around the calcium ion in geometrical arrangements that permit the formation of calcium–oxygen bonds with bond lengths ranging from 230 to 285 pm. Three general types of calcium coordination polyhedra have been found. In most cases, calcium is coordinated to eight oxygen atoms, following the geometry dictated by a square–antiprism arrangement. Seven-fold and nine-fold coordination have also been observed. Obviously, tetra- and pentadentate chelation are rare [83Dh]. For pentadentate complexation, calcium α-D-allopyranosyl α-D-allopyranoside is a good example [78Ol].

Calcium complexes of L-arabinose have been the subject of numerous investigations. A structural analysis of Ca(α-L-arabinose)Cl_2.4H_2O has shown that the calcium ion is eight-coordinated in this complex, binding through two symmetry-related L-arabinose molecules (via O-3 and O-4 in the first, and O-1 and O-5 in the second), as well as to four water molecules [78Te]. The structure of L-arabinose, with the numbering of the atoms, is shown below:

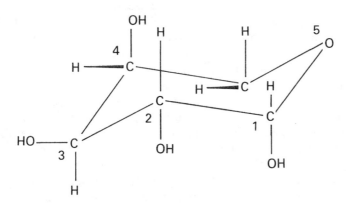

The eight oxygen atoms form a slightly distorted, square–antiprism around the calcium ion, without any direct interaction between calcium and chloride. All sugar hydroxy groups and all water molecules, as well as the chloride ions, are involved in a strong, inter- and intramolecular hydrogen-bonding network [78Te]. The results of FT-IR spectroscopic investigations have supported these results. It was also demonstrated that the sugar molecule crystallizes as the α-anomer in these calcium complexes. The chloride and bromide salts of the calcium arabinose complexes are isomorphous, in terms of octacoordination and common binding sites [84Ta1]. The magnesium-L-arabinose complex with similar overall composition was shown to have a different structure [84Ta2]. The magnesium ion in the latter compound is hexacoordinated via four oxygens atoms of the two sugar moieties and two coordinated water molecules. The strontium and barium complexes with analogous composition have structure similar to that of the calcium complex (but different from that of the magnesium one). In the former compounds, the metal ions are bound to two arabinose molecules (via O-1 and O-5 of the first, and O-3 and O-4 of the second), and also to four water-oxygens [86Ta1]. The β-anomer conformation is known be predominant in free L-arabinose. Complex formation altered the intermolecular sugar network, leading to the α-anomer conformation. The same can be said for the L-arabinose complexes of zinc(II), cadmium(II), mercury(II). [86Ta2], and uranyl(II) [87Ta1].

The complex-forming ability of β-D-fructose has been studied by means of X-ray diffraction, FT-IR spectroscopy, and molar conductivity measurements. X-ray powder diagrams of the magnesium–fructose compound [86Ta3] exhibited marked similarities with those of the calcium–fructose complex of known structure, [74Cr1,

76Co, 74Cr2], but also differences due to the lower coordination number of magnesium(II), i.e. 6 [83Ta1], in contrast with the 7 [74Cr1] or 8 [74Cr2] for calcium. The strong hydrogen-bonding network of free D-fructose undergoes a typical rearrangement upon reaction of the sugar with magnesium(II) [86Ta3], uranyl(2 +) [87Ta2], zinc(II), cadmium(II), and mercury(II) ions [88Ta1].

The crystalline sugar is in the β-D-fructopyranose form. The calcium ion is coordinated by the β-fructopyranose isomer, whereas magnesium(II), zinc(II), cadmium(II), mercury(II), and uranyl(II) cations may be bound to either the β-D-fructopyranose or the β-D-fructofuranose form [88Ta1]. The binding mode of the latter metal ions is coordination to two D-fructose molecules through oxygens O-2 and O-3 of the first sugar and O-4 and Q-5 of the second, as well as to two water oxygens, resulting in six-coordination geometry around each metal ion, except for the mercury ion, which is four-coordinated (no bonded H_2O molecules). Conductivity measurements suggested that there is no direct coordinative interaction between the metal ions and the inorganic anions in the compound.

H-NMR spectroscopy was employed to study the interaction between sucrose and the calcium ion by Poncini [79Po2]. Surprisingly, he did not find any evidence of complexation in this system. The chemical shifts were explained in terms of a bulk solvent effect, where the calcium ion rapidly exchanges with the solvent and the solute in a typical solute–solvent interaction. Interactions of sucrose with metal ions have been reviewed generally by Poncini [79Po3]. However, recent investigations have shown the existence of saccharose complexes in the crystalline state with the compositions $Na(saccharose)Cl \cdot 2H_2O$, $Na(saccharose)Br \cdot 2H_2O$, $Na_3(saccharose)_2I_3 \cdot H_2O$ [86Ta9], $Mg(saccharose)Cl_2 \cdot 4H_2O$, $Mg(saccharose)_2Br_2 \cdot 4H_2O$, $Ca(saccharose)Cl_2 \cdot 2H_2O$, and $Ca(saccharose)_2Br_2 \cdot 2H_2O$ [87Ta7]. X-ray structural information shows that the sodium ion is six-coordinated in the structurally identified bromide [46Be] and in the isomorphous analogous chloride compound, but differs in the iodide. The sodium ion binds *via* O-4, O-6, and O-6′ to the sugar, to two water molecules and to a halide ion. In the 1:2 metal–sugar compounds, the magnesium(II) ion is possibly six-coordinated, binding to two saccharose molecules via O-4 and O-6 of the first and O-6′ of the second saccharose, and to three water molecules. The calcium(II) ion is considered to be seven-coordinated in the 1:1 adduct, binding to the sugar in the same way as magnesium (II). In every case, the strong hydrogen-bonding sugar network is rearranged because of metal ion complexation. The formation of $Mg(D\text{-}glucose)X_2 \cdot 4H_2O$, $Ca(D\text{-}glucose)X_2 \cdot 4H_2O$ and $Ca(D\text{-}glucose)_2X_2 \cdot 4H_2O$ ($X = Cl^-$ and Br^- has been reported by Tajmir-Riahi [88Ta3]. The magnesium(II) ion may be six-coordinated, while the calcium(II) ion is seven- or eight-coordinated. Both metal ions are bound to D-glucose through donor atoms O-1 and O-2 of the α-anomeric form of the ligand.

The EXAFS (Extended X-ray Absorption Fine Structure) method seems to be suitable for determination of the local structure of iron(III) in its sugar complexes in solution and in the solid state [89Na3]. The EXAFS method provides structural information relating to the radial distribution of atom pairs in a system: the number of neighbouring atoms around a central atom (coordination number), the interatomic distances between them and their root mean square deviation [81Sa, 84Ya].

The results of EXAFS measurements on iron(III)-sugar complexes are collected in Table 6.1. The average Fe-O distance in the iron(III) complexes formed with

Table 6.1 — Results of curve-fitting analysis for iron(III)–sugar complexes in aqueous solution (aq) and in the solid state (pw). The values in parentheses are X-ray crystallographic data; r, and n denote the interatomic distance, the root mean square displacement and the coordination number, respectively [89Na3]

Ligand		Fe-O			Fe...C			Fe...Fe		
		r/pm	σ/pm	n	r/pm	σ/pm	n	r/pm	σ/pm	n
A. Mannitol	aq	194	9.4	6.5	287	5.5	5.2			
B. Mannitol	pw	198	7.3	6.1	285	5.0	4.4			
C. Fructose	aq	192	9.8	6.0	277	6.9	$4^{\ddagger}$	310	6.9	0.5
D. Fructose	pw	194	10.1	5.7	285	9.3	$4^{\ddagger}$	304	9.3	0.7
E. Gluconate	aq	195	8.3	6.3	285	5.2	4.8			
F. Gluconate	pw	196	8.7	6.3	283	5.0	4.5			
G. Water	pw	196	8.7	$6^{\ddagger}$						
		$190^{\S}$								
		$(198.6^{\dagger})$								
H. Fe foil								249	6.88	$8^{\ddagger}$

† Ref. 77 Ha; ‡ fixed; § EXAFS result in ref. 85As.

different sugar-type ligands is 195 pm, irrespective of the nature of the ligand. This seems to indicate that the Fe-O (hydroxide ion), Fe-O (sugar ligand), and Fe-O (water molecule) bond lengths are not distinguishable by EXAFS analysis. The Fe-Fe distance of 310 pm is in good agreement with the value found in [76Th].

The Fe K-edge EXAFS and XANES measurements were complemented by Mössbauer and EPR investigations [85Wo, 86Na2], and a model has been proposed for the structure of binuclear iron(III)–sugar complexes: two deprotonated alcoholic oxygen atoms (in the case of a sugar acid, one alcoholic and one carboxylate oxygen atom) coordinate to the iron(III) centre. Two free coordination sites of iron(III) are occupied by hydroxide ions, which form bridges between two iron(III) ions. In all the complexes, the iron(III) ion has a coordination number of six (Table 6.1). Consequently, two water molecules or two nondeprotonated alcoholic oxygens should be coordinated to fill the remainder of the coordination sites of the iron(III) ions, forming an elongated octahedral arrangement.

Similarly, K-EXAFS and XANES methods have been applied to determine the structures of the copper(II) complexes of D-ribose and D-glucosamine [90Na1]. All data demonstrated that, in aqueous solution at a high pH, 1:1 copper(II)–sugar complexes are formed. The EPR spectra of the two complexes indicated the presence of two copper(II) ions interacting via hydroxide oxygen atoms in the solution. An analysis of the EXAFS spectra indicated that the oxygen coordination geometry around the copper(II) (distorted octahedral) and the Cu-O bond lengths (Cu-O_{eq} 192 pm, Cu-O_{ax} 230 pm (within the complexes are not very different from those for the hexaaquacopper(II) ion (Cu-O_{eq} 195 pm, Cu-O_{ax} 228 pm). EXAFS Fourier transform curves clearly showed a peak ascribed to non-bonded Cu...Cu and Cu...C distances (274 pm), indicating the formation of sugar-ligand chelate rings around the copper(II) ion. The most likely structure of the copper(II)–sugar complexes is similar to that proposed for the iron(III)–sugar complexes [89Na3].

Optical rotatory dispersion (ORD) and circular dichroism (CD) studies have also been used to characterize optically active carbohydrate complexes. ORD and CD investigations of molybdenum(VI) complexes with different sugars [60Bo, 64An, 66Ve, 66Ba1, 68Vo] have indicated that the sugars react in the pyranose form and that three hydroxy groups are chelated by a metal ion. These investigations are of interest because of the suggestion that sugars are possible complexing sites for molybdenum in certain enzymes [66Ba2]. Molybdenum(V) also forms complexes with several sugar molecules, but in a much broader range (pH 5.0–9.5) than for molybdenum(VI) (pH 4.5–6.5) [70Br]. The signs of the observed Cotton effects in the molybdenum(VI) complexes of D-mannose, D-xylose, D-rhamnose, D-ribose, L-sorbose, and D-fructose depend on the conformation of the pyranose ring, which can give rise to three hydroxy groups so placed that they can occupy three adjacent sites in an octahedral complex.

Optically active chromium(III)–sugar complexes have also been prepared [86Br].

Cyclodextrins are cyclic oligosaccharides consisting of six or more α-D-gluco-pyranose units. They form inclusion compounds with various molecules that fit into their 5-8 Å cavity. The direct coordination of cyclodextrins to metal ions has been found only in the copper(II) [75Ma], manganese(III) [83Na], and (recently) cobalt (III) [89Ya] complexes, because of the poor coordinating ability of their hydroxy groups. The proposed structure for the complex $[Co(\alpha\text{-CDX})$ or $(\beta\text{-CDX})^{2-}(en)_2]^+$ is shown in Fig. 6.3.

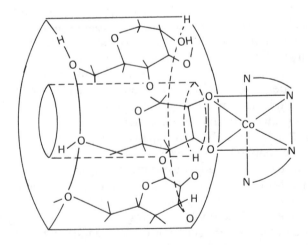

Fig. 6.3 — Proposed structure for $[CoLen_2]^+$ complexes, where L denotes α or β cyclodextrin [89Ya].

Interactions of organotin compounds with carbohydrates have attracted considerable interest [85Pa], because the presence of carbohydrate ligands modifies the biological properties of the organotin group [82Ba]. In spite of this, the chemical and electronic structures of the reaction products remained largely unexplored.

Dialkylstannylenes are easily prepared by the reaction of vicinal diols with dialkyltin oxide [86Da, 82Ho, 74Wa, 84Ru, 87Do] or dialkyltin diethoxide [65Po, 76Po, 65Me]. Molecular weight measurements in solution indicated that the products are dimeric [65Me, 66Co, 79Do, 68Po]. The high-field ^{119}Sn NMR chemical shifts unambiguously showed that the complexes contained five-coordinated tin [72Sm]. David et al. [79Da] showed by X-ray diffraction studies that glucose derivatives of dibutyl tin(IV)-oxide has a dimeric structure, Holzapfel et al. reported [82Ho] that the corresponding mannose derivatives have a pentameric structure containing two five-coordinated and three six-coordinated tin atoms. Recently Davis et al. [86Da] indicated on the basis of Mössbauer spectra of 2,2-dibutyl-mannose-stannolane that this compound contains five- and six-coordinated tin atoms in 3:2 ratio. The X-ray measurements have shown average Sn-O bond lengths of 204 pm, within each monomer unit and 251 pm between the units. The endocyclic O-Sn-O angle was shown to be 79,0°, and the exocyclic C-Sn-C angle 138,6°.

Mössbauer spectroscopy has also been used to investigate the structure of dialkylstannylene derivatives of carbohydrates in solid state [87Do, 72Sm, 84He]. The magnitude of the QS($\sim 2,90$ mm/s) indicated a coordination number greater than four, with tin(IV) in five- or six-coordinated surroundings.

The comparison of experimental QS values with those calculated on the basis of the PQS concept revealed that the complexes are of the three types with central tin(IV) present (a) in purely trigonal bipyramidal surroundings, (b) in purely octahedral surroundings, and (c) in both octahedral and trigonal bipyramidal arrangements in approximately 1:1 ratio [90Na2].

Eight different kind of diethyltin(IV) complexes with carbohydrate ligands (aldoses, polyalcohols, sugar acids, sugar amines, and disaccharides) were prepared. The analytical data showed that all these complexes contained the diethyltin moiety and carbohydrate ligand in 1:1 molar ratio. Their local structure has been determined by extended X-ray absorption fine structure (EXAFS) techniques in solid state. Dioxastannolane units are associated into an infinite ribbon polymer, in which the tin is bound by two carbon atoms and three or four oxygen atoms in a highly distorted octahedral or trigonal bipyramidal arrangement. This observation agreed with results published earlier [90Na2]. Within each unit with average Sn-O, C bond lengths of 200 pm, and with Sn . . . C distances in the seconds shell of 278 pm [90Na3].

3.2 Sugar acid complexes

The biological importance of D-glucuronic acid is mainly due to its presence in specific bacterial polysaccharides [33Go].

The structure of potassium-D-glucuronate.2H$_2$O is known from X-ray measurements [63Gu]. The infrared spectra of crystalline α-D-glucuronic acid and its sodium; potassium, and rubidium salts reveal that the strong hydrogen-bonding systems that is present in the free acid is preserved in the salt formation process, and the sugar moiety crystallizes as the α-anomer. The three alkali metal salts are isomorphous, in terms of common coordination geometry and similar binding sites. The cations interact slightly with the COO^-, CO, and OH groups of the D-glucuronate anion [84Ta3].

D-Glucuronic acid forms complexes of two different compositions with alkaline earth metal ions. These are M(D-glucuronate)X. nH_2O and M(D-glucuronate)$_2$ $4H_2O$, where M = Ca^{2+}, Mg^{2+}, Sr^{2+}, or Ba^{2+}, X = Cl^- or Br^-, and $n = 2$–4 [83Ta2, 85Ta1]. These metal ions have been shown to be eight-coordinated in the complexes with the former composition (except for magnesium, which is six-coordinated), while hexa coordination geometry has been proposed for them in the metal gluconate salts. In every case, the strong hydrogen bonding within the free ligand is weakened by complex formation, and the sugar moiety crystallizes as the α-anomer. Just the opposite is true for the uranyl cation in its D-glucuronate complex. The β-anomer is predominant in the latter compound [86Ta4].

The coordination number found for the zinc(II), cadmium(II) [86Ta5], and uranyl(II)–glucuronic acid complexes is six. The latter cation is bound to two sugar moieties via oxygens O-5 and O-6 in the first (ionized), and O-6' and O-4 in the second (neutral) molecule, further two-sites are occupied by uranyl oxygens [86Ta4].

Spectroscopic (FT-IR and NMR) evidence indicates that the silver ion is bound to two glucuronic acids via oxygens O-6' of the non-ionized carboxy group in Ag(D-glucuronic acid)NO_3, resulting in linear coordination with no direct interaction between the inorganic anion and the silver. On the other hand, Ag(glucuronate) has been shown to have a dimeric structure with linear geometry involving coordination number 2 around the silver(I), as expected on the basis of sp hybridization in silver complexes [86Ta6]. A similar structure is displayed by Ag(D-gluconate) [86Ta7]. The silver(I) ion in Ag(D-gluconolactone)$NO_3 \cdot H_2O$ is also two-coordinated, being bound to the lactone via the carbonyl oxygen atom and to the nitrate ion. Nitrate and perchlorate stabilize the lactone form in the silver complex, while acetate and carbonate facilitate the sugar-to-acid conversion in aqueous solution [87Ta3].

D-Glucurono-1,4-lactone has biological importance because it is converted in animal and human organisms into vitamin C. Since magnesium(II) and calcium(II) seem to participate in this process, the coordination of the latter ions by β-D-glucurono-1,4-lactone has been studied [88Ta2]. The results demonstrated the formation of several complexes in both ethanolic and aqueous solutions. FT-IR and NMR investigations have indicated the binding mode of the ligands to the metal ions. The metal ions are bound to two lactone molecules via O-1-H and O-2-H of the first sugar, and O-6 and O-5-H (five-membered ring chelation) of the second, and to four H_2O molecules in the calcium complex, and to two H_2O molecules in the magnesium complex, resulting in eight-coordination around the calcium(II) and six-coordination around the magnesium(II) in the systems.

Sodium, magnesium, barium, and tetraethylammonium copper-gluconates and barium copper-galacturonates have been prepared and characterized by different methods [81Pa2]. The copper(II) complex formed by the slime of the bacterium *Pseudomonas aeruginosa* has been isolated and its structure studied in comparison with those of model copper(II) complexes of uronic acids of glucose. The local structure of the complexes was determined by using EXAFS [83Sa]. It was shown that the coordination number in all these complexes is four, and square–planar symmetry was suggested. The Cu-O distances (196 pm) in the copper glucuronates and in the complex from *Pseudomanas aeruginosa* were found to be similar to those in copper acetylacetonate (194 pm). The proposed structure was rather surprising, however. In spite of the presence of the carboxylic group in the ligands, the authors concluded that the copper is coordinated through the ether oxygen of C-1 and the aldehydic oxygen, or through the C-3 and C-4 hydroxy oxygens. A reinvestigation of the system is needed to confirm these suggestions or rather to determine the correct structure.

3.3   Complexes of nitrogen-containing sugar derivatives

Small glycopeptides model biologically important carbohydrate-containing molecules. This may be the reason for the increasing interest in their complexation behaviour. The binding of gadolinium(III) and mangangse(II) to tri-O-D-galacto-sylated hexapeptide (Gly-L-Ser-L-Thr-L-Thr-Gly) through the carbohydrate residues appears to be weak and to differ somewhat for the two metals [83Di1]. On the other hand, the metal ion binding to the β-D-galactose groups of di-O-β-D-galacto-sylated tripeptide (Gly-L-Thr-L-Thr) indicates a strong binding site near O-6′, and possible several weak ones near O-3′ and Thr O-3. The modes of interaction of gadolinium(III) and manganese(II) are similar for the glycopeptide containing only α-D-galactose groups, but may differ for the glycopeptide containing β-D-galactose groups [83Di2]. The interaction of manganese(II) with methyl-D-galactopyranoside [84Da2] is similar to those found for the D-galactosylated peptides discussed above [83Di1, 83Di2].

Comparison of the gadolinium(III) binding of glycosylated polypeptides with that of the free polypeptides shows that the peptide with free carboxylate and amino groups has a higher affinity towards the metal ion than that of the glycopeptide without these free functional groups. NMR studies prove the significant, but weak, metal binding to the glycosyl group of the glycopeptide.

D-Glucosamines can readily be prepared through the reaction, in methanolic solution, of an equimolar mixture of D-glucono-1,5-lactone and the appropriate amine or amino acid [64Is]. ^{13}C NMR spectroscopic studies have revealed that the Gd^{3+}-binding ability of these compounds is specific, requiring the C-5, C-4, and C-3 oxygen as donor atoms, whereas the Mn^{2+} binding appears to be nonspecific, involving only one or a pair of oxygen atoms [85Di]. A similar difference in metal ion (gadolinium(III) vs manganese(II)) binding has been observed for *epi*-inositol [84Da1]. From studies of alditol–europium systems [74An1, 74An2], it was concluded that the strongest lanthanide-binding structure occurs when three vicinal alcohols have the *threo-threo* configation. This observation holds for the D-glucosamines too.

In recent years, Yano *et al*. have made extensive studies of the synthesis and characterization of novel nickel(II) complexes containing glycosylamines prepared by the reaction of a diamine and a monosaccharide [88Ya].

These complexes are formed quite easily, when $Ni(diamine)_3^{2+}$ salts in methanolic solution are refluxed with an excess of ketoses. Purification of the resulting blue solution was carried out on a Sephadex LH-20 gel permeation column. Finally, the solution was evaporated to give blue crystals or powder complexes with relatively low yields [80Ta, 85Ya1, 85Ts, 84Ts, 83Ts].

The   molecular   structures   of   [Ni(en)(D-fructose-en)]$^{2+}$ [85Ts]

[Ni(en)(L-sorbose-en)]$^{2+}$ [84Ts], and Ni(ampr)(L-sorbose-ampr)$^{2+}$ [85Ts] (where en = ethylenediamine, and ampr = aminomethylpyrrolidine) were determined by X-ray crystallography. The coordination geometry is very similar in these complexes, being pseudooctahedral in each case. Four coordination positions are occupied by a tetradentate glycosylamine ligand, and the other two coordination sites by a bidentate diamine. The Ni-N bond distances of approximately 200 pm are typical for octahedral nickel(II) complexes. On the other hand, the average Ni-O bond distances are considerably longer than those generally reported for carboxylate complexes. The results indicate that the coordinate bonds between nickel and the hydroxy oxygens are comparatively weak. The proposed structures of the ketose–diamine complexes prepared from D-fructose, L-sorbose, D-tagatose, and D-psicose are depicted in Fig. 6.4.

Aldoses react with [Ni(en)$_2$]$^{2+}$ or [Ni(tn)]$^{2+}$ (tn = trimethylenediamine), similarly to ketoses forming blue octahedral complexes of [Ni(N-glucoside)$_2$]$^{2+}$ composition [80Ta]. The molecular structures of latter complexes have been determined by X-ray crystallography (Fig. 6.5). The N-glucoside was shown to coordinate to the nickel ion through the sugar C-2 oxygen and the two N atoms of the diamine. Mannose reacts with the nickel complex of N,N'-dimethylethylenediamine to give a blue-green binuclear μ-manno furanoside binuclear nickel(II) complex [85Ta2]. Analogous complexes are formed with other sugar-type ligands. The X-ray crystal structure of the octahedral [Ni(L-rham-tn)$_2$] Br$_2$·2H$_2$O·CH$_3$OH complex (rham = rhamnose) [83Sh], shows that the two N-glycoside molecules form a pseudo octahedral coordination sphere around nickel in meridional mode; the complex has an approximately C$_2$ symmetry. Each pyranoside ring of the sugar moieties has the β-^{1}C$_4$ chair conformation, while the absolute configuration of the two coordination chiral nitrogen atoms is S.

Tris(ethylenediamine) nickel(II) was shown to react with the hydrochloride salts of D-glucosamine, D-galactosamine or D-mannosamine to give blue-violet, paramagnetic bis (tridentate) octahedral nickel(II) complexes [85Ya2]. The nickel atom is surrounded by six nitrogen atoms at the apices of a distorted octahedron. Each N-glycoside ligand is coordinated to the metal at three points, through the sugar amino group on C-2 and through the two nitrogen atoms of the en residue; the complex has approximately C$_2$ symmetry. The pyranose ring of the sugar moiety is in the usual β-^{4}C$_1$ chair conformation (Fig. 6.5).

Tsubomura *et al.* [86Ts1] reported that Ni(β-alanine)$_2$·(H$_2$O)$_2$ reacted with aldoses (D-glucose, D-galactose, D-xylose, D-ribose, 4,6,-O-benzylidene-D-glucose, and 3-O-methyl-D-methyl-D-glucose) to give blue or green compounds. The ligands are tridentate in these complexes with coordination through the β-alanine carboxylate and aldose C-2 hydroxy oxygens, and through the nitrogen atom of β-alanine (Fig. 6.5).

Both C-2 epimeric pairs of aldoses (D-glucose-D-mannose, D-quinobose-L-rhamnose, D-galactose-D-talose, D-arabinose-D-ribose and D-xylose-D-lysose) were shown to react with Ni^{2+}-N,N,N'-trimethylethylendiamine (tmen) to give the corresponding green, very hygroscopic complexes containing only mannose-type sugars [87Ta4]. The local structures of some of the complexes have been determined by EXAFS. The results indicated polynuclear structure with the N-mannofuranosides bridging the nickel(II) centres. The mechanism of the above mentioned

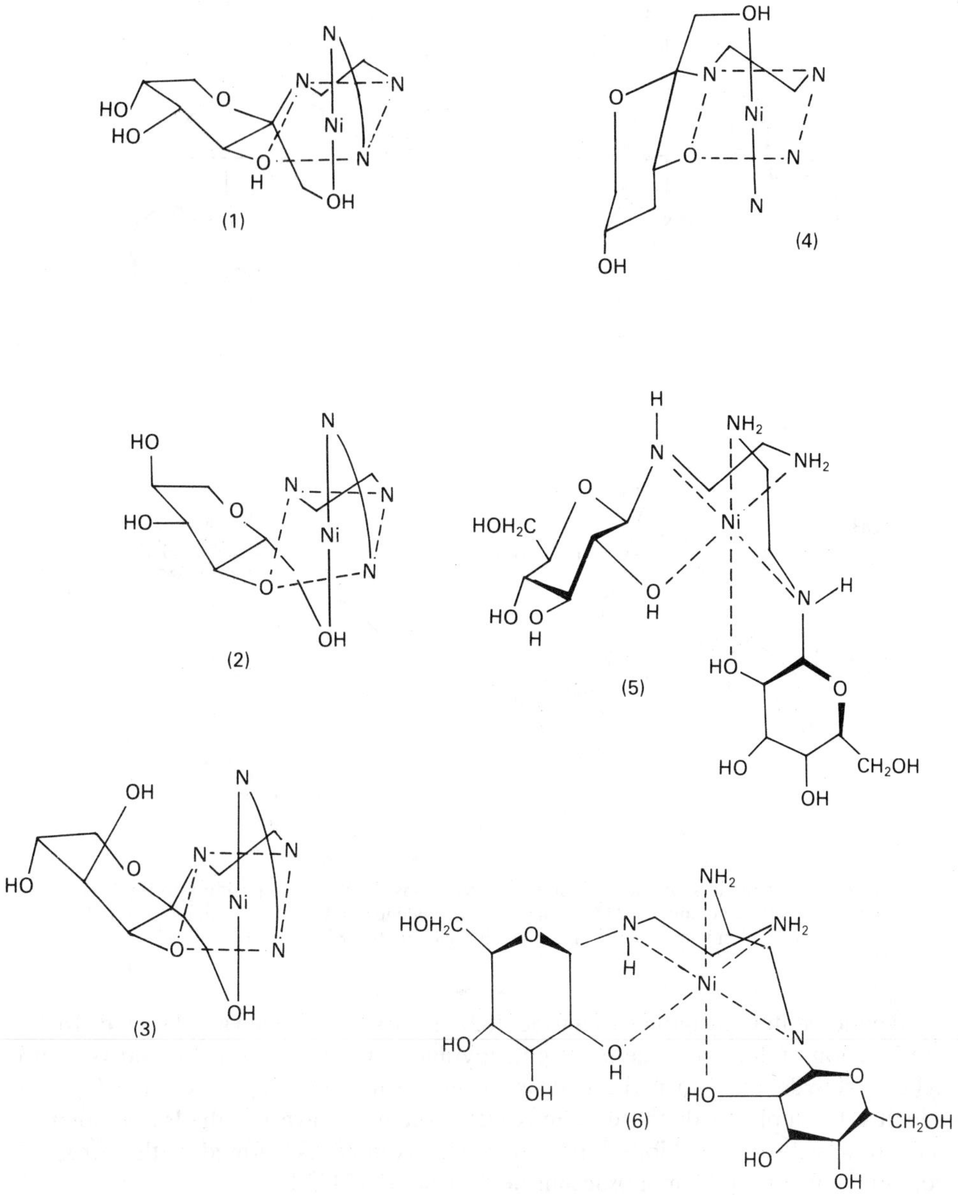

Fig. 6.4 — Structures of ketose and aldose–diamine complexes derived from (1) D-fructose, (2) L-sorbose, (3) D-tagatose, (4) D-picose, (5) D-glucose, and (6) D-mannose [80Ta, 85Ts].

epimerization reaction has been reported in [87Ta5]. The molybdate-catalysed C-2 epimerization of xylose has been decribed by Taylor & Waters [81Ta2]. The complex formed in the reaction of ammonium molybdate and D-xylose contains two molybdates and one sugar moiety (D-xylose) in an unusual furanose form.

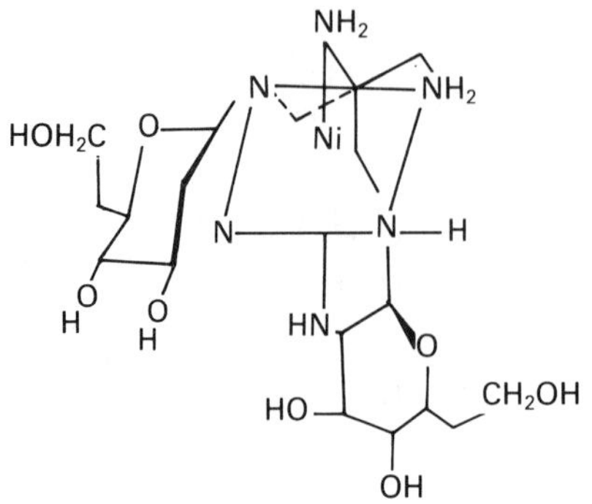

Fig. 6.5 — Structures of the nickel(II) complex ions of NiL$_2$ composition formed with D-glucose-*N*-ethylendiamine (1), D-galactose-*N*-ethylenediamine (2), D-mannose-*N*-ethylendiamine (3), and β-alanine-glucose (4) [85Ya2, 86Ts].

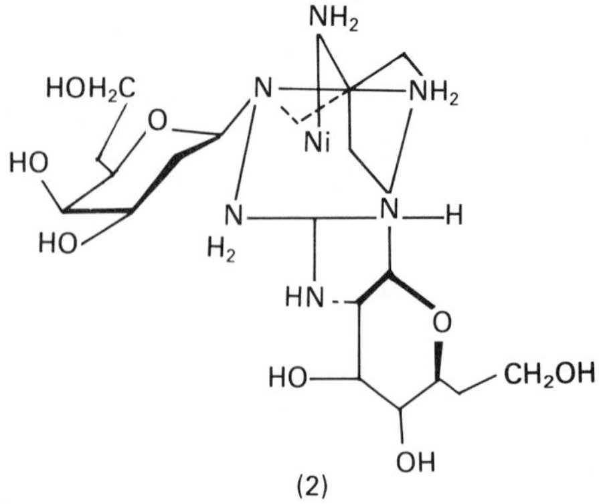

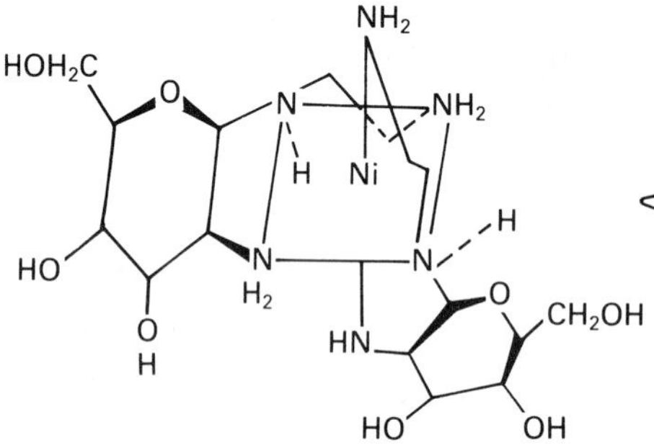

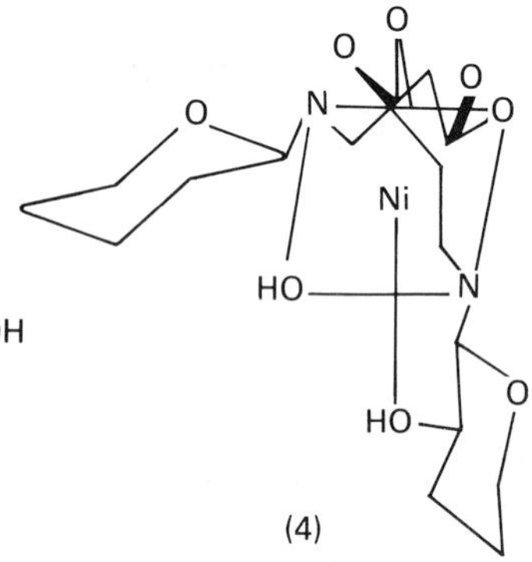

Amino-containing monosaccharides [82Ad, 79Ad] have been synthesized also by the reaction of 2-amino-2-deoxy-D-glucopyranose and some of its derivatives with salicylaldehyde and 3-formyl-2-hydroxybenzoic acid. The copper(II), zinc(II), and cobalt(II) complexes of these ligands were found to have tetrahedral geometry. Similar structure is exhibited by copper(II) complexes formed with nitrogen-containing heterocyclic monosaccharide derivatives [74Zh].

Yano *et al.* [86Is] reported the first successful synthesis, isolation, and characterization of substitution-inert cobalt(III) complexes containing an *N*-glycoside derived from ethylenediamine and D-mannose (1), L-rhamnose (2), and D-ribose (3), respectively. The results of optical measurements suggested the Δ configuration for (1) and the Λ configuration for (2) and (3).

Substitution-inert cobalt(III) complexes containing an *N*-glycoside formed from ethylenediamine and an aldose were synthesized by the oxidation of cobalt(II) to cobalt(III) in the presence of the diamine. It was found that

$[(en)_2Co(O_2)(OH)Co(en)_2]^{3+}$ is the reactive species toward aldoses in the reaction. The results of semiempirical AMI calculations demonstrate that the D-mannose and L-rhamnose units take the pyranose form with β-$^3S_5$ skew-boat conformation in the cobalt(III) complexes, while the D-ribose unit adopts the furanose form with α-^2E envelope conformation in the complex [89Is].

Similarly, two diastereomers (Δ and Λ) have been separated by ion-exchange chromatography from the complexes $[CoL(en)_2]^+$, (where L denotes D-gluconate or L-mannoate) [90Ts]. In latter cases, the carboxylate group and one deprotonated hydroxy group on C-2 are coordinated to the cobalt(III) centre, resulting in a five-membered chelate ring. The preparation of tetraammine complexes of cobalt(III) containing D-ribose, L-sorbose or D-glucosoamine as secondary ligand is described in [84Da1]. It was established that the ligands (except for L-sorbose) are preferably bound to cobalt(III) through the hydroxy groups on C-1 and C-2. L-Sorbose is probably coordinated through the hydroxy groups on C-2 and C-3. The interactions between $[Co(NH_3)_5ClCl_2]$ or $[Co(NH_3)_4Cl_2]Cl$ and L-ascorbic acid led to the formation of the complexes $[Co(NH_3)_5ascorbate]Cl_2 \cdot H_2O$ and $[Co(NH_3)_4ascorbate]Cl_2 \cdot H_2O$. The ascorbate anion is coordinated monodentately in the former (pentaammine) complex via ionized O-3, and bidentately in the latter (tetrammine) one through ionized O-1 and O-4, in both cases leading to pseudo-octahedreal hexacoordinated cobalt(III) [86Ta8]. The complex formation results in the rearrangement of the strong intermolecular hydrogen bounding network of ascorbic acid.

Platinum(II) complexes containing amino sugars [methyl-2,3-diamino-2,3-dideoxy-α-D-mannopyranoside (D-MeManNN)], [methyl-2,3-diamino-2,3-dideoxy-β-D-glucopyranoside (D-MeGlcNN)], and [2,3-diamino-2,3-dideoxy-D-glucose (D-GlcNN)] were prepared [86Ts2] because of their suggested antitumour activity. The structures of these complexes are presented in Fig. 6.6. Intramolecular hydrogen bonds are suggested between the solvate water molecules, chloride ions, hydroxy groups, and amino groups, but no intermolecular interactions are observed between the platinum atoms.

FT-IR, H-NMR spectroscopic, X-ray diffraction, and other evidence indicates that the D-glucuronic or D-gluconic acid is bound monodentately via the dissociated carboxy group in $trans$-$PtL_2(NH_3)_2 \cdot H_2O$ and bidentately through the carboxylate group and a sugar oxygen in cis-$PtL(NH_3)_2L \cdot H_2O$. The strong intermolecular hydrogen-bonding sugar network is altered by the interaction with the platinum ammine, resulting in simple sugar-OH . . . $NH_3(H_2O)$ . . . OH sugar H-bond interactions [87Ta6].

3.4 Dimeric and oligomeric complexes

The presence of metal ion-containing dimeric units in inorganic and metal organic complexes formed with ligands is a well-known phenomenon [64Ka, 70Ko, 74Sm]. When the transition metal ion centres in such compounds are paramagnetic, their magnetic properties reflect the dimeric moiety because of the magnetic interactions between the metal ions in question. Among the complexes formed with sugar-type ligands, few examples of this type of compound have been reported.

Manganese(II) has been shown to form complexes with gluconate [76Bo1], saccharose [86Na1], and maltitol [88Na] that contain such dimeric units. For the

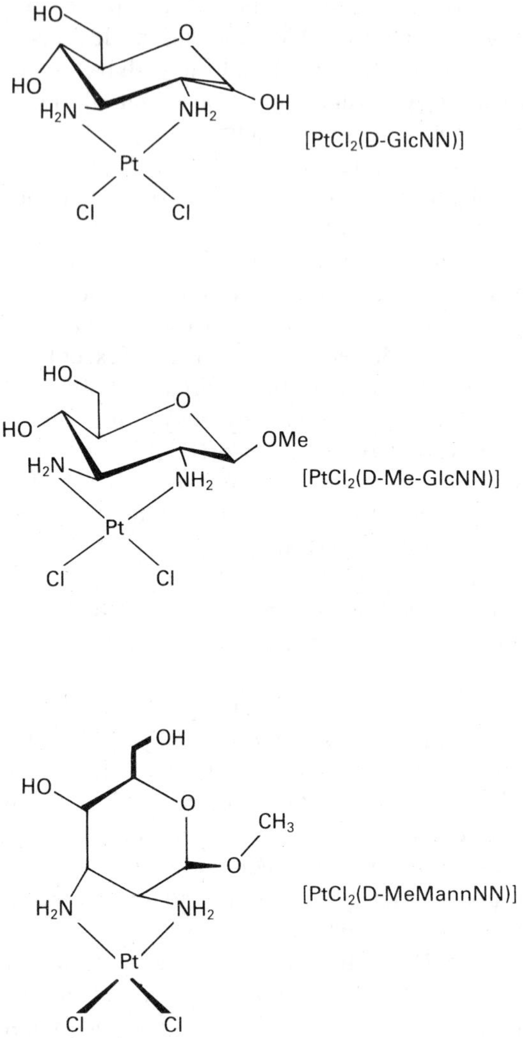

Fig. 6.6 — Structures of complexes PtCl₂(D-GlcNN), PtCl₂(D-Me-GlcNN), PtCl₂(D-Me-MannNN) [86Ts2].

latter two ligands, the corresponding manganese(III) complexes have a similar dimeric structure. The magnetic moment of a manganese(III)-β-cyclodextrin complex, measured by the Evans method, yielded a value of 3.51 μ_B per Mn at 302 K, which decreased to 3.38 μ_B at 224 K. This is well below the spin-only value (4.9 μ_B) for mononuclear manganese(III) species, suggesting a weak antiferromagnetic coupling of the manganese spins. This could be explained by the bis μ-hydroxo-bridged structure [83Na].

EPR measurements have been made on the copper(II) complexes of saccharic, lactobionic, and gluconic acids. Copper(II)-copper(II) interactions indicating dimer formation have been confirmed for the complexes by the observation of $\Delta M = 2$ and in certain cases also $\Delta M = 1$ signals. The calculated copper–copper distances (4.0 Å) are somewhat larger than the copper–copper separation of 3.1 Å observed in the dimeric copper complex of citric acid. The increased distance between the copper ions could be accounted for by a bridging arrangement involving the terminal carboxylate group and a deprotonated alcoholic hydroxy group on the β-carbon atom [71To].

A series of copper(II) complexes formed with polyalcohols, sugar acids, and di- and trisaccharides were prepared by Nagy *et al.* [89Na2]. The analytical results demonstrated that deprotonated alcoholic hydroxy groups participated in the complex formation in all systems. In the case of the sugar acids, the carboxylate group was also coordinated to the metal ion. The analytical data indicated the dimeric or oligomertic composition of some of the complexes. However, in most of the systems the EPR spectra revealed the presence of non-interacting copper(II) ions, indicating either that the complexes are mononuclear, or that the copper(II) centres are connected (and thus magnetically separated) by the carbohydrate ligands in dimeric or oligomeric species [89Na2]. Interacting copper centres have been detected in gluconic acid and dulcitol complexes. It was suggested that in the latter complexes two hydroxide ions form bridges between the two copper ions, resulting in a strong magnetic interaction between them, similarly to that in copper(II)-cyclo-dextrin complexes [75Ma]. In the latter case, two pairs of C_2 and C'_3 secondary hydroxy groups of contiguous glucose units are crosslinked by a $Cu(OH^-)_2Cu$ bridge in the α-cyclodextrin complex and by a $Cu(OH^-)(O^{2-})Cu$ bridge in the β-cyclo-dextrin one.

The direct isolation of the water-soluble iron(III)–fructose complex was first reported by Saltman *et al.* [63Ch]. Elemental analysis indicated that the complex formed at pH 9.0 contains two iron(III) ions per two fructose ligands. The pH-dependence of the ESR signal intensity, the NMR line width, and the magnetic susceptibility demonstrated a minimum at pH 7 in the presence of excess fructose. The results are consistent with the formation of a dimer, which breaks up at a higher pH [64Aa]. The Mössbauer results reflected the similar structures of the iron(III) complexes formed with L-sorbose, D-fructose [82To], and some other pentoses, hexoses, and sugar acids [74Kr]. For the complexes formed with reducing sugars (glucose, galactose, mannose, and lactose) the Mössbauer spectra indicated the presence of a certain amount of iron(II) in the systems [74Kr, 85Wo].

With a view to the acquisition of information on the factors influencing the compositions, magnetic interactions, and structure of iron(III)–sugar complexes, a series of iron(III) complexes with aldoses, ketoses, polyalcohols, sugar acids, and di- and trisaccharides were prepated [86Na2, 87Kü, 89Na3]. The compositions of the complexes were determined by standard analytical methods (Table 6.2). In contrast with the results obtained in ref. [85Wo] it was shown that deprotonated alcoholic hydroxy groups participate in the complex formation, and that polynuclear complex anionic species are formed.

The Mössbauer parameters (IS = 0.47 mm/s, QS = 0.72–0.92 mm/s) reveal the presence of high-spin iron(III) central atoms.

Table 6.2 — Compositions and EPR data on the complexes [86Na2]

Number	Ligand (L)	L:Fe ratio in initial solution	pH	Composition	Molecular weight		Negative charge per L	$I_{ia}:I_{is}$
					exp.	calc.		
I	mannitol	8.0	11.0	$Fe_4L_3(OH)_4(H_2O)_2Na_2$	916	916	3.3	∞
II	sorbitol	3.2	10.0	$Fe_3L_2(OH)_4(H_2O)_8Na$	752	764	3	∞
III	dulcitol	3.0	11.3	$Fe_3L_2(OH)_3Na$	590	602	3.5	∞
IV	1,2-propylene glycol	10.0	14.0	$Fe_2L_2(OH)_2(H_2O)_4$	360	368	2	∞
V	glycerol	10.0	14.0	$Fe_2L_2(OH)_2(H_2O)_4$	388	400	2	∞
VI	deoxy-D-ribose	20.0	12.5	$Fe_2L(OH)_4Na$	340	334	3	2700
VII	raffinose	4.0	12.5	$Fe_3L(OH)_6(H_2O)_5Na_2$	992	994	5	900
VIII	deoxy-D-glucose	4.8	12.5	$Fe_2L_2(OH)_4Na_2$	603	606	3.5	850
IX	L-arabinose	8.0	10.3	$Fe_2L_2(OH)_6(H_2O)_3Na$	642	645	2	700
X	Na-saccharate	2.0	10.3	$Fe_2L(OH)_5(H_2O)_3Na_3$	700	680	2	700
XI	D-mannose	8.0	10.0	$Fe_2L_2(OH)_2(H_2O)Na$	532	544	2.5	350
XII	fructose	8.0	10.0	$Fe_2L_2(OH)_2(H_2O)_4Na$	600	599	2.5	330
XIII	L-xylose	5.2	11.5	$Fe_2L_2(OH)_2(H_2O)Na$	486	484	2.5	170
XIV	D-xylose	8.0	12.0	$Fe_2L_2(OH)_4(H_2O)_6Na$	630	609	1.5	17
XV	Na-gluconate	1.6	10.0	$Fe_2L_2(OH)_4(H_2O)_2Na_2$	640	651	2	105
XVI	Na-lactobionate	1.3	10.3	$Fe_2L_2(OH)_2(H_2O)_{14}Na_2$	1154	1158	3	75
XVII	lactose	2.5	12.5	$Fe_2L_2(OH)_6(H_2O)_6Na_3$	1198	1164	3	22
XVIII	glucose	2.0	12.0	$Fe_2L_2(OH)_2(H_2O)_4Na$	594	599	2.5	12

The EPR spectra of most of the complexes consist of two components:

(1)   There is a broad pattern with g ~ 2, with peak-to-peak widths between 35 and 250 mT, indicative of strong interactions between the paramagnetic high-spin iron(III) centres in the polynuclear species. The shape and peak-to-peak width depend on the dipole–dipole and exchange interactions. A dominance of the former results in Gaussian curves, while that of the latter leads to Lorentzian curves. The spectra of these complexes point to a mixture of the two, with some asymmetry.

(2)   A narrow component with g' ~ 4.3, with peak-to-peak widths between 5 and 30 mT, reflects the presence of isolated high-spin iron(III) centres in the compounds. The magnetic susceptibility of the isolated iron(III) atoms of these complexes follows the Curie law. Toward lower temperatures, the susceptibility curves point to an antiferromagnetic interaction [87Kü]. On the assumption that the compounds are dimeric, the exchange integrals could be fitted to the Van Vleck equation [85Wo]. The values of $-J$ $(11–20\,cm^{-1})$ are much smaller than those for μ-oxo-bridged iron(III) complexes [74Mu] $(\sim 100\,cm^{-1})$, but resemble those for the dihydroxo-bridged complexes of iron(III) $(\sim 15\,cm^{-1})$ [76Th, 83Bo].

The intensity ratio $(I_{ia}:I_{is})$ of the EPR spectral components characterizing the interacting (ia) and isolated (is) iron(III) centres was determined for each spectrum. The results are presented in Tables 6.2 and 6.3. From the data, it may be concluded that the concentration ratio of the two types of species depends on (1) the nature (structure) of the organic ligand (the presence of carboxylate oxygen is needed for the appearance of isolated iron(III) centres, the latter oxygens being stronger electron pair donors than those of alcoholic hydroxy groups, (2) the mode of preparation of the compounds; an increase of the ligand-to-iron ratio in the initial solution and an increase of the pH of this solution favour the formation of isolated iron(III) centres.

Recent potentiometric and spectroscopic studies reflected the interaction of oxovanadium(IV) with D-galacturonic acid [89Mi, 89Br]. The carboxylate group initiates the coordination at pH 3, and one or two deprotonated sugar hydroxyl groups participate in the process. ESR and ENDOR spectra indicated in alkaline concentrated solutions the formation of a dimeric VO(IV)-D-galacturonic acid complex, which exhibits spectroscopic properties consistent with a metal-metal distance of about 4.8 Å similar to that found for VO(IV)-tartarate dimers [90Br].

All these investigations show that sugar-type ligands permit or even promote the formation of polynuclear species. The transition metal centres are connected either by inorganic (hydroxide ion or oxo) bridges or by bridging organic (sugar) ligands, or possibly by both. The strong hydration of polyalcohol moieties has the consequence that even sugar complexes with high metal contents can be water-soluble. This has made possible the preparation of iron(III) mixed complexes containing 100 mg iron/cm^3 in aqueous solution at physiological pH [83Bu] which could therefore be used in human and veterinary iron therapy.

3.5   Redox behaviour

Complex formation is known to influence the redox behaviour of metal ions, which is of importance in a great number of life processes. This initiated research activities in connection with the redox behaviour of carbohydrate complexes. The most thorough

Table 6.3 — Effects of composition of the reaction mixture on complexes [86Na2]

Number	Ligand (L)	L:Fe ratio in initial solution	pH	Composition	Molecular weight		Negative charge per L	$I_{ia}:I_{is}$
					exp.	calc.		
XIX	mannitol	1.0	12.0	$Fe_2L_2(OH)_2(H_2O)_4Na$	620	602	2.5	∞
XX	mannitol	2.0	12.0	$Fe_2L_2(OH)_2Na$	528	531	2.5	2000
XXI	mannitol	2.0	14.0	$FeL(OH)_3(H_2O)Na_2$	358	350	2	11
X	Na-saccharate	2.0	10.3	$Fe_2L_2(OH)_5(H_2O)_3Na_3$	698	680	2	700
XXII	Na-saccharate	2.0	12.5	$Fe_2L_2(OH)_6Na_4$	683	666	2	300
XXIII	lactose	1.0	12.5	$Fe_4L_3(OH)_9(H_2O)_5Na_5$	1598	1660	2.6	45
XVII	lactose	2.5	12.5	$Fe_3L_2(OH)_6(H_2O)_6Na_3$	1198	1164	3	22
XXIV	lactose	1.0	14.0	$FeL(OH)_5(H_2O)_5Na_4$	663	681	2	2
XXV	glucose	1.0	12.0	$Fe_2L_2(OH)_2Na$	522	526	2.5	45
XVIII	glucose	2.0	12.0	$Fe_2L_2(OH)_2(H_2O)_4Na$	594	598	2.5	12

investigations have been reported for manganese complex systems. These are summarized below.

Manganese is known to play a significant role in biological redox systems, e.g. in photosystem II in green plant photosynthesis [79He, 70Ch2] and in mitochondrial superoxide dismutase [70Ke]. The $+2$, $+3$, and $+4$ oxidation states of manganese are believed to be involved in these processes, although their role has not been explained completely. A systematic search has therefore been made for complexing agents (among them sugar-type compounds) to stabilize these oxidation states of manganese.

The stabilization of manganese in different oxidation states in alkaline gluconate solutions and the interaction of this system with hydrogen peroxide and with dioxygen have been discussed by Sawyer *et al.* [75Sa, 76Bo1]. They also described the evolution of oxygen from a solution of the manganese(IV)–gluconate complex. The results demonstrated [76Bo1] that stable complexes of manganese(II), manganese(III), and manganese(IV) are formed, and their apparent formulae are: $[\text{Mn}^{\text{II}}(\text{GH}_3)_2]^{2-}$, $[\text{Mn}^{\text{III}}(\text{GH}_3)_2(\text{OH})]^{2-}$, and $[\text{Mn}^{\text{IV}}(\text{GH}_3)_2(\text{OH})_3]^{3-}$ (in which GH_3^{2-} denotes the dianion of D-gluconic acid). Manganese(III) also forms a second complex, with the possible formula $[\text{Mn}^{\text{III}}(\text{GH}_3)_3]^{3-}$. Electrochemical and magnetic measurements indicated that the manganese(II) complex dimerizes at high pH. A similar observation was made on the manganese(II) complexes with saccharose [86Na1], and maltitol [88Na], but there is no evidence for the formation of dimeric manganese complexes with sorbitol, mannitol [76Ve3], or lactobionic acid [85Na]. The polyols do not form appreciable amounts of soluble manganese(II) complexes, but they do produce stable manganese(III) and manganese(IV) complexes [78Ma1].

The redox behaviour of manganese–gluconate complexes can be expressed by the following equations [76Sa]:

$$[\text{Mn}^{\text{II}}_2(\text{GH}_3)_4(\text{H}_2\text{O})_2]^{4-} + 2\text{H}_2\text{O} + 4\text{e} \xrightarrow{-1.73\,\text{V}} 2\,\text{Mn} + 4\text{GH}_4^- + 4\text{OH}^-$$

$$[\text{Mn}^{\text{II}}_2(\text{GH}_3)_4(\text{H}_2\text{O})_2]^{4-} + 2\text{OH}^- \xrightarrow{-0.54\,\text{V}} [\text{Mn}^{\text{III}}_2(\text{GH}_3)_4(\text{OH})_2]^{4-} + 2\text{H}_2\text{O} + 2\text{e}$$

$$[\text{Mn}^{\text{III}}_2(\text{GH}_3)_4(\text{OH})_2]^{4-} + 4\text{OH}^- \xrightarrow{-0.28\,\text{V}} [\text{Mn}^{\text{IV}}_2(\text{GH}_3)_4\text{O}_2(\text{OH})_2]^{6-} + 2\text{H}_2\text{O} + 2\text{e}$$

At higher gluconate concentrations, the dimeric manganese(III) complex rearranges to give a tris complex, with an apparent equilibrium constant of 0.13:

$$[\text{Mn}^{\text{III}}_2(\text{GH}_3)_4(\text{OH})_2]^{4-} + 2\text{GH}_4^- \xrightarrow{\text{slow}} 2[\text{Mn}^{\text{III}}(\text{GH}_3)_3]^{3-} + 2\text{H}_2\text{O}$$

The latter species is reduced at much more negative potentials, in an irreversible process:

$$2[\text{Mn}^{\text{III}}(\text{GH}_3)_3]^{2-} + 4\text{H}_2\text{O} + 2\text{e}^- \xrightarrow{-1.07\,\text{V}} [\text{Mn}^{\text{II}}_2(\text{GH}_3)_4(\text{H}_2\text{O})_2]^{4-} + 2\text{GH}_4^- + 2\text{H}_2\text{O}$$

Below pH 6, dioxygen (molecular oxygen) does not oxidize the manganese(II)-gluconate complex; between pH 6 and 12, the stable oxidation products are manganese(III) gluconate complexes. Manganese(IV)–gluconate complex formation starts above pH 12. At pH 14 and above, oxidation of the manganese(II) complex by dioxygen to the corresponding manganese(IV) complex is stoichiometric.

When it is in the form of the binuclear bisgluconatomanganese(III) complex, manganese(III) is oxidized only to the manganese(IV) complex by dioxygen. The results indicate that the manganese(IV) complex reacts with hydrogen peroxide to produce molecular oxygen. It should be emphasized that the redox potential of the Mn(IV)/Mn(III) couple at pH 10 is only $+0.08$ V vs SCF which is not positive enough to oxidise water [76Sa, 76Bo2].

In lactobionate complexes [85Na] the ligand does not stabilize the manganese(III) oxidation state, but reduces it to manganese(II).

Polarographic, potentiometric, spectrophotometric, and ESR studies on the manganese complexes of maltitol (H-O-α-glucopyranosyl-D-sorbitol) in aqueous alkaline medium have shown that they participate in electrode reactions according to the following equations [88Na]:

(a) The oxidation of manganese(II) to manganese(III), and of manganese(III) to manganese(IV):

$$[Mn^{III}_2L_3(OH)_4]^{6-} + 20H^- \xrightarrow{-0.625\,V} [Mn^{III}_2L_3(OH)_6]^{6-} + 2e$$

$$[Mn^{III}_2L_3(OH)_6]^{6-} + 20H^- \xrightarrow{-0.347\,V} [Mn^{IV}L(OH)_4]^{2-} + L^{2-} + 2e$$

(b) The reduction of the manganese(II) complex to metallic manganese:

$$[Mn^{II}_2L_3(OH)_4]^{6-} + 4e \xrightarrow{-1.75\,V} 2Mn^0 + 3L^{2-} + 4OH^-$$

(where L denotes the maltitol anion).

Oxidation of the manganese(II)–maltitol complex by molecular oxygen gives stable manganese(III) and manganese(IV) complexes. The UV–visible absorption and ESR spectra of the manganese(IV) compounds confirm that this complex has a distorted octahedral structure [88Na].

EPR measurements revealed that saccharose forms dimeric complexes with manganese(II) in alkaline solution [86Na1].

Oxidation of the latter by molecular oxygen (or electrochemically) led first to a mixed valence manganese(II,III) complex, which could be oxidized to the manganese(III) complex. Polarographic, potentiometric, and analytical chemical results showed the presence of manganese in both oxidation states ($+$II and $+$III) in the dimeric mixed oxidation state complex [86Na1]. The manganese–saccharose complex system undergoes the following electrode reactions:

(a) Oxidation of the manganese(II) complex to the mixed valence complex (violet):

$$[Mn^{II}_2L_2(OH)_4] + 2OH^- \xrightarrow{-0.493\,V} [Mn^{II}L(OH)_2Mn^{III}L(OH)_4]^- + e \ .$$

(b) Oxidation of the latter complex to the manganese(III) complex (brown):

$$[Mn^{II}L(OH)_2Mn^{III}L(OH)_4]^- + 2OH^- \rightarrow [Mn^{III}_2L_2(OH)_8]^{2-}_+ e \ .$$

(c) Oxidation of the manganese(II) complex to the corresponding manganese(III) complex:

$$[Mn^{II}_2L_2(OH)_4] + 4OH^- \xrightarrow{-0.447\,V} [Mn^{III}_2L_2(OH)_8]^{2-} + 2e \ .$$

The oxidation of the manganese(II,III) mixed valence saccharose complex to the corresponding manganese(III) complex, and the reduction of the latter, proved to be reversible processes. The mixed valence complex has an absorption maximum at 490 nm and a minimum at 380 nm (Fig. 6.7, curve 1). Gradual oxidation of the complex by dioxygen to the manganese(III) complex, with periodic recording of the spectra during this process, disclosed an isosbestic point at 608 nm, indicating the equilibrium of two saccharose complexes (Fig. 6.7). After completion of the

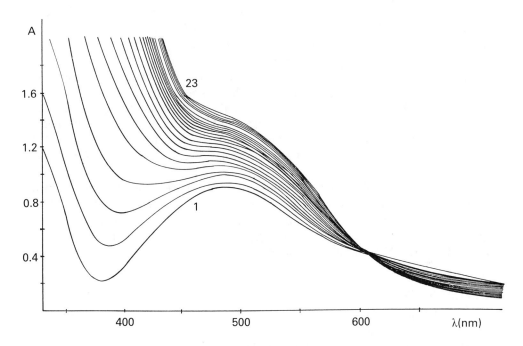

Fig. 6.7 — Oxidation of manganese(II,III)-saccharose to the corresponding manganese(III) complex by molecular oxygen. Curves 1–23: the duration of oxygenation gradually increases [86Na1].

oxidation, when the solution containing the manganese(III)–saccharose complex was stored in an oxygen-free nitrogen atmosphere for 12 h, the complex underwent spontaneous reduction to the parent mixed valence complex. The whole procedure proved to be completely reversible and could be repeated several times [86Na1]. This reaction seems to mimic the thermal step in the Kok model proposed for photosystem II [78La].

Studies have been made on the coordination chemistry and redox behaviour of the manganese–sorbitol complexes, and also in their reactions with molecular oxygen and hydrogen peroxide [79Ri]. The results showed that stable complexes of manganese(III) and manganese(IV) are formed in aqueous alkaline solutions. The measurements indicated that one simple form of the manganese(IV) complex predominates, but that manganese(III) complexes with three different compositions are formed. A dimeric manganese(III)$_A$ complex appears to be the initial product in the reaction of molecular oxygen with the manganese(II) complex, and is in rapid equilibrium with the apparently monomeric manganese(III)$_A$ and manganese(III)$_B$ complexes at higher hydroxide ion and sorbitol concentrations.

A reasonable mechanism for the formation and reaction pathways of the various manganese–sorbitol complexes and their interaction with dioxygen is presented in Fig. 6.8.

A series of polarographic studies by Dolezal have shown that polyhydroxy ligands effectively stabilize the $+3$ and $+4$ oxidation states of manganese in aqeous alkaline solution [72Do, 75Ch, 74Do1, 74Do2, 76Do]. The goal of this research was to find the best experimental conditions for the polarographic determination of manganese in the presence of different cations.

The redox behaviour of carbohydrate complexes with metal ions other than manganese has been investigated much less thoroughly.

The reduction reactions of iron(III) to iron(II) and of vanadium(V) to vanadium (IV) by polygalacturonic acid were studied by Micera et al. [83Ge]. The results suggested that the reaction mechanism involves oxidation of the terminal units of the polymeric chain in a four-electron process, one mole of formic acid being formed.

Iron injections containing a high-spin iron(III)-acetate-lactobionate mixed ligand complex [83Bu] may undergo reductive decomposition during preparation or storage [85Za]. In response to heat treatment, the lactobionate ligand decomposes to gluconic acid and galactose, with reducing properties. The latter reduces the central atom of the iron(III) complex to iron(II).

REFERENCES

04Li   Lippmann, E. O.: *Die Chemie der Zuckerarten*; Friedrich Vieweg und Sohn, Braunschweig (1904).

13Mi   Michales, L.: *Ber.* **46**, 3683 (1913).

29Hi   Hirsch, P. & Schlage, R.: *Z. Phys. Chem., Abt. A* **141**, 387 (1929).

33Go   Goebel, W. F. & Barber, F. H.: *J. Biol. Chem.* **100**, 573 (1933).

46Be   Beevers, C. A. & Cochran, W.: *Proc. Roy. Soc., London* **109**, 257 (1946).

50So   Souchay, R. & Schaal, R.: *Bull. Soc. Chim. France* 819 (1950).

52Th   Thamsen, J.: *Acta Chem. Scand.* **6**, 270 (1952).

55Pe1   Pecsok, R. L. & Sandera, J.: *J. Am. Chem. Soc.* **77**, 1489 (1955).

55Pe2   Pecsok, R. L. & Juvet, R. S.: *J. Am. Chem. Soc.* **77**, 202 (1955).

56Pe   Pecsok, R. L. & Juvet, R. S.: *J. Am. Chem. Soc.* **78**, 3967 (1956).

59We   Weitzel, G., Engelmann, J., & Fretzdorff, A.: *Hoppe-Seyler's Z. Physiol. Chem.* **315**, 236 (1959).

60Bo   Bourne, E. J., Hudson, D. H., & Weigel, H.: *J. Chem. Soc.* 4252 (1960).

60Cr   Croon, I.: *Sven. Papperstidn* **63**, 247 (1960).

60Le   Lenz, R. W.: *J. Am. Chem. Soc.* **82**, 182 (1960).

61Mi   Mills, J. A.: *Biochem. Biophys. Res. Commun.* **6**, 418 (1961).

62Sa1   Sawyer, D. T. & Ambrose, R. T.: *Inorg. Chem.* **1**, 296 (1962).

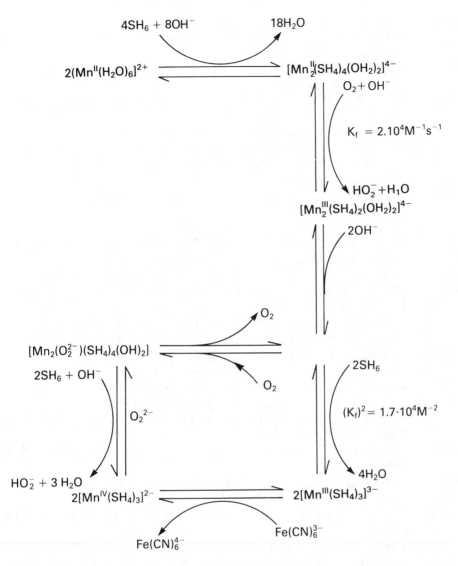

Fig. 6.8 — Redox and solution equilibria for the manganese(II), manganese(III) and manganese(IV) complexes formed by polyhydroxy ligands in aqueous alkaline solution (SH_6 = sorbitol) [76Ri].

62Sa2   Sawyer, D. T. & Kula, R. J.: *Inorg. Chem* **1**, 303 (1962).
62St   Stacey, M. & Barker, S. A.: *Carbohydrates of living tissues*; Van Nostrand, London (1962) Chap. 2.
63Ch   Charley, P. J., Sarkar, B., Stitt, C. F., & Saltman, P.: *Biochim. Biophys. Acta* **69**, 313 (1963).
63Gu   Gurr, G. E.: *Acta Crystallogr.* **16**, 690 (1963).

63We   Weigel, H.: *Adv. Carbohydr. Chem.* **18**, 61 (1963).
63Zo   Zolotukhin, V. K., Linok, Sz. V., Verbljan, N. I., & Balabasz, Sz. I.: *Ukr. Khim. Zhurn.* **29**, 3 (1963).
64Aa   Aasa, R., Malmström, B., Saltman, P., & Vänngärd, T.: *Biochim. Biophys. Acta* **80**, 430 (1964).
64An   Angus, H. J. F., Bourne, E. S., Searle, F., & Weigel, H.: *Tetrahedron Lett.* 55 (1964).
64Is   Ishikawa, T.: *Nippon Kagaku Zasshi* **85**, 897 (1964).
64Ka   Kato, M., Jonassen, H. B., & Fanning, J. C. *Chem. Rev.* **64**, 99 (1964).
64Ma   Mathews, M. B.: *Arch. Biochem. Biophys.* **104**, 394 (1964).
64Mu   Murto, J.: *Acta Chem. Scand.* **18**, 1043 (1964).
64Sa1   Sarkar, B., Saltman, P., Benson, S., & Adamson, A.: *J. Inorg. Nucl. Chem.* **26**, 1551 (1964).
64Sa2   Sawyer, D. T.: *Chem. Rev.* **64**, 633 (1964).
65El   Eliel, E. L., Allinger, N. L., Angyal, S. J., & Morrison, G. A.: *Conformational analysis*; Interscience, New York, (1965), pp. 186–187.
65Jo   Joyce, L. G. & Pickering, W. F.: *Aust. J. Chem.*, **18**, 783 (1965).
65Me   Mehrota, R. C. & Gupta, V. D.: *J. Organomet. Chem.* **4**, 145 (1965).
65Po   Pommier, J. C. & Valade, J.: *Bull. Soc. Chim. France* 1257 (1965).
65Ta1   Tamura, Z. & Miyazaki, M.: *Chem. Pharm. Bull.* **13**, 333 (1965).
65Ta2   Tamura, Z. & Miyazaki, M.: *Chem. Pharm. Bull.* **13**, 345 (1965).
65Ta3   Tamura, Z. & Miyazaki, M.: *Chem. Pharm. Bull.* **13**, 387 (1965).
66Ba1   Bayer, E. & Voelter, W.: *Annalen* **696**, 194 (1966).
66Ba2   Bayer, E. & Voelter, W.: *Biochim. Biophys. Acta* **113**, 632 (1966).
66Co   Considine, W. J.: *J. Organomet. Chem.* **5**, 263 (1966).
66Iz   Izatt, R. M. Rytting, J. H., Hansen, L. D., & Christensen, J. J.: *J. Am. Chem. Soc.* **88**, 2641 (1966).
66Mi   Miyazaki, M., Senshu, T., & Tamura, Z.: *Chem. Pharm. Bull.* **14**, 114 (1966).
66Re   Rendleman, J. A.: *Adv. Carbohydr. Chem.* **21**, 209 (1966).
66Ry   Rys, P. & Zollinger, H.: *Helv. Chim. Acta* **49**, 1406 (1966).
66Ve   Velluz, L. & Legrand, M.: *C.r. hebd Séanc. Acad. Sci. Paris C.* **263**, 1429 (1966).
68Po   Pommier, J. C. & Valade, J.: *J. Organomet. Chem.* **12**, 433 (1968).
68Ur   Urist, M. R., Speer, D. P., Ibsen, K. J., & Strates, B. S.: *Calcif. Tissue Res.* **2**, 253 (1968).
68Vi   Vicedomini, M.: *Gazz. Chim. Ital.* **98**, 1161 (1968).
68Vo   Voelter, W., Bayer, E., Records, R., Bunnenberg, E., & Djerassi, C.: *Annalen* **718**, 238 (1968).
69Ma   Manning, G. S.: *J. Chem. Phys.* **51**, 924 (1969).
69Pe   Peterson, P. J.: *J. Expt. Bot.* **20**, 863 (1969).
70Br   Brown, D. H. & MacPherson, J.: *J. Inorg. Nucl. Chem.* **12**, 3309 (1970).
70Ch1   Christensen, J. J., Rytting, J. H., & Izatt, R. M. *J. Chem. Soc.* (*B*), 1646 (1970).
70Ch2   Cheniae, G. M.: *Ann. Rev. Plant. Physiol* **21**, 467 (1970).
70Cl   Cleland, R. L. & Wang, J. L.: *Biopolymers* **9**, 799 (1970).

70Ke   Keele, B. B., McCord, J. M., & Fridovich, I.: *J. Biol. Chem.* **24**, 6176 (1970).
70Ko   Kokoska, G. F. & Duerst, R. W.: *Coord. Chem. Rev.* **5**, 209 (1970).
70Ve   Veis, A.: *Biological polyelectrolytes*; Marcel Dekker, New York (1970).
71An   Angyal, S. J. & Davies, K. P.: *Chem. Commun.* **10**, 500 (1971).
71Ar   Armitage, I. & Hall, L. D.: *Can. J. Chem.* **49**, 2770 (1971).
71Bu   Butterworth, R. F., Pernet, H. G., & Hanessian, S.: *Can. J. Chem.* **49**, 981 (1971).
71De   Degani, Ch.: *Carbohydr. Res.* **18**, 329 (1971).
71Ho   Horton, D. & Thomson, J. K.: *Chem. Commun.* 1389 (1971).
71Ro   Roberts, E. J., Wade, C. P., & Rowland,, S. P.: *Carbohydr. Res.* **17**, 393 (1971).
71Le   Leppart, G. G. & Colrin, J. R.: *J. Polymer Sci.* **36C**, 321 (1971).
71Mc   McGregor, E. A. & Bownes, J. M.: *Can. J. Biochem.* **49**, 417 (1971).
71Oo   Oosawa, F.: *Polyelectrolytes*; Marcel Dekker, New York, (1971).
71To   Toy, A. D. & Smith, T. D.: *J. Chem. Soc.* (*A*) 2925 (1971).
72An   Angyal, S. J.: *Aust. J. Chem.* **25**, 1957 (1972).
72Ar   Armitage, I. & Hall, L. D.: *Carbohydr. Res.* **24**, 221 (1972).
72Do   Dolezal, J. & Landmyhr, F. J.: *Anal. Chim. Acta* **61**, 73 (1972).
72Fe   Fedorov, A. A. & Pavlinova, A. V.: *Zh. Vses. Khim. Obschest.* **17**, 352 (1972).
72Pa   Pavlinova, A. V. & Demyanchuk, L. S.: *Izv. Vyssh. Ucheb. Zaved., Khim. Khim. Tekhnol.* **15**, 1152 (1972).
72Pr   Preuss, F. & Rosenhahn, L.: *J. Inorg. Nucl. Chem.* **34**, 1691 (1972).
72Sm   Smith, P. J., White, R. F. M., & Smith, L.: *J. Organomet. Chem.* **40**, 341 (1972).
72Yo   Yoshimura, J., Kobayashi, K., Sato, K., & Funabashi, M.: *Bull. Chem. Soc. Japan* **25**, 1806 (1972).
73Al   Albersheim, P.: *The primary cell wall*, in: *Plant carbohydrate biochemistry*, Ed. J. B. Pridham, Academic Press, London (1973).
73An1   Angyal, S. J. *Pure Appl. Chem.* **35**, 131 (1973).
73An2   Angyal, S. J.: *Carbohydr. Res.* **26**, 271 (1973).
73Au   Authonsen, Cf. T., Larsen, B., & Smidsrod, O.: *Acta Chem. Scand.* **27**, 2671 (1973).
73Mu   Muzzarelli, R. A. A.: *Natural chelating polymers*; Pergamon Press, Oxford, (1973).
73Re   Rendleman, J. A. Jr.: *Adv. Chem. Ser.* **117**, 51 (1973).
74Al   Albersheim, P.: *Plant carbohydrate biochemistry*; Academic Press, London (1974).
74An1   Angyal, S. J., Greeves, D., & Mills, J. A.: *Aust. J. Chem.* **27**, 1447 (1974).
74An2   Angyal, S. J.: *Tetrahedron* **30**, 1695 (1974).
74Ca   Cape, J. N., Cook, D. H., & Williams, D. R.: *J. Chem. Soc., Dalton Trans.* 1849 (1974).
74Ch   Chernyshev, V. A. & Shishniashvili, M. E. *Helaty Met. Prir. Soedin. Ikh. Primen.* **1**, 34 (1974).

74Cr1   Craig, D. C., Stephenson, N. C., & Stevens, J. D.: *Cryst. Struct. Commun.* **3**, 277 (1974).

74Cr2   Craig, D. C., Stephenson, N. C. & Stevens, D., *Cryst. Struct. Commum.* **3**, 195 (1974).

74Do1   Donoso, G., Dolezal, J., & Zyka, J., *J. Electroanal. Chem., Interfacial Electrochem.* **49**, 461 (1974).

74Do2   Dolezal, J., Juláková, E., Cerny, M., & Kopanica, M.: *J. Electroanal. Chem., Interfacial Electrochem.* **52**, 261 (1974).

74Ko   Konunova, Ts. B. & Kachkar, L. S. *Zh. Neorg. Khim.* **19**, 2279 (1974).

74Kr   Krapov, V. U., Kulakov, V. N., Shafransky, V. H., Sztukan, R. A., & Sztanko, V. I.: *Izotopenpraxis* **6**, 227 (1974).

74Mu   Murray, K. S.: *Coord. Chem. Rev.* **12**, 1 (1974).

74Od   Odilabadze, L. N., Shishniashvili, M. E., Demetrashvili, E. V., & Kashiya, L. D.: *Helaty Met. Prir. Soedin. Ikh. Primen.* **1**, 45 (1974).

74Sm   Smith, T. D. & Pilbrow, J. R.: *Coord. Chem. Rev.* **13**, 173 (1974).

74Wa   Wagner, D., Verheyden, J. P. H., & Moffat, J. G.: *J. Org. Chem.* **39**, 74 (1974).

74Zh   Zhdanov Yu. A., Osipov, O. A., Grigoriev, A. D., Garnovsky, A. D., Alexeev, Yu. E., Alexeeva, V. G., Gontmacher, N. M., & Perov, P. A.: *Carbohydr. Res.* **38**, Cl (1974).

75An1   Angyal, S. J., Bodkin, C. L., & Parrish, F. W.: *Aust. J. Chem.* **28**, 1541 (1975).

75An2   Angyal, S. J. & Hickman, R. J.: *Aust. J. Chem.* **28**, 1279 (1975).

75Ch   Chughtai, N., Dolezal, J., & Zyka, J.: *Microchem. J.* 20 (1975).

75Fe   Fedorov, A. A.: *Izv. Vyssh, Uchebn. Zaved., Khim i Khim. Teknol.* **18**, 1360 (1975).

75Gr   Grasdalen, H., Anthoussen, T., Larsen, B., & Swidsrod, O.: *Acta Chem. Scand.* **B29**, 99 (1975).

75Gu   Guss, J. M., Hukins, D. W. L., Smith, P. J. C., Winter, W. T., Arnott, S., Moorhouse, R., & Rees, D. A.: *J. Mol. Biol.* **95**, 395 (1975).

75Ha   Hall, L. D. & Preston, C. H.: *Carbohydr. Res.* **41**, 53 (1975).

75Ki   Kieboom, A. P. G., Spoormarker, T., Sinnema, A., van der Toorn, J. M., & van Bekkum, H.: *Recl. Trav. Chem. Pay-Pas* **94**, 53 (1975).

75Ma   Matsui, Y., Kurita, T., Yagi, M., Okayama, T., Mochida, J. & Date, Y.: *Bull. Chem. Soc. Japan* **48**, 2187 (1975).

75Pa   Parrish, F. W., Angyal, S. J., Evans, M. E., & Mills, J. A.: *Carbohydr. Res.* **45**, 73 (1975).

75Sa   Sawyer, D. T. & Bodini, M. E.: *J. Am. Chem. Soc.* **97**, 6588 (1975).

75Ve   Velasco, J. G., Ortega, J., & Sancho, J.: *An. Quim.* **71**, 706 (1975).

75Wi   Winter, W. T., Smith, P. J. C., & Arnott, S.: *J. Mol. Biol.* **99**, 219 (1975).

76An1   Angyal, S. J. & Greeves, D.: *Aust. J. Chem.* **29**, 1223 (1976).

76An2   Angyal, S. J., Greeves, D., & Littlemore, L.: *Aust. J. Chem.* **29**, 1231 (1976).

76Bo1   Bodini, M. E., Willis, L. A., Riechel, T. L., & Sawyer, D. T.: *Inorg. Chem.* **15**, 1538 (1976).

76Bo2   Bodini, M. E. & Sawyer, D. T.: *J. Am. Chem. Soc.* **98**, 8366 (1976).

76Co   Cook, W. J. & Bugg, C. E.: *Acta Crystallogr.* **B32**, 656 (1976).

76Do   Dolezal, J. & Kekulova, H.: *J. Electroanal Chem., Interfacial Electrochem.* **69**, 239 (1976).

76Go   Good, R. and Sawyer, D. T.: *Inorg. Chem.* **15**, 1427 (1976).

76Ha   Harvey, J. M., Naftalin, R. J., & Symons, M. C. R.: *Nature* **261**, 435 (1976).

76Le   Lenkinski, R. E. & Renben, J.: *J. Am. Chem. Soc.* **98**,, 3089 (1976).

76Pa   Panda, C. & Patnaik, R. K.: *Indian J. Chem., Sect. A* **14A**, 446 (1976).

76Po   Pommier, J. C. & Pereyre, M.: *Adv. Chem. Ser.* **157**, 82 (1976).

76Sa   Sawyer, D. T., Bodini, M. E., Willis, L. A., Riechel, T. L., & Magers, K. D.: *Bioinorg. Chem.* **II**, p. 330 (1976).

76Th   Thich, J. A., Chih-Ou, C., Powers, D., Vassilion, B., Mastropaolo, D., Potenza, J. A., & Schugar, H. J.: *J. Am. Chem. Soc.* **98**, 1425 (1976).

76Ve1   Velasco, J. G., Ortega, J. & Sancho, J.: *An. Quim.* **72**, 743 (1976).

76Ve2   Velasco, J. G., Ortega, J., & Sancho, J.: *J. Inorg. Nucl. Chem.* **38**, 889 (1976).

76Ve3   Velikov, B. L. & Dolezal, J.: *J. Electroanal. Chem.* **71**, 91 (1976).

77An   Angyal, S. J., Bodkin, C. L., Mills, J. A. & Pojer, P. M.: *Aust. J. Chem.* **30**, 1259 (1977).

77Ba   Banon, M. L., Ortega, J., & Sancho, J.: *J. Electroanal. Chem.* **78**, 173 (1977).

77Ch   Chakrabarti, B.: *Arch. Biochem. Biophys.* **180**, 146 (1977).

77Cl   Cleland, R. L.: *Arch. Biochem. Biophys.* **180**, 57 (1977).

77Co   Cook, W. J. & Bugg, C. E., In: Pullman, B. & Goldblum, W. (Eds.), *Metal–ligand interactions in organic chemistry and biochemistry*, Vol. 2, Reidel, Dordrecht, Holland (1977) pp. 231–256.

77Cr   Czarniecki, M. F. & Thornton, E. R.: *Biochem. Biophys. Res. Commun.* **74**, 553 (1977).

77Ev   Evans, W. J. & Frampton, V. L.: *Carbohydr. Res.* **59**, 571 (1977).

77Fa   Farago, M. E. & Pitt, M. J.: *Inorg. Chim. Acta* **24**, 127 (1977).

77Fi   Figueroa, N., Nagy, B., & Chakrabarti, B.: *Biochem. Biophys. Res. Commun.* **74**, 460 (1977).

77Ha   Hair, N. J. & Beattie, J. K.: *Inorg. Chem.* **16**, 245 (1977).

77Ja   Jaques, L. W., Brown, E. B., Barrett, J. M., Brey, Jr., W. S., & Weltner, W. Jr.: *J. Biol. Chem.* **252**, 4533 (1977).

77Ki   Kieboom, A. P. G., Sinnema, A., van Toorn, J. M., & van Bekkum, H.: *Recl. Trav. Chim. Pays-Bas* **96**, 35 (1977).

77Ma1   Makridon, C., Croner-Morin, M., & Scharff, J. P.: *Bull. Soc. Chim. France* 59 (1977).

77Ma2   Matthews, M. B. & Decker, L.: *Biochem. Biophys. Acta* **498**, 259 (1977).

77Pa   Panda, C., & Patnaik, R. K.: *J. Inst. Chem.* **49**, 297 (1977).

77Ra   Ramamoorthy, R.: *J. Theor. Biol,* **66**, 527 (1977).

78Al   Alföldi, J., Petrus, L., & Bilik, V.: *Collect. Czech. Chem. Commun.* **43**, 1476 (1978).

78An   Angus, H. J. F., Briggs, J., Sufi, N. A., & Weigel, H.: *Carbohydr. Res.* **66**, 25 (1978).

78Ar   Arance, B. & Velasco, J. G.: *An. Quim.* **74**, 547 (1978).

78Co1   Comper, W. D. & Laurent, T. C.: *Physiol. Rev.* **58**, 255 (1978).

78Co2   Coccioli, F. & Vicedomini, M.: *J. Inorg. Nucl. Chem.* **40**, 2103 (1978).

78Co3   Coccioli, F. & Vicedomini, M.: *J. Inorg. Nucl. Chem.* **40,** 2106 (1978).

78Fi   Figureoa, N. & Chakrabarti, B.: *Biopolymers* **17**, 2415 (1978).

78Gu   Gustavsson, H., Siegel, G., Lindman, B. & Fransson, L.: *FEBS Lett.* **86**, 127 (1978).

78Ko   Kohn, R. & Kovác, P.: *Chem. Zvesti* **32**, 478 (1978).

78La   Lawrence, G. D. & Sawyer, D. T.: *Coord. Chem. Rev.* **27**, 173 (1978).

78Ma1   Magers, K..D., Smith, C. G., & Sawyer, D. T.: *Inorg. Chem.* **17**, 517 (1978).

78Ma2   Manning, G. S.: *Q. Rev. Biophys.* **11**, 179 (1978).

78Mu   Mukherjee, D. C., Park, J. W., & Chakrabarti, B.: *Arch. Biochem. Biophys.* **191**, 393 (1978).

78Ol   Ollis, J., James, V. J., Angyal, S. J., & Pojer, P. M.: *Carbohydr. Res.* **60**, 219 (1978).

78Pa   Parkash, O., Bhasin, S. K., & Jain, D. S.: *J. Less-Common. Met.* **60**, 179 (1978).

78Te   Tersis, A.: *Cryst. Struct. Commun.*, **7**, 95 (1978).

79Ad   Adam, M. J. & Hall, L. D.: *J. Chem. Soc., Chem. Commun.* 234 (1979).

79Cl   Cleland, R. L.: *Biopolymers*, **18**, 2673 (1979).

79Da   David, S., Pascard, C., & Cesario, M.: *Nouv. J. Chim.* **3**, 63 (1979).

79Do   Domazetis, G., Magee, R. G., & James, B. D.: *J. Inorg. Nucl. Chem.* **41**, 1546 (1979).

79Go   Gonzales Velasco, J., Ayllon, S., & Sancho, J.: *J. Inorg. Nucl. Chem.* **41**, 1075 (1979).

79He   Heat, R. L.: *Int. Rev. Cytol.* **34**, 49 (1979).

79Ki   Kieboom, A. P. G., Buurmans, H. M. A., van Leeuwen, L. K., & Benschop, H. J.: *Recl. Trav. Chim. Pays-Bas* **98**, 393 (1979).

79Ma   Magdalenat, H., Turq, P., Tivant, P., Chemla, M., Menez, R., & Drifford, M.: *Biopolymers* **18**, 187 (1979).

79Mi   Miyazaki, M., Nishimura, S., Yoshida, A., & Okubo, N.: *Chem. Pharm. Bull.* **27**, 532 (1979).

79Po1   Poonia, N. S. & Bajaj, A. V.: *Chem. Rev.* **79**, 389 (1979).

79Po2   Poncini, L.: *Indian J. Chem., Sect. A* **18**, 167 (1979).

79Po3   Poncini, L.: *Int. Sugar J.* **81**, 36 (1979).

79Ri   Riechens, D. T., Smith, C. G., & Sawyer, D. T.: *Inorg. Chem.* **18**, 706 (1979).

80An1   Angyal, S. J.: *Chem. Soc. Rev.* **9**, 415 (1980).

80An2   Angyal, S. J. & Kondo, Y.: *Aust. J. Chem.* **33**, 1013 (1980).

80Ag   Agarwal, L. K. & Jain, D. S.: *J. Indian Chem. Soc.* **57**, 309 (1980).

80De   Deiana, S., Erre, L., Micera, G., Piu, P., & Gessa, C.: *Inorg. Chim. Acta* **46**, 249 (1980).

80Löl   Lönnberg, H., Vesala, A., & Käppi, R.: *Carbohydr. Res.* **86**, 137 (1980).

80Lö2   Lönnberg, H. & Vesala, A.: *Carbohydr. Res.* **78**, 53 (1980).

80Ro   Rowland, S. P.: *Cellul. Chem. Technol.* **14**, 423 (1980).

80Ta   Takaziwa, S., Sugita, H., Yano, S., & Yoshikawa, S.: *J. Am. Chem. Soc.* **102** , 7969 (1980).

81Ar   Aruga, R.: *Bull. Chem. Soc. Japan* **54**, 1233 (1981).

81Br   Briggs, J., Finch, P., Matulewicz, M. C., & Weigel, H.: *Carbohydr. Res.* **97**, 181 (1981).

81Fr   Fridman, Ya. D., Dolgashova, N. V. & Rustemova, A. G.: *Zh. Neorg. Khim.* **26**, 2775 (1981).

81Jo   Joshi, Y. M. & Kwak, J. C. T.: *Biophys. Chem.* **13**, 65 (1981).

81Ma   Mattai, J. & Kwak, J. C. T.: *Biophys. Chem.* **14**, 55 (1981).

81Mc   McArdle, P., O'Reilly, J. P., Simmie, J., & Lee, E. E.: *Carbohydr. Res.* **90**, 165 (1981).

81Mi   Micera, G., Deiana, S., Gessa, C., & Petrera, M.: *Inorg. Chim. Acta* **56**, 109 (1981).

81Pa1   Parrish, R. F. & Fair, W. R.: *Bicohem. J.* **193**, 407 (1981).

81Pa2   Payne, R., Magee, R. J., Sarode, P. R., & Rao, C. N. R.: *Inorg. Nucl. Chem. Letters* **17**, 125 (1981).

81Sa   Sandstrom, D. R., Stubbs, B. R., & Greegar, R. B.: 'Structural evidence for solutions from EXAFS measurements' in *EXAFS spectroscopy*; ed. by Teo, B. K. & Joy, D. C., Plenum Press, New York (1981) Chap. 9., pp. 139–157.

81Ta1   Talarico, F. & Vicedomini, M.: *Ann. Chim.* **71**, 97 (1981).

81Ta2   Taylor, G. E. & Waters, J. M.: *Tetrahedron Lett.* **22**, 1277 (1981).

81Ve   Vesala, A. & Lönnberg, H.: *Acta Chem. Scand. Ser. A* **35**, 123 (1981).

81Vi   Vicedomini, M.: *Ann. Chim. (Rome)*, 213 (1981).

81Wr   Wright, W. M., Feigan, J., Denny, W., Leupin, W., & Klarus, D. R.: *J. Magn. Reson.* **45**, 514 (1981).

82Ad   Adam, M. J. & Hall, L. D.: *Can. J. Chem.* **60**, 2229 (1982).

82Ba   Barbieri, R., Pellerito, L., Ruisi, G., & Lo Giudice, M. T.: *Inorg. Chim. Acta* **66**, 39 (1982).

82Cl   Cleland, R. L., Wang, J. L. & Detweiler, D. M.: *Macromol.* **15**, 386 (1982).

82Da   Daman, M. E. & Dill, K., *Carbohydr. Res.* **102**, 47 (1982).

82Ho   Holzapfel, C. W., Kockemoer, J. M., Marais, C. M., Kruger, G. J., & Pretorius, J. A.: *S. Afr. J. Chem.* **35**, 81 (1982).

82Ma   Mattai, J. & Kwak, J. C. T.: *J. Phys. Chem.* **86**, 1026 (1982).

82Pa   Payne, P. & Magee, R. J.: *Proc. Indian Acad. Sci.* **91**, 31 (1982).

82Sy   Symons, M. C. R., Benbow, J. A., & Pelmore, H.: *J. Chem. Soc., Faraday Trans.* 1. **78**, 3671 (1982).

82To   Tonkovic, M., Music, S., Hadzija, O., Nagy-Czakó, I., & Vértes, A.: *Acta Chim. Hung.* **110**, 197 (1982).

82Ve   Vesala, A., Lönnberg, H., Käppi, R., & Arpalahti, J.: *Carbohydr. Res.*, **102**, 312 (1982).

83Ba1   Balt, S., De Bolster, M. W. G., Booij, M., & van Herk, A. M.: *Inorg. Biochem.* **19**, 213 (1983).

83Ba2   Balt, S., DeBolster, M. W. G., & Visser-Luirik, G.: *Inorg. Chim. Acta* **78**, 121 (1983).

83Bo   Borer, L., Thalken, L., Ceccarelli, C., Glick, M., Zhang, J. H., & Reiff, W. M.: *Inorg. Chem.* **22**, 1719 (1982).

83Bu   Burger, K., Zay, I., & Takácsi Nagy, G.: *Inorg. Chim. Acta* **80**, 231 (1983) and references therein.

83Da1   Davies, A., Gormally, J., Wyn-Jones, E., Wedlock, D. J., & Phillips, G. O.: *Biochem. J.* **213**, 363 (1983).

83Da2   Damand, M. E. & Dill, K., *Carbohydr. Res.* **111**, 205 (1983).

83De   Delville, A. & László, P.: *Biophys. Chem.* **17**, 119 (1983).

83Dh   Dheu-Andries, M. L. & Perez, S.: *Carbohydr. Res.* **124**, 324 (1983).

83Di1   Dill, K., Daman, M. E., Batstone-Cunningham, R. L., Denarie, M., & Pavia, A. A.: *Carbohydr. Res.* **123**, 137 (1983).

83Di2   Dill, K., Daman, M. E., Bastone, R. L., Batstone-Cunningham, R. L., Denarie, M., & Pavia, A. A.: *Carbohydr. Res.* **124**, 11 (1983).

83Ge   Gessa, C., De Cherchi, M. L., Dessi, A., Deiana, S., Micera, G.: *Inorg. Chim. Acta.* **80**, L53 (1983).

83Ma   Mario, V.: *J. Coord. Chem.* **12**, 307 (1983).

83Na   Nair, B. U. & Dismukes, G. C.: *J. Am. Chem. Soc.* **105**, 124 (1983).

83Sa   Sarode, P. R., Sankar, G., & Rao, C. N. R.: *Proc. Indian Acad. Sci.* **92**, 527 (1983).

83Sh   Shiai, H., Yano, S., Toriumi, K., Ito, T., & Yoshikawa, S.: *J. Chem. Soc., Chem. Commun.* 201 (1983).

83Ta1   Tajmir-Riahi, H. A. & Lotfipoor, M.: *Spectrochim. Acta* **39A**, 167 (1983).

83Ta2   Tajmir-Riahi, H. A.: *Carbohydr. Res.* **122**, 241 (1983).

83Ts   Tsubomura, T., Yano, S., Yoshikawa, S., Toriumi, K., & Ito, T.: *Polyhedron* **2**, 123 (1983).

83Ve1   Vesala, A., Käppi, R., & Lehikoinen, P.: *Finn. Chem. Lett.* 45 (1983).

83Ve2   Vesala, A., Käppi, R., & Lönnberg, H.: *Carbohydr. Res.* **119**, 25 (1983).

83Za   Zay, I., Gaizer, F., & Burger, K.: *Inorg. Chim. Acta* **80**, L9 (1983).

84Co   Cowman, M. K., Nakaniski, K., & Balázs, E. A.: *Arch. Biochem. Biophys.* **230**, 203 (1984).

84Da1   Daman, M. E. & Dill, K.: *J. Magn. Reson.* **60**, 118 (1984).

84Da2   Daman, M. E. & Dill, K.: *Carbohydr. Res.* **132**, 335 (1984).

84Gr   Grebenyikov, V. N., Vüszkrebcov, V. B., & Budarin, L. I.: *Zh. Neorg. Khim.* **29**, 51 (1984).

84He   Herber, R. H., Shanzer, A., & Libman, J.: *Organometallics* **3**, 586 (1984).

84Li   Limaye, S. N. & Saxena, M. C.: *J. Indian Chem. Soc.* **61**, 842 (1984).

84Ma1   Mattai, J. & Kwak, J. C. T.: *J. Phys. Chem.* **88**, 2625 (1984).

84Ma2   Manzini, G., Cesaro, A., Delben, F., Paoletti, S., & Reisenhofer, E.: *Bioelectrochem. Bioenerg.* **12**, 443 (1984).

84Mo   Motekaitis, R. J. & Martell, A. E.: *Inorg. Chem.* **23**, 18 (1984).

84Pi   Picullel, L., Lindman, B., & Einarsson, R.: *Biopolymers* **23**, 1683 (1984).

84Re   Reisenhofer, E., Cesaro, A., Delben, F., Manzini, G., & Paoletti, S.: *Bioelectrochem. Bioenerg.* **12**, 455 (1984).

84Ru   Ruisi, G., Lo Guidice, M. T., & Pellerito, L.: *Inorg. Chim. Acta* **93**, 161 (1984).

84Sy   Symons, M. C. R., Benbow, J. A., & Pelmore, H.: *J. Chem. Soc., Faraday Trans.* 1. **80**, 1999 (1984).

84Ta1   Tajmir-Riahi, H. A.: *Carbohydr. Res.* **127**, 1 (1984).

84Ta2   Tajmir-Riahi, H. A.: *J. Inorg. Biochem.* **22**, 55 (1984).

84Ta3   Tajmir-Riahi, H. A.: *Carbohydr. Res.* **125**, 13 (1984).

84Ts   Tsubomura, T., Yano, S., Yoshikawa, S., Toriumi, K., & Ito, T.: *Bull. Chem. Soc. Japan* **57**, 1833 (1984).
84Ya   Yamaguchi, T., Lindquist, O., Boyce, J. B., and Clason, T.: *EXAFS and near edge structure III*, ed. by K. O. Hodgson, B. Hedmann, & J. E. Pennen-Hahn, Springer Verlag, Berlin, (1984) pp. 417–419.
85As   Asakura, K., Nomura, M., & Kuroda, H.: *Bull. Chem. Soc. Japan.* **58**, 1543 (1985).
85Bu   Bunel, S. & Ibarra, C.: *Polyhedron* **4**, 1537 (1985).
85Ca   Carter, R. D. & Dill, K.: *Inorg. Chim. Acta* **108**, 83 (1985).
85Di   Dill, K., Daman, M. E., Decoster, E., Lacombe, J. M., & Pavia, A. A.: *Inorg. Chim. Acta.* **106**, 203 (1985).
85Fr   Fridman, Ya. D., Dzhusueva, M. S., & Dolgashova, N. V.: *Zh Neorg. Khim.* **30**, 2286 (1985).
85Ko   Kosmus, W. & Schmut, O.: *Carbohydr. Res.* **145**, 141 (1985).
85Ll   Llopis Jover, E., Ramirez, B., & Jose, A.: *Transition Met. Chem.* **10**, 405 (1985).
85Mi   Micera, G., Deiana, S., Dessi, A., Decock, P., Dubois, B., & Kozlowski, H.: *Inorg. Chim. Acta.* **107**, 45 (1985).
85Mo   Morel, J-Pierre, & Lhermet, C.: *Can. J. Chem.* **63**, 2639 (1985).
85Na   Nagy, L., Horváth, I., & Burger, K.: *Inorg. Chim. Acta* **107**, 179 (1985).
85Pa   Patel, A. & Poller, C.: *Rev. Silicon, Germanium, Tin, Lead Compl.* **8**, 263 (1985).
85St   Sterk, H., Braun, M., Schmut, O., & Feichtinger, H.: *Carbohydr. Res.* **145**, 1 (1985).
85Ta1   Tajmir-Riahi, H. A.: *J. Inorg. Biochem.* **24**, 127 (1985).
85Ta2   Tanase, T., Kurihara, K., Yano, S., Kobayashi, K., Sakurai, T., & Yoshikava, S.: *J. Chem. Soc., Chem. Commun.* 1562 (1985).
85Ts   Tsubomura, Y., Yano, S., Toriumi, K., Ito, T., & Yoshikawa, S.: *Inorg. Chem.* **24**, 3218 (1985).
85Ya1   Yano, S., Takizawa, S., Sugita, H., Takahoshi, T., Shioi, H., Tsubomura, T., & Yoshikawa, S.: *Carbohydr. Res.* **142**, 179 (1985).
85Ya2   Yano, S., Sakai, Y., Toriumi, K., Ito, T., Ito, H., & Yoshikawa, S.: *Inorg. Chem.* **24**, 498 (1985).
85Wo   Wolowiec, S. & Drabent, K.: *J. Radioanal. Nucl. Chem. Letters* **95**, 1 (1985).
85Za   Zay, I., Vértes, A., Takácsi Nagy, G., Suba, M., & Burger, K.: *J. Radioanal. Nucl. Chem.* **88**, 343 (1985).
86Br   Brown, D. H., Smith, W. E., El-Shahawi, M. S., & Wazir, M. F. K.: *Inorg. Chim. Acta* **124**, L25 (1986).
86Ca   Carter, R. D. & Dill, K.: *Inorg. Chim. Acta* **123**, L9 (1986).
86Co   Cook, I. B., Magee, R. J., Payne, R., & Ternai, B.: *Aust. J. Chem.* **39**, 1307 (1986).
86Da   Davies, A. G., Price, A. J., Dawes, H. M., & Hursthouse, M. B.: *J. Chem. Soc., Faraday Trans.* **2**, 297 (1986).
86Fr   Franks, F., Hall, J. R., Irish, D. E., & Norris, K.: *Carbohydr. Res.* **157**, 53 (1986).

86Is   Ishida, K., Yano, S., & Yoshikawa, S.: *Inorg. Chem.* **25**, 3552 (1986).

86Le1   Lerivrey, J., Dubois, B., Decock, P., Micera, G., Urbanska, J., & Kozlowski, H.: *Inorg. Chim. Acta.* **125**, 187 (1986).

86Le2   Lerivrey, J., Decock, P., Dubois, R., Urbanska, J., & Kozlowski, H.: *Inorg. Chim. Acta.* **124**, L11 (1986).

86Ll   Llopis Jover, E., Ramirez, B., & Jones, A.: *Transition Met. Chem.* **11**, 489 (1986).

86Ma   Manolov, St., Pjartman, A. K., & Gentshev, M.: *Z. Phys. Chem.* **267**, 1017 (1986).

86Mo   Morel, J-Pierre, Lhermet, C., & Morel-Desrosiers: *Can. J. Chem.* **64**, 996 (1986).

86Na1   Nagy, L., Gajda, T., Páli, T., & Burger, K.: *Inorg. Chim. Acta* **123**, 35 (1986).

86Na2   Nagy, L., Burger, K., Kürti, J., Mostafa, M. A., Korecz, L., & Kiricsi, I.: *Inorg. Chim. Acta* **124**, 55 (1986).

86Pe   Pekka, H., Seppo, P., & Kari, L.: *Finn. Chem. Lett.* **13**, 93 (1986).

86Ta1   Tajmir-Riahi, H. A.: *Biophys. Chem.* **25**, 271 (1986).

86Ta2   Tajmir-Riahi, H. A.: *J. Inorg. Biochem.* **27**, 65 (1986).

86Ta3   Tajmir-Riahi, H. A.: *Biophys. Chem.* **23**, 223 (1986).

86Ta4   Tajmir-Riahi, H. A.: *Inorg. Chim. Acta.* **119**, 227 (1986).

86Ta5   Tajmir-Riahi, H. A.: *J. Inorg. Biochem.* **26**, 23 (1986).

86Ta6   Tajmir-Riahi, H. A.: *Inorg. Chim. Acta* **125**, 43 (1986).

86Ta7   Tajmir-Riahi, H. A.: *J. Inorg. Biochem.* **27**, 205 (1986).

86Ta8   Tajmir-Riahi, H. A.: *Biophys. Chem.* **25**, 37 (1986).

86Ta9   Tajmir-Riahi, H. A.: *J. Coord. Chem.* **15**, 95 (1986).

86Ts1   Tsubomura, T., Yano, S., & Yoshikawa, S.: *Inorg. Chem.* **25**, 392 (1986).

86Ts2   Tsubomura, T., Yano, S., Kobayashi, K., Sakurai, T., & Yoshikawa, S.: *J. Chem. Soc., Chem. Commun.* **459** (1986).

87Al   Alvarez, A. M., Desrosiers, N. M., & Morel, J-Pierre: *Can. J. Chem.* **65**, 2656 (1987).

87Ce   Cesaro, A., Delben, F., Flaibani, A., & Paoletti, S.: *Carbohydr. Res.* **160**, 355 (1987).

87Do   Donaldson, J. D., Grimes, S. M., Pellerito, L., Girasolo, M. A., Smith, P. J., Cambria, A., & Fama, M.: *Polyhedron* **6**, 383 (1987).

87Kü   Kürti, J., Mostafa, M. A., Korecz, L., Nagy, L., & Burger, K.: *J. Radioanal. Nucl. Chem. Letters* **118**, 437 (1987).

87Pu   Pusino, A., Droma, D., Decock, P., Dubois, B., & Kozlowski, H.: *Inorg. Chim. Acta* **138**, 31 (1987).

87Se   Seppo, P., Kari, L., & Pekka, H.: *Finn. Chem. Lett.* **14**, 1 (1987).

87Ta1   Tajmir-Riahi, H. A.: *Monatsh. Chem.* **118**, 245 (1987).

87Ta2   Tajmir-Riahi, H. A.: *Inorg. Chim. Acta* **135**, 67 (1987).

87Ta3   Tajmir-Riahi, H. A.: *Inorg. Chim. Acta* **136**, 93 (1987).

87Ta4   Tanase, T., Shimizu, F., Yano, S., Hidai, M., Yoshikawa, S., & Asakura, K.: *Nippon Kagaku Kaishi* **3**, 322 (1987).

87Ta5   Tanase, T., Shimizu, F., Kuse, M., Yano, S., Yoshikawa, S., & Hidai, M.: *J. Chem. Soc., Chem. Commun.* **659** (1987).

87Ta6   Tajmir-Riahi, H. A.: *Biophys. Chem.* **27**, 243 (1987).

87Ta7   Tajmir-Riahi, H. A.: *J. Inorg. Biochem.* **31**, 255 (1987).
87Vé   Véber, M., Trogmayer-Málik, K., Horváth, I., & Burger, K.: *Bioelectroanal. 1. Symp., Akad. Kiadó, Budapest*, 421 (1987).
88Hu   Hunter, G. K., Wong, K. S., & Kim, J. J.: *Arch. Biochem. Biophys.* **260**, 161 (1988).
88Na   Nagy, L., Gajda, T., Páli, T., & Burger, K.: *Acta Chim. Hung.* **125**, 403 (1988).
88Ra   Radowska, B., Kozlowski, H., Decock, P., Dubois, B., & Micera, G.: *J. Inorg. Biochem.* **33**, 153 (1988).
88Ta1   Tajmir-Riahi, H. A.: *Carbohydr. Res.* **172**, 1 (1988).
88Ta2   Tajmir-Riahi, H. A.: *Carbohydr. Res.* **172**, 327 (1988).
88Ta3   Tajmir-Riahi, H. A.: *Carbohydr. Res.* **183**, 35 (1988).
88Ur   Urbanska, J., Kozlowski, H., Delannoy, A., & Hennion, J.: *Anal. Chim. Acta* **207**, 85 (1988).
88Ve   Verchere, J. F. & Sauvage, J. P.: *Bull. Soc. Chim. France* 263 (1988).
88Ya   Yano, S.: *Coord. Chem. Rev.* **92**, 113 (1988).
89An   Angyal, S. J.: *Advances in Carbohydrate Chemistry and Biochemistry* **47**, (1989).
89Br   Branca, M., Micera, G., Dessi, A., & Kozlowski, H.: *J. Chem. Soc., Dalton Trans.* 1283 (1989).
89Is   Ishida, K., Nonoyama, S., Hirano, T., Yano, S., Hidai, M., & Yoshikawa, S.: *J. Am. Chem. Soc.* **111**, 1599 (1989).
89Ko   Kozlowski, H., Decock, P., Oliver, I., Micera, G., Pusino, A. & Pettit, L. D.: *Carbohydr. Res.* (1989) (in press).
89Mi   Micera, G., Dessi, A., Kozlowski, H., Radowska, B., Urbanska, J., Decock, B., & Oliver, I.: *Carbohydr. Res.* **188**, 25 (1989).
89Na1   Nagy, L. & Burger, K.: (unpublished results) (1989).
89Na2   Nagy, L., Zsikla, L., Burger, K., Rockenbauer, A., & Kiss, J. T.: *J. Crystallogr. and Spectrosc. Res.* **19**, 911 (1989).
89Na3   Nagy, L., Yamaguchi, T., Nomura, M., & Ohtaki, H.: *Inorg. Chim. Acta* **159**, 201 (1989).
89Ve   Verchere, J. F. & Chappelle, S.: *Polyhedron* **8**, 333, (1989).
89Ya   Yamanari, K., Nakamichi, M., & Shimura, Y.: *Inorg. Chem.* **28**, 248 (1989).
90Br   Branca, M., Micera, G., Sanna, D., Dessi, A., & Kozlowski, H.: *J. Chem. Soc., Dalton Trans.* (1990) (in press).
90Ga1   Gajda, T., Nagy, L., & Burger, K.: *J. Chem. Soc. Dalton Trans.* (in press).
90Ga2   Gajda, T., Nagy, L., Rockenbauer, A., & Burger, K.: *J. Chem. Soc. Dalton Trans* (in press) (1990).
90Na1   Nagy, L., Yamaguchi, T., Nomura, M., Páli, T., & Ohtaki, H.: Accepted for publication in *Inorg. Chim. Acta* (1990).
90Na2   Nagy, L., Korecz, L., Kiricsi, I., Zsikla, L., & Burger, K.: *Struct. Chem.* (1990) (in press).
90Na3   Nagy, L., Yamashita, S., Yamaguchi, T., Wakita, H., & Burger, K.: to be published in *J. Chem. Soc., Dalton Trans.* (1990).
90Ts   Tsubomura, T., Yano, S., & Yoshikawa, S.: *Bull. Chem. Soc. Japan*; submitted for publication.
90Ur   Urbanska, J. & Kozlowski, H.: *J. Coord. Chem.* (1990) (in press).

VII

Proton and metal ion interaction with nucleic acid bases, nucleosides, and nucleoside monophosphates

Harri Lönnberg
Department of Chemistry, University of Turku, SF-20500 Turku, Finland

1. INTRODUCTION

Nucleic acids are biopolymers that contain the genetic information in the cell nucleus (DNA), transfer it to ribosomes (mRNA), and participate there in protein syntheses as carriers of amino acid residues (tRNA). Chemically they are chain-like macro-molecules consisting of monomeric nucleoside units linked by phosphodiester bonds (Fig. 7.1). Nucleosides are, in turn, N-glycosides of two different types of hetero-aromatic nitrogen bases: purines and pyrimidines. The carbohydrate moiety is a β-D-ribofuranosyl group in ribonucleic acids (RNA) and 2-deoxy-β-D-ribofuranosyl group in deoxyribonucleic acids (DNA). The pyrimidine bases are cytosine and thymine in DNA, and cytosine and uracil in RNA, the purine bases being adenine and guanine in both nucleic acids. The structures of these bases are indicated in Fig. 7.1. The phosphodiester bonds linking the nucleoside units are invariably of $3',5'$-type.

Another central biological function of nucleosides is the participation of their phosphoric acid esters, nucleotides (Fig. 7.2) in phosphate transfer reactions, which form the basis of cellular energy transfer. The energy released in exothermic metabolic processes is coupled with formation of nucleoside $5'$-triphosphates, the hydrolysis of which to nucleoside $5'$-monophosphates and inorganic pyrophosphate constitutes the driving force for endothermic biosynthetic reactions. Moreover, nucleotides are involved as coenzymes in numerous other enzyme catalysed processes.

Owing to the vital role of nucleic acids in molecular biology, their coordination chemistry has become one of the most extensively studied subjects of bioinorganic chemistry. As a matter of fact, nucleic acids and their monomeric constituents are

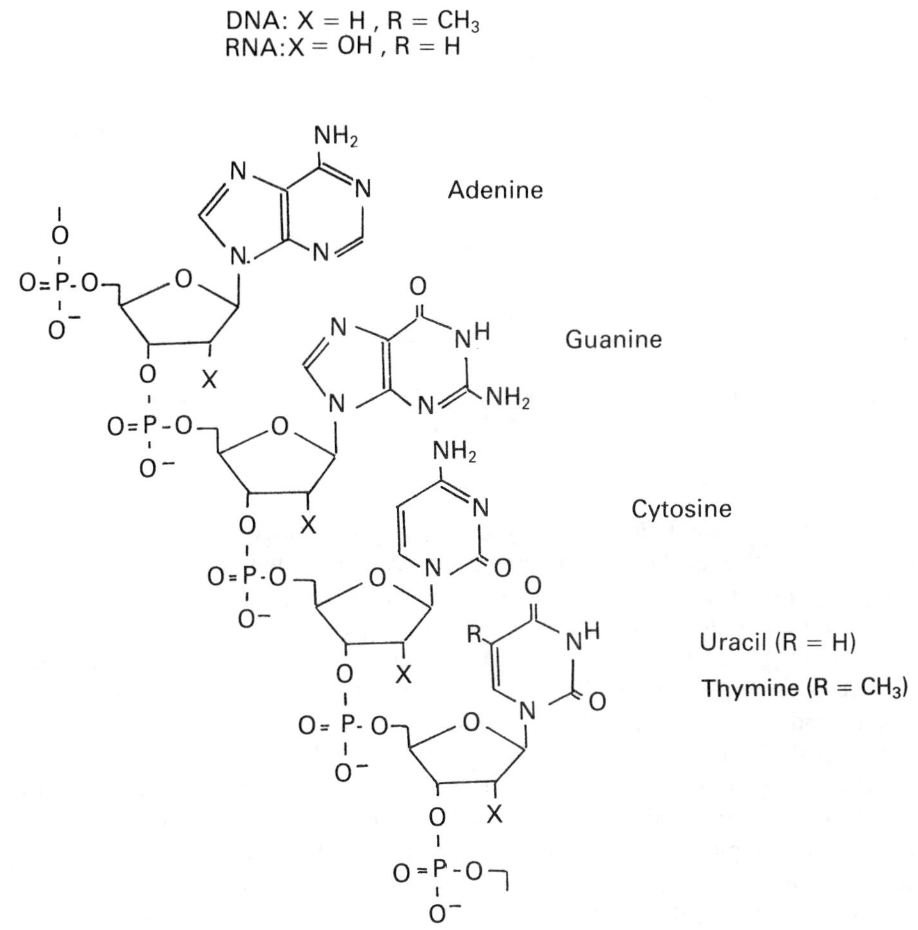

DNA: X = H , R = CH$_3$
RNA:X = OH , R = H

NH$_2$

Adenine

Guanine

NH$_2$

Cytosine

Uracil (R = H)

Thymine (R = CH$_3$)

O=P-O

O$^-$

O=P-O

O$^-$

O=P-O

O$^-$

O=P-O

O$^-$

O=P-O

O$^-$

Fig. 7.1 — Structure of ribonucleic acids.

ambidentate ligands, which may participate in the biological process as metal ion complexes. The potential binding sites include oxygen and nitrogen atoms of the base moieties, hydroxyl functions of the sugar moieties, and oxygen atoms of the phosphate groups. It has been well established that enzymes catalysing reactions which involve nucleic acid constituents, for example using nucleotides as coenzymes, require the presence of metal ions [77Oc, 87Ka, 87Mi1]. For example, DNA polymerase is activated by a Mg(II) ion, which probably binds the nucleoside triphosphate substrate to the enzyme protein [83Wu]. Accordingly, metal ions do not only participate in the catalysis itself, but may also be involved in selectivity phenomena. It has even been assumed that they might promote interactions between nucleic acids and proteins [81He]. Metal ions also affect the stability of the helical tertiary structure that nucleic acids adopt in solution [82Ma1, 88Ko]; phosphate coordination usually stabilizes, and base coordination destabilizes, this native

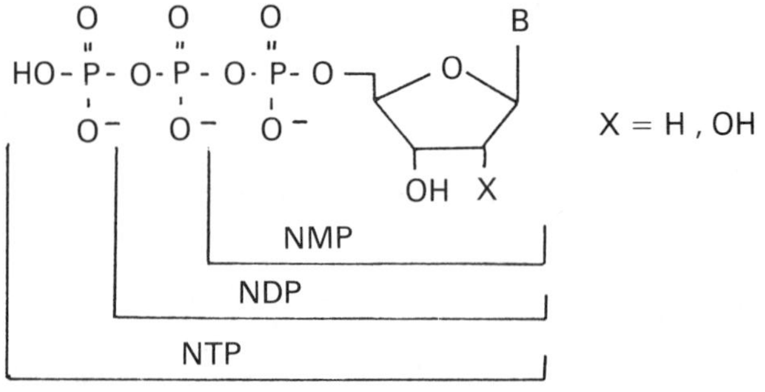

Fig. 7.2 — Structure of nucleotides: nucleoside 5'-mono- (NMP), 5'-di- (NDP) and 5'-tri-phosphates (NTP).

conformation [80Ma1]. *In vitro* they induce a transformation of a normal right-handed double helical DNA structure (B-DNA) to a left-handed Z-DNA [84Sa].

Metal ions that are not normally present in cells may also bind to nucleic acids and hence interfere with the transformation of genetic information. These metal ions are either toxic [80Ba, 80Ma1], like organomercurials, or may exhibit antiviral or anticarcinogenic properties. The breakthrough of the coordination chemistry of nucleic acids in the early 1970s was actually a consequence of the discovery of *cis*-$(NH_3)_2PtCl_2$ as an anticarcinogenic agent [69Ro]. Since then the biological influences of numerous metal complexes have been studied, but only a few of them have showed a comparable activity. Besides complexes of platinum group metals, the ammine complexes of ruthenium [80Cl1] and metallocene dihalides of titanium, vanadium, and neobium [87Kö] have been established as anticarcinogens. More-over, some evidence of anticarcinogenic activity of organotin compounds [84Cr] and gold [81Si] and rhodium [80Cl2] complexes has been reported.

Apart from biological influences, metal ions play an important role in the field of structural and synthetic nucleic acid chemistry. Illustrative examples of the numer-ous applications currently in use are: (i) paramagnetic [80Ma1] or fluorescent [80Ri, 86Ro] metal ions as structural probes, (ii) metal ions in chemical sequencing of DNA [87Iv], (iii) metal ion promoted cleavage of phosphodiester bonds [87Ba], (iv) structural studies with the aid of site-specific DNA binding metal complexes [88Ch, 89Ba, 89Sa], (v) elucidation of the structure of DNA–protein complexes with hydroxyl radicals generated by metal ion redox systems [87Tu], (vi) usage of metal ions as activators and protecting groups in synthetic nucleic acid chemistry [86Mi1], and (vii) polymerization of nucleotides with metal ions [88Sa1].

For the reasons described above, detailed knowledge of the structures and stabilities of metal ion complexes of nucleosides and nucleotides is desirable in attempting to advance the understanding of molecular biological phenomena and developing the usage of metal ions as tools of nucleic acid chemistry. Several review articles on these subjects have been published during the last two decades, the majority of them dealing mainly with the structures of the complexes, either in the solid state [77Ho, 77Ma1, 79Ge3, 79Sw] or in solution [73Ei, 73Tu, 75Da, 77Ma2, 79Ma1, 81Ma1, 80Pe]. By contrast, less attention has been paid to the thermodynamics of complex formation, the excellent survey of Izatt *et al.* [71Iz] being already almost 20 years old. More recent review articles are focused on special topics like the dichotomy between N1- and N7-binding with purine nucleosides [85Ma1], macrochelate formation of purine nucleoside triphosphates [87Si], and the kinetics of complexing 3d transition metal ions [84Di]. The very rapidly developing field of substitution inert platinum [87Re1, 87Sh1] and non-platinum group [87Kö] metal antitumor agents has been reviewed quite recently.

The aim of the present review is to survey the literature dealing with the interactions of substitution labile metal ions with nucleic acid bases, nucleosides, and nucleoside 5'-monophosphates in aqueous solutions. Most attention is paid to the thermodynamics of the complex formation and solution structures of the complexes. The tautomeric and protolytic equilibria are also briefly discussed to provide a background necessary for understanding the metal ion complexation of these amibidentate ligands. By contrast, the solid state structures and results of theoretical calculations are referred to only when these data help to evaluate the binding sites. Complexing the nucleoside di- and tri-phosphates has been omitted, since this subject has been thoroughly discussed in the extensive series of papers by Sigel [86Sc].

2.   COMPLEXING OF PURINE AND ITS 9-SUBSTITUTED DERIVATIVES

2.1   Protolytic equilibria

9-(β-D-Ribofuranosyl)purine (Ia), called purine riboside in the following, may be regarded as a parent compound of purine nucleosides. For this reason its coordination chemistry has been a subject of considerable interest, although the purine riboside itself is not a constituent of nucleic acids. The most basic site in this molecule and other 9-substituted purines is the N1 atom, as evidenced by ^{15}N NMR shift measurements with 9-methylpurine (Ic) in D_2O [82Go, 83Sc1], and with 9-(2-deoxy-β-D-*erythro*-pentofuranosyl)purine (Ib) in DMSO-d_6 [87Re2]. The N1 signal is shifted upfield by about 80 ppm upon protonation while the other nitrogen signals exhibit hardly detectable downfield shifts. Although the ^{15}N resonance shifts may be influenced by base-stacking and/or hydrogen bonding at the high substrate concentrations employed, it is apparent that N1 protonation predominates. On these bases, it is rather surprising that protonation of 9-methylpurine affects almost equally on the ^{1}H resonance shifts of H2, H6, and H8 [87Fr]. Protonation, however, promoted the J(H2, H6) coupling and increased the T_1 values of H2 and H6 more than that of H8, consistent with N1 protonation [87Fr]. The pK_a values for the monocations of Ia and Ic have been reported to be 2.46 and 3.05, respectively (Table 7.1), the higher basicity of Ic most probably resulting from the electropositive nature of the methyl group compared with the ribofuranosyl group.

Table 7.1 — Logarithmic stability constants, $\log(K/M^{-1})$, for the proton and 3d transition metal complexes of methylated purines and purine riboside at 25°C[†]

	Purines				Purine riboside
	9-Me	2,9-di-Me	6,9-di-Me	8,9-di-Me	
H^+	3.05	3.84	3.55	3.51	2.46
Co^{2+}	1.04	0.80	<0.2	0.78	1.00
Ni^{2+}	1.56	1.26	<0.2	1.28	1.31
Cu^{2+}	1.88	1.78	0.74	1.62	1.50
Zn^{2+}	0.9	0.7	<0.2	0.6	0.7

[†] From Ref. 80Lö and 83Ar2. The ionic strength adjusted to 1.0 M with $NaClO_4$.

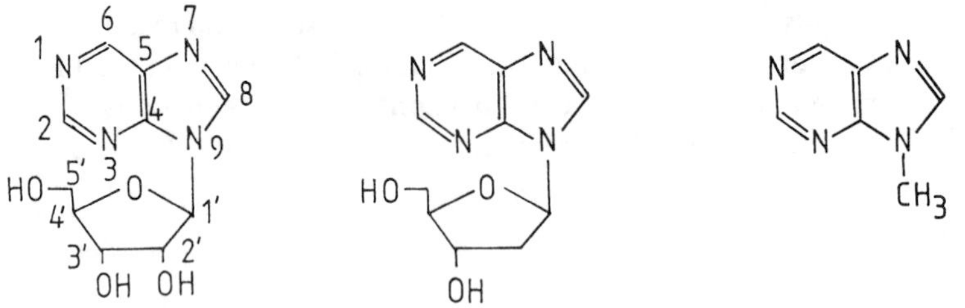

Ia   Ib   Ic

Formation of di- and tri-cations has been studied with the free purine base, only. Under neutral aqueous conditions this base exists as an equimolar mixture of N9H (IIa) and N7H tautomers (IIb) [70Pu, 71Pu1, 75Ch, 80Ch, 82Go, 82Sc, 83Sc1]. The macroscopic pK_a value for the dissociation of the labile proton is 8.75 ($I = 0.05$ M, $T = 25°C$) [69Re1]. In the crystalline state the N7H tautomer prevails [65Wa]. The monocation formation takes place at N1 [71Pu1, 75Ch, 80Ch, 82Go, 82Sc, 83Sc1], as with the purine riboside, the macroscopic pK_a value being almost equal to that of 9-methylpurine [54Al, 54Be, 82Go]. Possibly the N1H,N9H tautomer prevails, since 9-methylpurine has been shown to be slightly more basic than its 7-isomer [83Ar1, 83Ar2]. The subsequent protonations take place at the imidazole ring and the N3 site, respectively, as indicated by ^{15}N and ^{13}C NMR shift changes in D_2O [83Sc1, 84Be1, 88Sl], and by ^1H shift changes under nonaqueous conditions [71Wa]. The pK_a value of the dication is about -1.6 [82Sc, 84Be1], and that of trication < -6 [84Be1].

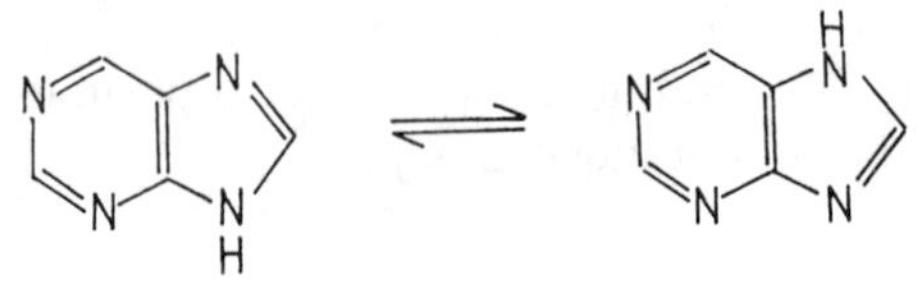

IIa   IIb

2.2 Metal ion complexes

While the purine riboside is protonated exclusively at N1, several metal ions seem to coordinate almost as favourably to N1 and N7 sites, or may even favour N7. For example, it has been shown by ^{1}H NMR relaxation time (T_1) measurements that Cu(II) and Mn(II) ions bind at neutral pH twice as firmly to N7 as to N1 of 9-methylpurine [87Fr]. A similar conclusion was earlier drawn by comparing the stabilities of the 3d transition metal complexes of methyl substituted purines [83Ar2]. The stability constants obtained with 2,9- and 8,9-dimethylpurines are only slightly smaller than those of 9-methylpurine, and one order of magnitude larger than the stabilities of the corresponding complexes of 6,9-dimethylpurine, despite the fact that the basicities of all dimethylpurines are comparable (Table 7.1). Evidently a methyl group at C6 sterically hinders the attachment of metal ions to both N1 and N7 atoms, while the methyl groups at positions 2 and 8 are able to retard coordination to one of these sites only. It has also been shown that the Zn(II) ion shifts all ^{1}H resonances of the base moiety of the purine riboside, the effect on H6 being slightly larger than on H2 and H8 [80Lö], but the significance of this observation may be questioned, since N1 protonation leads to very similar results. The concept of competitive attachment of metal ions to N1 and N7 sites also receives some support from the studies on substitution inert (dien)Pt(II) complexes; ^{1}H NMR spectroscopic measurements have indicated that both of these atoms are coordinated simultaneously upon mixing purine riboside and [(dien)PtCl]Cl in D_2O [74Ko1]. A solid Cu(II) complex of 9-methylpurine, obtained at pH 4, has been shown to exhibit N1,N7 bridging [79Sl].

The data on the binding behavior of unsubstituted purine in solution are limited to the ^{1}H and ^{15}N NMR shift and relaxation measurements, according to which $CuCl_2$ $ZnCl_2$ [66 Wa], and $HgCl_2$ [70Ka] interact with the N7 site of a neutral purine ligand in DMSO-d_6 and the Cu(II) ion with the N7 and N9 sites of a purinium cation in D_2O [88Sl]. The binary Cu(II) complex of a neutral purine is approximately as stable as that of 9-methylpurine, $\log(K/M^{-1}) = 1.9$ ($I = 0.05M$, $T = 25°C$), whereas the N7/N9 monoanion binds Cu(II) ion almost 10^4 times more strongly [69Re1].

In the solid state several different binding modes have been established for a neutral purine. These include N7-bonding in a 1:1 adduct with [Rh(PPh$_3$)$_2$(CO)]PF$_6$ [83Ab1,] and N1,N7-bonding in a 2:1 (M:L) adduct with $(\eta^3 - C_3H_5)Pd(Cl)$ [85Ro]. Moreover, several polymeric complexes have been prepared, but the binding sites were unambiguosly determined by X-ray in one case only. Accordingly, a polymeric Cu(II) complex, crystallized from aqueous ethanol, is known to be bridged through the N7 and N9 atoms of neutral purine bases, which bear a proton at N1 [81Ve1]. A similar binding mode has later been accepted for several other polymeric complexes, containing both neutral [82Mi1, 83Mi1, 83Mi2] and anionic purine ligands [82Mi2, 83Mi3], although in some earlier reports N1,N9 or N3,N9 bridging was preferred [80Sp1, 80Sp2]. N3 binding has been established for a solid dimeric copper(II) chloride complex of a purinium cation [81Sh], and N7 binding from a mononuclear trichloro-(purinium)–zinc complex [82Sh]. The X-ray structure of a purine dication Cu(II) chloro complex has revealed N3 coordination [81Sh].

3.   COMPLEXING OF ADENOSINE AND ADENINE

3.1   Protolytic equilibria

Adenosine (III), the 6-amino derivative of purine riboside, exists as an amino

III

tautomer in solution [61An, 69Wo], solid state [61An, 67Lo1, 82Ta] and gas-phase [78No], as also predicted by molecular orbital calculations [71Pu2, 83Sy]. In fact, the imino population is negligible in aqueous solution, less than 0.01% [69Wo]. The neutral adenosine molecule may thus be treated as a single tautomeric structure.

Adenosine like the purine riboside undergoes N1 protonation, as shown by UV [63Jo], IR [67Lo1], and ^{15}N NMR [77Ma3, 83Go, 87Re2] spectroscopic measurements. Predominance of N1 protonation is also consistent with the X-ray structure of adenosine hydrochloride [73Sh1] and with molecular orbital predictions [83De, 83Mi4]. In all likelihood the N1 site is considerably more basic than the other nitrogen atoms, in spite of the fact that both H2 and H8 resonances of 9-methyladenine undergo comparable shifts upon protonation [86Ta]. Accordingly, the macroscopic pK_a value of the adenosine monoaction, which is 3.6 at 25°C ($I = 0.1$ M) [71Iz, 76Og, 82Ra, 87Tr], may be regarded as the microscopic acidity constant of N1 deprotonation. The higher basicity of N1 of adenosine compared with the purine riboside may be accounted for by the electron-donating ability of the adjacent amino group. The microscopic pK_a value of the N7 monoprotonated species has been estimated on the basis of linear relationships prevailing between acidities and metal complex stabilities of imidazole type ligands. The values obtained fall in the range from -0.5 to 1.1 [84Ki1, 89Ki], in good agreement with the suggested predominance of N1 protonation. However, the fluorescence from adenosine in acidic solution may be regarded as a direct indication of the presence of the N7-monocation as a minor tautomer [63Bö, 67Bö, 82Kn].

The second protonation of adenosine takes place at N7, as shown by ^{13}C NMR spectroscopy [84Be1] and the X-ray structure of 9-methyladenine dihydrobromide [62Br1]. The macroscopic pK_a value of the dication is -1.5 [84Be1, 84Ki1]. In other words, N1 protonation retards N7 protonation by two orders of magnitude. Finally, it is important to note that the primary amino group at C6 is not basic. In fact, no indication of protonation of this site has been detected. The pK_a for the deprotonation of the 6-amino function has been estimated to be 17 [81Ta1].

The tautomeric equilibria of the free adenine base are more complicated than those of adenosine due to annular tautomerism. In inert gas matrices at low temperatures adenine has been shown to be a mixture of N7H (IVb) and N9H/amino (IVa) tautomers [85St]. In aqueous solution the predominant species is the N9H/amino tautomer, the mole fraction of the minor N7H tautomer being about 0.2 [69Ea, 75Ch, 75Dr, 77Dr, 83Go]. Accordingly, the pK_a value of neutral adenine, for which values range from 9.4 to 10.0 ($I = 0 - 0.1$ M, $T = 25°C$) [71Iz, 72Zi, 81Gh, 82Ra, 83Ta1, 88Ga], must be regarded as a macroscopic constant referring to deprotonation of both N7H and N9H tautomers. The acidity of this labile proton is one order of magnitude lower than that with unsubstituted purine, most probably due to electron donation from the 6-amino group, which increases the electron density of the aromatic nucleus.

NH₂ ... IVa ⇌ NH₂ H ... IVb

Protonation behaviour of adenine differs from that of adenosine in that, while N1 protonation prevails [83Go, 71Wa], the N3 site also appears to be rather basic. The ^{15}N NMR signals of both N1 and N3 were observed to shift upfield upon protonation, the influence on N1 resonance being larger (51 ppm) than on N3 resonance (11 ppm) [83Go, 88Sc]. The protonation pattern of adenine resembles that of 7-ethyladenine in this respect [83Go], and the increased basicity of N3 may thus be a feature typical for the N7H tautomer. The macroscopic pK_a value of adenine monocation is somewhat higher than that of adenosine, *viz.* 4.3 at 25°C ($I = 0.1$ M) [71Iz, 72Zi, 81Gh, 82Ra, 83Ta1, 88Be1, 88Ga]. The second protonation of neutral adenine takes place at the imidazole ring, and the third at N3, as shown by ^{1}H [71Wa] and ^{13}C [84Be1] NMR spectroscopy. The pK_a values of the di- and tri-cations have been reported to be -0.4 [74Bu, 88Ga] and -4.2 [84Be1], respectively. The X-ray structures determined for adenine mono- [51Co, 74Ki1] and dications [74Ki2, 78La] are consistent with the suggested protonation order of the nitrogen atoms.

3.2 Metal ion complexes of adenosine

Hard metal ions, like alkali metal [76Pl] and alkaline earth metal ions [66Ha, 80Ma2, 82Bu] do not interact with the base moiety of adenosine, whereas 3d transition metal ions form reasonably stable complexes (Table 7.2). The dichotomy between their binding at N1 and N7 has been one of the most intensively studied subjects in the coordination chemistry of nucleic acid constituents. While crystalline 3d transition metal complexes appear to display almost invariably N7 coordination [74Sl1, 75Sl, 75Sz1, 76Ki, 78Gu], or N1,N7-bridging in polymeric complexes [73Me1, 76Mc, 78Be, 85Mi1, 86Mi2], the importance of N1-binding in solution has been emphasized [84Ki1]. Solid state evidence for this binding mode has been obtained only in one case; the Zn(II) complex of 9-methyladenine displays N1 binding [75Mc]. The absence of N3 coordination among the potential binding modes most probably results from the steric hindrance of the bulky substituent at N9.

Table 7.2 — Logarithmic stability constants for 1:1 metal ion complexes of adenosine (L)[†]

Metal ion	T/°C	I/M	$\log(K/M^{-1})$	Method	Ref.
$M^{z+} + L = ML^{z+}$					
Co^{2+}	25.0	$1.0(NaClO_4)$	0.2(2)	pH stat.	80Lö
Ni^{2+}	25.0	$1.0(NaClO_4)$	0.4(2)	pH stat.	80Lö
	25.0	$0.1(NaClO_4)$	0.3(1)	Potent.	76Ta
	23	$3(NaClO_4)$	0.3	UV	79Ma4
Cu^{2+}	25.0	$1.0(NaClO_4)$	0.96(5)	pH stat.	80Lö
	25.0	$1.5(NaClO_4)$	0.85	T-jump	72Bo
	20.0	$1.0(NaNO_3)$	0.7(3)	Potent.	65Fi
Zn^{2+}	25.0	$1.0(NaClO_4)$	0.2(2)	pH stat.	80Lö
Pb^{2+}	20.0	$1.0(NaNO_3)$	< -0.5	Potent.	65Fi
Ag^+	25.0	$10^{-3}(NO_3^-)$	2.02(2)	Potent.	68Ph
$MeHg^+$	‡	variable	3	UV	64Si1
$(dien)Pd^{2+}$	34.0	$0.5(KNO_3)$	3.9[§]	^{1}H NMR	84Ki2
			4.5[¶]		
$M^{z+} + LH^+ = ML^{(z-1)+}$					
Cu^{2+}	21.0	$1.0(NaClO_4)$	0.16	UV	84Ki1
Zn^{2+}	21.0	$1.0(NaClO_4)$	-0.89	UV	84Ki1
$(dien)Pd^{2+}$	34.0	$0.5(KNO_3)$	2.0	^{1}H NMR	84Ki2

[†]See also Ref. 64Sc and 70He.
[‡]Temperature not indicated.
[§]For N7 binding.
[¶]For N1 binding.

Although N1 is the most basic atom in adenosine, the basicity of the N7 site is only 2 or 3 orders of magnitude lower [84Ki1, 89Ki]. Kim & Martin [84Ki1] concluded on the basis of linear stability–basicity correlations observed for pyridine type and imidazole type nitrogens, that Cu(II) and Ni(II) ions prefer N1 over N7, and the Zn(II) ion coordinates as readily to both sites. While N1 coordination undoubtedly plays a significant role in solution, the predominance of this binding mode is not in complete agreement with all experimental data currently available. In particular, several NMR spectroscopic observations in DMSO-d_6 seem to be consistent with N7 coordination. Firstly, Zn(II) ions induce considerably larger upfield shifts on the ^{15}N resonance of N7 than on that of N1 [82Bu]. Secondly, Cu(II) ions broaden specifically ^{13}C NMR signals of C4, C5 and C8, leaving those of C2 and C6 unchanged [74Fr, 78Ma1]. Thirdly, longitudinal ^{1}H resonance relaxation studies have indicated twice as efficient binding to N7 as to N1 [75Ma1]. Finally, ^{1}H signal shift [86Ta] and broadening studies [66Ei, 71Be1, 78Ma1, 79Fa], though shown to be susceptible to misinterpretation [74Es, 76Es1, 76Es2], agree with N7-binding. Having an extra methyl group at C2 or C8 of 9-methyladenine affects the stabilities of the Ni(II) and Cu(II) complexes in a manner consistent with the attachment of the metal ions predominantly at N7 [85Ar1]. For example, the logarithmic stability constants obtained for the Ni(II) complexes of 9-methyl, 2,9-dimethyl and 8,9-dimethyl-adenine were 0.7(2), 0.6(2) and <0.1 ($I = 1.0$ M, $T = 25$°C), respectively. Accordingly, the influence of the 8-methyl group is much larger than that of the 2-methyl group, suggesting that this substituent is situated adjacent to the main coordination site and hence sterically retards complex formation. Possibly the approach of Kim & Martin overestimates the complexing ability of the N1 site, since the stability–basicity correlations employed for this site were based on *m-* and *p-* substituted pyridines. However, in adenosine there is an amino group in the *ortho* position to the donor atom. Since the size of an *o*-substituent plays a more important role in complexing than in protonation, the stability constant obtained for the N1-complex may be too large. The importance of the steric hindrance caused by the 6-amino group is clearly demonstrated by the fact that adenosine, though considerably more basic than the purine riboside, forms less stable complexes with 3d transition metal ions (see Tables 7.1 and 7.2). The finding that N^6,N^6-dimethyladenosine does not form a detectably stable complex with Ni(II) ion may also be taken as a strong indication of the relevance of steric factors in complex formation [85Ar1]. Sigel [89Ki] has recently applied the approach of Kim & Martin by taking the steric hindrance of the 6-amino group into account. The results obtained suggest that N7 coordination is favoured, the microscopic $\log(K/M^{-1})$ values of N1 and N7 coordination being 0.28 and 0.59 with Ni(II), and 0.59 and 0.87 with Cu(II).

According to early observations $ZnCl_2$ induced large downfield shifts on the ^1H resonance of the amino protons of adenosine in DMSO-d_6, which led to the proposal of a direct involvement of the N^6 atom in metal ion binding [68Wa, 73Sh2]. However, crystal structures of Cu(II) [74Sl1, 75Sl, 76Ki] and Zn(II) complexes [76Mc] of 9-substituted adenines display only indirect chelate formation through hydrogen-bonding of the 6-amino group to aquo or chloride ligands of the N7-coordinated metal ion. IR analysis of the solid Cu(II) complex of adenosine also argues against metal ion bonding to N^6 [79Ne1]. The large shifts of amino protons that accompany the addition of $ZnCl_2$ have later been attributed to interactions with weakly solvated chloride ions [74Ch, 76Pl], consistent with the solid state structures. The fact that adenosine forms less stable complexes with 3d transition metal ions than unsubstituted purine riboside strongly argues against chelate formation through the N^6 and N7 atoms [80Lö]. A chelate formation would be expected to increase the complex stability. By comparison, the complexes of 1,N^6-ethenoadenosine (V), for which such a chelate structure has been suggested, are several orders of magnitude more

V

stable than those of the purine riboside [87Si]. ^1H resonance studies on the structure of Cu(II) complexes have given contradictory results; Berger & Eichhorn [71Be1] excluded the involvement of N^6 atom, whereas Fazakerley *et al.* [79Fa] interpreted their data in terms of a relatively strong N^6, N7-chelate formation. The results of *ab initio* SCF calculations are also conflicting. While Pullman *et al.* [79Pu] considered N^6,N7-chelation as the preferred binding mode of the Zn(II) ion, the results of Anwander *et al.* [87An] were in better agreement with unidentate binding of the Zn(II) ion, either to N1 or N3. To summarize, the evidence against a direct chelate formation seems to be stronger than that presented to support this binding mode. In all likelihood the predominance of N7 binding does not result from the interaction of the N7 coordinated metal ion with the amino substituent. It is worth noting that the distribution of metal ions between the N1 and N7 sites is rather similar in adenosine and purine riboside [87Fr], although the latter does not have a substituent at C6.

As seen from Table 7.2, the complex formation is markedly enhanced on going from 3d transition metal ions to softer Lewis acids, like Ag(I), Hg(II), and Pd(II) ions, which usually prefer binding to nitrogen ligands. While 3d transition metal ions favour N7 over N1 coordination, the opposite seems to be the case with soft metal ions. The early UV spectroscopic studies of Simpson [64Si1] showed that the MeHg(II) ion binds to N1 of adenosine under slightly acidic solutions and displaces a proton from the 6-amino group under basic conditions. Moreover, an N1,N^6-dimercurated species was formed at high MeHg(II) ion concentrations. In fact, binding of MeHg(II) ion to N1 facilitates the displacement of the N^6 proton by almost one order of magnitude. The assignment of N1 as a binding site has later been corroborated by ^{15}N [82Bu] and ^{13}C resonance [75Je, 80Ma2] shift studies in DMSO-d_6 and by Raman spectroscopic measurements [74Ma1, 79Ma2] in aqueous solution at pH 3.5. Formation of an N1,N^6-dicoordinated MeHg(II) complex has, in turn, been verified by ^{1}H NMR spectroscopy [79Ho]. N1-coordinated [80Ol], N^6-coordinated [81Ta1], and N1,N^6-dicoordinated [79Pr, 82Ol] MeHg(II) complexes of 9-methyladenine have all been prepared in the solid state. Moreover, a crystalline N^6,N^6-di-MeHg(II) complex of 9-methyladenine is known [85Ch1], and the corresponding N^6-MeHg(II) complex has been shown to disproportionate under nonaqueous conditions to this species and to uncomplexed 9-methyladenine [85Ch2].

Pd(II) complexes of adenosine have been studied rather extensively as models of substitution inert Pt(II) complexes, which exhibit a marked anticarcinogenic activity. The ligand exchange reactions of Pd(II), although several orders of magnitude faster than those of Pt(II), are still so slow that complexes with different binding modes give separate patterns of ^{1}H NMR signals, enabling a quantitative estimation of their proportions in solution. (dien)Pd(II) [76Li1, 84Ki2], (en)Pd(II) [83Hä1], and (gly-L-asp)Pd(II) ions [77Ko] have been shown to form two mono-nuclear complexes with the metal ion bound at N1 or N7 of adenosine, and a binuclear complex with both sites occupied. Under conditions where adenosine is present as an uncharged molecule, the N1 site is favoured over the N7 site, the microscopic log (K/M^{-1}) values being 4.7 and 3.9, respectively (see Table 7.2). Coordination of one (dien)Pd(II) ion at either the N1 or N7 site reduces the binding affinity of the remaining site by one log unit [84Ki2]. In acidic solutions protonation of N1 prevents binding to this site and lowers the stability of the N7 coordinated species by two orders of magnitude. At high pH the (en)Pd(II) ion has been shown to displace a proton of the 6-amino group with the concomitant formation of a binuclear dimer, $[(en)Pd]_2L_2^{2+}$, where each of the Pd(II) ions is bound to N1 of one adenosine ligand and to deprotonated N^6 of the other [83Hä1]. For several solid complexes both N1- [77Et], N7- [83Ca, 83Pn, 88Pn] and N1,N7-coordination [80Pn, 82Pn, 84Pn1] have been suggested on the basis of ^1H and ^{13}C resonance shift measurements in DMSO-d_6.

The data on binding modes of other soft substitution labile metal ions are limited to solid state structures of Cd(II), Ag(I), Rh(I), and Rh(II) complexes, which all show different binding behaviour. The Cd(II) complex of 9-methyladenine was crystallized from a mixture of DMSO and DMF as a chloride-bridged polymer, consisting of N7 coordinated mononuclear units [82Gr]. The Ag(I) complex was obtained as a N1,N7-bridged chain-like polymer from aqueous solution [75Ga, 77Ga1]. $[(PPh_3)_2Rh(CO)]^+$ ion was suggested to be N1 coordinated [83Ab1] and $[Rh(CO)_2]^+$ ion N7 coordinated [83Si] in their solid adenosine complexes, whereas dimeric rhodium(II) acetate formed N1,N7-bridged polymeric structures [79Pn, 80Fa].

3.3   Metal ion complexes of adenine

The complexes of neutral adenine are considerably more stable than those of adenosine (Tables 7.2 and 7.3), most probably owing to availability of an additional binding site, *viz.* N9 of the N7H tautomer (IVb). Deprotonation of the imidazole ring (pK_a 9.5) enhances the affinity of this site to metal ions, as can be seen from the stability being 4 orders of magnitude larger for the complexes of the adeninate anion compared with those of neutral adenine (Table 7.3).

The assumed N9 coordination receives considerable support from the solid state structures of adenine complexes. In particular, the structures of Cu(II) complexes have been extensively studied by X-ray crystallography, and two alternative binding modes have been established. Firstly, the metal ion may display monodentate binding to the N9 atom of either the adeninate anion [78Sa], neutral (N7H)adenine [73To] or the (N1H,N7H)adeninium cation [71Me1, 73Me2, 77Br1]. Secondly, a binuclear 2:2 complex may be formed, where the Cu(II) ions are bound to N9 of one ligand molecule and to N3 of the other. The bridging ligand is either the adeninate anion [67Sl, 69Sl1], neutral (N7H)adenine [71Me2, 71Me3, 73Te] or the (N1H,N7H)adeninium cation [72Me]. The logarithmic stability constants, $\log(K/M^{-3})$, of the binuclear Cu(II) complex of the adeninate anion, $Cu_2L_2^{2+}$, has been reported to be 17.9 ($I = 0.15$ M, $T = 37°C$) [86Ta]. A solid trinuclear Cu(II) complex of the (N1H,N7H)adeninium cation, $Cu_3Cl_8(LH_2)_2$, has also been prepared, and shown to exhibit N3,N9-bridging [70Me].

Among the other solid complexes studied, the Co(II) complex of neutral (N7H)adenine [73Me3, 73Me4], and the $(PPh_3)Au(I)$ [85Bo, 85Ro] and *trans*-(n-Bu$_3$P)Pd(II) [79Be] complexes of the adeninate anion have been established to display a monodentate N9-coordination, while Cd(II) [81Ch] and Ag(I) [77Ga1] ions form dimeric 2:2 complexes via N3,N9-bridging of the (N1H,N7H)adeninium cation. Ag(I) ion has also been suggested to form in dilute aqueous solution chain-like structures *via* N1,N9-bridging of the adeninate anion [84Ma1, 85Ma2]. The only well-documented case where the N9 site does not participate in binding, seems to be the trichloroadeniniumzinc(II) complex, where the metal ion is attached to the N7 site [70Sr, 73Ta].

Table 7.3 — Logarithmic stability constants for 1:1 metal ion complexes of adenine (LH)[†]

Metal ion	T/°C	I/M	$\log(K/M^{-1})$	Method	Ref.
$M^{z+} + LH = MLH^{z+}$					
Mg^{2+}	35.0	$0.1(KNO_3)$	2.71	Potent.	83Ta1
Ca^{2+}	35.0	$0.1(KNO_3)$	1.86	Potent.	83Ta1
Co^{2+}	35.0	$0.1(KNO_3)$	1.90	Potent.	83Ta1
Ni^{2+}	25.0	$0.1(NaCl)$	1.38	UV	82Or
	35.0	$0.1(KNO_3)$	2.45	Potent.	83Ta1
Cu^{2+}	25.0	$0.05(NaClO_4)$	2.7	Potent.	69Re1
	26.0	$0.5(NaClO_4)$	2.70	T-jump	73Bo
	35.0	$0.1(KNO_3)$	2.84	Potent.	83Ta1
	37.0	$0.15(NaNO_3)$	2.25(6)	Potent.	86Ta
Zn^{2+}	35.0	$0.1(KNO_3)$	2.17	Potent.	83Ta1
	37.0	$0.15(NaNO_3)$	0.6(3)	Potent.	86Ta
$M^{z+} + L^- = ML^{(z-1)+}$					
Be^{2+}	30.0	$0.1(KNO_3)$	6.77	Potent.	72Na
Al^{3+}	30.0	$0.1(KNO_3)$	7.56	Potent.	72Na
Fe^{3+}	30.0	$0.1(KNO_3)$	10.4	Potent.	72Na
Co^{2+}	25.0	$0.005(ClO_4^-)$	4.2	Potent.	58Ha
Ni^{2+}	25.0	$0.005(ClO_4^-)$	4.8	Potent.	58Ha
	25.0	$0.1(KNO_3)$	5.3	T-jump	71Ka
	25.0	$0.1(KNO_3)$	4.52	Potent.	80Gh
	20.0	var., $<10^{-3}$	4.37	Potent.	60Al
	30.0	$0.1(KNO_3)$	4.54	Potent.	72Na
Cu^{2+}	25.0	$0.005(ClO_3^-)$	7.1	Potent.	58Ha
	25.0	$0.05(NaClO_4)$	6.99	Potent.	69Re1
	25.0	$0.1(KNO_3)$	6.77(2)	Potent.	81Gh
	30.0	$0.1(KNO_3)$	6.56	Potent.	72Na
Th^{2+}	35.0	$0.1(KNO_3)$	10.3(1)	Potent.	82Ra
Uo_2^{2+}	30.0	$0.1(KNO_3)$	7.53	Potent.	72Na
	35.0	$0.1(KNO_3)$	8.38(3)	Potent.	82Ra

[†]See also Refs 71Iz, 71Ta, and 77Ta.

Formation of a Ni(II) complex of the adeninate anion is abnormally slow [71Ka, 75Ca]. It has been assumed that the metal ion initially attacks neutral adenine and the complex then undergoes deprotonation and subsequent rate-limiting ring-closure to a bidentate chelate [71Ka]. Alternatively, initial hydrogen bonding of neutral adenine through the N1 atom to a water molecule of the inner sphere of Ni(II) ion has been suggested to result in a misorientation, which slows the reaction [75Ca].

N9 coordination also predominates in complexing of soft metal ions with unsubstituted adenine, as has been demonstrated by X-ray structures of a series of MeHg(II) complexes prepared under aqueous conditions. These extensive studies clearly indicate that the adeninate anion is consecutively coordinated at N9 [81Pr1, 84Ch], N7 [79Pr, 80Be1, 82Pr, 86Ch1] and N3 [80Hu1, 80Hu2, 86Ch1], when treated with $MeHgClO_4$ or $MeHgNO_3$. Under alkaline conditions the amino protons may be displaced yielding N9,N[6]-dimercurated [87Ch], N9,N[6],N[6]-trimercurated [83Ch,

86Ch2] or N3,N9,N^6,N^6-tetramercurated [86Ch2] complexes. Analogously, N^6,N^6-dialkyladenine binds MeHg$^+$ cations at N9 and N3 [88Gr]. The lack of N1 among the potential binding sites is rather surprising, since this binding mode prevails with 9-methyladenine. It has been suggested that displacement of N9H by MeHg$^+$ tends to promote subsequent coordination at N7 instead of N1 [79Pr], and that mercuration of the primary amino group sterically prevents coordination to N1 and N7 sites [86Ch2]. Still, it is noteworthy that mercuration of N9 and N7 is followed by attachment of the third MeHg$^+$ ion at N3 and not at N1.

4. COMPLEXING OF 6-OXOPURINES AND THEIR NUCLEOSIDES

4.1 Protolytic equilibria

6-Oxo-substituted purine nucleosides, inosine (VI) and guanosine (VII), exist as oxo and oxo/amino tautomers in solution [58Mi, 61Mi, 67Lo1, 69Ts, 71Me4, 71Ps, 72Wo, 73Pi, 73Wo], as well as in the solid state [61An, 82Ta]. Accordingly, the neutral forms of these nucleosides bear a dissociable proton at N1, the pK_a value of which is reported to range from 8.8 to 9.0 ($I = 0 - 0.1$ M, $T = 25.0°C$) with inosine [70Ch1, 71Ps, 83Ra] and from 9.0 to 9.7 ($I = 0 - 0.1$ M, $T = 25.0°C$) with guanosine [70Ch1, 76Og, 81Gh, 88Kh]. The other annular tautomers, bearing a proton at N3 or N7, are much less stable. For example, the intrinsic tendency to bind a proton at N1 over N7, has been estimated to be 5.6 and 5.1 log units with inosine and guanosine, respectively [84Ki1]. In other words, the microscopic pK_a values of N7H tautomers are 3.2 (inosine) and 4.1 (guanosine) [84Ki1]. However, in inert matrices the enol form prevails, the oxo–enol equilibrium constant reported for 9-methylguanine being 5.9 [87Sz].

VI   VII

Neutral 6-oxopurine nucleosides are protonated at N7 with the formation of an N1H,N7H monocation, as shown by IR [63Mi1, 88Pa], UV [61Pf], and ^{15}N NMR [77Ma3] spectroscopy, and X-ray crystallography [64So1, 74To, 75Ma2]. The macroscopic pK_a value of the guanosine monocation range from 1.9 to 2.3 at $I = 0 - 1.0$ M ($T = 25.0°C$) [66Bu, 70Ch1, 76Og, 81Lö], while inosine is about one order of magnitude less basic (pK_a 1.39 for VI and 2.33 for VII at $I = 1.0$ M, $T = 25°C$ [81Lö]. A comparison of these data with the microscopic pK_a values of the N7H tautomers of neutral inosine and guanosine reveals that protonation of N1 reduces the basicity of N7 by almost 2 log units. The influence is thus comparable to that observed with adenosine. The pK_a value of the guanosine dication, which probably bears protons at N3 and N7 in addition to N1, has been reported to be -2.4 [70Zo]. As with adenosine, the primary amino substituent of guanosine is not a potential protonation site. The pK_a value for its deprotonation is about 15 [81Ta1].

The 6-oxo substituted purine bases, hypoxanthine and guanine, have been crystallized as N1H,N9H oxo/amino tautomers, (VIIIa) and (IXa), [61An, 68Bu, 68Sl, 71Th], but in solution they exist as mixtures of N1H,N7H (VIIIb,IXb), and N1H,N9H oxo/amino tautomers [65Cl, 68Sh, 72Li, 75Ch, 78De], consistent with quantum chemical predictions [71Pu2]. In the gas phase the N1H,N7H oxo/amino tautomer prevails [80Li], while in inert gas matrices the N7H/enol tautomer has been suggested to exist besides the N1H,N7H, and N1H,N9H oxo/amino forms [85Sh1, 87Sz, 87Sh2]. The protonation pattern of 6-oxopurines is analogous to that of the corresponding nucleosides. In other words, the basicity of the nitrogen atoms decreases in the order N7/N9(dianion) > N1(monoanion) > N7/N9(neutral species) > N3(monocation) [63Mi1, 65Ib, 68Sh, 69Sl2, 71Wa, 75Bo, 83De, 85Be1]. The macroscopic pK_a values for the consecutive protonation steps are 11.3 [88Ga], 9.0 [88Ga], 3.3 [48Ta], and -1.1 [74Bu, 85Be1, 88Ga] with guanine and 12.1 [54Al, 70Ch1], 8.9 [69Re1, 70Ch1, 81Gh], 1.9 [70Wo, 76Li2], and -3.7 [85Be1] with hypoxanthine. Accordingly, the basicity difference between the base and the corresponding nucleoside is considerably larger than with adenosine and purine riboside.

VIIIa   VIIIb   IXa   IXb

4.2   Metal ion complexes of guanosine and inosine

Neutral guanosine and inosine molecules interact only weakly, if at all, with alkali [76Pl] or alkaline earth metal cations [81Lö, 82Bu, 82Ma2, 83Bu1] in solution. The early ^{1}H NMR shift observations [72Jo, 73Sh2] on complexing of alkaline earth metal chlorides with guanosine in DMSO-d_6, have later been attributed to hydrogen-bonding of free chloride ion to the N1H and N^{2}H$_2$ sites [74Ch, 75Yo, 76Pl]. Similarly, the relatively large stability constants ($\log(K/M^{-1}) \sim 2$) reported for the alkaline earth metal complexes of guanosine in aqueous solution [88Kh] have later been criticized, since these values do not fall on the basicity–stability correlation line based on related ligands [89Ki]. It is noteworthy, however, that solid Li(I) [88Pa] and Mg(II) [85Ba] ion complexes of neutral 6-oxopurine derivatives have been detected to display N7 binding.

The 3d-transition metal complexes of neutral guanosine and inosine are approximately as stable as those of adenosine and purine riboside (Table 7.4). In these complexes N1 remains protonated, while the metal ion is attached to N7, as shown by ^{1}H [68Wa], ^{13}C [82Ma2] and ^{15}N [82Bu, 83Bu1] NMR shift measurements with

Table 7.4 — Logarithmic stability constants for 1:1 metal ion complexes of guanosine and inosine (LH)[†]

Metal ion	T/°C	I/M	$\log(K/M^{-1})$	Method	Ref.
$M^{z+} + LH = MLH^{z+}$					
Guanosine:					
Co^{2+}	25.0	$1.0(NaClO_4)$	1.0(1)	pH stat.	81Lö
Ni^{2+}	25.0	$1.0(NaClO_4)$	1.4(1)	pH stat.	81Lö
Cu^{2+}	25.0	$1.0(NaClO_4)$	1.9(1)	pH stat.	81Lö
	20.0	$1.0(NaNO_3)$	2.15(4)	Potent.	65Fi
Zn^{2+}	25.0	$1.0(NaClO_4)$	0.8(1)	pH stat.	81Lö
	21.0	$1.0(NaClO_4)$	0.87	UV	84Ki1
Pb^{2+}	20.0	$1.0(NaNO_3)$	0.5(2)	Potent.	65Fi
$MeHg^+$	‡	variable	4.5	UV	64Si1
Th^{4+}	20.0	$1.0(NaNO_3)$	0.9(1)	Potent.	65Fi
UO_2^{2+}	20.0	$1.0(NaNO_3)$	0.7(2)	Potent.	65Fi
Inosine:					
Co^{2+}	25.0	$1.0(NaClO_4)$	0.8(2)	pH stat.	81Lö
Ni^{2+}	25.0	$1.0(NaClO_4)$	1.1(2)	pH stat.	81Lö
	15.0	$0.1(NaClO_4)$	1.10	T-jump	81Na
	15.0	$1.0(NaClO_4)$	1.15	UV	81Na
	10.0	$0.4(NaClO_4)$	1.27	st.-flow	75Ca
Cu^{2+}	25.0	$1.0(NaClO_4)$	1.3(1)	pH stat.	81Lö
Zn^{2+}	25.0	$1.0(NaClO_4)$	0.7(1)	pH stat.	81Lö
	21.0	$1.0(NaClO_4)$	0.34	UV	84Ki1
$MeHg^+$	‡	variable	3.7	UV	64Si1
$(dien)Pd^{2+}$	25.0	$0.2(NaClO_4)$	5.2	st.-flow	88Me
	34.0	$0.5(KNO_3)$	5.34	^1H NMR	84Ki2
$M^{z+} + L^- = MI^{(z-1)+}$					
Guanosine:					
Co^{2+}	25.0	$0.1(KNO_3)$	3.25	Potent.	80Gh
Ni^{2+}	25.0	$0.1(KNO_3)$	3.90	Potent.	80Gh
C^{2+}	25.0	$0.1(KNO_3)$	5.32(2)	Potent.	81Gh
	20.0	$1.0(NaNO_3)$	4.3(4)	Potent.	65Fi
	25.0	$0.05(NaClO_4)$	5.58[§]	Potent.	69Re2
Pb^{2+}	20.0	$1.0(NaNO_3)$	3.5	Potent.	65Fi
$MeHg^+$	‡	variable	8.1	UV	64Si1
Inosine:					
Co^{2+}	25.0	$1.0(NaClO_4)$	2.1	pH stat.	81Lö
Ni^{2+}	25.0	$1.0(NaClO_4)$	2.8	pH stat.	81Lö
Cu^{2+}	25.0	$1.0(NaClO_4)$	4.5	pH stat.	81Lö
Zn^{2+}	25.0	$1.0(NaClO_4)$	2.4	pH stat.	81Lö
$MeHg^+$	‡	variable	8.2	UV	64Si1
$(dien)Pd^{2+}$	34.0	$0.5(KNO_3)$	8.3[¶]	^1H NMR	84Ki2
			6.8[††]		

[†]See also Ref. 82Or, 83Ra, 84Ma2, 87Ra and 88Kh.
[‡]Temperature not indicated.
[§]For 2′-Deoxyguanosine.
[¶]N1 coordination.
[††]N7 coordination.

Zn(II) ion, ^{13}C NMR relaxation time measurements with Mn(II) [74Fr] and Cu(II) [74Fr, 78Ma2, 81Ma1] ions, ^{1}H NMR relaxation time measurements with Cu(II) ion [66Ei, 70Fr1, 70Fr2, 71Be2, 78Ma2, 79Fa, 78Ma3, 81Ma1]. This binding mode has also been verified by crystal structures of the Cu(II) complexes of 9-methylhypox-anthine [71Sl, 75Sl2, 77Sl1] and 9-methylguanine [76Sl1, 77Sl2], which display a monodentate N7 coordination and hydrogen-bonding to O^6 through one of the water molecules in the inner sphere of the Cu(II) ion. Most probably the N7,O^6 chelate formation, if present at all, is mediated by a water molecule also in solution. As mentioned above, the complexes of 6-oxopurine nucleosides are only slightly more stable than those of the unsubstituted purine riboside. A more marked stabilization could be expected, if both the N7 and O^6 atoms were in the inner coordination sphere of the metal ion, since the N7 atom of 6-oxopurine nucleosides is at least as basic as the corresponding site in purine riboside (compare the discussion on the structure and stability of proton complexes). Moreover, only unspecific broadening of the carbonyl stretch took place upon addition of Cu(II) ions in a D_2O solution of guanosine [68Fr, 71Fr] or inosine [70Fr1]. The early observations on the significantly broadened carbonyl band of the Cu(II) complex [68Tu] has more recently been accounted for by interaction of the metal ion with the N1 site [78Ma2, 78Ma3]. Accordingly, the situation in solution seems similar to that in the solid state; the cehlate formation is indirect, if at all. Nevertheless, one should bear in mind that according to *ab initio* SCF calculations N7,O^6 chelation is the energetically favoured binding mode [77Pe1, 79Pu, 83Sa1, 84De, 87An].

The possible chelate formation between the N7 and O^6 sites has also been approached by studying the interactions of the Cu(II) ion with some structural analogues of guanosine and inosine, but no firm evidence for the direct participation of the O^6 atom in the complex formation has been obtained. For example, ^{1}H resonance relaxation measurements indicated that the Cu(II) ion coordinates to N7 of xanthosine (X), but the carbonyl group stretching band remained unchanged in the freeze-dried samples [78Ma4]. Some mixed ligand complexes involving a theophyllinate anion (XI) as a primary ligand have been shown to exhibit a roughly square planar geometry about Cu(II) ion, where the equatorial positions are occupied by N7 of theophylline and three donor atoms of the secondary ligand, while the O^6 atom is loosely bound to an axial position [75Ki1, 76Sz, 78Ki]. Even in these complexes the $Cu-O^6$ bond is much longer than the $Cu-N7$ bond. In a $CuCl_2$ complex of neutral theophylline no clear indication of $Cu-O^6$ interaction could be obtained [79Ci, 83Ci]. The X-ray structure of the N7-coordinated Cd(II) complex of the theophyllinate anion displayed an indirect water mediated chelation to O^6 [85Bu1].

The 3d transition metal complexes of the monoanions of guanosine and inosine are from 2 to 3 orders of magnitude more stable than those of the corresponding neutral nucleosides. While N7-coordination prevails under conditions where the N1 atom remains protonated, attachment of metal ions to the deprotonated ligands takes place at N1. The gradual change from N7-binding to N1-binding with increasing pH has been rather extensively documented by ^{1}H NMR [71Be2, 78Ma2, 81Ma1] and EPR [77Ch1, 85Ma3] spectroscopic measurements. It has also been shown by ^{13}C NMR spectroscopy that addition of triethylamine in the DMSO-d_6 solution of guanosine changes the coordination site of the Zn(II) ion from N7 to N1 [82Ma2], and that the Zn(II) ion prefers the N1 atom over the O^6 atom in complexing with 7-methylinosine monoanion (XII) [85Sh2]. IR data on solid Cu(II) complexes of 6-oxopurine nucleosides obtained under basic conditions are consistent with the attachment of the metal ion at the N1,O^6-region, as well [79Ne1, 79Ne2, 80Ne1]. Accordingly, the site of coordination is pH-dependent. With inosine the crossover pH values have been reported to be: Ni(II) 7.1, Cu(II) 6.1 and Zn(II) 6.7 [84Ki1]. With guanosine the values are slightly higher: Ni(II) 7.8, Cu(II) 6.9 and Zn(II) 7.5.

MeHg(II) ion complexes with 6-oxopurine nucleosides much more strongly than 3d transition metal ions (Table 7.4), but the binding sites remain unchanged. In other words, N1 coordination predominates under neutral conditions and N7 coordination under acidic conditions, as indicated by UV [64Si1], Raman [75Ma3] and [1]H [75Ma3], [13]C [75Je, 82Ma2], and ^{15}N [82Bu, 83Bu1] NMR spectroscopy. However, N1 binding is enhanced more markedly than N7 binding, compared with 3d transition metal ions. Accordingly, the crossover pH is considerably lower than with Ni(II), Cu(II), and Zn(II) ions, *viz.* 4.3 and 5.6 in the case of inosine and guanosine, respectively. At high concentrations of MeHg(II) ion a N1,N7-dimercurated complex is formed [75Ma3]. Furthermore, the MeHg(II) ion may displace one of the N^2H_2 protons of guanosine under alkaline conditions giving an N1,N^2-dimercurated species [64Si1]. All these binding modes have been verified by preparing solid complexes at an appropriate pH and metal ion concentration, and characterizing their structures by IR and ^1H and ^{13}C NMR spectroscopy [79Ca, 79Pe, 80Ca, 81Bu, 81Ta1, 85Ba]. In addition X-ray structures have been reported for the N7-MeHg(II) complex of 9-methylguanine [80Ca], N7-Hg(II) complex of guanosine [78Au] and N1,N7-bis-MeHg(II) complex of inosine N1 monoanion [82Be]. No evidence for N7,O^6-chelate formation could be detected either in these complexes or in the MeHg(II) [84No1, 84No2] or (β-methoxyethyl)Hg(II) [84Ca1] complexes of the theophyllinate anion.

The binding behaviour of the Ag(I) ion has been repeatedly suggested to differ from that of MeHg(II) and Hg(II) ions. On the basis of IR [66Tu, 84Ma2], UV [84Ma2], and ^1H NMR [77Ci] spectroscopic data a cyclic binuclear structure has been proposed, in which each metal ion is coordinated to N7 of one ligand and to enolized O^6 of the other. However, 9-methylhypoxanthine was crystallized from aqueous solution at pH 4–5 as a 2:1 (L/M) complex, which displayed N7-coordination without involvement of the carbonyl group [84Ao1, 80Be2]. For solid Au(I) complexes both unidentate N7 coordination [84Ca2, 81Ha] and bidentate N1,N7-bridging has been reported [85Ro].

^1H, ^{13}C, and ^{15}N resonance shift measurements unequivocally indicate that the (en)Pd(II) ion binds, like 3d transition metal ions and the MeHg(II) ion, to the deprotonated N1 atom of anionic 6-oxopurine nucleosides [76Ne, 83Hä1, 89Uc] and to the N7 atom of neutral ligands [80Ro, 83Hä1]. In a wide pH range around neutrality both binding modes occur [83Hä1], leading possibly to formation of a 4:4 cyclic structure [89Uc]. With the monofunctional (dien)Pd(II) ion the crossover pH for N1 and N7 coordination has been reported to be 6.1 in the case of inosine [81Sc]. The intrinsic tendency of binding to N1 and N7 sites is thus similar to that of 3d transition metal ions, although the complexes are much more stable (Table 7.4). Binding of one (dien)Pd(II) ion to the N7 site increases the acidity of the N1H by 1.5 log units and retards coordination of another (dien)Pd(II) ion to the deprotonated N1 by 1 log unit [84Ki2]. Replacement of the N1H by (dien)Pd(II) facilitates, in turn, the N7 coordination by 0.5 log unit.

Several solid Pd(II) complexes of guanosine and inosine have been suggested to exhibit N7,O^6 chelation [77Pn, 78Ha, 78Pn, 83Ca, 83Pn, 84PC2, 88Pn], although dimeric and polymeric structures have not always been rigorously excluded [83Pn, 84Pn2] and sometimes even regarded as the most probable binding mode [82Pn, 84Pn1]. The proposed chelation is not, however, in a complete agreement with the easy formation of 2:1 (L/M) (en)Pd(II) complexes in neutral solutions, which shows that N7 coordinated (en)Pd(II) prefers interligand N7 binding over intraligand chelate formation [76Ne, 83Hä1]. The complexes obtained under acidic solution exhibit N7 coordination [77De, 77Pn, 78Ha, 80Pn, 85Ro, 88Pn], consistent with the results in solution. For solid Rh(I) complexes prepared in organic solvents both N1 [83Ab1, 83Ab2,] and N7 [83Si, 87Pn] coordination has been suggested.

4.3   Metal ion complexes of guanine and hypoxanthine

The metal ion complexes of both neutral hypoxanthine and its monoanion are approximately one order of magnitude more stable than those of neutral and anionic inosine (Table 7.5). For guanine complexes no stability constants are available.

Table 7.5 — Logarithmic stability constants for 1:1 metal ion complexes of hypoxanthine $(LH_2)^\dagger$

Metal ion	T/°C	I/M	$\log(K/M^{-1})$	Method	Ref.
$M^{z+} + LH_2 = MLH_2^{z+}$					
Cu^{2+}	25.0	$0.05(NaClO_4)$	1.85	Potent.	69Re1
$M^{z+} + LH^- = MLH^{(z-1)+}$					
Mn^{2+}	25.0	$0.1(KNO_3)$	2.49	Potent.	80Gh
	25.0	$0.1(NaClO_4)$	3.50	Potent.	79Ra1
Co^{2+}	25.0	$0.1(KNO_3)$	4.00	Potent.	80Gh
	25.0	$0.1(NaClO_4)$	3.67	Potent.	79Ra1
Ni^{2+}	25.0	$0.1(KNO_3)$	4.82	Potent.	80Gh
	25.0	$0.1(NaClO_4)$	4.10	Potent.	76Li2
	25.0	$0.1(NaClO_4)$	4.55	Potent.	79Ra1
Cu^{2+}	25.0	$0.1(NaClO_4)$	6.54	Potent.	69Re1
	25.0	$0.1(KNO_3)$	5.80(1)	Potent.	81Gh
	25.0	$0.1(NaClO_4)$	6.22	Potent.	79Ra1
Zn^{2+}	25.0	$0.1(NaClO_4)$	5.20	Potent.	79Ra2
Cd^{2+}	25.0	$0.1(NaClO_4)$	4.44	Potent.	79Ra2
Hg^{2+}	25.0	$0.1(NaClO_4)$	7.32	Potent.	79Ra2

†See also Ref. 71Ta, 82Or, 83Ta1 and 84Pn1.

Experimental data on predominant binding sites in these complexes are limited to characterization of their solid state structures. X-ray analyses have revealed 3 different binding modes for mononuclear complexes. Firstly, Cu(II) [70Ca, 71De, 71Su1], and Zn(II) [70Sr] ions display a monodentate N9 binding in their complexes with guaninium cation. Secondly, Co(II) and Ni(II) ions display a monodentate N7 binding in their complexes with neutral hypoxanthine [87Du]. Thirdly, Cu(II) ion forms a N3,N9-chelate with neutral hypoxanthine [70Sl]. N9-binding has also been suggested for several guanine complexes, *viz.* $Cu(LH_3)Cl_3$ [77Vi], $Zn(LH_3)_2Cl_4$ [78Sh1], $[Fe(LH_2)_2(EtOH)_2(ClO_4)_2]ClO_4$ [83Mi5], and $Pd(LH_2)Cl_2 H_2O$ [86Ba] (guanine = LH_2), on the basis of spectroscopic observations. The versatility of the binding behaviour is further increased by various bonding modes detected in polymeric complexes. The polymeric Cu(II) complex of neutral hypoxanthine is N3,N7-bridged [87Du], whereas N7,N9-bridging has been proposed, although not definitely proved, for a number of polymeric guanine complexes [85Mi2, 85Mi3, 86Bi]. Accordingly, the binding behaviour is less clearcut than with free adenine, and no firm conclusions concerning the complex structures in solution can be drawn.

5.   COMPLEXING OF CYTIDINE AND CYTOSINE

5.1   Protolytic equilibria

As with purine nucleosides, the predominant tautomer of cytidine (XIII) is the oxo/

amino form [58Mi, 59Mi, 61Mi, 63Ja, 63Mi2, 63Ul, 65Be, 67Lo1, 73Pi, 73Wo, 82Ta, 84Sz]. Of the two possible annular tautomers of free cytosine the N1H oxo-amino form (XIVa)) prevails [61An, 62Br2, 63Ja, 63Je, 63Ka, 63Mi3, 64Ba, 64Ma, 65Be, 67Lo1, 71Su2, 72Bl, 72Ed, 73Mc, 77St], the mole fraction of the N3H oxo/amino

XIII   XIVa   XIVb

species (XIVb) being less than $3 \cdot 10^{-3}$ in aqueous solution [76Dr]. The pK_a value for the N1H deprotonation is 12.2 ($I = 0$, $T = 25.0°C$) [52Sh, 67Ch, 85Gn, 88Ga].

While the N1H oxo/amino tautomer of cytosine appears to prevail in solution, the situation is less clear in the gas phase. Microwave spectroscopic measurements have revealed the existence of three different tautomers, assigned as the N1H oxo/amino, hydroxy/amino, and N1H,N3H oxo/imino forms [89Br1]. By matrix isolation techniques both N1H oxo/amino and hydroxy/amino tautomers have been detected [84Ra, 88Ku]. In contrast, the earlier results of UV photoelectron studies were adequately simulated on the basis of the N1H oxo/amino tautomer [78Yu]. The results of quantum mechanical calculations contradict each other. MINDO/3 [79Cz] and recent *ab initio* calculations [84Sc] are consistent with the predominance of the N1H oxo/amino form, whereas earlier *ab initio* [75Go1, 83Pa], CNDO/2 [69Fu, 71Br, 73So], and MNDO [83Bu2] methods have predicted the hydroxy/amino form as the most stable tautomer.

Both cytosine and cytidine undergo N3 protonation under mildly acidic conditions, as shown by UV [63Ka], and ^{1}H [63Ja, 63Ka, 63Mi3, 70Wa], ^{13}C [86Be], and ^{15}N [65Be, 65Ro] NMR spectroscopy in solution and by X-ray crystallography in the solid state [82Ta]. The assignment of N3 as the preferential site of protonation also receives support from *ab initio* SCF calculations [77Pu, 83De]. The pK_a values of the N3 monocations of cytosine and cytidine are 4.5 ($I = 0$, $T = 25.0°C$) [52Sh, 61We, 67Ch, 69Re1, 88Ga] and 4.1 ($I = 0$, $T = 25.0°C$) [52Fo, 53Fo, 67Ch, 70Wr], respectively. The lower basicity of the nucleoside most probably results from the electron-withdrawing effect of the ribofuranosyl group which reduces the electron density in the cytosine ring. For comparison, the pK_a value of 1-methylcytosine is even slightly higher than that of cytosine [52Fo]. The second protonation of cytosine takes place at O^2 [70Wa, 86Be] with a pK_a value of -6.4 [86Be]. As with purine nucleosides, the primary amino group is not a site of protonation, but undergoes deprotonation under extremely basic conditions (pK_a 15.5 [81Ta1]). The difference between the successive acidity constants of cytosine varies from 8 to 10 log units, thus being considerably larger than with purines, which may be protonated alternately on the pyrimidine and imidazole ring.

5.2 Metal ion complexes

Neutral ligands cytidine and cytosine from detectably stable complexes with hard metal ions (Tables 7.6 and 7.7). The $\log(K/M^{-1})$ values reported by Taqui Khan *et*

Table 7.6 — Logarithmic stability constants for 1:1 metal ion complexes of cytidine (L)[†]

Metal ion	T/°C	I/M	$\log(K/M^{-1})$	Method	Ref.
$M^{z+} + L = ML^{z+}$					
Mg^{2+}	35.0	$0.1(KNO_3)$	2.42	Potent.	85Ra1
Ca^{2+}	35.0	$0.1(KNO_3)$	2.47	Potent.	85Ra1
Mn^{2+}	35.0	$0.1(KNO_3)$	2.51	Potent.	85Ra1
Co^{2+}	35.0	$0.1(KNO_3)$	2.74	Potent.	85Ra1
Ni^{2+}	21.0	$1.0(NaClO_4)$	0.95	UV	84Ki1
	35.0	$0.1(KNO_3)$	2.94	Potent.	85Ra1
Cu^{2+}	21.0	$1.0(NaClO_4)$	2.04	UV	84Ki1
	20.0	$1.0(NaNO_3)$	1.59(3)	Potent.	65Fi
	35.0	$0.1(KNO_3)$	3.19	Potent.	85Ra1
Zn^{2+}	21.0	$1.0(NaClO_4)$	0.56	UV	84Ki1
	35.0	$0.1(KNO_3)$	2.60	Potent.	85Ra1
Pb^{2+}	20.0	$1.0(NaNO_3)$	0.96(5)	Potent.	65Fi
$MeHg^+$	‡	variable	4.6	UV	64Si1
$(dien)Pd^{2+}$	25.0	$0.2(NaClO_4)$	5.29	UV	84Me
	34.0	$0.5(KNO_3)$	5.37(5)	^{1}H NMR	81Sc
$(en)Pd^{2+}$	34.0	$0.5(KNO_3)$	6.0	^{1}H NMR	83Hä2
La^{3+}	35.0	$0.1(KNO_3)$	2.90	Potent.	86Ra1
Pr^{3+}	35.0	$0.1(KNO_3)$	2.96	Potent.	86Ra1
Nd^{3+}	35.0	$0.1(KNO_3)$	3.03	Potent.	86Ra1
Sm^{3+}	35.0	$0.1(KNO_3)$	3.10	Potent.	86Ra1
Gd^{3+}	35.0	$0.1(KNO_3)$	3.13	Potent.	86Ra1
Dy^{3+}	35.0	$0.1(KNO_3)$	3.17	Potent.	86Ra1
Er^{3+}	35.0	$0.1(KNO_3)$	3.21	Potent.	86Ra1
Th^{4+}	35.0	$0.1(KNO_3)$	4.62(2)	Potent.	82Ra
UO_2^{2+}	35.0	$0.1(KNO_3)$	3.5(2)	Potent.	82Ra

[†]See also Ref. 81Ta2.
[‡]Temperature not indicated.

al. [83Ta1] and Rabindra Reddy *et al.* [85Ra1, 86Ra1] for the complexes of alkaline earth and rare earth metal ions vary from 1.7 to 2.5 and from 2.9 to 3.2, respectively. The observed strengthening of the interaction on going from a divalent to a trivalent ion is expected, but the absolute values of the stability constants appear rather high. In all likelihood they should be regarded as an upper limit to the complex stability, since the stability constants reported by the same authors for the 3d transition metal complexes tend to be considerably higher than those determined by Fiskin & Beer [65Fi], Reinert & Weiss [69Re1], or Kim & Martin [84Ki1].

The data on binding sites of hard metal ions are open to various interpretations. While O^2 has been assigned as the binding site on the basis of ^{13}C NMR and Raman difference spectroscopy in DMSO-d_6 [78Ma5, 80Ma2], the results of ^{7}Li [76Pl] and

Table 7.7 — Logarithmic stability constants for 1:1 metal ion complexes of cytosine (LH)

Metal ion	T/°C	I/M	$\log(K/M^{-1})$	Method	Ref.
$M^{z+} + LH = MLH^{z+}$					
Mg^{2+}	35.0	$0.1(KNO_3)$	1.76	Potent.	83Ta1
Ca^{2+}	35.0	$0.1(KNO_3)$	2.14	Potent.	83Ta1
Co^{2+}	35.0	$0.1(KNO_3)$	2.31	Potent.	83Ta1
Ni^{2+}	35.0	$0.1(KNO_3)$	2.31	Potent.	83Ta1
Cu^{2+}	25.0	$0.05(NaClO_4)$	1.40	Potent.	69Re1
	35.0	$0.1(KNO_3)$	2.73	Potent.	83Ta1
Zn^{2+}	35.0	$0.1(KNO_3)$	2.23	Potent.	83Ta1
Hg^{2+}	†	$0.1(NaClO_4)$	1.9	Potent.	61Fe
Th^{4+}	35.0	$0.1(KNO_3)$	5.52(2)	Potent.	82Ra
UO_2^{2+}	35.0	$0.1(KNO_3)$	3.7(2)	Potent.	82Ra
$M^{z+} + L^- = ML^{(z-1)+}$					
Mg^{2+}	45.0	$0.1(KNO_3)$	2.71(1)	Potent.	74Ta
Ca^{2+}	45.0	$0.1(KNO_3)$	2.5(1)	Potent.	74Ta
Ni^{2+}	35.0	$0.1(NaClO_4)$	5.42(5)	Ionoph.	87Mi3
Zn^{2+}	35.0	$0.1(NaClO_4)$	5.12(11)	Ionoph.	87Mi3
Th^{4+}	35.0	$0.1(KNO_3)$	12.4(1)	Potent.	82Ra
UO_2^{2+}	35.0	$0.1(NaClO_4)$	7.25(9)	Ionoph.	87Mi3
	35.0	$0.1(KNO_3)$	10.42(1)	Potent.	82Ra

†Temperature not indicated.

[13]C [76Sh] resonance relaxation time (T_1) measurements have been interpreted to support N3-binding. Consistent with the former proposal, Ba(II) ion was found not to affect the ^{15}N resonance shifts [82Bu]. According to *ab initio* SCF calculations the energetically favoured binding mode would be a simultaneous attachment of the metal ion to both of these sites with concomitant formation of a N3,O^2 chelate [77Pe1, 83Sa1, 84De]. This kind of structure has been suggested for a solid Ca(II) complex of cytosine [80Sh] and solid Ce(III) and La(III) complexes of cytidine [84Te], which exhibit the $C=O^2$ and $C=N3$ stretching bands at lower frequencies than the uncomplexed ligand.

The 3d transition metal complexes of cytidine and cytosine are approximately as stable as those of adenosine. Binding to the N3 site has been unequivocally established. [1]H [66Ei, 75Mc, 79Ma3] and [13]C [74Fr, 74Ko2, 78Ma5, 79Ma3] resonance relaxation studies with Cu(II) ion, [1]H [66Wa, 68Wa, 75Sz2] and [15]N [82Bu] resonance shift measurements with Zn(II) ion, and IR [79Ma3], and Raman [74Ma2, 78Ma5] spectral changes upon addition of Cu(II) ions are all consistent with this binding mode. Simultaneous interaction of the N3-coordinated metal ion with the adjacent O^2 atom has been suggested on the bases of theoretical calculations [79Pu, 87An] and decreased carbonyl stretching upon formation of a solid Cu(II) complex of cytidine [80Ne2]. However, Cu(II)–cytosine complexes show only minor spectral changes that may be consistent with, but do not definitely prove, O^2 involvement [78Sh1, 80Go, 87Mi2]. The crystal structures also lend only mild additional support for N3,O^2 chelation. Binding of (glygly)Cu(II) [74Sa, 75Sz3, 77Sz] and Cd(II) [79Ga] to N3 of cytidine or cytosine is probably enhanced by a weak axial metal-O^2 interaction, though the bond length to O^2 is in both cases considerably larger than that to N3. Similarly, a weak interaction with O^2 has been detected in several cytosine (LH) complexes: Cu(LH)Cl$_2$ [71Su1], (N-salicylidene-N'-methylethylenediamine)Cu(LH)NO$_3$ H$_2$O [75Sz2], Cu(LH)$_2$Cl$_2$ [68Ca], and (glygly)Cu(LH) 2H$_2$O [75Ki2]. In contrast, the O^2 atom is not significantly involved in binding of ZnCl$_2$ to 1-methylcytosine [84Be2]. In any case, the possible M–O^2 interaction is too weak to affect markedly the complex stabilities. The 3d transition metal ions exhibit only slightly higher affinity to N3 of cytidine than to N1 of adenosine, in spite of the fact that adenosine does not contain a neighbouring oxo substituent. The basicities of both sites are comparable and they are in the *ortho* position to an amino substituent.

Free cytosine may also act as an anionic ligand. These complexes are naturally more stable than those of neutral cytosine (Table 7.7), but the data on their binding modes are scanty. Actually, it is known only that the (Ph$_3$P)Au(I) complex of the cytosinate anion is N1 bonded [85Bo].

MeHg(II) and Hg(II) ions coordinate, like 3d transition metal ions, to N3 of cytidine and cytosine. The binding site has been established by UV [64Si1], Raman [67Lo2], and ^{1}H [70Ka, 70Te], ^{13}C [75Je, 80Ma2] and ^{15}N [82Bu] NMR spectroscopy in aqueous and/or DMSO-d_6 solution, and by X-ray crystallography [77Au] in the solid state. The X-ray structure of the N3 coordinated HgCl$_2$ complex of 1-methylcytosine displayed a simultaneous weak interaction with the O^2 atom [77Au]. Under alkaline conditions one of the N^{4}H$_2$ protons may be displaced by a MeHg(II) ion, as shown by UV [64Si1], Raman [65Fi], and ^{1}H NMR [81Ta1] spectroscopy in solution, and verified by X-ray crystallography [81Pr2]. Methylmercuration of N3 facilitates the displacement by almost one order of magnitude. Substitution of both of the amino protons has been demonstrated with the crystal structure of the N3,N^4,N^4-(HgMe)$_3$ complex of cytosine [83Ch].

A rather unusual binding mode has been reported for the Ag(I) ion [77Ma4, 79Ki]. The crystals obtained from acidic solutions of AgNO$_3$ and 1-methylcytosine were 2:2 dimers, in which the ligand molecules were crosslinked by two Ag(I) ions through the N3 and O^2 atoms. Participation of O^2 in complexation has been verified by IR dichroic measurements in aqueous solution [86No]. In contrast, from slightly alkaline solutions a solid complex was obtained, for which a polymeric N3,N^4 bridged structure was suggested on the basis of ^{1}H NMR spectroscopic data [78Ki]. ^{13}C NMR spectroscopic evidence has been presented for binding of a Au(I) ion to N3 of cytosine [77Br2].

The ^{1}H [76Co, 80Ma3, 80Ro, 82Je, 83Hä2] and ^{13}C [76Co, 76Ne, 80Me3, 81Ko, 82Je] NMR spectra observed for PdCl$_2$, (en)Pd(II), (dien)Pd(II), and several (peptide)Pd(II) complexes of cytidine and cytosine indicate the N3 atom as the binding site. The equilibria prevailing in equimolar solutions of cytidine and (en)Pd(II) ion have been studied in detail by ^{1}H NMR spectroscopy [83Hä2]. At pH $<$ 5 the N3 coordinated 1:1 complex predominates, the logarithmic stability constant being 6.0 ($I = 0.5$ M, $T = 34.0$°C), but under slightly alkaline conditions one of the N^{4}H$_2$ protons may additionally be displaced with concomitant formation of a dimeric species, [(en)Pd(L$^-$)]$_2$, where two anionic cytosine rings bridge two (en)Pd(II) ions through N3 and deprotonated N^4 atoms. The latter complex has been shown to exist in three distinct isomeric forms, *viz.* in two head-to-tail and one head-to-head arrangements. The macroscopic equilibrium constant for their formation reaction,

$$2\,(en)Pd^{2+} + 2\,L = [(en)Pd(LH_{-1})]_2^{2+} + 2\,H^+ \ ,$$

is $1.3\ 10^3\ M^{-1}$ ($I = 0.5$ M, $T = 34.0°C$). Moreover, three different hydroxo complexes are present in neutral and alkaline solutions, but their mole fractions remain small compared to that of the dimeric species. In an excess of cytidine a 1:2 complex (M/L) prevails over a wide pH range, the logarithmic equilibrium constant for the attachment of the second ligand molecule being only slightly smaller than that for the 1:1 complex formation, *viz.* 5.6 ($I = 0.5$ M, $T = 34.0°C$). The situation is rather similar with the monofunctional (dien)Pd(II) ion. The N3 coordinated complex, $(dien)PdL^{2+}$, predominates at pH < 7, a N3,N^4-dimetallated species, $[(dien)Pd]_2(LH_{-1})^{3+}$, at pH 10–12 and a N^4-coordinated complex, $(dien)Pd(LH_{-1})^+$, at pH > 12 [87Me]. Equilibrium constants for displacement of chloro ligands from $PdCl_4^{2-}$ and $(en)PdCl_2$ have also been determined [84Et]. The log K values obtained for the replacement of the first and second chloride ions were 4.49 and 3.45 with $PdCl_4^{2-}$, and 3.32 and 2.56 with $(en)PdCl_2$ ($I = 1.0$ M, $T = 25.0°C$).

Crystal structures of the complexes Pd(1-methylcytosine)Cl$_2$ [77Si] and (gly tyr)Pd(cytidine) [83Sa2] have been shown to display N3 coordination without marked bonding to O^2. A weak interaction with one of the N^{4}H$_2$ hydrogens has been suggested on the basis of the X-ray structure of the former complex [77Si]. The spectral properties of several solid Pd(II) complexes of cytosine and its N1 substituted derivatives have been reported to agree with monodentate N3 binding [77De, 78Ha, 78Pn, 80Pn, 82Pn, 83Ca, 84Pn1, 85Ro]. Solid Rh(I) complexes of cytosine [83Si] and cytidine [83Ab3, 83Si, 87Pn] appear also to be N3 coordinated.

6. COMPLEXING OF 4-OXOPYRIMIDINES AND THEIR NUCLEOSIDES

6.1 Protolytic equilibria

The dioxo forms have been established to be the predominant tautomers of uridine (XV) and uracil (XVI) by UV [62Ka], IR [58 Mi, 61An], Raman [67Lo1], and ^{15}N NMR [65Ro] spectroscopy in solution, by IR spectroscopy [61An, 61K] and X-ray crystallography [82Ta] in the solid state, by UV, IR [78No, 81Sh] and microwave [88Br] spectroscopy in the gas phase, and by IR spectroscopy in argon and nitrogen matrices [83Sz, 85Sz]. The same tautomers prevail with their 5-methyl derivatives, thymidine (XVII) and thymine (XVIII) [58Mi, 61An, 65Wi, 74Su, 80Fe, 82Ta, 84Ma3, 89Br2, 89No].

XV   XVI   XVII   XVIII

It has been estimated that with uracil the mole fractions of the rare $O^4H,N1H$ and $O^2H,N3H$ tautomers are about $3 \ 10^{-4}$ and $1 \ 10^{-5}$ [75Po]. The results of CNDO/2 [69Fu, 73So, 76Ab], MINDO/3 [79Cz] and recent *ab initio* [84Sc] calculations agree qualitatively with the observed predominance of the dioxo forms, whereas MINDO/2 [78Zi] and MNDO [83Bu2] and early *ab initio* [75Go1] methods have predicted dihydroxy forms as the most stable tautomers.

The pK_a values reported for N3H of uridine and thymidine range from 9.1 to 9.3 [52Fo, 64Si1, 67Ch, 69Re2, 76Og, 87Ta] and from 9.6 to 9.8 [52Fo, 61Fe, 67Ch, 87Mc] $(I = 0 - 0.1 \ M, T = 25.0°C)$, respectively. The macroscopic pK_a values of the corresponding free bases are of the same magnitude, 9.4–9.5 with uracil [52Sh, 67Ch, 69Re1, 83Ta1] and 9.8–9.9 [52Sh, 61Fe, 67Ch, 83Ta1, 84Pe1, 85Gn, 88Ga] with thymine $(I = 0$–$0.1 \ M, T = 25.0°C)$. It was previously assumed that these values mainly refer to deprotonation of N3H [71Sh, 75De], as with their N1 substituted derivatives, but the contribution of N1H deprotonation has more recently shown to be significant [79Li]. In less polar media the latter process may even become the predominant one [71Sh]. According to LCAO-SCF calculations the N3 monoanion of thymine in the gas phase is 80 kJ mol^{-1} more stable than its N1 counterpart [70Sn]. For the deprotonation of the uracil monoanion a pK_a value of 14.2 $(I = 1.0 \ M, T = 25.0°C)$ has been reported [75De].

4-Oxo substituted pyrimidine nucleosides and bases are all protonated at O^4, as shown by UV [62Ka, 72Gu, 72Sh], ^1H [70Wa, 72Sh, 86Be], and ^{13}C [86Be] NMR spectroscopy and the X-ray structure of 1-methyluracil hydrobromide [64So2]. The result is also consistent with *ab initio* SCF calculations [83De]. The macroscopic pK_a values determined for the monocations vary from -2.2 to -3.0 with uracil [62Ka, 72An, 75Po, 86Be] and from -3.0 to -5.2 with thymine [86Be, 88Ga]. The microscopic pK_a value of O^2 monocation has been estimated to be 1.5 log units lower than that of the O^4 protonated species [75Po]. The second protonation takes place at O^2 [86Be], the pK_a value of the dication being -7.3 with uracil and -6.8 with thymine [75Fr].

6.2   Metal ion complexes

2,4-Dioxopyrimidine nucleosides exhibit two alternative binding modes. In neutral and basic solutions metal ions may compete with protons for the N3 atom, whereas under acidic conditions deprotonation of the N3H site is impeded, hence the available binding sites in the base moiety are restricted to the carbonyl oxygens. According to theoretical calculations the preferred binding site of alkali [77Pe1,

83Sa1, 84De], alkaline earth metal [76Pe, 83Sa1, 83Sa3], and 3d transition metal ions [79Pu, 87An] of unionized uracil or thymine rings is the O^4 atom. No stability constants for these neutral ligand complexes are available, but most probably the interaction in solution is very weak, since Ba(II) and even La(III) ions did not appreciably affect the Raman and ^{13}C NMR spectra of uridine in DMSO-d_6 [80Ma2]. Several solid 3d transition metal complexes of 1,3-dimethyluracil [78Go1], unionized uracil [77Go, 83Sa4, 86Fi], and uridine [78Go2, 86Fi] have been prepared, and coordination to O^4 is suggested, on the basis of IR data. In the corresponding thymine complexes O^2 has been tentatively assigned as the binding site [77Go]. This change in binding behaviour might result from steric hindrance caused by the 5-methyl group, thus restricting attachment of metal ions to the adjacent O^4 site.

The complexes of the 4-oxopyrimidine nucleoside monoanions are approximately as stable as those of anionic 6-oxopurine nucleosides, which bind metal ions at N1, that is, a site comparable to N3 of uridine and thymidine (Tables 7.4, and 7.8).

Table 7.8 — Logarithmic stability constants for 1:1 metal ion complexes of uridine and thymidine (LH)

Metal ion	T/°C	I/M	$\log(K/M^{-1})$	Method	Ref.
$M^{z+} + L^- = ML^{(z-1)+}$					
Uridine:					
Mg^{2+}	25.0	$0.1(KNO_3)$	3.14(3)	Potent.	87Ta
	35.0	$0.1(KNO_3)$	2.71	Potent.	86Ra3
Ca^{2+}	25.0	$0.1(KNO_3)$	3.05(3)	Potent.	87Ta
	35.0	$0.1(KNO_3)$	2.62	Potent.	86Ra3
Mn^{2+}	25.0	$0.1(KNO_3)$	3.44(3)	Potent.	87Ta
	35.0	$0.1(KNO_3)$	3.20	Potent.	86Ra3
Co^{2+}	25.0	$0.1(KNO_3)$	3.79(3)	Potent.	87Ta
	35.0	$0.1(KNO_3)$	3.43	Potent.	86Ra3
Ni^{2+}	25.0	$0.1(KNO_3)$	3.90(3)	Potent.	87Ta
	35.0	$0.1(KNO_3)$	3.57	Potent.	86Ra3
Cu^{2+}	25.0	$0.1(KNO_3)$	4.03(3)	Potent.	87Ta
	20.0	$1.0(NaNO_3)$	4.2(2)	Potent.	65Fi
	35.0	$0.1(KNO_3)$	5.90	Potent.	86Ra3
Zn^{2+}	25.0	$0.1(KNO_3)$	3.67(3)	Potent.	87Ta
	35.0	$0.1(KNO_3)$	3.57	Potent.	86Ra3
Pb^{2+}	20.0	$1.0(NaNO_3)$	3.4(3)	Potent.	65Fi
$MeHg^+$	†	variable	9.0	UV	64Si1
$(en)Pd^{2+}$	25.0	$0.5(KNO_3)$	8.65(1)	Potent.	81Li
	34.0	$0.5(KNO_3)$	9.1	^1H NMR	83Hä2
$(dien)Pd^{2+}$	25.0	$0.5(KNO_3)$	8.08(1)	Potent.	81Li
	34.0	$0.5(KNO_3)$	8.60(5)	^1H NMR	81Sc
Thymidine:					
Cu^{2+}	20.0	$1.0(NaNO_3)$	4.7(1)	Potent.	65Fi
Pb^{2+}	20.0	$1.0(NaNO_3)$	3.7(5)	Potent.	65Fi
$(en)Pd^{2+}$	25.0	$0.5(KNO_3)$	8.84(1)	Potent.	81Li
$(dien)Pd^{2+}$	25.0	$0.5(KNO_3)$	8.31(1)	Potent.	81Li
	34.0	$0.5(KNO_3)$	8.67	^1H NMR	81Sc

†Temperature not included.

Binding of 3d transition metal ions to the deprotonated N3 atom has been demonstrated by ^{13}C resonance relaxation measurements in D_2O. Cu(II) [74Ko] and Mn(II) [73Ko1] ions were observed to exert similar influences on the ^{13}C resonances of uridine and cytidine. Since the complexes of cytidine were known to be N3 coordinated, a similar binding mode was proposed for the uridine monoanion. The Cu(II) ion has also been shown to form in aqueous solution an octanuclear structure with uridine trianion, involving $O^{2'}$ and $O^{3'}$ bonding in addition to N3 coordination [87Ga].

4-Oxopyrimidines complex with 3d transition metal ions as efficiently as their nucleosides (Table 7.9), but the site of coordination is possibly different, since the X-

Table 7.9 — Logarithmic stability constants for 1:1 metal ion complexes of uracil and thymine (LH_2)

Metal ion	T/°C	I/M	$\log(K/M^{-1})$	Method	Ref.
$M^{z+} + LH^- = MLH^{(z-1)+}$					
Uracil:					
Be^{2+}	31.0	0.1(KCl)	7.41	Potent.	87Se
	30.0	0.1(NaClO$_4$)	6.52(2)	Potent.	78Sr
Mg^{2+}	45.0	0.1(KNO$_3$)	2.6(1)	Potent.	74Ta
Ca^{2+}	45.0	0.1(KNO$_3$)	2.4(1)	Potent.	74Ta
Mn^{2+}	25.0	0.1(KNO$_3$)	3.35	Potent.	83Ta1
Co^{2+}	25.0	0.1(KNO$_3$)	3.76	Potent.	83Ta1
Ni^{2+}	25.0	0.1(KNO$_3$)	3.85	Potent.	83Ta1
	24.0	0.5(NaNO$_3$)	2.01	Potent.	73Yu
Cu^{2+}	25.0	0.1(KNO$_3$)	6.14	Potent.	83Ta1
	24.0	0.5(NaNO$_3$)	3.51	Potent.	73Yu
Zn^{2+}	25.0	0.1(KNO$_3$)	5.13	Potent.	83Ta1
Ag^+	24.0	0.5(NaNO$_3$)	4.03	Potent.	73Yu
	25.0	0.1(KNO$_3$)	6.0	Potent.	75De
Hg^{2+}	30.0	0.1(NaClO$_4$)	7.30(2)	Potent.	78Sr
$HgCl_2$	31.0	0.1(KCl)	5.04	Potent.	87Se
(en)Pd^{2+}	25.0	0.5(KNO$_3$)	8.59(1)	Potent.	81Li
(dien)Pd^{2+}	25.0	0.5(KNO$_3$)	8.01(1)	Potent.	81Li
Thymine:					
Be^{2+}	31.0	0.1(KCl)	7.81	Potent.	87Se
	30.0	0.1(NaClO$_4$)	7.01(3)	Potent.	78Sr
Mg^{2+}	25.0	0.1(KNO$_3$)	2.96	Potent.	83Ta1
Ca^{2+}	25.0	0.1(KNO$_3$)	2.92	Potent.	83Ta1
Mn^{2+}	25.0	0.1(KNO$_3$)	3.52	Potent.	83Ta1
Co^{2+}	25.0	0.1(KNO$_3$)	4.30	Potent.	83Ta1
Ni^{2+}	25.0	0.1(KNO$_3$)	4.38	Potent.	83Ta1
Cu^{2+}	25.0	0.1(KNO$_3$)	6.72	Potent.	83Ta1
Zn^{2+}	25.0	0.1(KNO$_3$)	5.32	Potent.	83Ta1
Hg^{2+}	25.0	0.1(NaClO$_4$)	10.65(3)	Potent.	84Pe1
	30.0	0.1(NaClO$_4$)	7.93(4)	Potent.	78Sr
$HgCl_2$	31.0	0.1(KCl)	5.45	Potent.	87Se
$M^{z+} + L^{2-} = ML^{(z-2)+}$					
Uracil:					
Ag^+	25.0	1.0(KNO$_3$)	8.2	Pot./Solub.	75De

ray structures of (dien)Cu(thyminato)Br 2H$_2$O [75Ki3] and (NH$_3$)$_2$Ni(uraliato)$_2$ (H$_2$O)$_2$ [80Lu] display N1 binding. However, N3 coordination [88Sa2, 88Sa3] or N3,O^2 chelation [84Gh, 88Sa3] has been preferred on the basis of IR spectroscopic characterization of several bis(uracilato) complexes. Moreover, a number of solid lanthanide complexes of the uracil dianion have been prepared, and involvement of both O^2 and O^4 assumed [81Bh].

Hg(II) and MeHg(II) ions are able to displace the N3 proton of uridine in neutral and even in mildly acidic aqueous solutions, as indicated by UV spectrophotometric [64Si1], potentiometric [63Ei], and IR [75Ma4] and Raman [74Ma2, 75Ma4] spectroscopic studies. In fact, the affinity of these ions to the N3 site is almost as high as that of a proton [64Si1]. According to Raman spectroscopic findings, 1-methylthymine behaves analogously [77Ch2, 83Gu]. Solid state studies lend further support for this binding mode. Accordingly, coordination to deprotonated N3 has been established by X-ray structures of Hg(1-methylthyminato)$_2$ [74Ko3], MeHg (1-methylthyminato) 1/2H$_2$O [82Gu], and Cd(uracilato)$_2$(H$_2$O)$_3$ [80Mu]. Furthermore, extensive spectroscopic data reported for solid mercury complexes of uridine [79Pe], thymidine [79Ca, 79Pe, 85Bu2] and 1-methylthymine [77Ch2] are consistent with N3 binding. At low pH, binding to carbonyl oxygens may occur. For example, the X-ray structure of the HgCl$_2$ complex of un-ionized uracil shows O^4 coordination [71Ca]. However, this interaction must be rather weak, since HgCl$_2$ did not appreciably affect the ^{13}C NMR shifts of uridine in DMSO-d_6 [80Ma2].

Ag(I) ion also exhibits a high affinity to deprotonated 4-oxopyrimidines and their nucleosides, though not as high as mercury ions. log(K/M^{-1}) values for the 1:1 complexes of uracil mono- and di-anions have been reported to be 6.0 and 8.2 (Table 7.9), respectively [75De]. On the basis of ^{13}C NMR data the N3 atom was assigned as the binding site in the former and the N1,O^2 region in the latter complex. Solid Ag(I) complexes of the monoanions of 1-methyluracil [84Ao3] and 1-methylthymine [79Gu] have been shown to display a polymeric structure where each ligand is coordinated to three metal ions through deprotonated N3 and the carbonyl oxygens. The solid Ph$_3$PAu(I) complexes of the monoanions of uridine [78We] and 1-methylthymine [87Fa] have been reported to be N3 coordinated.

The (en)Pd(II) [81Li, 83Hä2] and (dien)Pd(II) [81Li, 81Sc] ions complex with monoanions of 4-oxopyrimidine bases and nucleosides as efficiently as the MeHg(II) ion. The interactions between (en)Pd(II) and uridine in D$_2$O have been studied in detail by ^{1}H NMR spectroscopy [83Hä2]. The N3 coordinated 1:1 and 1:2 complexes of the uridine monoanion prevail in a wide pH range, the log(K/M^{-1}) values for the consecutive complexing steps being 9.1 and 6.1. Moreover, a hydroxo bridged 2:2 complex is formed under slightly alkaline conditions, the maximal mole fraction of this species being about 0.3 in equimolar solutions.

7. BINDING AT THE SUGAR MOIETY OF RIBONUCLEOSIDES

The sugar moiety of ribinucleosides, that is, the β-D-ribofuranosyl group (XIX), contains a vicinal *cis*-diol arrangement at C2′ and C3′, which affords a binding site for metal ions in addition to the heteroatoms of the base moiety. The 2′-OH function is exceptionally acidic as an aliphatic hydroxyl group. The pK_a values obtained with adenosine (12.35; $I = 0$, $T = 25.0°$C [65 Iz]), guanosine (12.33 [70Ch1]), inosine

(12.36 [70Ch1]), cytidine (12.5 [70Ch2]), and uridine (12.59 [67Ch]) are almost identical. The high acidity may partly be attributed to electron withdrawal by the aromatic base moiety, ring oxygen, and neighbouring 3'-OH, and partly to stabilization of the 2'-oxoanion by hydrogen bonding with the 3'-OH. Deprotonation of 2'-OH increases its affinity to metal ions and makes the *cis*-2',3'-diol system a potential binding centre under slightly alkaline conditions.

It has been known since the early 1970s that $Cu(OAc)_2$ broadens the ribose hydroxyl proton resonances of ribonucleosides in DMSO-d_6, with concomitant reduction of the VIS absorbance of the copper acetate dimer [71Be2]. In contrast, 2'-deoxyribonucleosides (XX) and 2',3',5'-tri-*O*-acetyl-ribonucleosides (XXI) do not produce similar changes in the VIS spectrum of dimeric copper acetate. These findings, together with the observed 2:1 (M:L) stoichiometry of the complex, led originally to the conclusion that one of the bridging acetate ligands in the copper acetate dimer is replaced by the ribose 2',3'-*cis*-diol system [71Be2, 72Be]. In other words, one Cu(II) ion in the dimer would be coordinated to 2'-OH and the other to 3'-OH, while three acetate molecules still bridge the metal ions (XXII). This proposal was later questioned and the data accounted for by disruption of the dimers to form bidentate monomeric Cu(II) complexes (XXIII) [75Br, 77Ki]. However, formation of the monomeric species was subsequently disputed on the basis of EPR and magnetic susceptibility data, thus the original concept of a binuclear structure was preferred [79Ch]. Finally, the discrepancy between various approaches was accounted for by two alternative kinds of behaviour for the *cis*-diols; uridine type ligands retain the dimeric structure of copper acetate upon complexing, while free ribose and its structural analogues destroy the dimer [80So]. Moreover, it was assumed that some *cis*-diols may react in both ways, and even switch their predominant mode of interaction with a change of experimental conditions. The spectrophotometric data obtained with uridine were then analyzed in terms of the species $Cu(OAc)_2$, $Cu_2(OAc)_4$, $Cu_2(OAc)_2L$, and $Cu_4(OAc)_4L_2$ (L = *cis*-diolate of uridine). In this manner a $\log(K/M^{-1})$ value of 2.0 was obtained for the reaction:

$$Cu_2(OAc)_4 + LH_2 = Cu_2(OAc)_2L + 2HOAc \ .$$

It has been shown by EPR measurements that the Cu(II) ion binds to the sugar moiety of ribonucleosides in alkaline water–DMSO mixtures, while at pH < 7 base coordination prevails [75Ma2, 77Ch1]. Formation of both mono- (XXIV) and dinuclear (XXV) complexes has been suggested, the dinuclear species prevailing at low alkalinities and the mononuclear ones under very basic conditions. Formation of two different 1:1 Cu(II)–adenosine complexes with ribose binding has also been demonstrated in aqueous solution by optical spectroscopy, one prevailing at pH 9 and the other at pH 11.5 [85On]. Similarly, ^{13}C NMR relaxation measurements have revealed the Mn(II) ion to interact with the 2'- and 3'-hydroxyl functions of ribonucleosides in aqueous alkali solutions [79Ku]. Furthermore, an octameric complex between the Cu(II) ion and uridine trianion has been prepared under aqueous conditions, where each uridine ligand is coordinated toward three Cu(II) ions through anionic N3, O2' and O3' sites [87Ga].

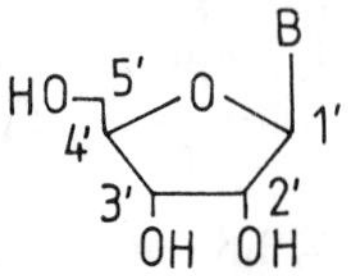

XIX

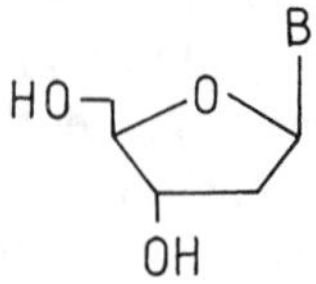

XX

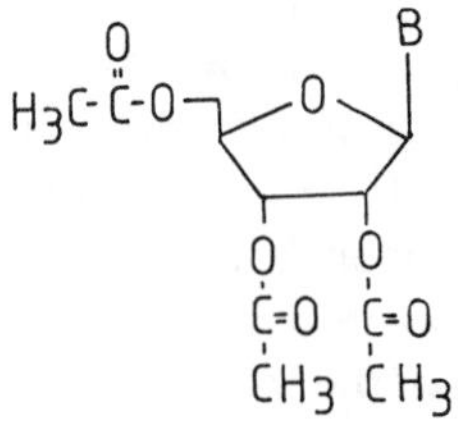

XXI

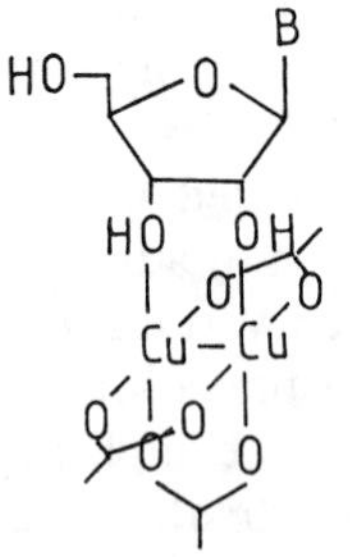

XXII

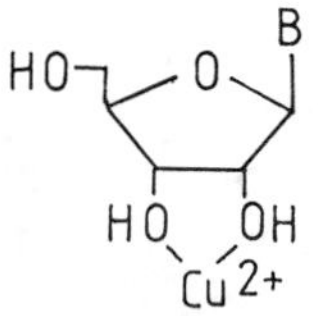

XXIII

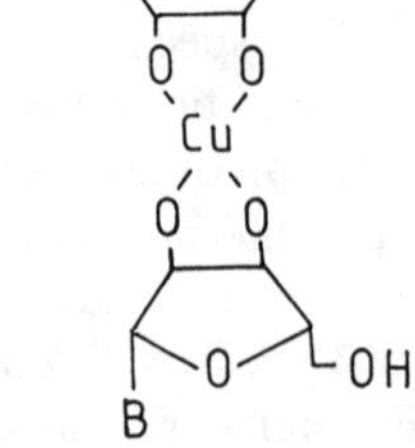

XXIV

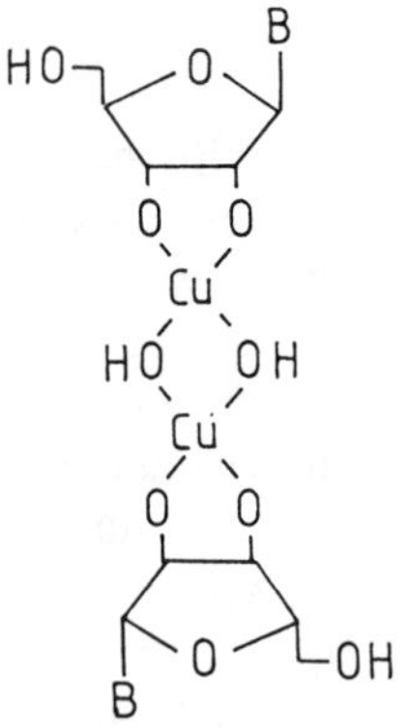

XXV

8. COMPLEXING OF NUCLEOSIDE 5'-MONOPHOSPHATES

5'-Monophosphates of nucleosides (XXVIa,b) may be regarded as monomeric constituents of nucleic acids. It should be noted, however, that internucleosidic 3',5'-phosphodiester groupings of polynucleotides bear only one negative charge at pH > 1, whereas the 5'-phosphate group of monomeric nucleoside monophosphates has two dissociable protons, the approximate pK_a values of which are 1 and 6 [88Ma]. Accordingly, nucleoside monophosphates may display binding modes, which cannot occur with their polymeric derivatives.

XXVIa   XXVIb

The 5'-monophosphate group does not markedly effect the basicity of the base moiety. As seen from Table 7.10, the pK_a values obtained with nucleoside monophosphates are equal to those of the corresponding nucleosides within 0.2 log units. Similarly, the acidity of the phosphate group does not depend on the structure of the base moiety, indicating that the intramolecular interaction between the base and phosphate moieties do not play an important role in solution. However, the 5'-monophosphate group reduces the acidity of 2' − OH by 0.7 log units [66Iz].

Hard Lewis acids, including alkali, alkaline earth, and rare earth metal ions, display almost equal affinities to all the nucleoside monophosphates studied (Table 7.11), suggesting that coordination takes place at the phosphate dianion without involvement of the base moiety. The complex stability is markedly increased on going from monovalent to di- and tri-valent metal ions, but remains rather constant within each group, consistent with the assumed electrostatic character of the interaction. However, it is worth noting that even alkali metal ions form reasonably stable complexes with nucleoside monophosphates [56Sm], hence attention should be paid to the nature of the electrolyte used to adjust the ionic strength in equilibrium measurements.

Structures of solid metal ion complexes of nucleoside monophosphates are only partly consistent with the phosphate bonding suggested to prevail in solution. This binding mode has been established for the Ca(II) complex of thymidine 5'-monophosphate (TMP) by X-ray crystallography [61Tr], whereas the Ca(II) complex of inosine 5'-monophosphate (IMP) was shown to exhibit N7 coordination and hydrogen bonding to the 5'-phosphate through a molecule of water [80Br]. The latter kind of structure was also described for the Mg(II) complexes of the 5'-monophosphates of adenosine (AMP) [83Ta2], guanosine (GMP) [83Ta3], 2'-deoxyguanosine (dGMP) [83Ta4], and IMP [85Ta] on the basis of IR spectroscopic characterization. Interaction with the phosphate group was suggested to be either a direct one or mediated through a molecule of water, depending on the conditions in which the complexes were prepared. The X-ray structure of the Ba(II) complex of AMP

Table 7.10 — pK_a values of nucleoside 5'-monophosphates (NMP) at $T = 25.0°C$
$(I = 0.1$ M$)$

NMP	Site	pK_a	Ref.
Base moiety:			
AMP	N1H$^+$	3.80–3.86	[62Ta, 74Ba, 76Og, 87Sh3, 87Tr, 88An, 88Si]
GMP	N7H$^+$	2.34	[76Og]
	N1H	9.46	[76Og]
IMP	N7H$^+$	1.3[†]	[81Sc]
	N1H	9.15[‡]	[81Na]
		9.27[†]	[81Sc]
CMP	N3H$^+$	4.33–4.36	[76Og, 80Or, 88Ma]
UMP	N3H	9.43–9.45	[76Og, 88Ma]
TMP	NSH	9.90	[88Ma]
Phosphate moiety:			
AMP	OPO$_3$H$^-$	6.10–6.30	[62Ta, 64Si2, 74Ba, 76Ta, 80Or, 87Sh3, 87Tr, 88An, 88Si]
GMP		6.2[§]	[81Ve2]
		6.23[†]	[81Sc]
IMP		5.93[‡]	[81Na]
		6.3[§]	[81Ve2]
		6.00[†]	[81Sc]
CMP		6.19–6.30	[80Or, 88Ma]
UMP		6.15	[84Si]
TMP		6.36	[88Ma]

[†] $I = 0.5$, M, $T = 34.0°C$.
[‡] $I = 0.2$ M, $T = 15.0°C$.
[§] $I = 0.2$ M, $T = 34.0°C$.

showed outer-sphere complexing only [76St]. Despite the versatility of solid state structures, the experimental evidence for only phosphate bonding in solution seems unequivocal.

3d transition metal complexes of nucleoside monophosphates are considerably more stable than those of the corresponding nucleosides under neutral conditions, that is, in solutions where the phosphate group is already dianionic, but possible deprotonation of the base moiety is impeded. Accordingly, the phosphate group must participate in the binding process. In fact, interaction of 3d transition metal ions with both the base and phosphate moieties has been established by extensive ^{1}H, ^{13}C and ^{31}P NMR relaxation measurements with AMP [65Sh, 66Ei, 66Si, 71Be1, 72Mi, 73Ko2, 73Ko3, 74We, 78Ma1, 86Le], GMP [73Ko1, 73Ko3, 74Fr, 78Ma3], IMP [73Ko1, 73Ko3, 78Ma2, 81Ma2], and pyrimidine nucleoside monophosphates [71Be3, 73Ko1, 74Ko2, 79Ma3]. The question that remains to be answered is whether binding to both sites takes place simultaneously with formation of a

Table 7.11 — Logarithmic stability constants for 1:1 metal ion complexes of nucleoside 5'-monophosphates[†]

Metal ion	T/°C	I/M	$\log(K/M^{-1})$	Method	Ref.
Binding of monoanions:					
AMP:					
Co^{2+}	8.0	$0.2(NaClO_4)$	1.32	Potent.	77Pe2
Ni^{2+}	25.0	$0.1(NaClO_4)$	1.11	Potent.	76Ta
	15.0	$0.1(KNO_3)$	1.18	Potent.	80Th
IMP:					
Ni^{2+}	15.0	$0.1(NaClO_4)$	0.9(1)	UV	81Na
Binding of nucleotide dianions:					
AMP:					
Li^+	25.0	$0.2(R_4NBr)$	0.61(4)	Potent.	56Sm
Na^+	25.0	$0.2(R_4NBr)$	0.41(7)	Potent.	56Sm
K^+	25.0	$0.2(R_4NBr)$	0.23(8)	Potent.	56Sm
Mg^{2+}	25.0	$0.1(NaNO_3)$	1.60(2)	Potent.	88Si
	25.0	$0.1(KNO_3)$	1.97(1)	Potent.	62Ta
	25.0	$0.1((NaClO_4)$	2.10	Potent.	87Sh3
	23.0	$0.1(NaCl)$	1.95	Distrib.	58Wa
	15.0	$0.1(KNO_3)$	1.80	Potent.	72Fr
Ca^{2+}	25.0	$0.1(NaNO_3)$	1.46(1)	Potent.	88Si
	25.0	$0.1(KNO_3)$	1.85(2)	Potent.	62Ta
	25.0	$0.1(NaClO_4)$	2.03	Potent.	87Sh3
	23.0	$0.1(NaCl)$	1.76	Distrib.	58Wa
Sr^{2+}	25.0	$0.1(NaNO_3)$	1.24(1)	Potent.	88Si
	25.0	$0.1(KNO_3)$	1.79(1)	Potent.	62Ta
Ba^{2+}	25.0	$0.1(NaNO_3)$	1.17(2)	Potent.	88Si
	25.0	$0.1(KNO_3)$	1.73(1)	Potent.	62Ta
Mn^{2+}	25.0	$0.1(NaNO_3)$	2.23(1)	Potent.	88Si
	25.0	$0.1(KNO_3)$	2.40(2)	Potent.	62Ta
	25.0	$0.2(R_4NBr)$	2.33(5)	Potent.	77Ra
	25.0	$0.1(KNO_3)$	2.35	Potent.	66Do
	23.0	$0.1(NaCl)$	2.31	Distrib.	58Wa
Co^{2+}	25.0	$0.2(NaNO_3)$	2.23(2)	Potent.	88Si
	25.0	$0.1(KNO_3)$	2.64(2)	Potent.	62Ta
	25.0	$0.1(KNO_3)$	2.57	Potent.	66Do
	23.0	$0.1(NaCl)$	2.58	Distrib.	58Wa
	7.9	$0.2(NaClO_4)$	2.35	Potent.	77Pe2
Ni^{2+}	25.0	$0.1(NaNO_3)$	2.49(2)	Potent.	88Si
	25.0	$0.1(NaCl)$	2.43	UV	82Or
	25.0	$0.1(KNO_3)$	2.84(1)	Potent.	62Ta
	25.0	$0.1(KNO_3)$	2.49(2)	Potent.	74Ba
	25.0	$0.1(NaClO_4)$	2.50(5)	Potent.	67Ta
	25.0	$0.1(KNO_3)$	2.61(5)	Potent.	80Or
	25.0	$0.1(KNO_3)$	2.67	Potent.	66Do
Cu^{2+}	25.0	$0.1(NaNO_3)$	3.14(1)	Potent.	88Si
	25.0	$0.1(NaClO_4)$	3.22(5)	Potent.	74Na
	25.0	$0.1(KNO_3)$	3.18(1)	Potent.	62Ta
	25.0	$0.1(KNO_3)$	3.6(1)	Potent.	88An
Zn^{2+}	25.0	$0.1(NaNO_3)$	2.38(7)	Potent.	88Si
	25.0	$0.1(KNO_3)$	2.72(2)	Potent.	62Ta
Cd^{2+}	25.0	$0.1(NaNO_3)$	2.68(2)	Potent.	88Si
Y^{3+}	25.0	$0.1(NaClO_4)$	4.35	Potent.	87Sh4
La^{3+}	25.0	$0.1(NaClO_4)$	3.78	Potent.	87Sh4
Nd^{3+}	25.0	$0.1(NaClO_4)$	4.09	Potent.	87Sh4
Eu^{3+}	25.0	$0.1(NaClO_4)$	4.50	Potent.	87Sh4

Table 7.11 — Contd.

Metal ion	T/°C	I/M	$\log(K/M^{-1})$	Method	Ref.
Binding of nucleotide dianions:					
AMP:					
Dy^{3+}	25.0	$0.1(NaClO_4)$	4.57	Potent.	87Sh4
Tm^{3+}	25.0	$0.1(NaClO_4)$	4.46	Potent.	87Sh4
IMP:					
Ni^{2+}	25.0	$0.1(NaCl)$	2.89	UV	82Or
	15.0	$0.1(NaClO_4)$	2.96(2)	UV	81Na
Cu^{2+}	25.0	$0.1(NaClO_4)$	3.3(1)	Potent.	74Na
CMP:					
Mg^{2+}	25.0	$0.1(NaNO_3)$	1.51(5)	Potent.	88Ma
	25.0	$0.1(KNO_3)$	1.75	Potent.	72Fr
Ca^{2+}	25.0	$0.1(NaNO_3)$	1.40(5)	Potent.	88Ma
Sr^{2+}	25.0	$0.1(NaNO_3)$	1.17(4)	Potent.	88Ma
Ba^{2+}	25.0	$0.1(NaNO_3)$	1.11(3)	Potent.	88Ma
Mn^{2+}	25.0	$0.1(NaNO_3)$	2.10(4)	Potent.	88Ma
Co^{2+}	25.0	$0.1(NaNO_3)$	1.86(5)	Potent.	88Ma
Ni^{2+}	25.0	$0.1(NaNO_3)$	1.94(6)	Potent.	88Ma
	25.0	$0.1(KNO_3)$	2.00(6)	Potent.	80Or
	15.0	$0.1(KNO_3)$	1.90	Potent.	72Fr
Cu^{2+}	25.0	$0.1(NaNO_3)$	2.84(6)	Potent.	88Ma
Zn^{2+}	25.0	$0.1(NaNO_3)$	2.06(5)	Potent.	88Ma
Cd^{2+}	25.0	$0.1(NaNO_3)$	2.40(8)	Potent.	88Ma
$(dien)Pd^{2+}$	25.0	$0.2(NaClO_4)$	2.63	UV	84Me
La^{3+}	35.0	$0.1(KNO_3)$	3.73(2)	Potent.	85Ra3
Pr^{3+}	35.0	$0.1(KNO_3)$	3.84(2)	Potent.	85Ra3
Nd^{3+}	35.0	$0.1(KNO_3)$	4.01(3)	Potent.	85Ra3
Sm^{3+}	35.0	$0.1(KNO_3)$	4.11(3)	Potent.	85Ra3
Gd^{3+}	35.0	$0.1(KNO_3)$	4.20(3)	Potent.	85Ra3
Dy^{3+}	35.0	$0.1(KNO_3)$	4.21(4)	Potent.	85Ra3
Er^{3+}	35.0	$0.1(KNO_3)$	4.35(4)	Potent.	85Ra3
UMP:					
Mg^{2+}	25.0	$0.1(NaNO_3)$	1.56(2)	Potent.	88Ma
	23.0	$0.1(NaCl)$	2.25	Distrib.	58Wa
Ca^{2+}	25.0	$0.1(NaNO_3)$	1.44(5)	Potent.	88Ma
Sr^{2+}	25.0	$0.1(NaNO_3)$	1.25(4)	Potent.	88Ma
Ba^{2+}	25.0	$0.1(NaNO_3)$	1.13(6)	Potent.	88Ma
Mn^{2+}	25.0	$0.1(NaNO_3)$	2.11(2)	Potent.	88Ma
Co^{2+}	25.0	$0.1(NaNO_3)$	1.87(5)	Potent.	88Ma
Ni^{2+}	25.0	$0.1(NaNO_3)$	1.97(3)	Potent.	88Ma
Cu^{2+}	25.0	$0.1(NaNO_3)$	2.77(6)	Potent.	88Ma
Zn^{2+}	25.0	$0.1(NaNO_3)$	2.02(7)	Potent.	88Ma
Cd^{2+}	25.0	$0.1(NaNO_3)$	2.38(4)	Potent.	88Ma
TMP:					
Mg^{2+}	25.0	$0.1(NaNO_3)$	1.55(2)	Potent.	88Ma
Ca^{2+}	25.0	$0.1(NaNO_3)$	1.40(6)	Potent.	88Ma
Sr^{2+}	25.0	$0.1(NaNO_3)$	1.19(6)	Potent.	88Ma
Ba^{2+}	25.0	$0.1(NaNO_3)$	1.11(4)	Potent.	88Ma
Mn^{2+}	25.0	$0.1(NaNO_3)$	2.11(5)	Potent.	88Ma
Co^{2+}	25.0	$0.1(NaNO_3)$	1.89(5)	Potent.	88Ma
Ni^{2+}	25.0	$0.1(NaNO_3)$	1.92(6)	Potent.	88Ma
Cu^{2+}	25.0	$0.1(NaNO_3)$	2.87(5)	Potent.	88Ma
Zn^{2+}	25.0	$0.1(NaNO_3)$	2.10(6)	Potent.	88Ma
Cd^{2+}	25.0	$0.1(NaNO_3)$	2.42(3)	Potent.	88Ma

†See also Ref. 64Si, 70He, 79Ma4, 84Si and 85Kr.

macrochelate structure (XXVII), or do the phosphate and base coordinated complexes separately contribute to the observed stability. Massoud & Sigel [88Ma] have shown by comparing the stabilities of metal ion complexes of nucleoside monophosphates and simple phosphate monoesters that the base moiety does not play any role in complexing of the 5'-monophosphates of cytidine (CMP), uridine (UMP), and thymidine. In other words, the complex stabilities are solely determined by the basicity of the phosphate group. This was also shown to be the case with 7-deazaadenosine 5'-monophosphate (XXVIII), a structural analogue of AMP in which the potential base moiety binding site, N7, has been replaced by carbon [88Si]. In contrast, the 3d transition metal complexes of AMP exhibited increased stabilities, which were attributed to simultaneous coordination of the metal ion to N7 and the phosphate group [88Si]. The extent of macrochelate formation was estimated to increase on going from Mn(II) (0.15) to Co(II) (0.49) and Ni(II) (0.71) ions, and then decrease on going to Cu(II) (0.46), Zn(II) (0.45), and Cd(II) (0.43) ions. The more facile macrochelate formation of the Ni(II) ion compared to the Cu(II) ion, which contrasts to their affinity toward the imidazole nitrogen donor, was explained by the different coordination geometry of these two metal ions; owing to Jahn–Teller distortion a phosphate coordinated Cu(II) ion would have fewer sterically suitable positions left to accept the electron pair of N7. Kinetic studies on complexing of Ni(II) ion with IMP [81Na, 84Di] and Co(II) ion with AMP [77Pe2] have lent additional support for macrochelate formation. Both reaction systems were observed to follow stepwise kinetics, consisting of initial outer-sphere association, subsequent inner-sphere phosphate binding, and final attachment of the phosphate coordinated metal ion to the base moiety.

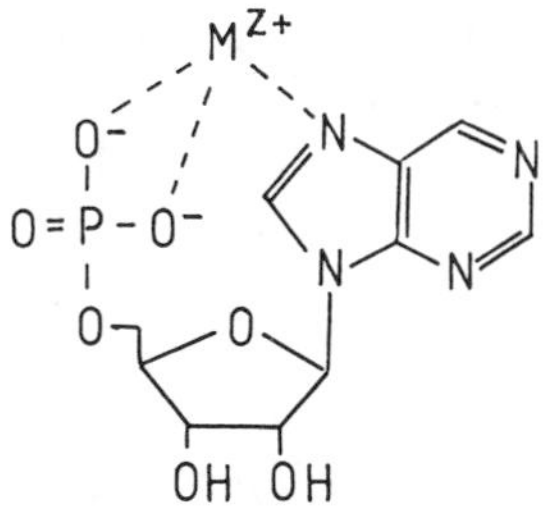

XXVII

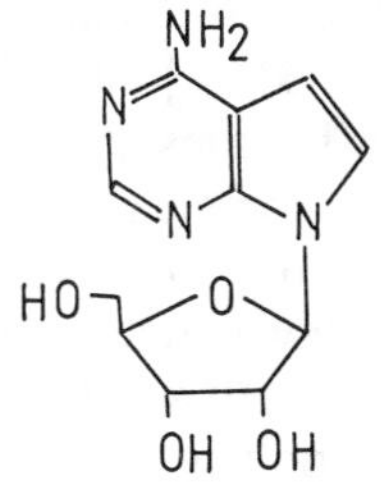

XXVIII

While the kinetic and thermodynamic data on complexing of purine nucleoside 5′-monophosphates with 3d transition metal ions suggest formation of a direct macrochelate structure (XXVII), the results of spectroscopic studies have been accounted for by indirect chelation. Accordingly, ^{1}H and ^{31}P resonance relaxation data on Co(II)-AMP and Co(II)-CMP systems were interpreted in terms of two indirect chelate structures (XXIXa,b) which are in equilibrium with each other [82Le, 86Le]. Structure XXIXa was suggested to prevail at low temperatures and XXIXb at high temperatures. The former type of chelate structure has also been accepted on the basis of UV resonance Raman spectroscopic measurements carried out with Ni(II)-dGMP system in dilute aqueous solutions [87Pe].

XXIXa   XXIXb

The interaction of 3d transition metal ions with both base and phosphate moieties of purine nucleoside monophosphates has been demonstrated in the solid state by IR [79Vi, 83Ta4, 85Gr] and Raman [88Ca] spectroscopy. It is worth noting that direct macrochelate formation is not usually encountered in crystalline complexes. Binary complexes of purine mononucleotides invariably exhibit N7 coordination with water mediated bonding to the phosphate group (XXIXa), as illustrated by the X-ray structures of the Ni(II) complex of AMP [75Co], Mn(II) [74Me1], Ni(II) [74Me2, 79Ge1, 88Be2], and Fe(II) [86Ca] complexes of GMP and/or dGMP, Mn(II) [86Ca], Ni(II) [74Cl, 75Ao], Co(II) [75Ao, 86Po] and Cu(II) [87Ma] complexes of IMP and/or dIMP, and Zn(II) complexes of GMP and IMP methyl esters [85Mi4]. When the binary complex is polymeric, the metal ions bridge between the N7 atom and the phosphate oxygens of different ligand molecules. This kind of structure has been reported for polymeric Cu(II) complexes of GMP [76Ao1, 76Sl2, 78Cl1] and IMP [79Cl], as well as for a polymeric Zn(II) complex of IMP [74Me3]. Ternary complexes of GMP [85Ma4, 87Ma], dGMP [88Be2], and IMP [78Ch] with (en)-Ni(II), (en)Cu(II), and/or (dien)Cu(II) ions display indirect chelate structures similar to those of mononuclear binary complexes.

The X-ray structures of the 3d transition metal complexes of pyrimidine nucleoside monophosphates display no indication of macrochelate formation, in agreement with the suggested solution structures. Binary Co(II) [75Cl, 78Cl2] and Zn(II) [76Ao2, 86Mi3] complexes of CMP are polymeric in nature and exhibit coordination of the central ion to N3 and the phosphate group of separate ligand molecules. The

Mn(II) complex is structurally similar, but O^2 rather than N3 constitutes the binding site in the cytosine ring [76Ao3]. The binary Co(II) complex of UMP is phosphate coordinated [77Ca].

Ternary complexes of 3d transition metal ions with nucleoside monophosphates and heteroaromatic amines usually display only phosphate binding. The list of illustrative examples comprises the (bpy)Cu(II) complexes of AMP [78Ao] and IMP [79Ge2, 87La], and (dpa)Cu(II) complexes of IMP [87Ci, 88Gi], CMP [79Ao] and UMP [77Fi, 78Fi]. Exceptions are the complexes,](bpy)Cu(IMP) $(H_2O)_2]NO_3 H_2O$ [77Ao] and $Na_2[(im)_{0.8}Cu(IMP)_2(H_2O)_{1.2}(H_2O)_2]$ 12.4 H_2O [85 Po], in which the metal ion is N7 coordinated and hydrogen bonded through water to the phosphate group. In the ternary complex, $[(phen)_4Cu_4(IMP)_2(H_2O)_4](NO_3)_2$, two Cu(II) ions bridge in between the N1 sites and phosphate groups of separate ligands [80Ge]. Cd(II) complexes are structurally similar to the corresponding 3d transition metal complexes, as demonstrated by the X-ray structures of binary GMP [76Ao4] and CMP [75Cl, 75Go2, 78Cl2, 78Sh2, 84Ao2] complexes and a ternary (dpa)Cd(II) complex of UMP [79Ao]. The IMP complex, however, displayed ribose binding in addition to the expected N7 coordination [75Go3].

Soft metal cations, including Hg(II), Ag(I), and Pd(II) ions, do not exhibit a marked affinity for the phosphate group, hence their interaction with nucleoside monophosphates is, in all likelihood, similar to that with the corresponding nucleosides. The similar binding of these metal ions to nucleosides and their 5'-monophosphates has been demonstrated by IR spectroscopy for complexes of Ag(I) with GMP and IMP [66Tu], by Raman spectroscopy for complexes of MeHg(II) with AMP [74Ma2, 75Ma5, 88Ta], GMP [74Ma1] and TMP [77Ch2], by ^1H, ^{13}C, and ^{15}N NMR spectroscopy for complexes of (en)Pd(III), (dien)Pd(II), and (peptide)Pd(II) with AMP [80Mo, 80Ro, 81Ve2, 81Ve3, 83Hä1], GMP [80Je, 80Ro, 81Ve2, 89Uc], IMP [80Mo, 89Uc], CMP and TMP [81Ve2], and by ^{15}N NMR spectroscopy for the complex of MeHg(II) with GMP [86Bu]. However, the 5'-phosphate group also has some influence on the binding behaviour of these metal ions. For example, binding of (dien)Pd(II) to N7 of purine nucleoside monophosphates is enhanced by 0.5–0.7 log units compared to nucleosides, whereas no influence on N1 binding is observed [84Ki2]. It has been assumed that the dien ligand hydrogen bonds to the phosphate group, since no similar enhancement of N7 binding can be detected, when the amino protons of dien have been replaced by methyl groups [84Ki2]. Consistent with this proposal, the second pK_a value of the phosphate group of purine mononicleotide is decreased by 0.3–0.5 log units upon coordination of (dien)Pd(II) ion at the N7 site, while the influence of N1 bound metal ion is negligible [84Ki2]. ^{31}P NMR and vibrational spectroscopic studies on the (en)Pd^{2+} complex of GMP have led to a similar conclusion [89Uc]. The higher affinity of the (dien)Pd(II) ion to CMP than to cytidine has also been explained by intracomplex hydrogen bonding [84Me]. However, binding of the (en)Pd(II) ion at N3 of CMP does not exert any effect on the acidity of the phosphate group [83Hä2].

9. THERMODYNAMICS OF PROTOLYTIC AND COMPLEXING EQUILIBRIA

Table 7.12 summarizes the enthalpy and entropy data for the protolytic reactions discussed in the foregoing sections. All the values listed are based on calorimetric enthalpy measurements. In addition, enthalpy and entropy values based on temperature-dependence of the acidity constant have been reported in the literature. These values are less reliable, however, since the temperature range employed is usually limited to 30°C, hence the $\Delta H^{\ominus}$ and $\Delta S^{\ominus}$ values obtained are rather susceptible to experimental errors.

Protonation of nucleic acid bases, as well as the base moieties of nucleosides and their 5′-monophosphates, exhibits negative reaction enthalpies and usually also positive entropies, irrespective of the charge type of the nitrogen atom protonated. In other words, these processes are both enthalpically and entropically favoured. The enthalpy changes expectedly correlate with the basicity of the donor atom, being approximately -10 kJ mol^{-1} for N7 protonation of 6-oxopurines (pK_a 2), -15 to -20 kJ Mol^{-1} for protonation of N1 of adenine and N3 of cytosine (pK_a 4), and -30 to -40 kJ mol^{-1} for protonation of anionic nitrogen atoms (pK_a 9). The entropy factor plays a minor role in monocation formation, but becomes more important in protonation of anionic nitrogen atoms. Accordingly, the entropy changes reported for protonation of neutral pyrimidine type nitrogen atoms fall in the range from $+10$ to $+15$ J K^{-1} mol^{-1}, while those observed for protonation of uncharged imidazole type nitrogen atoms are even slightly negative. Protonation of negatively charged nitrogen atoms, in turn, exhibits entropies from $+50$ to $+90$ J K^{-1} mol^{-1}. Protonation of the 2′-oxyanion is also characterized by a negative enthalpy change (-45 kJ mol^{-1}) and a markedly positive entropy term ($+90$ J K^{-1} mol^{-1}), whereas the thermodynamics of the protolytic equilibria of the 5′-phosphate group are different. Protonation of the phosphate dianion displays a small positive reaction enthalpy, which is overcompensated by a large positive entropy term ($+150$ kJ mol^{-1}). Consequently, the phosphate moiety pK_a values are rather insensitive to temperature, while deprotonation of the base moieties is markedly enhanced at elevated temperatures, which may affect the behaviour of nucleoside monophosphates as ambidentate ligands.

The thermodynamics of monocation formation have also been studied in DMSO [85Be2, 88Be1], a solvent frequently used in spectroscopic studies of complexing equilibria. Protonation in this medium is approximately one order of magnitude more exothermic than in water, owing to a stronger interaction of the monocation with basic DMSO. However, this enthalpy gain is overcompensated by an entropy change of 30 J K^{-1} mol^{-1}. Consequently, the pK_a values obtained in DMSO are usually smaller than those obtained in water, the difference being less than 1 unit.

Calorimetric data on complexing of metal ions with nucleosides and their 5′-monophosphates are extremely scarce. The enthalpies reported for binding of Co(II), Ni(II), Cu(II), and Zn(II) ions at N1 and N7 of 9-methylpurine range from -15 to -20 kJ mol^{-1} [83Ar3], being of the same order of magnitude as those of protonation. However, the entropy change of complex formation is unfavourable (-35 J K^{-1} mol^{-1}). The $\Delta H^{\ominus}$ and $\Delta S^{\ominus}$ values calculated by van't Hoff's equation for complexing of Ag(I) ion with adenosine are qualitatively similar [68Ph]. Coordi-

Table 7.12 — Enthalpies and entropies for protonation of nucleic acid bases, nucleosides and nucleoside 5′-monophosphates at 25.0°C ($I = 0$ if not otherwise stated).

Site of protonation	$H^{\ominus}$/kJ Mol^{-1}	$S^{\ominus}$/J K^{-1} mol^{-1}	Ref.
Protonation of uncharged N:			
N1 purine	− 8.8(4)[†]	+ 16	85Be2
N1 adenine	− 20.1(1)	+ 13.3(3)	70Ch1
	− 16.7(8)[‡]		60Ra
N1 adenosine	− 16.4(1)	+ 12.2(3)	70Ch1
	− 15.9(3)[‡]		60Ra1
	− 15[†]		85Be2
N1 2′-deoxyadenosine	− 16.2(3)[‡]		60Ra
N1 AMP	− 17.6(4)	+ 12.6	62Ch
N1 dAMP	− 11.0(5)[‡]		60Ra
N7 guanosine	− 13.4(8)	− 9(3)	70Ch1
	− 4(1)[‡]		60Ra
N7 2′-deoxyguanosine	− 8.0(4)[‡]		60Ra
N7 dGMP	− 0.6(4)[‡]		60Ra
N7 hypoxanthine	− 12.3(3)	− 7.1(1)	70Ch1
	− 10.5(6)	− 1.3(8)	70Wo
	− 20(2)[§]	− 28	88Be1
N3 cytosine	− 21.5(1)	+ 15.5(4)	67Ch
	− 18.7(8)[‡]		60Ra
N3 cytidine	− 20.2(1)	+ 10.5(4)	70Ch2
N3 2′-deoxycytidine	− 18.0(2)[‡]		60Ra
N3 dCMP	− 17.9(3)		60Ra
Protonation of negatively charged N:			
N7/N9 adenine	− 40.4(2)	+ 53.9(2)	70Ch1
N1 guanosine	− 32.0(1)	+ 69.9(4)	70Ch1
N1 hypoxanthine	− 33.0(2)	+ 60.2(8)	70Ch1
	− 34(1)	+ 56(8)	70Wo
N9 hypoxanthine	− 39.9(1)	+ 97.5(4)	70Ch1
	− 42(4)	+ 90(10)	70Wo
N1 inosine	− 27.2(3)	+ 80.3(8)	70Ch1
N1 cytosine	− 48.1(4)	+ 71(1)	67Ch
N3 uracil	− 32.8(1)	+ 70.7(4)	70Ch2
N3 uridine	− 31.8(6)	+ 72(2)	67Ch
N3 thymine	− 34.1(4)	+ 75.3(4)	70Ch2
N3 thymidine	− 30.6(1)	+ 84.5(4)	70Ch2
Protonation of 2′-oxyanion:			
2′-O$^-$ adenosine	− 40.6(2)	+ 100(1)	66Iz
2′-O$^-$ AMP	− 46(2)	+ 98(6)	66Iz
2′-O$^-$ guanosine	− 45.4(1)	+ 83.7(4)	70Ch1
2′-O$^-$ inosine	− 44.6(1)	+ 86.2(4)	70Ch1
2′-O$^-$ cytidine	− 44.8(4)	+ 85(3)	70Ch2
Protonation of 5′-phosphate dianion:			
AMP	+ 7.5(4)	+ 151	62Ch
	+ 2.1(4)[¶]	+ 126(17)	88An
	− 3.3(1)[§]	+ 132(1)	87Sh3

[†] $I = 0.1$ M (KCl).
[‡] $I = 0.1$ M (NaCl).
[§] $I = 0.1$ M (Et$_4$NClO$_4$).
[¶] $I = 0.1$ M (KNO$_3$).

nation of alkaline earth metal ions to the phosphate group of mononucleotides is, in turn, a slightly endothermic reaction ($\Delta H^{\ominus} = +7$ kJ mol^{-1}), the driving force of which is a favourable entropy change ($+60$ J K^{-1} mol^{-1}) [73Sa, 87Sh3]. The complex formation is thus thermodynamically analogous to protonation of the phosphate dianion. In contrast, complexing of Cu(II) ion with AMP is an exothermic process ($\Delta H^{\ominus} = -26$ kJ mol^{-1}) retarded by an unfavourable entropy change (-21 kJ mol^{-1}) [88An]. The dissimilar thermodynamic behaviour of Cu(II) and alkaline earth metal ions may well reflect different binding modes. As presented in section 8, Cu(II) ion has been suggested to form a macrochelate strcuture, whereas alkaline earth metal cations exhibit phosphate bonding.

10. STABILITY–BASICITY CORRELATIONS

The ability of various nitrogen atoms in ribonucleosides to bind metal ions, decreases in the order uridine(N3$^-$) ~ thymidine(N3$^-$) ~ guanosine(N1$^-$) ~ inosine(N1$^-$) > cytidine(N3) ~ guanosine(N7) > inosine(N7) > adenosine(N7) = adenosine(N1). As seen from Fig. 7.3, the complexing ability parallels the basicity order of the donor

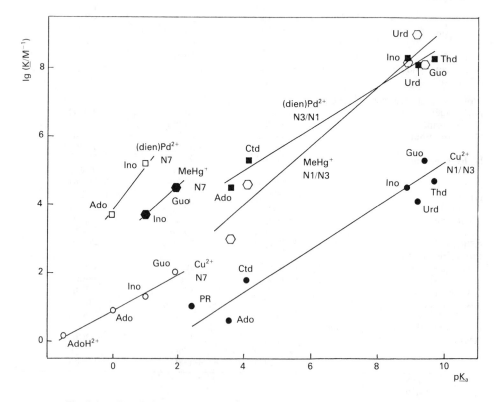

Fig. 7.3 — Correlation between the complexing-ability and basicity of nucleosides.

atoms, although the correlation is a rather poor one. In particular the points referring to binding at the imidazole type nitrogen atoms, fall about 1 log unit above the correlation line given by the pyrimidine bonded complexes. The reason for this difference in binding behaviour remains obscure. It should also be noted that the N1 atom of 6-oxopurine nucleosides and N3 atom of 4-oxopyrimidine nucleosides undergo protonation at pH 9, hence coordination to these sites is impeded under neutral and acidic conditions.

The affinity of various metal ions towards nitrogen atoms of nucleic acid bases or nucleosides decreases in the order (dien)Pd(II) > MeHg(II) > Ag(I) > Cu(II) > Ni(II) > Co(II) > Zn(II) ~ La(III) > Mn(II) > Mg(II), irrespective of the site of coordination. The difference between the affinities of the first and last member in this series appears to be more than 5 orders of magnitude. As seen from Fig. 7.4,

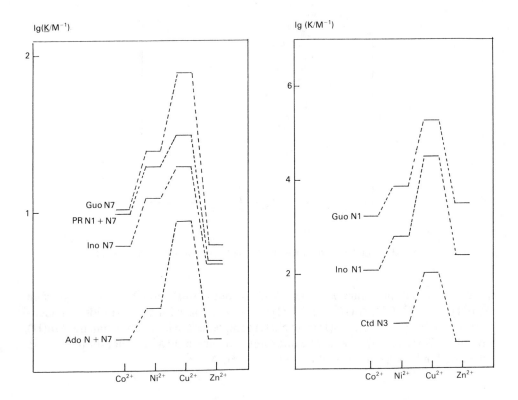

Fig. 7.4 — Stabilities of the 3d transition metal complexes of nucleosides.

complexing of 3d transition metal ions obeys expectedly the Irving–Williams order, the magnitude of ligand field stabilization energy being rather independent of the charge type or position of the donor atom in the ligand. The affinity toward the phosphate group of nucleoside monophosphates, in turn, decreases in the order La(III) > Cu(II) > (dien)Pd(II) > Zn(II) > Ni(II) > Co(II) ~ Mn(II) > Mg(II).

However, phosphate coordination is considerably less susceptible to the nature of the metal ion than the nitrogen binding, the La(III) complexes being only two orders of magnitude more stable than the Mg(II) complexes. The ligand field stabilization energy is somewhat lower than with nitrogen bonded complexes (Fig. 7.5). It is also

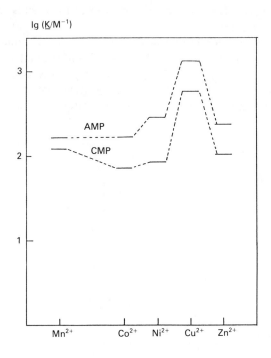

lg ($\underline{K}$/M^{-1})

Fig. 7.5 — Stabilities of the 3d transition metal complexes of nucleoside 5′-monophosphates.

noteworthy that the complexation of AMP is continuously enhanced on going from Mn(II) ion to Co(II), Ni(II), and Cu(II) ions, whereas with the pyrimidine nucleoside monophosphates the Co(II) and Ni(II) complexes are less stable than the Mn(II) complex. This dissimilarity may be interpreted to lend mild additional support to the suggested macrochelate formation in the case of AMP.

11. INTERLIGAND INTERACTIONS

The 1:2 (M:L) metal ion complexes of anionic nucleosides and nucleic acid bases are expectedly less stable than the corresponding 1:1 complexes, the difference between the consecutive stability constants, $K_1 = [ML]/[M][L]$ and $K_2 = [Ml_2]/[M][ML]$, ranging from 0.1 to 1 log unit [69Re1, 72Na, 78Sr, 79Ra1, 80Gh, 87Mi3, 87Ta]. In contrast, the data on relative stabilities of 1:1 and 1:2 complexes of neutral ligands are conflicting. While $\log(K_1/K_2)$ values of 0.2 and 0.4 were reported for the Ag(I) complex of adenosine [68Ph] and (en)Pd(II) complex of cytidine [83Hä2], the 1:2

complexes of neutral nucleic acid bases [83Ta1] and guanosine [88Kh] were suggested to be more stable than the 1:1 complexes. The enhanced stability was attributed to a more negative reaction enthalpy, and this enthalpy gain, which was more marked with purines than with cytosine, was explained to result from interligand stacking interactions [83Ta1]. It remains obscure, however, how a parallel overlapping of the interacting ligands with a distance of 3.4 Å, which is needed for efficient stacking, may be obtained in these complexes. For comparison, the mixed ligand complexes of nucleic acid bases and bipyridyl or o-phenantrholine do not show an extra stabilization compared to binary metal ion–nucleic acid base systems [78Ta, 80Gh, 83Ta5].

The possible interligand interactions in mixed ligand complexes of nucleosides and nucleotides has also been the subject of considerable interest, since metal ions in enzymic systems may be simultaneously coordinated to amino acid residues of proteins and base and/or phosphate moities of nucleic acid constituents. The complexes of mononucleotides with $(bpy)Cu^{2+}$ and/or $(o\text{-}phen)Cu^{2+}$ have been shown to be from 0.2 to 0.5 log units more stable than those with Cu^{2+} [74Na, 89Ma]. With 7-deazaadenosine 5'-monophosphate (XXVIII) the stability difference is even greater, about one order of magnitude [89Ma]. The increased stability has been attributed partly, but not entirely, to stacking interactions between the aromatic ligand and the base moiety of the mononucleotide, coordinated to the central ion through the phosphate group. The electron back donation from Cu(II) ion to the bipyrimidyl ligand may increase the effective positive charge at the central ion and hence promote binding of the nucleotide ligand. For example, the ternary $(bpy)Cu^{2+}$ and $(o\text{-}phen)Cu^{2+}$ complexes of D-ribose 5'-monophosphate are slightly more stable (0.05 log units) than the binary Cu(II) complexes, although no stacking interactions can be present. With mononucleotides the stability difference between the ternary and binary complexes is, however, considerably greater, and this additional stabilization has been regarded as an indication of intracomplex stacking with possible involvement of a hydrophobic contribution [89Ma].

The concept of intracomplex stacking has later been applied to ternary metal ion complexes of adenine [83Ta6], cytosine [86Ra2], cytidine[85Ra1, 86Ra1], inosine [85Ra2], and CMP [86Ra2] with aromatic amino acids, though it is not always quite clear what kind of a geometry could enable intracomplex stacking. As mentioned above, mixed ligand complexes involving oxygen and aromatic nitrogen as donor atoms may display enhanced stability due to electron back donation, even when no stacking interaction occur. In contrast, the ternary 3d transition metal complexes of the uridine monoanion and aromatic amino acids do not exhibit any extra stabilization [86Ra3].

While stacking interactions may possibly enhance complex formation when both interacting molecules are coordinated to the same metal ion, stacking as such appears to retard complexation of nucleosides with metal ions. It has been shown that association of 9-methylpurine with N^6,N^6-dimethyladenosine, which does not form detectably stable complexes with 3d transition metal ions, lowers its affinity to Ni(II) ion by 0.6 log units [85Ar2]. In contrast, binary metal ion complexes of nucleotides may stack somewhat more efficiently then the uncomplexed ligands, owing to diminution of electrostatic repulsion [74Na, 83Sc2]. In summary, the effects that interligand interactions exert on the complex stability remains small, though undoubtedly significant.

ACKNOWLEDGEMENT

The invaluable assistance of Mrs Marjo Lahti, MSc, during the preparation of the manuscript is gratefully acknowledged.

REFERENCES

48Ta   Taylor, H. F. W.: *J. Chem. Soc.* 765 (1948).

51Co   Cochran, W.: *Acta Cryst.* **4**, 81 (1951).

52Fo   Fox, J. J. & Shugar, D.: *Biochim. Biophys. Acta* **9**, 369 (1952).

52Sh   Shugar, D. & Fox, J. J.: *Biochim. Biophys. Acta* **9**, 199 (1952).

53Fo   Fox, J. J., Cavalieri, L. F., & Chang, N.: *J. Am. Chem. Soc.* **75**, 4315 (1953).

54Al   Albert, A. & Brown, D. J.: *J. Chem. Soc.* 2060 (1954).

54Be   Bendich, A., Russell, R. J., Jr., & Fox, J. J.: *J. Am. Chem. Soc.* **76**, 6073 (1954).

56Sm   Smith, R. M. & Alberty, R. A.: *J. Phys. Chem.* **60**, 180 (1956).

58Ha   Harkins, T. R. & Freiser, H.: *J. Am. Chem. Soc.* **80**, 1132 (1958).

58Mi   Miles, H. T.: *Biochim. Biophys. Acta* **27**, 46 (1958).

58Wa   Walaas, E.: *Acta Chem. Scand.* **12**, 528 (1958).

59Mi   Miles, H. T.: *Biochim. Biophys. Acta* **35**, 274 (1959).

60Al   Albert, A. & Serjeant, E. P.: *Biochem. J.* **76**, 621 (1960).

60Ra   Rawitscher, M. & Sturtevant, J. M.: *J. Am. Chem. Soc.* **82**, 3739 (1960).

61An   Angell, C. L.: *J. Chem. Soc.* 504 (1961).

61Fe   Ferreira, R., Ben-Zvi, E., Yamane, T., Vasilevskis, J., & Davidson, N.: *Advances in the chemistry of the coordination compounds*, Kirschner, S., ed., Macmillan, New York (1961), p. 457.

61Mi   Miles, H. T.: *Proc. Natl. Acad. Sci. USA* **47**, 791 (1961).

61Pf   Pfleiderer, W.: *Liebigs Ann. Chem.* **647**, 167 (1961).

61Tr   Trueblood, K. N., Horn, P., & Luzzati, V.: *Acta Cryst.* **14**, 965 (1961).

61We   Wempen, I., Duschinsky, R., Kaplan, L., & Fox, J. J.: *J. Am. Chem. Soc.* **83**, 4755 (1961).

62Br1   Bryan, R. F. & Tomita, K.-I.: *Acta Cryst.* **15**, 1179 (1962).

62Br2   Brown, D. J. & Lyall, J. M.: *Aust. J. Chem.* **15**, 851 (1962).

62Ch   Christensen, J. J. & Izatt, R. M.: *J. Phys. Chem.* **66**, 1030 (1962).

62Ka   Katritzky, A. R. & Waring, A. J.: *J. Chem. Soc.* 1540 (1962).

62Ta   Taqui Khan, M. M. & Martell, A. E.: *J. Am. Chem. Soc.* **84**, 3037 (1962).

63Bö   Börresen, H. C.: *Acta Chem. Scand.* **17**, 921 (1963).

63Ei   Eichhorn, G. L. & Clark, P.: *J. Am. Chem. Soc.* **85**, 4020 (1963).

63Ja   Jardetzky,. O., Pappas, P., & Wade, N. G.: *J. Am. Chem. Soc.* **85**, 1657 (1963).

63Je   Jeffrey, G. A. & Kinoshita, Y.: *Acta Cryst.* **16**, 20 (1963).

63Jo   Jones, J. W. & Robins, R. K.: *J. Am. Chem. Soc.* **85**, 193 (1963).

63Ka   Katritzky, A. R. & Waring, A. J.: *J. Chem. Soc.* 851 (1963).

63Mi1   Miles, H. T., Howard, F. B., & Frazier, J.: *Science* **142**, 1458 (1963).
63Mi2   Miles, H. T.: *J. Am. Chem. Soc.* **85**, 1007 (1963).
63Mi3   Miles, H. T., Bradley, R. B., & Becker, E. D.: *Science* **142**, 1569 (1963).
63Ul   Ulbricht, T. V. L.: *Tetrahedron Lett.* 1027 (1963).
64Ba   Barker, D. L. & Marsh, R. E.: *Acta Cryst.* **17**, 1581 (1964).
64Ma   Mathews, F. S. & Rich, A.: *Nature* **201**, 179 (1964).
64Sc   Schneider, P. W., Brintzinger, H., & Erlenmeyer, H.: *Helv. Chim. Acta* **47**, 992 (1964).
64Si1   Simpson, R. B.: *J. Am. Chem. Soc.* **86**, 2059 (1964).
64Si2   Sigel, H. & Brintzinger, H.: *Helv. Chim. Acta* **47**, 1701 (1964).
64So1   Sobell, H. M. & Tomita, K.: *Acta Cryst.* **17**, 126 (1964).
64So2   Sobell, H. M. & Tomita, K.: *Acta Cryst.* **17**, 122 (1964)..
65Be   Becker, E. D., Miles, H. T., & Bradley, R. B.: *J. Am. Chem. Soc.* **87**, 5575 (1965).
65Cl   Clark, L. B. & Tinoco, I., Jr.: *J. Am. Chem. Soc.* **87**, 11 (1965).
65Fi   Fiskin, A. M. & Beer, M.: *Biochemistry* **4**, 1289 (1965).
65Ib   Iball, J. & Wilson, H. R.: *Proc. Roy. Soc.* **A288**, 418 (1965).
65Iz   Izatt, R. M., Hansen, L. D., Rytting, J. H., & Christensen, J. J.: *J. Am. Chem. Soc.* **87**, 2760 (1965).
65Ro   Roberts, B. W., Lambert, J. B., & Roberts, J. D.: *J. Am. Chem. Soc.* **87**, 5439 (1965).
65Sh   Shulman, R. G., Sternlicht, H., & Wyluda, B. J.: *J. Chem. Phys.* **43**, 3116 (1965).
65Wa   Watson, D. G., Sweet, R. M., & Marsh, R. E.: *Acta Cryst.* **19**, 573 (1965).
65Wi   Wierzchowski, K. L., Litonska, E., & Shugar, D.: *J. Am. Chem. Soc.* **87**, 4621 (1965).
66Bu   Bunville, L. G. & Schwalbe, S. J.: *Biochemistry* **5**, 3521 (1966).
66Do   Doody, B. E., Tucci, E. R., Scruggs, R., & Li, N. C.: *J. Inorg. Nucl. Chem.* **28**, 833 (1966).
66Ei   Eichhorn, G. L., Clark, P., & Becker, E. D.: *Biochemistry* **5**, 245 (1966).
66Ha   Happe, J. A. & Morales, M.: *J. Am. Chem. Soc.* **88**, 2077 (1966).
66Iz   Izatt, R. M., Rytting, J. H., Hansen, L. D., & Christensen, J. J.: *J. Am. Chem. Soc.* **88**, 2641 (1966).
66Si   Sigel, H.: *Experientia* **22**, 497 (1966).
66Tu   Tu, A. T. & Reinosa, J. A.: *Biochemistry* **5**, 3375 (1966).
66Wa   Wang, S. M. & Li, N. C.: *J. Am. Chem. Soc.* **88**, 4592 (1966).
67Bö   Börresen, H. C.: *Acta Chem. Scand.* **21**, 2463 (1967).
67Ch   Christensen, J. J., Rytting, J. H., & Izatt, R. M.: *J. Phys. Chem.* **71**, 2700 (1967).
67Lo1   Lord, R. C. & Thomas, G. J., Jr.: *Spectrochim. Acta* **A23**, 2551 (1967).
67Lo2   Lord, R. C. & Thomas, G. J., Jr.: *Biochim. Biophys. Acta* **142**, 1 (1967).
67Sl   Sletten, E.: *J. Chem. Soc., Chem. Commun.* 1119 (1967).
67Ta   Taqui Khan, M. M. & Martell, A. E.: *J. Am. Chem. Soc.* **89**, 5585 (1967).
68Bu   Bugg, C. E., Thewalt, U., & Marsh, R. E.: *Biochem. Biophys. Res. Commun.* **33**, 436 (1968).
68Ca   Carrabine, J. A. & Sundaralingam, M.: *J. Chem. Soc., Chem. Commun.* 746 (1968).

68Fr   Fritzsche, H. & Zimmer, C.: *Eur. J. Biochem.* **5**, 42 (1968).

68Ph   Phillips, R. & George, P.: *Biochim. Biophys. Acta* **162**, 73 (1968).

68Sh   Shapiro, R.: *Progr. Nucleic Acid Res. Mol. Biol.* **8**, 73 (1968).

68Sl   Sletten, J., Sletten, E., & Jensen, L. H.: *Acta Cryst.* **B24**, 1692 (1968).

68Tu   Tu, A. T. & Friedrich, C. G.: *Biochemistry* **7**, 4367 (1968).

68Wa   Wang, S. M. & Li, N. C.: *J. Am. Chem. Soc.* **90**, 5069 (1968).

69Ea   Eastman, J. W.: *Ber. Bunsenges. Phys. Chem.* **73**, 407 (1969).

69Fu   Fujita, H., Imamura, A., & Nagata, C.: *Bull. Chem. Soc. Jpn.* **42**, 1467 (1969).

69Re1   Reinert, H. & Weiss, R.: *Hoppe-Seyler's Z. Physiol. Chem.* **350**, 1310 (1969).

69Re2   Reinert, H. & Weiss, R.: *Hoppe-Seyler's Z. Physiol. Chem.* **350**, 1321 (1969).

69Ro   Rosenberg, B., Van Camp, L., Trosko, J. E., & Mansour, V. H.: *Nature* **222**, 385 (1969).

69Sl1   Sletten, E.: *Acta Cryst.* **B25**, 1480 (1969).

69Sl2   Sletten, J. & Jensen, L. H.: *Acta Cryst.* **B25**, 1608 (1969).

69Ts   Ts'O, P. O. P., Schweizer, M. P., & Hollis, D. P.: *Ann. N. Y. Acad. Sci.* **158**, 256 (1969).

69Wo   Wolfenden, R. V.: *J. Mol. Biol.* **40**, 307 (1969).

70Ca   Carrabine, J. A. & Sundaralingam, M.: *J. Am. Chem. Soc.* **92**, 369 (1970).

70Ch1   Christensen, J. J., Rytting, J. H., & Izatt, R. M.: *Biochemistry* **9**, 4907 (1970).

70Ch2   Christensen, J. J., Rytting, J. H., & Izatt, R. M.: *J. Chem. Soc. B* 1643 (1970).

70Fr1   Fritzsche, H. & Tresselt, D.: *Stud. Biophys.* **24/25**, 299 (1970).

70Fr2   Fritzsche, H.: *Biochim. Biophys. Acta* **224** 608 (1970).

70He   Heller, M. J., Jones, A. J., & Tu, A. T.: *Biochemistry* **9**, 4981 (1970).

70Ka   Kan, L. S. & Li, N. C.: *J. Am. Chem. Soc.* **92**, 4823 (1970).

70Me   de Meester, P., Goodgame, D. M. L., Price, K. A., & Skapski, A. C.: *J. Chem. Soc., Chem. Commun.* 1573 (1970).

70Pu   Pullman, B., Berthod, H., Bergmann, F., & Neiman, Z.: *Tetrahedron* **26**, 1483 (1970).

70Sl   Sletten, E.: *Acta Cryst.* **B26**, 1609 (1970).

70Sn   Snyder, L. C., Shulman, R. G., & Neuman, D. B.: *J. Chem. Phys.* **53**, 256 (1970).

70Sr   Srinivasan, L. & Taylor, M. R.: *J. Chem. Soc., Chem. Commun.* 1668 (1970).

70Te   Tewari, K. C., Lee, J., & Li, N. C.: *Trans. Faraday Soc.* **66**, 2069 (1970).

70Wa   Wagner, R. & von Phillipsborn, W.: *Helv. Chim. Acta* **53**, 299 (1970).

70Wo   Woolley, E. M., Wilton, R. W., & Hepler, L. G.: *Can. J. Chem.* **48**, 3249 (1970).

70Wr   Wrobel, A., Rabczenko, A., & Shugar, D.: *Acta Biochim. Pol.* **17**, 339 (1970).

70Zo   Zoltewicz, J. A., Clark, D. F., Sharpless, T. W., & Grahe, G.: *J. Am. Chem. Soc.* **92**, 1741 (1970).

71Be1   Berger, N. A. & Eichhorn, G. L.: *Biochemistry* **10**, 1847 (1971).

71Be2   Berger, N. A. & Eichhorn, G. L.: *J. Am. Chem. Soc.* **93**, 7062 (1971).
71Be3   Berger, N. A. & Eichhorn, G. L.: *Biochemistry* **10**, 1857 (1971).
71Br   Breen, D. L. & Flurry, R. L.: *Theor. Chim. Acta* **13**, 138 (1971).
71Ca   Carrabine, J. A. & Sundaralingam, M.: *Biochemistry* **10**, 292 (1971).
71De   Declercq, J. P., Debbaudt, M., & Van Meerssche, M.: *Bull. Soc. Chim. Belg.* **80**, 527 (1971).
71Fr   Fritzsche, H., Tresselt, D., & Zimmer, C.: *Experimentia* **27**, 1253 (1971).
71Iz   Izatt, R. M., Christensen, J. J., & Rytting, J. H.: *Chem. Rev.* **71**, 439 (1971) and references therein.
71Ka   Karpel, R. L., Kustin, K., & Wolff, M. A.: *J. Phys. Chem.* **75**, 799 (1971).
71Me1   de Meester, P., Goodgame, D. M. L., Price, K. A., & Skapski, A. C.: *Biochem. Biophys. Res. Commun.* **44**, 510 (1971).
71Me2   de Meester, P. & Skapski, A. C.: *J. Chem. Soc.* (*A*) 2167 (1971).
71Me3   de Meester, P., Goodgame, D. M. L., Price, K. A., & Skapski, A. C.: *Nature* **229**, 191 (1971).
71Me4   Medeiros, G. C. & Thomas, G. J., Jr.: *Biochim. Biophys. Acta* **238** 1 (1971).
71Ps   Psoda, A. & Shugar, D.: *Biochim. Biophys. Acta* **247**, 507 (1971).
71Pu1   Pugmire, R. J. & Grant, D. M.: *J. Am. Chem. Soc.* **93**, 1880 (1971).
71Pu2   Pullman, B. & Pullman, A.: *Adv. Heterocyclic Chem.* **13**, 77 (1971).
71Sh   Shapiro, R. & Kang, S.: *Biochim. Biophys. Acta* **232**, 1 (1971).
71Sl   Sletten, E.: *J. Chem. Soc., Chem. Commun.* 558 (1971).
71Su1   Sundaralingam, M. & Carrabine, J. A.: *J. Mol. Biol.* **61**, 287 (1971).
71Su2   Susi, H., Ard, J. S., & Purcell, J. M.: *Spectrochim. Acta* **A27**, 1549 (1971).
71Ta   Taqui Khan, M. M. & Krishnamoorthy, C. R.: *J. Inorg. Nucl. Chem.* **33**, 1417 (1971).
71Th   Thewalt, U., Bugg, C. E., & Marsh, R. E.: *Acta Cryst.* **B27**, 2358 (1971).
71Wa   Wagner, R. & von Phillipsborn, W.: *Helv. Chim. Acta* **54**, 1543 (1971).
72An   Antonovskii, V. L., Gokovskaya, A. S., Prokopeva, T. M., & Sukhorukov, B. I.: *Dokl. Akad. Nauk SSSR* **205**, 461 (1972).
72Be   Berger, N. A., Tarien, E., & Eichhorn, G. L.: *Nature* **239**, 237 (1972).
72Bl   Blinc, R., Mali, M., Osredkar, R., Prelesnik, A., Seliger, J., Zupancic, I., & Ehrenberg, L.: *J. Chem. Phys.* **57**, 5087 (1972).
72Bo   Boivin, G. & Zador, M.: *Can. J. Chem.* **50**, 3117 (1972).
72Ed   Edmonds, D. T. & Speight, P. A.: *J. Magn. Res.* **6**, 265 (1972).
72Fr   Frey, C. M. & Stuehr, J. E.: *J. Am. Chem. Soc.* **94**, 8898 (1972).
72Gu   Gukovskaya, A. S., Sukharakov, B. I., Prokopeva, T. M., & Antonovskii, V. L.: *Bull. Akad. Sci. USSR* **21**, 2614 (1972).
72Jo   Jordan, F. & McFarquhar, B. Y.: *J. Am. Chem. Soc.* **94**, 6557 (1972).
72Li   Lichtenberg, D., Bergmann, F., & Neiman, Z.: *Isr. J. Chem.* **10**, 805 (1972).
72Me   de Meester, P. & Skapski, A. C.: *J. Chem. Soc., Dalton Trans.* 2400 (1972).
72Mi   Missen, A. W., Natusch, D. F. S., & Porter, L. J.: *Aust. J. Chem.* **25**, 129 (1972).
72Na   Nayan, R. & Dey, A. K.: *Z. Naturforsch.* **B27**, 688 (1972).
72Sh   Shapiro, R. & Danzig, M.: *Biochemistry* **11**, 23 (1972).

72Wo   Wong, Y. P., Wong, K. L., & Kearns, D. R.: *Biochem. Biophys. Res. Commun.* **6**, 1580 (1972).

72Zi   Zimmer, S. & Biltonen, R.: *J. Solution Chem.* **1**, 291 (1972).

73Bo   Boivin, G. & Zador, M.: *Can. J. Chem.* **51**, 3322 (1973).

73Ei   Eichhorn, G. L.: *Inorganic biochemistry*, Eichhorn, G. L., Ed., Elsevier, Amsterdam (1973), p. 1191.

73Ko1   Kotowycz, G. & Suzuki, O.: *Biochemistry* **12**, 3434 (1973).

73Ko2   Kotowycz, G. & Hayamizu, K.: *Biochemistry* **12**, 517 (1973).

73Ko3   Kotowycz, G. & Suzuki, O.: *Biochemistry* **12**, 5325 (1973).

73Mc   McClure, R. J. & Craven, B. M.: *Acta Cryst.* B29, 1234 (1973).

73Me1   de Meester, P., Goodgame, D. M. L., Skapski, A. C., & Warnke, Z.: *Biochim. Biophys. Acta* **324**, 301 (1973).

73Me2   de Meester, P. & Skapski, A. C.: *J. Chem. Soc., Dalton Trans.* 424 (1973).

73Me3   de Meester, P. & Skapski, A. C.: *J. Chem. Soc., Dalton Trans.* 1596 (1973).

73Me4   de Meester, P., Goodgame, D. M. L., Richman, D. J., & Skapski, A. C.: *Nature* **242**, 257 (1973).

73Pi   Pieber, M., Kroon, P. A., Prestegard, J. H., & Chan, S.: *J. Am. Chem. Soc.* **95**, 3408 (1973).

73Sa   Sari, J. C. & Belaich, J. P.: *J. Am. Chem. Soc.* **95**, 7491 (1973).

73Sh1   Shikata, K., Ueki, T., & Mitsui, T.: *Acta Cryst.* **B29**, 31 (1973).

73Sh2   Shimokawa, S., Fukui, H., Sohma, J., & Hotta, K.: *J. Am. Chem. Soc.* **95**, 1777 (1973).

73So   Sorarrain, O. M. & Castro, E. A.: *Chem. Phys. Lett.* **19**, 422 (1973).

73Ta   Taylor, M.: *Acta Cryst.* **B29** 884 (1973).

73Te   Terzis, A., Beauchamp, A. L., & Rivest, R.: *Inorg. Chem.* **12**, 1166 (1973).

73To   Tomita, K., Izuno, T., & Fujiwara, T.: *Biochem. Biophys. Res. Commun.* **54**, 96 (1973).

73Tu   Tu, A. T. & Heller, M. J.: *Met. Ions Biol. Syst.* **1**, 1 (1973).

73Wo   Wong, Y. P.: *J. Am. Chem. Soc.* **95**, 3511 (1973).

73Yu   Yulin Liu Tan & Beck, A.: *Biochim. Biophys. Acta* **299**, 500 (1973).

74Ba   Banyasz, J. L. & Stuehr, J. E.: *J. Am. Chem. Soc.* **96**, 6481 (1974).

74Bu   Budo, G. & Tomasz, J.: *Acta Biochim. Biophys. Acad. Sci. Hung.* **9**, 217 (1974).

74Ch   Chang, C. H. & Marzilli, L. G.: *J. Am. Chem. Soc.* **96**, 3656 (1974).

74Cl   Clark, G. P. & Orbell, J. D.: *J. Chem. Soc. Chem. Commun.* 139 (1974).

74Es   Espersen, W. G., Hutton, W. C., Chow, S. T., & Martin, R. B.: *J. Am. Chem. Soc.* **96**, 8111 (1974).

74Fr   Fritzsche, H., Arnold, K., & Krusche, R.: *Stud. Biophys.* **45**, 131 (1974).

74Ki1   Kistenmacher, T. J. & Shigematsu, T.: *Acta Cryst.* **B30**, 166 (1974).

74Ki2   Kistenmacher, T. J. & Shigematsu, T.: *Acta Cryst.* **B30**, 1528 (1974).

74Ko1   Kong, P.-C. & Theophanides, T.: *Inorg. Chem.* **13**, 1981 (1974).

74Ko2   Kotowycz, G.: *Can. J. Chem.* **52**, 924 (1974).

74Ko3   Kosturko, L. D., Folzer, C., & Stewart, R. F.: *Biochemistry* **13**, 3949 (1974).

74Ma1   Mansy, S. & Tobias, R. S.: *J. Am. Chem. Soc.* **96**, 6874 (1974).

74Ma2   Mansy, S., Wood, T. E., Sprowles, J. C., & Tobias, R. S.: *J. Am. Chem. Soc.* **96**, 1762 (1974).

74Me1	de Meester, P., Goodgame, D. M. L., Jones, T. J., & Skapski, A. C.: *Biochem. J.* **139**, 791 (1974).
74Me2	de Meester, P., Goodgame, D. M. L., Skapski, A. C., & Smith, B. T.: *Biochim. Biophys. Acta* **340**, 113 (1974).
74Me3	de Meester, P., Goodgame, D. M. L., Jones, T. J., & Skapski, A. C.: *Biochim. Biophys. Acta* **353**, 392 (1974).
74Na	Naumann, C. F. & Sigel, H.: *J. Am. Chem. Soc.* **96**, 2750 (1974).
74Sa	Saito, K., Terashima, R., Sakaki, T., & Tomita, K.: *Biochem. Biophys. Res. Commun.* **61**, 83 (1974).
74Sl1	Sletten, E. & Thorstensen, B.: *Acta Cryst.* **B30**, 2438 (1974).
74Sl2	Sletten, E.: *Acta Cryst.* **B30**, 1961 (1974).
74Su	Susi, H. & Ard, J. S.: *Spectrochim. Acta* **A30**, 1843 (1974).
74Ta	Taqui Khan, M. M. & Krishnamoorthy, C. R.: *J. Inorg. Nucl. Chem.* **36**, 711 (1974).
74To	Tougard, P. & Chantot, J.-F.: *Acta Cryst.* **B30**, 214 (1974).
74We	Weser, U., Strobel, G.-J., & Voelter, W.: *FEBS Lett.* **41**, 243 (1974).
75Ao	Aoki, K.: *Bull. Chem. Soc. Japan* **48**, 1260 (1975).
75Bo	Bonnaccorsi, R., Scrocco, E., Tomasi, J., & Pullman, A.: *Theor. Chim. Acta* **36**, 339 (1975).
75Br	Brun, G., Goodgame, D. M. L., & Skapski, A. C.: *Nature* **253**, 127 (1975).
75Ca	Casper, A. & Fazakerley, G. V.: *J. Chem. Soc., Dalton Trans.* 1977 (1975).
75Ch	Chenon, M.-T., Pugmire, R. J., Grant, D. M., Panzica, R., & Townsend, L. B.: *J. Am. Chem. Soc.* **97**, 4636 (1975).
75Cl	Clark, G. R. & Orbell, J. D.: *J. Chem. Soc.* 697 (1975).
75Co	Collins, A. D., de Meester, P., Goodgame, D. M. L., & Skapski, A. C.: *Biochim. Biophys. Acta* **402**, 1 (1975).
75Da	Daune, M.: *Met. Ions Biol. Syst.* **3**, 1 (1975).
75De	DeMember, J. R. & Wallace, F. A.: *J. Am. Chem. Soc.* **97**, 6240 (1975).
75Dr	Dreyfus, M., Dodin, G., Bensaude, O., & Dubois, J. E.: *J. Am. Chem. Soc.* **97**, 2369 (1975).
75Fr	Frederick, G. D. & Poulter, C. D.: *J. Am. Chem. Soc.* **97**, 1797 (1975).
75Ga	Gagnon, C. & Beauchamp, A. L.: *Inorg. Chim. Acta* **14**, L52 (1975).
75Go1	Goddard, J. D., Mexey, P. G., & Csizmadia, I. G.: *Theor. Chim. Acta* **39**, 1 (1975).
75Go2	Goodgame, D. M. L., Jeeves, I., Reynolds, C. D., & Skapski, A. C.: *Biochem. J.* **151**, 467 (1975).
75Go3	Goodgame, D. M. L., Jeeves, I., Reynolds, C. D., & Skapski, A. C.; *Nucleic Acids Res.* **2**, 1375 (1975).
75Je	Jennette, K. W., Lippard, S. J., & Ucko, D. A.: *Biochim. Biophys. Acta* **402**, 403 (1975).
75Ki1	Kistenmacher, T. J., Szalda, D. J., & Marzilli, L. G.: *Inorg. Chem.* **14**, 1686 (1975).
75Ki2	Kistenmacher, T. J., Szalda, D. J., & Marzilli, L. G.: *Acta Cryst.* **B31**, 2416 (1975).
75Ki3	Kistenmacher, T. J., Sorrell, T., & Marzilli, L. G.: *Inorg. Chem.* **14**, 2479 (1975).

75Ma1   Marzilli, L. G., Trogler, W. C., Hollis, D. P., Kistenmacher, T. J., Chang, C.-H., & Hollis, B. E.: *Inorg. Chem.* **14**, 2568 (1975).

75Ma2   Mandel, G. S. & Marsh, R. E.: *Acta Cryst.* **B31**, 2862 (1975).

75Ma3   Mansy, S. & Tobias, R. S.: *Biochemistry* **14**, 2952 (1975).

75Ma4   Mansy, S. & Tobias, R. S.: *Inorg. Chem.* **14**, 287 (1975).

75Ma5   Mansy, S., Frick, J. P., & Tobias, R. S.: *Biochim. Biophys. Acta* **378**, 319 (1975).

75Mc   McCall, M. J. & Taylor, M. R.: *Biochim. Biophys. Acta* **390**, 137 (1975).

75Po   Poulter, C. D. & Frederick, G. D.: *Tetrahedron Lett.* 2171 (1975).

75Sl   Sletten, E. & Ruud, M.: *Acta Cryst.* **B31**, 982 (1975).

75Sz1   Szalda, D. J., Kistenmacher, T. J., & Marzilli, L. G.: *Inorg. Chem.* **14**, 2623 (1975).

75Sz2   Szalda, D. J., Marzilli, L. G., & Kistenmacher, T. J.: *Inorg. Chem.* **14**, 2076 (1975).

75Sz3   Szalda, D. J., Marzilli, L. G., & Kistenmacher, T. J.: *Biochem. Biophys. Res. Commun.* **63**, 601 (1975).

75Yo   Yokono, T., Shimokawa, S., & Sohma, J.: *J. Am. Chem. Soc.* **97**, 3827 (1975).

76Ab   Abdulnur, S.: *J. Theor. Biol.* **58**, 165 (1976).

76Ao1   Aoki, K., Clark, G. R., & Orbell, J. D.: *Biochim. Biophys. Acta* **425**, 369 (1976).

76Ao2   Aoki, K.: *Biochim. Biophys. Acta* **447**, 379 (1976).

76Ao3   Aoki, K.: *J. Chem. Soc., Chem. Commun.* 748 (1976).

76Ao4   Aoki, K.: *Acta Cryst.* **B32**, 1454 (1976).

76Co   Coletta, F., Ettore, R., & Gambaro, A.: *J. Magn. Res.* **22**, 453 (1976).

76Dr   Dreyfus, M., Bensuade, O., Dodin, G., & Dubois, J. E.: *J. Am. Chem. Soc.* **98**, 6338 (1976).

76Es1   Espersen, W. G. & Martin, R. B.: *J. Am. Chem. Soc.* **98**, 40 (1976).

76Es2   Espersen, W. G. & Martin, R. B.: *J. Phys. Chem.* **80**, 161 (1976).

76Ki   Kistenmacher, T. J., Marzilli, L. G., & Szalda, D. J.: *Acta Cryst.* **B32**, 186 (1976).

76Li1   Lim, M. C. & Martin, R. B.: *J. Inorg. Nucl. Chem.* **83** 1915 (1976).

76Li2   Linder, P. W., Stanford, M. J., & Williams, D. R.: *J. Inorg. Nucl. Chem.* **38**, 1847 (1976).

76Mc   McCall, M. J. & Taylor, M. R.: *Acta Cryst.* **B32**, 1687 (1976).

76Ne   Nelson, D. J., Yeagle, P. L., Miller, T. L., & Martin, R. B.: *Bioinorg. Chem.* **5**, 353 (1976).

76Og   Ogasawara, N. & Inoue, Y.: *J. Am. Chem. Soc.* **98**, 7048 (1976).

76Pe   Perahia, D., Pullman, A., & Pullman, B.: *Theor. Chim. Acta* **42**, 23 (1976).

76Pl   Plaush, A. C. & Sharp, R. R.: *J. Am. Chem. Soc.* **98**, 7973 (1976).

76Sh   Shimokawa, S., Yokono, T., & Sohma, J.: *Biochim. Biophys. Acta* **425**, 349 (1976).

76Sl1   Sletten, E. & Flogstad, N.: *Acta Cryst.* **B32**, 461 (1976).

76Sl2   Sletten, E. & Lie, B.: *Acta Cryst.* **B32**, 3301 (1976).

76St   Sternglanz, H., Subramanian, E., Lacey, J. C., Jr., & Bugg, C. E.: *Biochemistry* **15**, 4797 (1976).

76Sz   Szalda, D. J., Kistenmacher, T. J., & Marzilli, L. G.: *J. Am. Chem. Soc.* **98**, 8371 (1976).

76Ta   Taylor, R. S. & Diebler, H.: *Bioinorg. Chem.* **6**, 247 (1976).

77Au   Authier-Martin, M. & Beauchamp, A. L.: *Can. J. Chem.* **55**, 1213 (1977).

77Ao   Aoki, K.: *J. Chem. Soc., Chem. Commun.* 600 (1977).

77Br1   Brown, D. B., Hall, J. W., Helis, H. M., Walton, E. G., Hodgson, D. J., & Hatfield, W. E.: *Inorg. Chem.* **16**, 2675 (1977).

77Br2   Bressan, M., Ettore, R., & Rigo, P.: *J. Magn. Res.* **26**, 42 (1977).

77Ca   Cartwright, B. A., Goodgame, D. M. L., Jeeves, I., & Skapski, A. C.: *Biochim. Biophys. Acta* **477**, 195 (1977).

77Ch1   Chao, Y.-Y. H. & Kearns, D. R.: *J. Am. Chem. Soc.* **99**, 6425 (1977).

77Ch2   Chrisman, R. W., Mansy, S., Peresie, H. J., Ranade, A., Berg, T. A., & Tobias, R. S.: *Bioinorg. Chem.* **7**, 245 (1977).

77Ci   Cini, R., Colamario, P., & Orioli, P. L.: *Bioinorg. Chem.* **7**, 345 (1977).

77De   Dehand, J. & Jordanov, J.: *J. Chem. Soc., Dalton Trans.* 1588 (1977).

77Dr   Dreyfus, M., Dodin, G., Bensaude, O., & Dubois, J. E.: *J. Am. Chem. Soc.* **99**, 7027 (1977).

77Et   Ettore, R.: *Inorg. Chim. Acta* **25**, L9 (1977).

77Fi   Fischer, B. E. & Bau, R.: *J. Chem. Soc., Chem. Commun.* 272 (1977).

77Ga1   Gagnon, C. & Beauchamp, A. L.: *Acta Cryst.* **B33**, 1448 (1977).

77Ga2   Gagnon, C., Hubert, J., Rivest, R., & Beauchamp, A. L.: *Inorg. Chem.* **16**, 2469 (1977).

77Go   Goodgame, M. & Johns, K. W.: *J. Chem. Soc., Dalton Trans.* 1680 (1977).

77Ho   Hodgson, D. J.: *Progr. Inorg. Chem.* **23**, 211 (1977).

77Ki   Kirchner, S. J., Fernando, Q., & Chvapil, M.: *Inorg. Chim. Acta* **25**, L45 (1977).

77Ko   Kozlowski, H.: *Inorg. Chim. Acta* **24**, 215 (1977).

77Ma1   Marzilli, L. G. & Kistenmacher, T. J.: *Acc. Chem. Res.* **10**, 146 (1977).

77Ma2   Marzilli, L. G.: *Progr. Inorg. Chem.* **23**, 255 (1977).

77Ma3   Markowski, V., Sullivan, G. R., & Roberts, J. D.: *J. Am. Chem. Soc.* **99**, 714 (1977).

77Ma4   Marzilli, L. G., Kistenmacher, T. J., & Rossi, M.: *J. Am. Chem. Soc.* **99**, 2797 (1977).

77Oc   Ochiai, E.-I.: *Bioinorganic chemistry: an introduction*, Allyn & Bacon, Boston (1977).

77Pe1   Perahia, D., Pullman, A., & Pullman, B.: *Theor. Chim. Acta* **43**, 207 (1977).

77Pe2   Peguy, A. & Diebler, H.: *J. Phys. Chem.* **81**, 1355 (1977).

77Pn   Pneumatikakis, G., Hadjiliadis, N., & Theophanides, T.: *Inorg. Chim. Acta* **22**, L1 (1977).

77Pu   Pullman, A. & Armbruster, A. M.: *Theor. Chim. Acta* **45**, 249 (1977).

77Ra   Ragot, M., Sari, J. C., & Belaich, J. P.: *Biochim. Biophys. Acta* **499**, 411 (1977).

77Si   Sinn, E., Flynn, C. M., Jr., & Martin, R. B.: *Inorg. Chem.* **16**, 2403 (1977).

77Sl1   Sletten, E. & Kaale, R.: *Acta Cryst.* **B32**, 158 (1977).

77Sl2   Sletten, E. & Ervik, G.: *Acta Cryst.* **B33**, 1633 (1977).

77St   Stolarski, R., Remin, M., & Shugar, D.: *Z. Naturforsch.* **C32**, 894 (1977).

77Sz   Szalda, D. J. & Kistenmacher, T. J.: *Acta Cryst.* **B33**, 865 (1977).

77Ta   Taqui Khan, M. M. & Jyoti, M. S.: *Indian J. Chem.* **A15**, 1002 (1977).

77Vi   Villa, J. F., Doyle, R., Nelson, H. C., & Richards, J. L.: *Inorg. Chim. Acta* **25**, 49 (1977).

78Ao   Aoki, K.: *J. Am. Chem. Soc.* **100**, 7106 (1978).

78Au   Authier-Martin, M., Hubert, J., Rivest, R., & Beauchamp, A. L.: *Acta Cryst.* **B34**, 273 (1978).

78Be   Behrens, N. B., Goodgame, D. M. L., & Warnke, Z.: *Inorg. Chim. Acta* **31**, 257 (1978).

78Ch   Chiang, C. C., Sorrell, T., Kistenmacher, T. J., & Marzilli, L. G.: *J. Am. Chem. Soc.* **100**, 5102 (1978).

78Cl1   Clark, G. R., Orbell, J. D., & Aoki, K.: *Acta Cryst.* **B34**, 2119 (1978).

78Cl2   Clark, G. R. & Orbell, J. D.: *Acta Cryst.* **B34**, 1815 (1978).

78De   Delabar, J. M. & Majoube, M.: *Spectrochim. Acta* **A34**, 129 (1978).

78Fi   Fischer, B. E. & Bau, R.: *Inorg. Chem.* **17**, 27 (1978).

78Go1   Goodgame, M. & Johns, K. W.: *Inorg. Chim. Acta* **30**, L335 (1978).

78Go2   Goodgame, M. & Johns, K. W.: *J. Chem. Soc., Dalton Trans.* 1294 (1978).

78Gu   Guichelaar, M. A. & Reedijk, J.: *Recl. Trav. Chim. Pays-Bas* **97**, 295 (1978).

78Ha   Hadjiliadis, N. & Pneumatikakis, G.: *J. Chem. Soc., Dalton Trans.* 1691 (1978).

78Ki   Kistenmacher, T. J., Szalda, D. J., Chiang, C. C., Rossi, M., & Marzilli, L. G.: *Inorg. Chem.* **17**, 2582 (1978).

78La   Langer, V., Huml, K.: *Acta Cryst.* **B34**, 1157 (1978).

78Ma1   Maskos, K.: *Acta Biochim. Pol.* **25**, 311 (1978).

78Ma2   Maskos, K.: *Acta Biochim. Pol.* **25**, 113 (1978).

78Ma3   Maskos, K.: *Acta Biochim. Pol.* **25**, 101 (1978).

78Ma4   Maskos, K.: *Acta Biochim. Pol.* **25**, 303 (1978).

78Ma5   Marzilli, L. G., Stewart, R. C., Van Vuuren, C. P., de Castro, B., & Caradonna, J. P.: *J. Am. Chem. Soc.* **100**, 3967 (1978).

78No   Nowak, J., Szczepaniak, K., Barski, A., & Shugar, D.: *Z. Naturforsch.* **C33**, 876 (1978).

78Pn   Pneumatikakis, G., Hadjiliadis, N., & Theophanides, T.: *Inorg. Chem.* **17**, 915 (1978).

78Sa   Sakaguchi, H., Anzai, H., Furuhata, K., Ogura, H., Iitaka, Y., Fukita, T., & Sakaguchi, T.: *Chem. Pharm. Bull.* **26**, 2465 (1978).

78Sh1   Shirotake, S. & Sakaguchi, T.: *Chem. Pharm. Bull.* **26**, 2941 (1978).

78Sh2   Shiba, J. K. & Bau, R.: *Inorg. Chem.* **17**, 3484 (1978).

78Sr   Srivastava, R. C. & Srivastava, M. N.: *J. Inorg. Nucl. Chem.* **40**, 1439 (1978).

78Ta   Taqui Khan, M. M. & Jyoti, M. S.: *J. Inorg. Nucl. Chem.* **40**, 1731 (1978).

78We   Wei, C.-Y., Fischer, B. E., & Bau, R.: *J. Chem. Soc., Chem. Commun.* 1053 (1978).

78Yu   Yu, C., Peng, S., Akiyama, I., Lin, K., & LeBreton, P. R.: *J. Am. Chem. Soc.* **100**, 2303 (1978).

79Ao   Aoki, K.: *J. Chem. Soc., Chem. Commun.* 589 (1979).

79Be   Beck, W. M., Calabrese, J. C., & Kottmair, N. D.: *Inorg. Chem.* **18**, 176 (1979).
79Ca   Canty, A. J. & Tobias, R. S.: *Inorg. Chem.* **18**, 413 (1979).
79Ch   Chalilpoyil, P. & Marzilli, L. G.: *Inorg. Chem.* **18**, 2328 (1979).
79Ci   Cingi, M. B., Lanfredi, A. M. M., Tiripicchio, A., & Camelli, M. T.: *Transition Met. Chem.* **4**, 221 (1979).
79Cl   Clark, G. R., Orbell, J. D., & Waters, J. M.: *Biochim. Biophys. Acta* **562**, 361 (1979).
79Cz   Czerminski, R., Lesyng, B., & Pohorille, A.: *Int. J. Quantum Chem.* **16**, 605 (1979).
79Fa   Fazakerley, G. V., Jackson, G. E., Phillips, M. A., & Niekerk, J. C.: *Inorg. Chim. Acta* **35**, 151 (1979).
79Ga   Gagnon, C., Beauchamp, A. L., & Tranqui, D.: *Can. J. Chem.* **57**, 1372 (1979).
79Ge1   Gellert, R. W., Shiba, J. K., & Bau, R.: *Biochem. Biophys. Res. Commun.* **88**, 1449 (1979).
79Ge2   Gellert, R. W., Foscher, B. E., & Bau, R.: *Biochem. Biophys. Res Commun.* **88**, 1443 (1979).
79Ge3   Gellert, R. W. & Bau, R.: *Met. Ions Biol. Syst.* **8**, 1 (1979).
79Gu   Guay, F. & Beauchamp, A. L.: *J. Am. Chem. Soc.* **101**, 6260 (1979).
79Ho   Hoo, D.-L. & McConnell, B.: *J. Am. Chem. Soc.* **101**, 7470 (1979).
79Ki   Kistenmacher, T. J., Rossi, M., & Marzilli, L. G.: *Inorg. Chem.* **18**, 240 (1979).
79Ku   Kupce, E. & Sekacis, I.: *Khim. Prir. Soedin* 565 (1979).
79Li   Lippert, B.: *J. Raman Spectrosc.* **8**, 274 (1979).
79Ma1   Martin, R. B. & Mariam, Y.: *Met. Ions Biol. Syst.* **8**, 57 (1979).
79Ma2   Mansy, S., Peticolas, W. L., & Tobias, R. S.: *Spectrochim. Acta* **A35**, 315 (1979).
79Ma3   Maskos, K.: *Acta Biochem. Pol.* **26**, 249 (1979).
79Ma4   Mariam, Y. H. & Martin, R. B.: *Inorg. Chim. Acta* **35**, 23 (1979).
79Ne1   Nelson, H. C. & Villa, J. F.: *J. Inorg. Nucl. Chem.* **41**, 1643 (1979).
79Ne2   Nelson, H. C. & Villa, J. F.: *Inorg. Chem.* **18**, 1725 (1979).
79Pe   Peringer, P.: *Z. Naturforsch.* **B34**, 1459 (1979).
79Pn   Pneumatikakis, G. & Hadjiliadis, N.: *J. Chem. Soc., Dalton Trans.* 596 (1979).
79Pr   Prizant, L., Olivier, M. J., Rivest, R., & Beauchamp, A. L.: *J. Am. Chem. Soc.* **101**, 2765 (1979).
79Pu   Pullman, A., Ebbesen, T., & Rholam, M.: *Thoer. Chim. Acta* **51**, 247 (1979).
79Ra1   Randhawa, S., Pannu, B. S., & Chopra, S. L.: *Thermochim. Acta* **32**, 111 (1979).
79Ra2   Randhawa, S., Pannu, B. S., & Chopra, S. L.: *Thermochim. Acta* **33**, 335 (1979).
79Sl   Sletten, E. & Våland, E.: *Acta Cryst.* **35**, 840 (1979).
79Sw   Swaminathan, V. & Sundaralingam, M.: *Crit. Rev. Biochem.* **6**, 245 (1979).

79Vi   Villa, J. F., Rudd, F. J., & Nelson, H. C.: *J. Chem. Soc.*, *Dalton Trans.* 110 (1979).
80Ba   Barton, J. K. & Lippard, S. J.: *Metal ions in biology*, Spiro, T. G., Ed., Vol. I, Wiley, NY (1980), p. 31.
80Be1 Beauchamp, A. L.: *J. Cryst. Mol. Struct.* **10**, 149 (1980).
80Be2 Belanger-Gariepy, F. & Beauchamp, A. L.: *J. Am. Chem. Soc.* **102**, 3461 (1980).
80Br   Brown, E. A. & Bugg, C. E.: *Acta Cryst.* **B36**, 2597 (1980).
80Ca   Canty, A. J., Tobias, R. S., Chaichit, N., & Gatehouse, B. M.: *J. Chem. Soc.*, *Dalton Trans.* 1693 (1980).
80Ch   Cheng, D. M., Kan, L. S., Ts'O, P. O. P., Giessner-Prettre, C., & Pullman, B.: *J. Am. Chem. Soc.* **102**, 525 (1980).
80Cl1 Clarke, M. J.: *Met. Ions Biol. Syst.* **11**, 231 (1980).
80Cl2 Cleare, M. L. & Hydes, P. C.: *Met. Ions Biol. Sysyt.* **11**, 1 (1980).
80Fa   Farrell, N.: *J. Chem. Soc.*, *Chem. Commun.* 1014 (1980).
80Fe   Ferro, D., Bencivenni, L., Teghil, R., & Mastromarino, R.: *Thermochim. Acta* **42**, 75 (1980).
80Ge   Gellert, R. W., Fischer, B. E., & Bau, R.: *J. Am. Chem. Soc.* **102**, 7812 (1980).
80Gh   Ghose, R. & Dey, A. K.: *Rev. Chim. Minerale* **17**, 492 (1980).
80Go   Goodgame, M. & Johns, K. W.: *Inorg. Chim. Acta* **46**, 23 (1980).
80Hu1 Hubert, J. & Beauchamp, A. L.: *Can. J. Chem.* **58**, 1439 (1980).
80Hu2 Hubert, J. & Beauchamp, A. L: *Acta Cryst.* **B36**, 2613 (1980).
80Je   Jezowska-Trzebiatowska, B., Kozlowski, H., & Wolowiec, S.: *Acta Biochim. Pol.* **27**, 99 (1980).
80Li   Lin, J., Yu, C., Peng, S., Akiyama, I., Li, K., Li Kao Lee, & LeBreton, P. R.: *J. Phys. Chem.* **84**, 1006 (1980).
80Lö   Lönnberg, H., & Arpalahti, J.: *Inorg. Chim. Acta* **55**, 39 (1980).
80Lu   Lumme, P. & Mutikainen, I.: *Acta Cryst.* **B36**, 2251 (1980).
80Ma1 Marzilli, L. G., Kistenmacher, T. J., & Eichhorn, G. L.: *Metal ions in biology*, Spiro, T. G., Ed., Vol. I, Wiley (1980), p. 179.
80Ma2 Marzilli, L. G., de Castro, B., Caradonna, J. P., Stewart, R. C., & van Vuuren, C. P.: *J. Am. Chem. Soc.* **102**, 916 (1980).
80Ma3 Matczak-Jon, E., Jezowska-Trzebiatowska, B., & Kozlowski, H.: *J. Inorg. Biochem.* **12**, 143 (1980).
80Ma4 Makrigiannis, G., Papagiannakopoulos, P., & Theophanides, T.: *Inorg. Chim. Acta* **46**, 263 (1980).
80Mo   Moller, M. R., Bruck, M. A., O'Conner, T., Armatis, F. J., Jr., Knolinski, E. A., Kottmair, N., & Tobias, R. S.: *J. Am. Chem. Soc.* **102**, 4589 (1980).
80Mu   Mutikainen, I. & Lumme, P.: *Acta Cryst.* **B36**, 2237 (1980).
80Ne1 Nelson, H. C. & Villa, J. F.: *J. Inorg. Nucl. Chem.* **42**, 133 (1980).
80Ne2 Nelson, H. C. & Villa, J. F.: *J. Inorg. Nucl. Chem.* **42**, 1669 (1980).
80Ol   Olivier, M. J. & Beauchamp, A. L.: *Inorg. Chem.* **19**, 1064 (1980).
80Or   Orenberg, J. B., Fischer, B. E., & Sigel, H.: *J. Inorg. Nucl. Chem.* **42**, 785 (1980).
80Pe   Pezzano, H. & Podo, F.: *Chem. Rev.* **80**, 365 (1980).
80Pn   Pneumatikakis, G.: *Inorg. Chim. Acta* **46**, 243 (1980).

80Ri   Ringer, D. P., Howell, B. A., & Kizer, D. E.: *Anal. Biochem.* **103**, 337 (1980).

80Ro   Rochon, F. D., Kong, P. C., Coulombe, B., & Melanson, R.: *Can. J. Chem.* **58**, 381 (1980).

80Sh   Shirotake, S.: *Chem. Pharm. Bull.* **28**, 956 (1980).

80So   Sóvágó, I. & Martin, R. B.: *Inorg. Chim. Acta* **46**, 91 (1980).

80Sp1   Speca, A. N., Mikulski, C. M., Iaconianni, F. J., Pytlewski, L. L., & Karayannis, N. M.: *Inorg. Chim. Acta* **46**, 235 (1980).

80Sp2   Speca, A. N., Mikulski, C. M., Iaconianni, F. J., Pytlewski, L. L., & Karayannis, N. M.: *Inorg. Chem.* **19**, 3491 (1980).

80Th   Thomas, J. C., Frey, C. M., & Stuehr, J. E.: *Inorg. Chem.* **19**, 505 (1980).

81Bh   Bhandari, A. M., Solanki, A. K., & Wadhwa, S.: *J. Inorg. Nucl. Chem.* **43**, 2995 (1981).

81Bu   Buncel, E., Norris, A. R., Racz, W. J., & Taylor, S. E.: *Inorg. Chem.* **20**, 98 (1981).

81Ch   Chin Hsuan Wei & Jacobson, K. B.: *Inorg. Chem.* **20**, 356 (1981).

81Gh   Ghose, R. & Dey, K.: *Acta Chim. Acad. Sci. Hung.* **108**, 9 (1981).

81Ha   Hadjiliadis, N., Pneumatikakis, G., & Basosi, R.: *J. Inorg. Biochem.* **14**, 115 (1981).

81He   Helene, C. & Maurizot, J.-C.: *Crit. Rev. Biochem.* **11**, 213 (1981).

81Ko   Kozlowski, H. & Matczak-Jon, E.: *Pol. J. Chem.* **55**, 2243 (1981).

81Li   Lim, M.-C.: *J. Inorg. Nucl. Chem.* **43**, 221 (1981).

81Lö   Lönnberg, H. & Vihanto, P.: *Inorg. Chim. Acta* **56**, 157 (1981).

81Ma1   Maskos, K.: *Acta Biochim. Pol.* **28**, 317 (1981).

81Ma2   Maskos, K. : *Acta Biochim. Pol.* **28**, 183 (1981).

81Na   Nagasawa, A. & Diebler, H.: *J. Phys. Chem.* **85**, 3523 (1981).

81Pr1   Prizant, L., Olivier, M. J., Rivest, R., & Beauchamp, A. L.: *Can. J. Chem.* **59**, 1311 (1981).

81Pr2   Prizant, L., Rivest, R., & Beauchamp, A. L.: *Can. J. Chem.* **59**, 2290 (1981).

81Sc   Scheller, K. H., Scheller-Krattiger, V., & Martin, R. B.: *J. Am. Chem. Soc.* **103**, 6833 (1981).

81Sh   Sheldrick, W. S.: *Acta Cryst.* **B37**, 945 (1981).

81Si   Simon, T. M., Kunishima, D. H., Vibert, G. J., & Lorber, A.: *Cancer Res.* **41**, 94 (1981).

81Ta1   Taylor, S. E., Buncel, E., & Norris, A. R.: *J. Inorg. Biochem.* **15**, 131 (1981).

81Ta2   Taqui Khan, B. & Raju, R. M.: *Indian J. Chem.* **A20**, 860 (1981).

81Ve1   Vestues, P. I. & Sletten, E.: *Inorg. Chim. Acta* **52**, 269 (1981).

81Ve2   Vestues, P. I. & Martin, R. B.: *J. Am. Chem. Soc.* **103**, 806 (1981).

81Ve3   Vestues, P. I. & Martin, R. B.: *Inorg. Chim. Acta* **55**, 99 (1981).

82Be   Belanger-Gariery, F. & Beauchamp, A. L.: *Cryst. Struct. Commun.* **11**, 991 (1982).

82Bu   Buchanan, G. W. & Stothers, J. B: *Can. J. Chem.* **60**, 787 (1982).

82Go   Gonella, N. C. & Roberts, J. D.: *J. Am. Chem. Soc.* **104**, 3162 (1982).

82Gr   Griffith, E. A. H., Charles, N. G., & Amma, E. L.: *Acta Cryst.* **B38**, 942 (1982).

82Gu Guay, F. & Beauchamp, A. L.: *Inorg. Chim. Acta* **66**, 57 (1982).

82Je Jezowska-Trzebiatowska, B., & Wolowiec, S.: *Biochim. Biophys. Acta* **708**, 12 (1982).

82Kn Knighton, W. B., Giskaas, G. O., & Callis, P. R.: *J. Phys. Chem.* **86**, 49 (1982).

82Le Leroy, J.-L. & Gueron, M.: *Biochimie* **64**, 297 (1982).

82Ma1 Manning, G. S.: *O. Rev. Biophys. II* **2**, 179 (1982).

82Ma2 Marzilli, L. G., de Castro, B., & Solorzano, C.: *J. Am. Chem. Soc.* **104**, 461 (1982).

82Mi1 Mikulski, C. M., Cocco, S., de Franco, N., & Karayannis, N. M.: *Inorg. Chim. Acta* **67**, 61 (1982).

82Mi2 Mikulski, C. M., Cocco, S., Mattucci, L., de Franco, N., Weiss, L., & Karayannis, N. M.: *Inorg. Chim. Acta* **67**, 173 (1982).

81Ol Olivier, M. J. & Beauchamp, A. L.: *Acta Cryst.* **B38**, 2159 (1982).

82Or Orenberg, J. B., Kjos, K. M., Winkler, R., Link, J., & Lawless, J. G.: *J. Mol. Evol.* **18**, 137 (1982).

82Pn Pneumatikakis, G.: *Inorg. Chim. Acta* **66**, 131 (1982).

82Pr Prizant, L., Olivier, M. J., Rivest, R., & Beauchamp, A. L.: *Acta Cryst.* **B38**, 88 (1982).

82Ra Ramalingam, K. & Krishnamoorthy, C. R.: *Inorg. Chim. Acta* **67**, 167 (1982).

82Sc Schumacher, M. & Gunther, H.: *J. Am. Chem. Soc.* **104**, 4167 (1982).

82Sh Sheldrick, W. S.: *Z. Naturforsch.* **B37**, 653 (1982).

82Ta Taylor, R. & Kennard, O.: *J. Mol. Struct.* **78**, 1 (1982) and references therein.

83Ab1 Abbott, D. W. & Woods, C.: *Inorg. Chem.* **22**, 597 (1983).

83Ab2 Abbott, D. W. & Woods, C.: *Inorg. Chem.* **22**, 1918 (1983).

83Ab3 Abbott, D. W. & Woods, C.: *Inorg. Chem.* **22**, 2918 (1983).

83Ar1 Arpalahti, J.: *Finn. Chem. Lett.* 111 (1983).

83Ar2 Arpalahti, J. & Lönnberg, H.: *Inorg. Chim. Acta* **78**, 63 (1983).

83Ar3 Arpalahti, J. & Lönnberg, H.: *Inorg. Chim. Acta* **80**, 25 (1983).

83Bu1 Buchanan, G. W. & Bell, M. J.: *Can. J. Chem.* **61**, 2445 (1983).

83Bu2 Buda, A. & Sygula, A.: *J. Mol. Struct.* **92**, 255 (1983).

83Ca Camboli, D., Besancon, J., Tirouflet, J., Gautheron, B., & Meunier, P.: *Inorg. Chim. Acta* **78**, L51 (1983).

83Ch Charland, J.-P., Simard, M., & Beauchamp, A. L.: *Inorg. Chim. Acta* **80**, L57 (1983).

83Ci Cigni, M. B., Landfredi, A. M. M., & Tiripicchio, A.: *Acta Cryst.* **C39**, 1523 (1983).

83De Del Bene, J. E.: *J. Phys. Chem.* **87**, 367 (1983).

83Go Gonnella, N. C., Nakanishi, H., Holtwick, J. B., Horowitz, J. B., Kanamori, K., Leonard, N. J., & Roberts, J. D.: *J. Am. Chem. Soc.* **105**, 2050 (1983).

83Gu Guay, F., Beauchamp, A. L., Gilbert, C., & Savoie, R.: *Can. J. Spectrosc.* **28**, 13 (1983).

83Hä1 Häring, U. K. & Martin, R. B.: *Inorg. Chim. Acta* **80**, 1 (1983).

83Hä2 Häring, U. K. & Martin, R. B.: *Inorg. Chim. Acta* **78**, 259 (1983).

83Mi1   Mikulski, C. M., Cocco, S., de Franco, N., & Karayannis, N. M.: *Inorg. Chim. Acta* **80**, L23 (1983).

83Mi2   Mikulski, C. M., Cocco, S., de Franco, N., & Karayannis, N. M.: *Inorg. Chim. Acta* **78**, L25 (1983).

83Mi3   Mikulski, C. M., Cocco, S., de Franco, N., & Karayannis, N. M.: *Inorg. Chim. Acta* **80**, L61 (1983).

83Mi4   Miertus, S. & Trebaticka, M.: *Collect. Czech. Chem. Commun.* **48**, 3517 (1983).

83Mi5   Mikulski, C. M., Mattucci, L., Smith, Y., Thu Ba Tran, & Karayannis, N. M.: *Inorg. Chim. Acta* **80**, 127 (1983).

83Pa   Palmer, M. H., Wheeler, J. R., Kwiatkowski, J. S., & Lesyng, B.: *J. Mol. Struct. (Theochem)* **92**, 283 (1983).

83Pn   Pneumatikakis, G.: *Inorg. Chim. Acta* **80**, 89 (1983).

83Ra   Rabinra Reddy, P., Venugopal Reddy, K., & Taqui Khan, M. M.: *Indian J. Chem.* **A22**, 999 (1983).

83Sa1   Sagarik, K. P. & Rode, B. M.: *Inorg. Chim. Acta* **76**, L209 (1983).

83Sa2   Sabat, M., Satyshur, K. A., & Sundaralingam, M.: *J. Am. Chem. Soc.* **105**, 976 (1983).

83Sa3   Sagarik, K. P. & Rode, B. M.: *Inorg. Chim. Acta* **78**, 177 (1983).

83Sa4   Sarkar, A. R. & Ghosh, P.: *Inorg. Chim. Acta* **78**, L39 (1983).

83Sc1   Schumacher, M. & Gunther, H.: *Chem. Ber.* **116**, 2001 (1983).

83Sc2   Scheller, K. H. & Sigel, H.: *J. Am. Chem. Soc.* **105**, 5891 (1983).

83Si   Singh, M. M., Rosopoulos, Y., & Beck, W.: *Chem. Ber.* **116**, 1364 (1983).

83Sy   Sygula, A. & Buda, A.: *J. Mol. Struct.* **92**, 267 (1983).

83Sz   Szczesniak, M., Nowak, M. J., Rostkowska, H., Szczepaniak, K., Person, W. B., & Shugar, D.: *J. Am. Chem. Soc.* **105**, 5969 (1983).

83Ta1   Taqui Khan, M. M., Satyanarayana, S., Jyoti, M. S., & Lincoln, C. A.: *Indian J. Chem.* **A22**, 357 (1983).

83Ta2   Tajmir-Riahi, H. A. & Theophanides, T.: *Inorg. Chim. Acta* **80**, 183 (1983).

83Ta3   Tajmir-Riahi, H. A. & Theophanides, T.: *Can. J. Chem.* **61**, 1813 (1983).

83Ta4   Tajmir-Riahi, H. A. & Theophanides, T.: *Inorg. Chim. Acta* **80**, 223 (1983).

83Ta5   Taqui Khan, M. M., Satyanarayana, S., Jyoti, M. S., & Purshotham Reddy, A.: *Indian J. Chem.* **A22**, 364 (1983).

83Ta6   Taqui Khan, M. M. & Satyanarayana, S.: *Indian J. Chem.* **A22**, 584 (1983).

83Wu   Wu, F. Y.-H. & Wu, C.-W.: *Met. Ions Biol. Syst.* **15**, 157 (1983).

84Ao1   Aoki, K. & Saenger, W.: *Acta Cryst.* **C40**, 772 (1984).

84Ao2   Aoki, K. & Saenger, W.: *J. Inorg. Biochem.* **20**, 225 (1984).

84Ao3   Aoki, K. & Saenger, W.: *Acta Cryst.* **C40**, 775 (1984).

84Be1   Benoit, R. L. & Frechette, M.: *Can. J. Chem.* **62**, 995 (1984).

84Be2   Beauchamp, A. L.: *Inorg. Chim. Acta* **91**, 33 (1984).

84Ca1   Caldwell, K., Deacon, G. B., Gatehouse, B. M., Lee, S. C., & Canty, A. J.: *Acta Cryst.* **C40**, 1533 (1984).

84Ca2   Calis, G. H. M. & Hadjiliadis, N.: *Inorg. Chim. Acta* **91**, 203 (1984).

84Ch   Charland, J.-P. & Beauchamp, A. L.: *Croat. Chem. Acta* **57**, 679 (1984).

84Cr   Crowe, A. J., Smith, P. J., & Atassi, G.: *Inorg. Chim. Acta* **93**, 179 (1984).

84De   Del Bene, J. E.: *J. Phys. Chem.* **88**, 5927 (1984).

84Di   Diebler, H.: *J. Mol. Catal.* **23**, 209 (1984).

84Et   Ettore, R.: *Inorg. Chim. Acta* **91**, 167 (1984).

84Gh   Ghosh, P., Mukhopadhyay, T. K., & Sarkar, A. R.: *Transition Met. Chem.* **9**, 46 (1984).

84Ki1   Kim, S.-H. & Martin, R. B.: *Inorg. Chim. Acta* **91**, 19 (1984).

84Ki2   Kim, S.-H. & Martin, R. B.: *Inorg. Chim. Acta* **91**, 11 (1984).

84Ma1   Matsuoka, Y., Norden, B., & Kurucsev, T.: *J. Chem. Soc., Chem. Commun.* 1573 (1984).

84Ma2   Matsuoka, Y., Norden, B., & Kurucsev, T.: *J., Phys. Chem.* **88**, 971 (1984).

84Ma3   Mathlouthi, M., Suvre, A. M., & Koenig, J. L.: *Carbohydr. Res.* **134**, 23 (1984).

84Me   Menard, R., Lachapelle, M., & Zador, M.: *Biophys. Chem.* **20**, 29 (1984).

84No1   Norris, A. R., Taylor, S. E., Buncel, E., Belanger-Gariery, F., & Beauchamp, A. L.: *Inorg. Chim. Acta* **92**, 271 (1984).

84No2   Norris, A. R., Kumar, R., Buncel, E., & Beauchamp, A. L.: *J. Inorg. Biochem.* **21**, 277 (1984).

84Pe   Petit-Ramel, M. M., Thomas-David, G., Perichet, G., & Pouyet, P.: *Can. J. Chem.* **62**, 22 (1984).

84Pn1   Pneumatikakis, G.: *Polyhedron* **3**, 9 (1984).

84Pn2   Pneumatikakis, G.: *Inorg. Chim. Acta* **93**, 5 (1984).

84Ra   Radchenko, E. D., Sheina, G. G., Smorygo, N. O., & Blagoi, Y. P.: *J. Mol. Struct.* **116**, 387 (1984).

84Sa   Saenger, W.: *Principles of nucleic acid structure*, Springer, NY, (1984) Ch. 9 and 12.

84Sc   Scanlan, M. J. & Hillier, I. H.: *J. Am. Chem. Soc.* **106**, 3737 (1984).

84Si   Sigel, H. & Scheller, K. H.: *Eur. J. Biochem.* **138**, 291 (1984).

84Sz   Szczesniak, M., Nowak, M. J., & Szczepaniak, K.: *J. Mol. Struct.* **115**, 221 (1984).

84Te   Temerek, Y. M., Kamal, M. M., Ahmed, M. E., & Abd-El-Hamid, M. I.: *Bioelectrochem. Bioenergetics* **12**, 475 (1984).

85Ar1   Arpalahti, J. & Ottoila, E.: *Inorg. Chim. Acta* **107**, 105 (1985).

85Ar2   Arpalahti, J. & Lönnberg, H.: *Inorg. Chim. Acta* **107**, 197 (1985).

85Ba   Bariyanga, J. & Theophanides, T.: *Inorg. Chim. Acta* **108**, 133 (1985).

85Be1   Benoit, R. L. & Frechette, M.: *Can. J. Chem.* **63**, 3053 (1985).

85Be2   Benoit, R. L., Boulet, D., Seguin, L., & Frechette, M.: *Can. J. Chem.* **63**, 1228 (1985).

85Bo   Bonati, F., Burini, A., & Pietroni, B. R.: *Z. Naturforsch.* **B40**, 1749 (1985).

85Bu1   Buncel, E., Kumar, R., Norris, A. R., & Beauchamp, A. L.: *Can. J. Chem.* **63**, 2575 (1985).

85Bu2   Buncel, E., Boone, C., Joly, H., Kumar, R., & Norris, A. R.: *J. Inorg. Biochem.* **25**, 61 (1985).

85Ch1   Charland, J.-P. & Beauchamnp, A. L.: *Acta Cryst.* **C41**, 505 (1985).

85Ch2   Charland, J.-P., Minh Tan Phan Viet, St-Jacques, M., & Beauchamp, A. L.: *J. Am. Chem. Soc.* **107**, 8202 (1985).

85Gn   Gningue, D. & Aaron, J.-J.: *Talanta* **32**, 183 (1985).

85Gr   Grigoratou, A. & Katsaros, N.: *J. Inorg. Biochem.* **24**, 147 (1985).

85Kr   Krishnamoorthy, C. R., Sunil, S., & Ramalingam, K.: *Polyhedron* **4**, 1451 (1985).

85Ma1   Martin, R. B.: *Acc. Chem. Res.* **18**, 32 (1985).

85Ma2   Matsuoka, Y., Norden, B., & Kurucsev, T.: *J. Cryst. Spectrosc. Res.* **15**, 545 (1985).

85Ma3   Maskos, K.: *J. Inorg. Biochem.* **25**, 1 (1985).

85Ma4   Mangani, S. & Orioli, P.: *J. Chem. Soc., Chem. Commun.* 780 (1985).

85Mi1   Mikulski, C. M., Minutella, R., de Franco, N., & Karayannis, N. M.: *Inorg. Chim. Acta* **106**, L33 (1985).

85Mi2   Mikulski, C. M., Mattucci, L., Weiss, L., & Karayannis, N. M.: *Inorg. Chim. Acta* **107**, 81 (1985) and references therein.

85Mi3   Mikulski, C. M., Mattucci, L., Weiss, L., & Karayannis, N. M.: *Inorg. Chim. Acta* **107**, 147 (1985) and references therein.

85Mi4   Miller, S. K., VanDerveer, D. G., & Marzilli, L. G.: *J. Am. Chem. Soc.* **107**, 1048 (1985).

85On   Onori, G. & Blidaru, D.: *Nouv. Chim.* **D5**, 339 (1985).

85Po   Poojary, M. D. & Manohar, H.: *Inorg. Chem.* **24**, 1065 (1985).

85Ro   Rosopulos, Y., Nagel, U., & Beck, W.: *Chem. Ber.* **118**, 931 (1985).

85Ra1   Rabindra Reddy, P. & Malleswara Rao, V. B.: *Polyhedron* **4**, 1603 (1985).

85Ra2   Rabindra Reddy, P., Marilatha Reddy, M., & Prasad Reddy, P. R.: *Proc. Indian Acad. Sci.* **95**, 547 (1985).

85Ra3   Rabindra Reddy, P. & Madhusudhan Reddy, B.: *Natl. Acad. Sci. Lett.* **8**, 205 (1985).

85Ra4   Rabindra Reddy, P. & Madhusudhan Reddy, B.: *Proc. Indian Acad. Sci.* **95**, 229 (1985).

85Sh1   Sheina, G. G., Radchenko, E. D., & Blagoi, Y. P.: *Dokl. Akad. Nauk SSSR* **282**, 1497 (1985).

85Sh2   Shinozuka, K., Wilkowski, K., Heyl, B., & Marzilli, L. G.: *Inorg. Chim. Acta* **100**, 141 (1985).

85Sz   Szczesniak, M., Nowak, M. J., Szczepaniak, K., Chin, S., Scott, I., & Person, W. B.: *Spectrochim. Acta* **A41**, 223 (1985).

85St   Stepanian, S. G., Sheina, G. G., Radchenko, E. D., & Blagoi, Y. P.: *J. Mol. Struct.* **131**, 333 (1985).

85Ta   Tajmir-Riahi, H. A. & Theophanides, T.:, *Can. J. Chem.* **63**, 2065 (1985).

86Ba   Basallote, M. G., Vilaplana, R., & Gonzales-Vilchez, F.: *Transition Met. Chem.* **11**, 232 (1986).

86Be   Benoit, R. L. & Frechette, M.: *Can. J. Chem.* **64**, 2348 (1986).

86Bi   Birdsall, W. J., Pfennig, B. W., & Toto, J. L.: *Polyhedron* **5**, 1357 (1986) and references therein.

86Bu   Buchanan, G. W. & Bell, M.-J.: *Magn. Res. Chem.* **24**, 493 (1986).

86Ca   Capparelli, M. V., Goodgame, D. M. L., Hayman, P. B., & Skapski, A. C.: *Inorg. Chim. Acta* **125**, L47 (1986).

86Ch1   Charland, J.-P., Britten, J. F., & Beauchamp, A. L.: *Inorg. Chim. Acta* **124**, 161 (1986).

86Ch2   Charland, J.-P. & Beauchamp, A. L.: *Inorg. Chem.* **25**, 4870 (1986).

86Fi   Fiol, J. J., Terron, A., & Moreno, V.: *Inorg. Chim. Acta* **125**, 159 (1986).

86Le   Leroy, J. L. & Gueron, M.: *J. Am. Chem. Soc.* **108**, 5753 (1986).
86Mi1   Mizuno, Y.: *The organic chemistry of nucleic acids*, Kodansha, Tokyo (1986).
86Mi2   Mikulski, C. M., Minutella, R., de Franco, N., Borges, G., Jr., & Karayannis, N. M.: *Inorg. Chim. Acta* **123**, 105 (1986).
86Mi3   Miller, S. K., Marzilli, L. G., Dörre, S., Kollat, P., Stigler, R.-D., & Stezowski, J. J.: *Inorg. Chem.* **25**, 4272 (1986).
86No   Norden, B., Matsuoka, Y., & Kurucsev, T.: *J. Cryst. Spectrosc. Res.* **16**, 217 (1986).
86Po   Poojary, M. D. & Manohar, H.: *J. Chem. Soc., Dalton Trans.* 309 (1986).
86Ra1   Rabindra Reddy, P. & Malleswara Rao, V. B.: *Inorg. Chim. Acta* **125**, 191 (1986).
86Ra2   Rabindra Reddy, P. & Madhusudhan Reddy, B.: *Polyhedron* **12**, 1947 (1986).
86Ra3   Rabindra Reddy, P. & Malleswara Rao, V. B.: *J. Chem. Soc., Dalton Trans.* 2331 (1986).
86Ro   Rosenthal, L. S. & Nelson, D. J.: *Inorg. Chim. Acta* **125**, 89 (1986).
86Sc   Scheller-Krattiger, V. & Sigel, H.: *Inorg. Chem.* **25**, 2628 (1986), and references therein.
86Ta   Tauler, R., Rainer, M. J. A., & Rode, B. M.: *Inorg. Chim. Acta* **123**, 75 (1986).
87An   Anwander, E. H. S., Probst, M. M., & Rode, B. M.: *Inorg. Chim. Acta* **137**, 203 (1987).
87Ba   Basile, L. A., Raphael, A. L., & Barton, J. K.: *J. Am. Chem. Soc.* **109**, 7557 (1987).
87Ch   Charland, J.-P.: *Inorg. Chim. Acta* **135**, 191 (1987).
87Ci   Cini, R. & Giorgi, G.: *Inorg. Chim. Acta* **137**, 87 (1987).
87Du   Dubler, E., Hanggi, G., & Bensch, W.: *J. Inorg. Biochem.* **29**, 269 (1987).
87Fa   Faggiani, R., Howard-Lock, H. E., Lock, C. J. L., & Turner, M. A.: *Can. J. Chem.* **65**, 1568 (1987).
86Fr   Froystein, N. Å. & Sletten, E.: *Inorg. Chim. Acta* **138**, 49 (1987).
87Ga   Galy, J., Mosset, A., Grenthe, I., Puigdomenech, I., Sjöberg, B., & Hulten, F.: *J. Am. Chem. Soc.* **109**, 380 (1987).
87Iv   Iverson, B. L. & Dervan, P.: *Nucleic Acids Res.* **15**, 7823 (1987).
87Ka   Kalbitzer, H. R.: *Met. Ions Biol. Syst.* **22**, 81 (1987).
87Kö   Köpf-Maier, P. & Köpf, H.: *Chem. Rev.* **87**, 1137 (1987).
87La   Laschi, F., Picchi, M. P., Rossi, C., & Cini, R.: *Inorg. Chim. Acta* **135**, 215 (1987).
87Ma   Mangani, S. & Orioli, P.: *J. Chem. Soc., Dalton Trans.* 2999 (1987).
87Mb   M'Boungou, R., Petit-Ramel, M., Thomas-David, G., Perichet, G., & Pouyet, B.: *Can. J. Chem.* **65**, 1479 (1987).
87Me   Menard, R., Phan Viet, M. T., & Zador, M.: *Inorg. Chim. Acta* **136**, 25 (1987).
87Mi1   Mildvan, A. S.: *Magnesium* **6**, 28 (1987).
87Mi2   Mikulski, C. M., Chung Ja Lee, Thu Ba Tran, & Karayannis, N. M.: *Inorg. Chim. Acta* **136**, L13 (1987).

87Mi3   Mishra, A. P., Mishra, S. K., & Yadava, K. L.: *Trans. SAEST* **22**, 247 (1987).
87Pe   Perno, J. R., Cwikel, D., Spiro, T. G.: *Inorg. Chem.* **26**, 400 (1987).
87Pn   Pneumatikakis, G., Markopoulos, J., & Yannopoulos, A.: *Inorg. Chim. Acta* **136**, L25 (1987).
87Ra   Rabindra Reddy, P., Prasad Reddy, P. R., & Harilatha Reddy, M.: *Indian Acad. Sci.* **99**, 297 (1987).
87Re1   Reedijk, J.: *Pure Appl. Chem.* **59**, 181 (1987).
87Re2   Remaud, G., Zhou, X.-X., Chattopadhyaya, J., Oivanen, M., & Lönnberg, H.: *Tetrahedron* **43**, 4453 (1987).
87Se   Sekhon, B. S.: *J. Indian Chem. Soc.* **64**, 308 (1987).
87Si   Sigel, H.: *Chimia* **41**, 11 (1987).
87Sh1   Sherman, S. E. & Lippard, S. J.: *Chem. Rev.* **87**, 1153 (1987).
87Sh2   Sheina, G. G., Stepanian, S. G., Radchenko, E. D., & Blagoi, Y. P.: *J. Mol. Struct.* **158**, 275 (1987).
87Sh3   Shanbhag, S. M. & Choppin, G. R.: *Inorg. Chim. Acta* **138**, 187 (1987).
87Sh4   Shanbhag, S. M. & Choppin, G. R.: *Inorg. Chim. Acta* **139**, 119 (1987).
87Sz   Szczepaniak, K. & Szczesniak, M.: *J. Mol. Struct.* **156**, 29 (1987).
87Ta   Taqui Khan, B., Raju, M. R., & Zakeeruddin, S. M.: *J. Coord. Chem.* **16**, 237 (1987).
87Tr   Tribolet, R. & Sigel, H.: *Eur. J. Biochem.* **163**, 353 (1987).
87Tu   Tullius, T. D.: *Recl. Trav. Chim. Pays-Bas* **106**, 190 (1987).
88An   Antonelli, M. L., Balzamo, S., Carunchio, V., Cernia, E., & Purrello, R.: *J. Inorg. Biochem.* **32**, 153 (1988).
88Be1   Benoit, R. L. & Frechette, M.: *Thermochim. Acta* **127**, 125 (1988).
88Be2   Begum, N. S., Poojary, M. D., & Manohar, H.: *J. Chem. Soc., Dalton Trans.* 1303 (1988).
88Br   Brown, R. D., Godfrey, P. D., McNaughton, D., & Pierlot, A. P.: *J. Am. Chem. Soc.* **110**, 2329 (1988).
88Ca   Carmona, P., Huong, P. V., & Gredilla, E.: *J. Raman Spectrosc.* **19**, 315 (1988).
88Ch   Chen, C. B. & Sigman, D. S.: *J. Am. Chem. Soc.* **110**, 6570 (1988).
88Ga   Garcia, B. & Palacios, J.: *Ber. Bunsenges. Phys. Chem.* **92**, 696 (1988).
88Gi   Giorgi, G. & Cini, R.: *Inorg. Chim. Acta* **151**, 153 (1988).
88Gr   Grenier, L., Charland, J.-P., & Beauchamp, A. L.: *Can. J. Chem.* **66**, 1663 (1988).
88Kh   Khan, B. T., Raju, M., & Zakeeruddin, S. M.: *J. Chem. Soc., Dalton Trans.* 67 (1988).
88Ko   Kornilova, S. V., Blagoi, Y. P., Moskalenko, I. P., Nikiforova, N. A., & Gladchenko, N. A.: *Stud. Biophys.* **123**, 77 (1988)..
88Ku   Kuczera, K., Szczesniak, M., & Szczepaniak, K.: *J. Mol. Struct.* **172**, 101 (1988).
88Ma   Massoud, S. S. & Sigel, H.: *Inorg. Chem.* **27**, 1447 (1988).
88Me   Menard, R. & Zador, M.: *Can. J. Chem.* **66**, 178 (1988).
88Pa   Parmentier, J., De Taeye, J., & Zeegers-Huyskens, T.: *Bull. Soc. Chim. Belg.* **97**, 893 (1988).

88Pn   Pneumatikakis,, G., Yannopoulos, A., Markopoulos, J., & Angelopoulos, C.: *Inorg. Chim. Acta* **152**, 101 (1988).

88Sa1   Sawai, H.: *J. Mol. Evol.* **27**, 181 (1988).

88Sa2   Sarkar, A. R., Bandyopadhyay, R. K., & Ghosh, P.: *J. Indian Chem. Soc.* **65**, 593 (1988).

88Sa3   Sarkar, A. R., Ghosh, P., & Bandyopadhyay, R. K.: *Indian J. Chem.* **A27**, 819 (1988).

88Sc   Schindler, M.: *J. Am. Chem. Soc.* **110**, 6623 (1988).

88Si   Sigel, H., Massoud, S. S., & Tribolet, R.: *J. Am. Chem. Soc.* **110**, 6857 (1988).

88Sl   Sletten, E., Sletten, J., & Froystein, N. Å.: *Acta Chem. Scand. A* **A42**, 413 (1988).

88Ta   Tajmir-Riahi, H. A., Langlais, M., & Savoie, R.: *Biochim. Biophys. Acta* **926**, 211 (1988).

89Ba   Barton, J. K.: *Pure Appl. Chem.* **61**, 563 (1989), and references therein.

89Br1   Brown, R. D., Godfrey, P. D., McNaughton, D., & Pierlot, A. P.: *J. Am. Chem. Soc.* **111**, 2308 (1989).

89Br2   Brown, R. D., Godfrey, P. D., McNaughton, D., & Pierlot, A. P.: *J. Chem. Soc., Chem. Commun.* 37 (1989).

89Ki   Kinjo, Y., Tribolet, R., Corfu, N. A., & Sigel, H.: *Inorg. Chem.* **28**, 1480 (1989).

89Ma   Massoud, S. S. & Sigel, H.: *Inorg. Chim. Acta* **159**, 243 (1989).

89No   Nowak, M. J.: *J. Mol. Struct.* **193**, 35 (1989).

89Sa   Sagripanti, J. L. & Kraemer, K. H.: *J. Biol. Chem.* **264**, 1729 (1989).

89Uc   Uchida, K., Toyama, A., Tamura, Y., Sugimura, M., Mitsumori, F., Furukawa, Y., Takeuchi, H., & Harada, I.: *Inorg. Chem.* **28**, 2067 (1989).

Index

(Keywords appearing in headings or subheadings in the Table of Contents are mostly omitted.)

Absorption spectra, 68, 73, 75, 80, 82, 83, 85, 89, 93, 102, 105, 141–143
Acid–base properties, 18, 59, 61, 136
Actinides, 69, 81, 84
Active site cavity of superoxide dismutase, 200–201
Adenine, 41, 42, 291
Adenine complexes, stability constants, 296
Adenosine, 290
Adenosine complexes, stability constants, 292
Alanine complexes, stability constants, 74
Albumin, 156, 158, 162
Alditol complexes, 241
Alkaline earth metals, 77, 81
Ambidentate, 85, 88, 91, 97
Amide deprotonation by metal ions in peptide complexes, 146, 174
Amino-acid complexes
  of cadmium(II), stability constants, 67
  of cobalt(II), stability constants, 63
  of copper(II), stability constants, 65
  of iron(III), structure, 72
  of nickel(II), stability constants, 64
  of silver(I), structure, 69–70
  of zinc(II), stability constants, 66
Amino acid rotamers, protonation scheme, 34
Amino acid protonation–deprotonation sites, 43–48, 59
Amino acids, 41, 57–58
  acid dissociation constants, 61
Aminophosphonic acid, dissociation constants, 105
Angiotensine, 28, 45–47, 158
Anion binding
  of carbonic anhydrase, 223–227
  of superoxide dismutase, 227–230
Apocarbonic anhydrase, metal ion binding, 188–193
Apocarboxypeptidase, metal ion binding, 188–193
Apotransferrin, metal ion binding, 208–209
L-Arabinose, complex structures, 253
Arsanilazo-Tyr-248-carboxypeptidase, 211

Aspartic acid
  deprotonation, 23, 38, 49–51
  metal complexes, 77–81
Auxiliary ligand, 27, 49
Azide adduct formation of carbonic anhydrase, 224–227

Basicity–complex stability correlation of nucleosides, 324
BFP (2-benzyl-3-formylpropionic acid), 216
  binding by carboxypeptidase, 216–217
Biological activity influencing factors, 12
Bis indole alcaloids, 12
Biuret reaction, 135, 141
BMBP(3-p-methoxybenzoyl)-2-benzylpropanoic acid, 214
  binding by carboxypeptidase, 214–216
Bulk constant, 36, 51

Calorimetric studies of sugar complexation, 82, 94, 161, 242
Carbohydrate complex formation in water, 237
Carbonic anhydrase
  anion binding, 223–227
  azide adduct formation, 224–227
  diamox binding, 218
  metal binding sites, 197
  structure, 196–197
  sulphanylamide binding, 217, 219
γ-Carboxyglutamic acid derivatives, 14, 27, 28, 79
Carboxypeptidase
  BMBP binding, 214–216
  BFP binding, 216–217
  substrate binding, 213–216
  zinc ion binding, 187, 196, 198
L-Carnosine, 160–162
Catecholate, 111
Chondroitin sulphate, 249, 250, 251
Cis-trans isomerism, 72, 75
Copper–copper interactions, 108–109, 203, 244, 265, 313–314
Copper(III) oligopeptides, 146
Cyclic peptides, 165
Coordination polyhedra of sugar complexes, 252

Corticotropin (ACTH)
    protonation–deprotonation equilibria, 21–22, 32
Cyclodextrin complexes, 256, 264, 265
Cytidine, 303–304
Cytidine complexes, stability constants, 305
Cytosine complexes, stability constants, 306

Duanorubicine, 29
Deoxyribonucleic acid (DNA), 284, 285
Dextran sulphate, 250, 251
Diamox (acetazolamide), binding by carbonic anhydrase, 218
Diastereomers, 150
Dipeptide complexes
    coordination mode, 141, 160–161, 167
    formation process, 140, 167, 169
    stability constants, 142, 148, 155, 156
Dipeptide diastereomers, proton pK values and copper(II) complex stability constants, 151
Dissaccharide, see Sugar
Dopa(3,4-dihydroxiphenylalanine) complexes, 86–87
Double isomerism, 36

Enthalpy values, 71, 72, 76, 80, 83, 84, 90
Epimerization, 237, 261
EXAFS study
    of iron(III) sugar complexes, 254–255
    of tin organic derivatives of sugars, 257

Fibrinopeptide A, 158
Five-membered carbohydrate rings, 242
Fructose complex structure, 253–254, 255

GHL(glycyl-histidine-L-lysine), 163
Gluconic acid complexes, 245–247, 255, 258–259
Glucosamine complexes, 248, 255, 259–261
Glucose, 240, 254, 260, 265
Glucuronic acid complexes, 244–245, 257–258
Glutathion, 171, 172
    microspecies, 40
Glycine, protonation scheme, 24–25
Glycopeptide complexation, 259
Glycosminoglycan complexes, 249–250
Glycosylamine mixed complexes, 259–261
Guanine, 298
Guanosine, 297
Guanosine complexes, stability constants, 299

H-bonding, 11–12, 14, 31, 69, 75, 76, 83, 84, 155, 238
Heparin, 249, 250, 251
^{1}H-NMR in sugar complexation studies, 240, 241, 244
Hydrophobic interactions, 151, 166
Hydroxy carbonic acid complexes, 246
Hydroxyproline complexes, stability constants, 74
Hystidin complexes, binding modes, 91–92
Hyaluronic acid complexes, 249–250
Hypoxanthine, 298
Hypoxanthine complexes, stability constants, 303

Inosine, 297
Inosine complexes, stability constants, 299
Inositol, 239, 241, 242, 259
Interactivity parameter, 25, 27, 32, 43–48
Iron(III)–iron(III) interactions, 266–268
Iron(III)–sugar complexes, EXAFS data, 255
Isoenzymes of carbonic anhydrase, 196

Ketose- and aldose-diamine complexes, structure, 261

Lactobionic acid, 243, 245, 265
Ligand–ligand interactions, 112, 113
Loop structures, 152, 153, 174
Lysine (and other diaminocarboxylate) complexes, 88–91
    binding modes, 89–90

Manganese complexes, redox behaviour, 269–271
Manning's theory, 251
Melanostatin (L-prolyl-L-leucylglycineamide), 152
Metal dependence of superoxide dismutase activity, 202
Metal ion–aromatic ring interaction, 83, 84
Metal ion binding
    effect on molecular structure, 12
    modes in nucleosides, 289, 291, 292–297
    of apocarbonic anhydrase, 188–193
    of apocarboxypeptidase A, 188–193
    of apotransferrin, 208–209
    of oligopeptides, binding mode, 138–139, 141, 143, 160, 163, 164–165, 167, 168
    strength in superoxide dismutase, 203–204
Metal ions in bioprocesses, 12
Mixed ligand complexes
    in amino acid chemistry, 95, 110, 114–116
        stability constants, 111, 112
    in carbohydrate chemistry, 244, 246, 259–261, 272
    in enzyme chemistry, 186, 196, 210, 218, 221, 222
    in nucleoside-, nucleotide-chemistry, 327
    of glycosylamine, 258–261
Mixed valence complexes, 101, 102, 270–271
Molybdenum(V) and (VI) carbohydrate complexes, 256
Monosaccharide, see Sugar

N-glycoside complexes, 262–263
Nickel(III) oligopeptides, 146
Nucleic acids, 40, 284, 285
Nucleic bases, protonation constants, 41–42
Nucleoside 5'-monophosphate complexes
    binding mode of metal ions, 319, 320
    stability constants, 317–318
Nucleoside 5'-monophosphates, pk_a values, 316
Nucleosides, 284, 285
Nucleotides, 284, 286

Octapeptides, 32
Olation equilibria, 246
Oligopeptide complexes, stability constants, 142, 153

Oligopeptides
   protonation p*K*-values, 136
   metal binding modes, 138–139, 141, 143, 160,
      163, 164–165, 167, 168
ORD and CD studies, 68, 75, 76, 80, 82, 89, 93,
   143, 165, 256
Organotin complexes of carbohydrates, 257
Overlapping protonation and complexation
   equilibria, 14–15
Oxygen uptake of Co(II) complexes, 95

Paper electrophoresis in sugar complexation
   studies, 239
Penicillamine, 56, 67, 68, 103
Peptides of
   α-alanine, 148
   β-alanine, 148, 149
   arginine, 148, 151
   asparagine, 157
   aspartic acid, 151, 157, 158, 174
   cysteine, 169
   glutamic acid, 157, 171
   glycine, 140
   histidine, 151, 159–165
   isoleucine, 148
   leucine, 148, 150
   lysine, 151, 153, 157, 158
   methionine, 166
   phenylalanine, 148, 149, 156
   proline, 151, 157
   sarcosine, 152, 153
   serine, 154
   thereonine, 154
   tyrosine, 153, 155
   valine, 148, 166
Peptide protonation–deprotonation sites, 43–48
Planar-octahedral equilibria, 90
Platinum(II) complexes of amino sugars, 263–264
2-(polyhydroxyalkyl;)-thiazolidin-L(R)-
   carboxylate complexes, 247
Polyelectrolyte theory, 249
Polynuclear sugar acid complexes, 245, 247
Proline complexes, stability constants, 74
Protonated amino acid complexes, 80
Protonation enthalpies and entropies of nucleic
   acid bases, 323
Protonation isomers, 25–27
Purine and derivatives
   metal complex stabilities, 288
   metal ion binding, 289
Purine nucleoside 5′-monophosphate complex
   structures, 320
Purine nucleosides, 287–288
Purine riboside, metal ion binding, 289

Reaction of metal complexes
   with enzymes, 193–194
   with macrocyclic ligands, 195
Ribofuranosyl group, 312, 314
Ribonucleic acid (RNA), 284, 285
Ribonucleosides, 313, 314
Ribose, 240, 242, 255, 256, 260, 261, 263

Rotamers, 33, 34

Saccharic acid, 243, 246, 247, 265
Saccharose
   complex structures, 254
   mixed valence Mn complex, 270–271
Serine complex structures, 75–76
Sialic acid complexes, 252
Stacking interactions, 327
Stereoselectivity, 76, 80, 84, 93
Substrate binding by carboxypeptidase, 213–216
Sugar (mono- and disaccharide) complexes
   copper coordination mode, 244
   coordination polyhedra, 252
Sugar (mono- and disaccharide) molecules
   acidity, 238–239
   complexing strength sequence, 240–241
   coordination sites, 240, 242, 247
Sugar-*N*-ethylenediamine complex, structure, 262
Sulphanilamide binding by carbonic anhydrase,
   217, 219
Superoxide dismutase
   active site cavity, 200–201
   anion binding, 227–230
   metal binding site, 200

Template effect, 11
Tetrapeptide complexes
   binding mode, 165
   stability constants, 142
Theophylline, 300, 301
Thiol-containing amino acid complexes, binding
   mode, 98, 100
Thymidine, 308–309
Thymidine complexes, stability constants, 310
Thymine, 308–309
Thymine complexes, stability constants, 311
Thymopoietine, 28, 32, 157–158, 159
Transferrin
   cyanide binding, 230
   iron binding site, 206–207
   structure, 205–207
TRF (L-pyroglutamyl-L-histidyl-L-prolineamide),
   163
Tripeptide complexes
   stability constants, 142, 150
   complexation mode, 144, 163, 170, 171–172

Uracil, 309–309
Uracil complexes, 311
Uranium complexes of sugar acids, 247, 258
Uridine, 308–309
Uridine complexes, stability constants, 310

Vanadate complexes of sugar acids, 247

Wilson's disease, 56

Xanthosine, 300, 301